Lecture Notes in Computer Science 15538

Advanced Research in Computing and Software Science

Subline of Lecture Notes in Computer Science

More information about this series at https://link.springer.com/bookseries/558

Rastislav Královič · Věra Kůrková
Editors

SOFSEM 2025: Theory and Practice of Computer Science

50th International Conference on Current Trends in Theory and Practice of Computer Science, SOFSEM 2025
Bratislava, Slovak Republic, January 20–23, 2025
Proceedings, Part I

Springer

Editors
Rastislav Královič
Comenius University in Bratislava
Bratislava, Slovakia

Věra Kůrková
Czech Academy of Sciences
Prague, Czech Republic

ISSN 0302-9743 ISSN 1611-3349 (electronic)
Lecture Notes in Computer Science
ISBN 978-3-031-82669-6 ISBN 978-3-031-82670-2 (eBook)
https://doi.org/10.1007/978-3-031-82670-2

This Springer imprint is published by the registered company Springer Nature Switzerland AG
The registered company address is: Gewerbestrasse 11, 6330 Cham, Switzerland

Preface

The *50th International Conference on Current Trends in Theory and Practice of Computer Science* (SOFSEM 2025), organized by the Slovak Society for Computer Science and the Faculty of Mathematics, Physics, and Informatics of the Comenius University in Bratislava, was held in Bratislava on January 20–23, 2025.

SOFSEM is an annual winter conference devoted to the theory and practice of computer science. It is focused on the latest results and developments in fundamental computer and artificial intelligence research. The series of SOFSEM (originally SOFtware SEMinar) conferences began in 1974 as a winter seminar for computer scientists in the former Czechoslovakia. Gradually, SOFSEM has transformed from a regional conference into an international one. Since 1995, its proceedings have been published in the LNCS series of Springer.

SOFSEM's scope has extended to include original research from all areas of foundations of computer science and artificial intelligence. Currently, it includes AI-based algorithms and techniques, nature-inspired computing, machine learning theory, multi-agent algorithms and games, neural network theory, parallel and distributed computing, quantum computing, computability, decidability, classical and non-classical models of computation, computational complexity, computational learning, cryptographic techniques and security, data compression, data and pattern mining methods, discrete combinatorial optimization, automata, languages, machine models, rewriting systems, efficient data structures, graph structures and algorithms, logics of computation, robotics, and other relevant theory topics in computing and AI.

The SOFSEM series of conferences was only interrupted in 2022 due to the COVID pandemic, while in 2021, it was organized as a virtual meeting. SOFSEM 2023 was again held as a physical meeting, its venue was Nový Smokovec in High Tatras in Slovakia. It was followed by SOFSEM 2024 held at the University of Trier in Cochem. SOFSEM maintains its tradition of a venue where researchers in all stages of their careers can share their achievements and insights.

For SOFSEM 2025, 109 full papers were submitted. The Program Committee, chaired by Rastislav Královič and Věra Kůrková, selected 48 papers after a thorough peer-review process. The evaluation of the papers was based on quality, originality, and relevance for the conference. The handling of the reviewing process was supported by the EasyChair system. Most papers received three single-blind reviews. The accepted papers are published in the two volumes of these proceedings. The conference included four invited talks:

- Paola Flocchini (University of Ottawa, Canada): *Distributed Computing by Mobile Robots: Exploring the Computational Landscape*
- Erik Jan van Leeuwen (Utrecht University, The Netherlands): *Open Problems and Recent Developments on a Complexity Framework for Forbidden Subgraphs*

– Paul Spirakis (University of Liverpool, UK): *Temporal Graph Realization Problems*
– Ivan Tyukin (King's College London, UK): *The Challenge of Stability, Accuracy, and Robustness in Data-Driven AI*

We are pleased to thank everyone who contributed to the event's success. In particular, we thank the invited speakers for presenting inspirational talks and sharing their insights in discussions with participants. We are grateful to the members of the Steering Committee for keeping the tradition of the high-level SOFSEM series and to all the members of the program committee and the reviewers for their efforts in the reviewing process. We thank the organizers led by Branislav Rovan, Dana Pardubská, and Jana Kostičová for perfect local arrangements. We also thank Springer for sponsoring the Best Paper Award and for publishing these proceedings in the Advanced Research in Computing and Software Science (ARCoSS) of the Lecture Notes in Computer Science (LNCS) series. Last but not least, we thank all authors who contributed to this volume to share their new ideas and results with the community of researchers in the rapidly developing field of foundations of computer science and artificial intelligence.

We hope that you will enjoy reading and find inspiration for your future work in the papers contained in these two volumes.

December 2024

Rastislav Královič
Věra Kůrková

Organization

Steering Committee

Henning Fernau (Chair)	Trier University, Germany
Leszek A. Gąsieniec	University of Liverpool, UK
Serge Gaspers	UNSW Sydney, Australia
Ralf Klasing	CNRS and University of Bordeaux, France
Tiziana Margaria	University of Limerick, Ireland
Mirosław Kutyłowski	NASK – National Research Institute, Poland
Branislav Rovan	Comenius University in Bratislava, Slovakia
Jan van Leeuwen	Utrecht University, The Netherlands
Július Štuller	Czech Academy of Sciences, Czech Republic

Program Committee

Amihood Amir	Bar-Ilan University, Israel and Georgia Tech, USA
Přemysl Brada	University of West Bohemia, Czech Republic
Tiziana Calamoneri	Sapienza University of Rome, Italy
Jérémie Chalopin	LIS, CNRS, Aix-Marseille Université, Université de Toulon, France
Marek Chrobak	University of California, Riverside, USA
Ivana Černá	Masaryk University, Czech Republic
Gianluca De Marco	University of Salerno, Italy
Stefan Dobrev	Slovak Academy of Sciences, Slovakia
Martin Drozda	Slovak University of Technology, Slovakia
Robert Ganian	Vienna University of Technology, Austria
Leszek Gąsieniec	University of Liverpool, UK
Cyril Gavoille	LaBRI, University of Bordeaux, France
Lucjan Hanzlik	CISPA Helmholtz Center for Information Security, Germany
Markus Holzer	Universität Giessen, Germany
Ling-Ju Hung	National Taipei University of Business, Taiwan
Petr Jančar	Palacky University Olomouc, Czech Republic
Galina Jiraskova	Slovak Academy of Sciences, Slovakia
Tomasz Jurdzinski	University of Wrocław, Poland
Petteri Kaski	Aalto University, Finland
Philipp Kindermann	Universität Trier, Germany

Dennis Komm	ETH Zurich, Switzerland
Rastislav Královič (Chair)	Comenius University in Bratislava, Slovakia
Danny Krizanc	Wesleyan University, USA
Věra Kůrková (Chair)	Czech Academy of Sciences, Czech Republic
Giuseppe Liotta	University of Perugia, Italy
Alexei Lisitsa	University of Liverpool, UK
Hsiang-Hsuan Liu	Utrecht University, The Netherlands
Alessio Mansutti	IMDEA Software Institute, Spain
Marco Mesiti	University of Milan, Italy
Xavier Muñoz Lopez	Universitat Politècnica de Catalunya, Spain
Vangelis Paschos	Université Paris Dauphine-PSL, France
Rajeev Raman	University of Leicester, UK
Peter Rossmanith	RWTH Aachen University, Germany
Pawel Sobocinski	Tallinn University of Technology, Estonia
Ulrike Stege	University of Victoria, Canada
Gerth Stölting Brodal	Aarhus University, Denmark

Additional Reviewers

Angara, Prashanti
Ayaziová, Paulína
Baheri, Betis
Bai, Tian
Bakhshi-Khaniki, Hessam
Banik, Aritra
Bednarczyk, Bartosz
Benes, Nikola
Bentert, Matthias
Bercea, Ioana
Berendsohn, Benjamin Aram
Binucci, Carla
Bouchard, Sébastien
Bournez, Olivier
Brötzner, Anna
Burjons, Elisabet
Butman, Moshe
Böckenhauer, Hans-Joachim
Böhm, Martin
Cavalleri, Emanuele
Chistikov, Dmitry
Corò, Federico
D'Elia, Marco
de Castro Mendes Gomes, Guilherme
Defrain, Oscar
Deligkas, Argyrios
Disser, Yann
Dondi, Riccardo
Dreier, Jan
Dudek, Bartlomiej
Döring, Michelle
Erlebach, Thomas
Fasoulakis, Michail
Felsner, Stefan
Fijalkow, Nathanaël
Filakovský, Marek
Fioravantes, Foivos
Förster, Henry
Galby, Esther
Ganty, Pierre
Gargano, Luisa
Garncarek, Paweł
Gehnen, Matthias
Gigante, Nicola
Glazenburg, Erwin
Grüne, Christoph

Haase, Carolina
Han, Yo-Sub
Hansen, Kristoffer Arnsfelt
Hlineny, Petr
Huang, Shang-En
Hörsch, Florian
Itzhaki, Michael
Jana, Satyabrata
Kawahara, Jun
Klemz, Boris
Kobayashi, Yasuaki
Korhonen, Tuukka
Kralovic, Richard
Krekelberg, Bob
Kryven, Myroslav
Král, Pavel
Labourel, Arnaud
Lagarde, Guillaume
Lecroq, Thierry
Lenc, Ladislav
Lin, Chuang-Chieh
Liskiewicz, Maciej
Liu, Fu-Hong
Martínek, Jiří
Masopust, Tomas
Mavronicolas, Marios
McKenzie, Pierre
Migler, Theresa
Mock, Daniel
Mráz, František
Nanoti, Saraswati
Ng, Timothy
Nõmm, Sven
Obdrzalek, Jan
Olchanyi, Maxim
Ortali, Giacomo
Osička, Petr
Paesani, Giacomo
Pashaeibarough, Ali
Perz, Daniel
Prigioniero, Luca
Prusa, Daniel
Rampersad, Narad
Reinhardt, Klaus
Rescigno, Adele
Rinaldi, Francesco
Roy, Shivesh K.
Rysgaard, Casper
S. Sankar, Govind
Sadhukhan, Arpan
Sahu, Abhishek
Salvo, Ivano
Sampaio, Rudini
Saumell, Maria
Sawa, Zdeněk
Schapire, Robert
Schou, Jens Kristian Refsgaard
Seki, Shinnosuke
Shapira, Dana
Sheth, Kshiteej
Sieper, Marie Diana
Skoviera, Martin
Stachowiak, Grzegorz
Stocker, Moritz
Suchý, Ondřej
Svenning, Rolf
Szykuła, Marek
Tantau, Till
Tao, Terence
Tappini, Alessandra
Tu, Ta-Wei
Udwani, Rajan
Unger, Walter
Valencia, Frank
Van Der Merwe, Brink
Vaszil, György
Walen, Tomasz
Walzer, Stefan
Whittington, Philip
Zaborniak, Tristan
Zhang, Qiankun
Zhu, Zixuan
Zink, Johannes

Short Invited Talks

Temporal Graph Realization Problems

George B. Mertzios[1] and Paul G. Spirakis[2]

[1] Department of Computer Science, Durham University, UK
george.mertzios@durham.ac.uk
[2] Department of Computer Science, University of Liverpool, UK
p.spirakis@liverpool.ac.uk

Abstract. In this talk, we introduce the *temporal graph realization* problem with respect to the fastest path durations among its vertices, while we focus on periodic temporal graphs. In the basic version of the problem, given an $n \times n$ matrix D and a $\Delta \in \mathbb{N}$, the goal is to construct a Δ-periodic temporal graph with n vertices such that the duration of a *fastest path* from v_i to v_j is equal to $D_{i,j}$, or to decide that such a temporal graph does not exist. The variations of the problem on static graphs have been well studied and understood since the 1960s, see e.g. [1, 2]. As it turns out, this basic version of the periodic temporal graph realization problem has a very different computational complexity behavior than its static (ie. non-temporal) counterpart [3].

First, the problem is NP-hard in general, but polynomial-time solvable if the so-called underlying graph is a tree. Building upon those results, we investigate its parameterized computational complexity with respect to structural parameters of the underlying static graph which measure the "tree-likeness". We prove a tight classification between such parameters that allow fixed-parameter tractability (FPT) and those which imply W[1]-hardness. We show that our problem is W[1]-hard when parameterized by the *feedback vertex number* (and therefore also any smaller parameter such as *treewidth*, *degeneracy*, and *cliquewidth*) of the underlying graph, while we show that it is in FPT when parameterized by the *feedback edge number* (and therefore also any larger parameter such as *maximum leaf number*) of the underlying graph.

Then, we focus on the *upper bound* variation of the problem where, given an $n \times n$ matrix D, the question is whether there exists a periodic temporal graph on n vertices such that the duration of the fastest temporal path from a vertex u to a vertex v is *at most* $D_{u,v}$ [4]. This constraint with respect to upper bounds appears naturally in *transportation network design* applications where, for example, a road network is given, and the goal is to appropriately schedule periodic travel routes, while not exceeding some desired upper bounds on the travel times. This approach is in contrast to verification applications of the graph realization problems, where *exact* values for the distances (respectively, fastest travel times) are given, following some kind of precise measurement. In this problem variation, we focus only on underlying *tree topologies*, which are fundamental in many transportation network applications.

As it turns out, the periodic upper-bounded temporal tree realization problem (TTR) has a very different computational complexity behavior than both (i) the classic graph realization problem with respect to shortest path distances in static graphs and (ii) the periodic temporal graph realization problem with *exact* given fastest travel times (which was recently introduced). First, we prove that, surprisingly, TTR is NP-hard, even for a constant period Δ and when the input tree G satisfies one of the following conditions: (a) G has a constant diameter, or (b) G has constant maximum degree. In contrast, when we are given exact values of the fastest travel delays, the problem is known to be solvable in polynomial time. Second, we prove that TTR is fixed-parameter tractable (FPT) with respect to the number of leaves in the input tree G, via a novel combination of techniques for totally unimodular matrices and mixed integer linear programming.

Keywords: Temporal graph · Periodic temporal labeling · Fastest temporal path · Graph realization · Temporal connectivity · Parameterized complexity

References

1. Erdős, P., Gallai, T.: Graphs with prescribed degrees of vertices. Mat. Lapok, **11**, 264–274 (1960)
2. Hakimi, S.L., Yau, S.S.: Distance matrix of a graph and its realizability. Q. Appl. Math. **22**(4), 305–317 (1965)
3. Klobas, N., Mertzios, G.B., Molter, H., Spirakis, P.G.: Temporal graph realization from fastest paths. In: Proceedings of the 3rd Symposium on Algorithmic Foundations of Dynamic Networks (SAND), pp. 16:1–16:18. Best Student Paper Award (2024)
4. Mertzios, G.B., Molter, H., Spirakis, P.G.: Realizing temporal transportation trees. CoRR, abs/2403.18513 (2024)

The Challenge of Stability, Accuracy and Robustness in Data-Driven AI

Ivan Tyukin

Department of Computer Science, King's College London, London, WC2R 2LS, UK
ivan.tyukin@kcl.ac.uk

Abstract. In this lecture, we will delve into the theoretical limitations of determining the guaranteed stability and accuracy of neural networks built from empirical data in classification tasks. We will show that there is a large family of tasks and settings in which computing and verifying stability and accuracy is extremely challenging. We will also discuss an intriguing connection of these results with adversarial data and examples and propose a potential way to remedy the issues by enabling the networks to adapt over time.

Keywords: Stability of AI · Accuracy of AI · Verifiability of AI

The problem of reliability and validity of artificial intelligence (AI) systems has been the focus of attention of the scientific community since the Dartmouth Conference of 1956, considered the starting point and place of birth of AI as a new independent discipline. The problem of reliability and validity of AI at that time was initially reflected in the assumption set out by McCarthy, Minsky, Rochester and Shannon that "... any aspect of learning or any other property of intelligence can in principle be so precisely described that a machine can simulate it" [1].

This vision determined the development of science in the field of AI for many decades. For example, formal languages were developed for the analysis and verification of AI systems built on rules, including Isabelle and Lean [2], which are now successfully used for automatic verification of formal statements and proofs of theorems. In non-deterministic cases, when it was not possible to compile a deterministic system of rules describing AI decisions, or when the problems themselves had a significant random component (for example, under the assumption of statistical data generation), the issue of reliability, provability and robustness was addressed in the context of statistical methods and approaches [3, 4].

The rapid development of technologies in the last two decades, coupled with the ever-growing demand for automation of increasingly complex tasks that were previously entirely within the competence of humans, has led to the identification of practical limitations of classical theories and methods that have traditionally been used to justify the reliability and robustness of AI systems. It turned out that many important problems often cannot be effectively solved by deterministic algorithms (a formal system of rules). Examples include the well-known halting problems (determine whether a program will end/stop) and the Traveling Salesman problem.

On the other hand, problems have also emerged in statistical settings, the reliable and robust solution of which faces serious and often insoluble practical problems. This class of problems includes classification tasks in high dimensions, in which the number of significant yet mutually dependent attributes is large. Such problems are common in medicine, finance, and computer vision. Obtaining complete information about the distribution in the absence of any assumptions for problems with 100 relevant attributes requires samples of the order of $2^{100} > 10^{30}$. This limits the application of standard and well-studied statistical approaches (e.g. Bayesian networks and trees, Bayesian decision theory, etc.).

Alternative approaches based on the worst-case analysis (e.g. using the Vapnik-Chervonenkis theory, Rademacher complexity, and covering numbers) do not require knowledge of distributions. However, their practical application presupposes the availability of data, the volume of which directly depends on the complexity of the model, which also creates barriers and limitations. On the one hand, the more complex and expressive the AI, the more accurate and precise the solution is potentially. On the other hand, the more data is required to guarantee and confidently achieve this accuracy. Moreover, for modern AI systems, which are described by millions and trillions of parameters (Chat GPT 4), estimates of provable reliability in the form of bounds on the risk of errors following from these generally accepted theories may require unacceptably large amounts of data, growing proportionally to the number of parameters (see, for example, [6], Theorems 5.2 and 8.9).

To circumvent these limitations, alternative approaches to the analysis and assessment of reliability have been proposed in the last decade. Among these approaches, it is worth noting "explainable AI" (XAI) and the approach of quantitative assessment of uncertainty, which is applied to already built systems (also known as uncertainty quantification). Unfortunately, to date, explainable AI is still far from solving the problem of trust and reliability [5, 7], and effective assessment of the uncertainty of modern large generative models, including calibration, is still in its infancy and is difficult to apply in practice (see, for example, [8], " ... the field is still in its infancy.").

In our work, we explore the possibility of ensuring both, stability and accuracy, for a large class of AI models –feed-forward neural networks. We show that this problem may bring new challenges. These challenges revolve around the computational complexity of producing rigorous stability guarantees in classical distribution-agnostic settings. Strikingly, the issues emerge for both low- and high-dimensional problems. High dimension, naturally, makes verification of stability more difficult. The validation challenge is inherent in static networks whose architecture does not change over time. Therefore a potential resolution of the issue may arise from enabling AI models to change their architecture dynamically in response to detected undesirable behaviors. An example of such changes could be the addition of error correctors [9–11]. This, however, may require new notions of stability and a new theory for assessing asymptotic temporal properties of such adaptive AI models.

References

1. McCarthy, J., Minsky, M., Rochester, N., Shannon C.E.: A Proposal for the Dartmouth Summer Research Project on Artificial Intelligence (1955). http://raysolomonoff.com/dartmouth/boxa/dart564props.pdf
2. Harrison, J., Urban, J., Wiedijk, F.: History of interactive theorem proving. In: Handbook of the History of Logic, vol. 9, pp. 135–214, North-Holland (2014)
3. Devroye, L., Györfi, L., Lugosi, G.: A Probabilistic Theory of Pattern Recognition, vol. 31. Springer Science and Business Media (1997)
4. Vapnik, V.: Statistical Learning Theory. John Wiley and Sons (1998)
5. de Bruijn, H., Warnier, M., Janssen, M.: The perils and pitfalls of explainable AI: strategies for explaining algorithmic decision-making. Gov. Inf. Q. **39**(2), 101666 (2022)
6. Antony, M., Bartlet, P: Neural Network Learning: Theoretical Foundations. Cambridge University Press, Cambridge (1999)
7. Bove, C., Laugel, T., Lesot, M.J., Tijus, C., Detyniecki, M.: Why do Explanations Fail? A Typology and Discussion on Failures in XAI (2024). arXiv preprint arXiv:2405.13474
8. He, W., Jiang, Z., Xiao, T., Xu, Z., Li, Y.: A Survey on Uncertainty Quantification Methods for Deep Learning (2023). arXiv preprint arXiv:2302.13425
9. Gorban, A., Grechuk, B., Tyukin, I.: Stochastic separation theorems: how geometry may help to correct AI errors. Not. Am. Math. Soc. (1), 25–33 (2023)
10. Gorban, A.N., Golubkov, A., Grechuk, B., Mirkes, E.M., Tyukin, I.Y.: Correction of AI systems by linear discriminants: probabilistic foundations. Inf. Sci. **466**, 303–322 (2018)
11. Tyukin, I.Y., Gorban, A.N., Sofeykov, K.I., Romanenko, I.: Knowledge transfer between artificial intelligence systems. Front. Neurorobotics **12**, 49 (2018)

Contents – Part I

Invited Talks

Contributed Papers

Contents – Part II

Contributed Papers

Invited Talks

Distributed Computing by Mobile Robots: Exploring the Computational Landscape (Extended Abstract)

Paola Flocchini(✉)

University of Ottawa, Ottawa, Canada
paola.flocchini@uottawa.ca

A large body of work in distributed computing has focused on the computational and complexity issues that arise in systems of mobile computational entities, called robots. The robots operate in Euclidean spaces (typically the plane) in `Look-Compute-Move` (`LCM`) cycles of activities: in each cycle, a robot perceives the positions of the other robots within its local coordinate system (`Look`), executes a set of deterministic rules common to all robots to compute a destination (`Compute`), and then moves toward that destination, not necessarily reaching it (`Move`). The main concern of the research in the field is on understanding the minimal conditions that would allow the robots to solve a given problem. Some of the tasks that have been extensively investigated in the literature are *gathering/convergence* (e.g., [1,2,7,16,22,28]), and *pattern formation* (e.g., [10,15,17,21,25,27,28]). See [14] and chapters therein for an account on the current research.

The Schedulers. One of the key factors influencing the computational power of these robots is the level of synchronization and the duration of each operation within the `LCM` cycle. In the synchronous setting, time is divided into discrete rounds; in each round, a subset of robots is activated and perform their `LCM` cycle simultaneously. In contrast, in the asynchronous setting, each robot operates independently, following its own cycle without any coordination with the others. In both cases, the activation of robots is governed by a fair adversary (called *scheduler*). Various types of schedulers have been proposed and studied within both the synchronous and asynchronous settings. The three classical ones are: the special fully synchronous scheduler `F` when all robots are activated at each cycle, the most general synchronous scheduler `S` (also called semi-synchronous), where the adversary can chose any subset of robots to be activated at each cycle, and the most general asynchronous scheduler `A` where the adversary has full power both in terms of activation time and of duration of each activity within the cycle[1].

The $\mathcal{OBLOT}$ Model. In the classical $\mathcal{OBLOT}$ model, the robots are characterized by *obliviousness*, *silence*, and *anonymity*: when activated, they have no memory of previous cycles; moreover, they lack explicit communication mechanisms, and they are identical in their appearance and behavior. One of the

[1] The only activity that takes an instantaneous amount of time is the `Look` activity.

R. Královič and V. Kůrková (Eds.): SOFSEM 2025, LNCS 15538, pp. 3–7, 2025.
https://doi.org/10.1007/978-3-031-82670-2_1

central research questions has been how internal weaknesses (like obliviousness and silence), and external factors (like synchrony and the scheduler type) impact the computational power of these robots.

Enhancement: Luminous Robots. Over the years, various enhancement of the basic model have been studied in regards to memory and communication under the different activation schedules (e.g., [5,6,8,9,11,19,20,23,26]).

A widely studied extension of $\mathcal{OBLOT}$ is the $\mathcal{LUMI}$ model [9], where each robot is equipped with a light showing a colour (from a small set of colours) that can change at every cycle. This light, which is visible to all the robots, can be used as a constant-size persistent memory, as well as a limited form of communication. Between the basic $\mathcal{OBLOT}$ model and the more powerful $\mathcal{LUMI}$ model are: $\mathcal{FSTA}$, where the robots have an internal light only, providing a constant-size persistent memory but no communication mechanism, and $\mathcal{FCOM}$, where their light is visible to others but not to themselves, allowing them to communicate a constant number of bits but not to remember. A natural question in this context is to determine the impact of lights on the overall capabilities of the robots and, in particular, to understand the relationship between memory and communication (is it better to remember or to communicate?).

The Landscape. The computational power of the robots depends on several factors; in particular on the adversarial scheduler, and on the model (i.e., the availability of light and their extent).

The relationship among $\mathcal{OBLOT}$, $\mathcal{FSTA}$, $\mathcal{FCOM}$ and $\mathcal{LUMI}$ has been thoroughly studied with respect to the three major schedulers F, S, and A, and over time a complete map of their computational relationship has been drawn [5,6,9,18,20].

Let $M \in \{\mathcal{OBLOT}, \mathcal{FSTA}, \mathcal{FCOM}, \mathcal{LUMI}\}$ be a model, and let $\mathtt{H} \in \{\mathtt{F}, \mathtt{S}, \mathtt{A}\}$ be a scheduler. Let M^H denote the set of problems solvable in model M under scheduler H. The relative computational power of the robots can be studied comparing all the combinations of $M_i^{H_j}$.

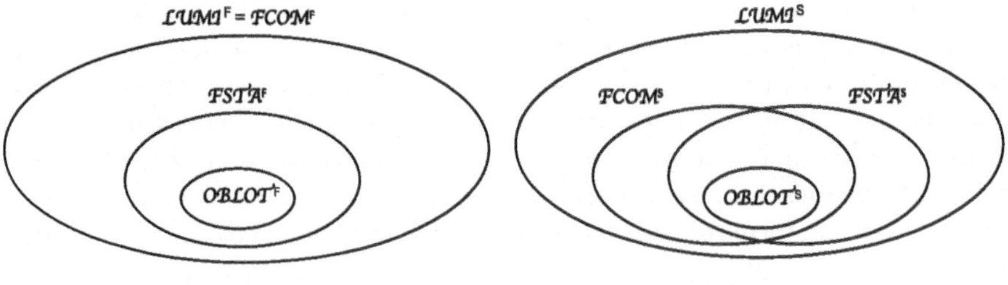

Fig. 1. Computational relationship between the models under scheduler F and S.

Among other things, it has been shown that $\mathcal{FSTA}^F \subset \mathcal{FCOM}^F$, that is, the little communication provided by $\mathcal{FCOM}$ is strictly more powerful than the little memory provided by $\mathcal{FSTA}$ if the scheduler is fully synchronous; on the other hand, the two models are orthogonal under the semi-synchronous and the asynchronous schedulers (see Fig. 1).

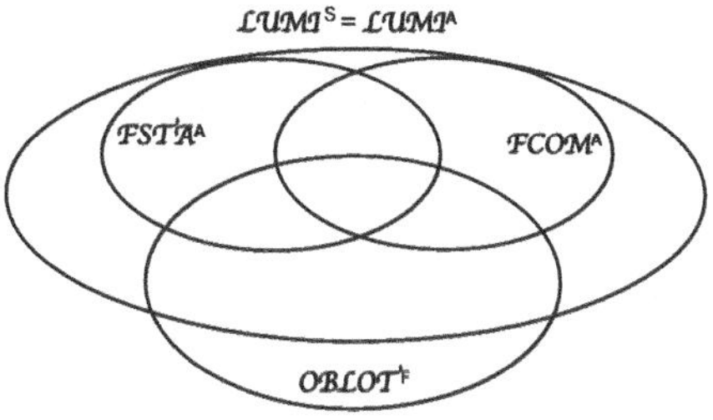

Fig. 2. Some orthogonality results.

Another very interesting orthogonality result concerns the least powerful $\mathcal{OBLOT}$ model under the strongest scheduler F, versus the most powerful $\mathcal{LUMI}$ under the weakest scheduler A: the classes of problems solvable under the two model's combinations are orthogonal, meaning that the power of lights and that of full synchronicity are incomparable (see Fig. 2).

Another notable aspect is the presence in the landscape of *separator* problems which provide a clear hierarchy of the computational power of the robots under the three main schedulers. In fact, with the interesting exception of $\mathcal{LUMI}$, in all other models the computational power of the robots under full synchrony is strictly stronger than the one under semi-synchrony, which is strictly stronger than the one under asynchrony (see Fig. 3).

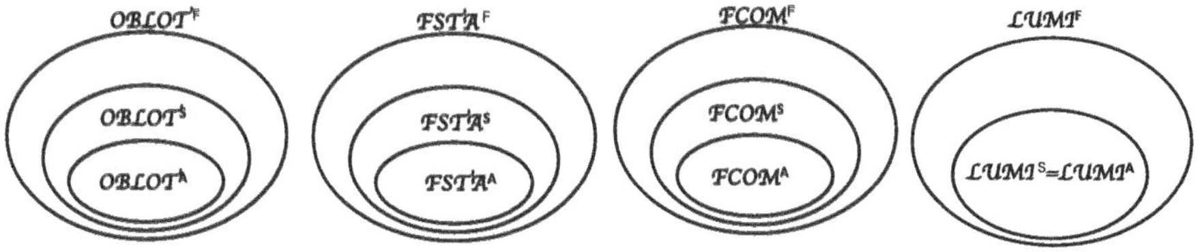

Fig. 3. Computational relationship within $\mathcal{OBLOT}$, $\mathcal{FSTA}$, $\mathcal{FCOM}$, $\mathcal{LUMI}$.

Enhancement: Robots with Ids. In recent years, the computational landscape has expanded to include new aspects, particularly those that are typically explored in other areas of distributed computing but have not yet been fully studied in this context. One such development is the introduction of identifiers (Ids) for robots, breaking with the traditional assumption of anonymity (e.g., [3,4,12,13,24]). Analogously to what is seen in other areas of distributed computing, a problem that is unsolvable by a team of anonymous robots might become solvable if the robots are equipped with Ids, even if the Ids are not necessarily all distinct. An interesting new research direction is to investigate the computational power of the robots in relation to the number of distinct Ids that they are assigned (e.g., [3,13]). Another research direction is the employment of different Ids to have different teams of robots that, while externally identical, are operating in the same space at the same time trying to accomplish different tasks (e.g., [4]).

References

1. Agmon, N., Peleg, D.: Fault-tolerant gathering algorithms for autonomous mobile robots. SIAM J. Comput. **36**(1), 56–82 (2006)
2. Ando, H., Oasa, Y., Suzuki, I., Yamashita, M.: A distributed memoryless point convergence algorithm for mobile robots with limited visibility. IEEE Trans. Rob. Autom. **5**, 818–828 (1999)
3. Asahiro, Y., Yamashita, M.: Minimum algorithm sizes for self-stabilizing gathering and related problems of autonomous mobile robots (extended abstract). In: 25th International Symposium on Stabilization, Safety, and Security of Distributed Systems (SSS), pp. 312–327 (2023)
4. Bhagat, S., Flocchini, P., Mukhopadhyaya, K., Santoro, N.: Weak robots performing conflicting tasks without knowing who is in their team. In: 21st International Conference on Distributed Computing and Networking (ICDCN), pp. 29:1–29:6 (2020)
5. Buchin, K., Flocchini, P., Kostitsyna, I., Peters, T., Santoro, N., Wada, K.: Autonomous mobile robots: refining the computational landscape. In: IEEE International Parallel and Distributed Processing Symposium Workshops, pp. 576–585 (2021)
6. Buchin, K., Flocchini, P., Kostitsyna, I., Peters, T., Santoro, N., Wada, K.: On the computational power of energy-constrained mobile robots: algorithms and cross-model analysis. In: 29th International Colloquium on Structural Information and Communication Complexity (SIROCCO), pp. 42–61 (2022)
7. Cieliebak, M., Flocchini, P., Prencipe, G., Santoro, N.: Distributed computing by mobile robots: gathering. SIAM J. Comput. **41**(4), 829–879 (2012)
8. Das, S., Flocchini, P., Prencipe, G., Santoro, N.: Forming sequences of patterns with luminous robots. IEEE Access **8**, 90577–90597 (2020)
9. Das, S., Flocchini, P., Prencipe, G., Santoro, N., Yamashita, M.: Autonomous mobile robots with lights. Theor. Comput. Sci. **609**, 171–184 (2016)
10. Das, S., Flocchini, P., Santoro, N., Yamashita, M.: Forming sequences of geometric patterns with oblivious mobile robots. Distrib. Comput. **28**(2), 131–145 (2015)
11. Feletti, C., Mambretti, L., Mereghetti, C., Palano,: Computational power of opaque robots. In: 3rd Symposium on Algorithmic Foundations of Dynamic Networks (SAND), pp. 13:1–13:19 (2024)
12. Flocchini, P., Pattanayak, D., Piselli, F., Santoro, N., Yamauchi, Y.: Asynchronous separation of unconscious colored robots. In: 16th International Workshop on Parallel and Distributed Algorithms and Applications (2024)
13. Flocchini, P., Pattanayak, D., Santoro, N., Yamashita, M.: The minimum algorithm size of k-grouping by silent oblivious robots. In: 35th International Workshop on Combinatorial Algorithms (IWOCA), pp. 472–484 (2024)

14. Flocchini, P., Prencipe, G., Santoro, N. (eds.): Distributed Computing by Mobile Entities. Springer, Heidelberg (2019)
15. Flocchini, P., Prencipe, G., Santoro, N., Viglietta, G.: Distributed computing by mobile robots: uniform circle formation. Theor. Comput. Sci. **30**(6), 413–457 (2017)
16. Flocchini, P., Prencipe, G., Santoro, N., Widmayer, P.: Gathering of asynchronous robots with limited visibility. Theor. Comput. Sci. **337**(1–3), 147–168 (2005)
17. Flocchini, P., Prencipe, G., Santoro, N., Widmayer, P.: Arbitrary pattern formation by asynchronous, anonymous, oblivious robots. Theor. Comput. Sci. **28**, 1–23 (2008)
18. Flocchini, P., Santoro, N., Sudo, Y., Wada, K.: On asynchrony, memory, and communication: separations and landscapes. In: 27th International Conference on Principles of Distributed Systems (OPODIS), pp. 28:1–28:23 (2023)
19. Flocchini, P., Santoro, N., Viglietta, G., Yamashita, M.: Rendezvous with constant memory. Theor. Comput. Sci. **621**, 57–72 (2016)
20. Flocchini, P., Santoro, N., Wada, K.: On memory, sommunication, and synchronous schedulers when moving and computing. In: 23rd International Conference on Principles of Distributed Systems (OPODIS), pp. 225:1–25:17 (2019)
21. Fujinaga, N., Yamauchi, Y., Ono, H., Kijima, S., Yamashita, M.: Pattern formation by oblivious asynchronous mobile robots. SIAM J. Comput. **44**(3), 740–785 (2016)
22. Kirkpatrick, D.G., Kostitsyna, I., Navarra, A., Prencipe, G., Santoro, N.: On the power of bounded asynchrony: convergence by autonomous robots with limited visibility. Distrib. Comput. (2024)
23. Nakai, R., Sudo, Y., Wada, K.: Asynchronous gathering algorithms for autonomous mobile robots with lights. In: 23rd International Symposium on Stabilization, Safety, and Security of Distributed Systems (SSS), pp. 410–424 (2021)
24. Seike, H., Yamauchi, Y.: Separation of unconscious colored robots. In: 25th International Symposium on Stabilization, Safety, and Security of Distributed Systems (SSS), pp. 328–343 (2023)
25. Suzuki, I., Yamashita, M.: Distributed anonymous mobile robots: formation of geometric patterns. SIAM J. Comput. **28**(4), 1347–1363 (1999)
26. Terai, S., Wada, K., Katayama, Y.: Gathering problems for autonomous mobile robots with lights. Theor. Comput. Sci. **941**, 241–261 (2023)
27. Yamashita, M., Suzuki, I.: Characterizing geometric patterns formable by oblivious anonymous mobile robots. Theor. Comput. Sci. **411**(26–28), 2433–2453 (2010)
28. Yamauchi, Y., Uehara, T., Kijima, S., Yamashita, M.: Plane formation by synchronous mobile robots in the three dimensional euclidean space. J. ACM **64**(3), 16:1–16:43 (2017)

Open Problems and Recent Developments on a Complexity Framework for Forbidden Subgraphs

Erik Jan van Leeuwen(✉)

Department Information and Computing Sciences, Utrecht University, Utrecht, The Netherlands
e.j.vanleeuwen@uu.nl

Abstract. For any finite set $\mathcal{H} = \{H_1, \ldots, H_p\}$ of graphs, a graph is $\mathcal{H}$-subgraph-free if it does not contain any of $H_1, \ldots, H_p$ as a subgraph. In this invited talk, I discuss a recently proposed algorithmic meta classification that precisely classifies if certain problems (sharing specific properties) are "efficiently solvable" or "computationally hard" for $\mathcal{H}$-subgraph-free graphs, depending on $\mathcal{H}$. For a broad set of classic graph problems, this framework yields a dichotomy (depending on $\mathcal{H}$) between polynomial-time solvability and NP-completeness. For other problems, like computing the diameter of a graph, it gives a dichotomy between almost-linear-time solvability and having no subquadratic-time algorithm (conditioned on some hardness hypotheses). This paper discusses this framework, highlights current developments and open problems, and surveys recent insights into the complexity on $\mathcal{H}$-subgraph-free graphs of problems that do not fall within the framework.

1 Introduction

Many graph problems are computationally hard. However, making assumptions on the structure of the input graph can often help in the design of efficient algorithms for such problems. This has led to significant research into algorithms for classes of graphs such as planar graphs, (subclasses of) perfect graphs, tree-structured graphs, and graphs with forbidden substructures. The extensive Information System on Graph Classes and their Inclusions [17] provides a glimpse into the wealth of research into this topic.

This article focusses on graphs that forbid certain substructures. All graphs in this article are simple and undirected. Let $\mathcal{H}$ be any set of graphs. Then a graph G is *$\mathcal{H}$-free* if G contains no graph of $H \in \mathcal{H}$ as an induced subgraph, i.e., no such H can be obtained by removing vertices (and their incident edges) from G. A graph G is *$\mathcal{H}$-subgraph-free* if G contains no graph of $H \in \mathcal{H}$ as a subgraph, i.e., no such H can be obtained by removing vertices and edges from G. A graph G is *$\mathcal{H}$-minor-free* if G contains no graph of $H \in \mathcal{H}$ as a minor, i.e., no such H can be obtained by removing vertices and edges from G and contracting edges of G. Recall that the edge contraction operation identifies the endpoints of an edge

R. Královič and V. Kůrková (Eds.): SOFSEM 2025, LNCS 15538, pp. 8–20, 2025.
https://doi.org/10.1007/978-3-031-82670-2_2

(and removes any loops and duplicate edges that may arise). Note that for any set $\mathcal{H}$, the class of $\mathcal{H}$-free graphs is a superclass of the class of $\mathcal{H}$-subgraph-free graphs, which in turn is a superclass of the class of $\mathcal{H}$-minor-free graphs.

Of these classes, the $\mathcal{H}$-minor-free graphs are perhaps the best understood. Not only do they generalize the well-studied class of planar graphs, but the seminal work of Robertson and Seymour [43] gave deep structural insights into the structure of $\mathcal{H}$-minor-free graphs that is highly useful in algorithm design (see, for example, the widely applicable bidimensionality framework for parameterized problems [18,23]).

A crucial tool in such algorithms is the notion of *treewidth*, which is the minimum width of any tree decomposition of a graph. Recall that a *tree decomposition* of a graph G is a tree T and a set of vertex *bags* $\{X_t \subseteq V(G) \mid t \in V(T)\}$ such that $V(G) = \bigcup_{t\in V(T)} X_t$; for any $v \in V(G)$, the bags containing v induce a subtree T_v of T; for any edge $uv \in E(G)$, the subtrees T_u and T_v intersect. The *width* of a tree decomposition is the largest size of any bag minus 1. Intuitively, treewidth is a measure of tree-likeness of a graph; in particular, the treewidth of a tree is 1, the treewidth of a cycle is 2, and the treewidth of an n-vertex complete graph is $n - 1$. There is substantial literature and active research on efficient algorithms for problems on graphs of bounded treewidth (see e.g. [3,8,14,16]).

Our broad understanding of treewidth enables many algorithmic results. Notably, it sometimes enables so-called *algorithmic meta classifications*, which state that all problems sharing some property are efficiently solvable when restricted to a particular class of inputs, but are computationally hard for all other classes of inputs. The results of Robertson and Seymour on $\mathcal{H}$-minor-free graphs directly imply the following algorithmic meta classification:

Theorem 1 [33,37,44]. *Consider a graph problem* Π *that is:*
(C1) efficiently solvable on every graph class of bounded treewidth;
(C4) computationally hard on planar graphs.
Then for any set $\mathcal{H}$ *of graphs, on the class of* $\mathcal{H}$*-minor-free graphs,* Π *is efficiently solvable if* $\mathcal{H}$ *contains a planar graph and is computationally hard otherwise.*

The notions of "efficiently solvable" and "computationally hard" in this theorem are left ambiguous on purpose, to retain broad applicability, but for now the reader may think of them as polynomial-time solvable and NP-complete respectively.

Theorem 1 is widely applicable, because we know that many problems satisfy (C1) and (C4). For example, this holds for Independent Set, Vertex Cover, Steiner Tree, Disjoint Paths, Feedback Vertex Set, 3-Coloring, and many others. Hence, a dichotomy between polynomial-time solvability and NP-completeness on $\mathcal{H}$-minor-free graphs for these problems, depending on whether $\mathcal{H}$ contains a planar graph, follows directly from this classification. Refer to Johnson et al. [33, Table 1] for proofs and more examples.

With Theorem 1 in hand, we may wonder about meta classifications for other graph classes with forbidden substructures, which this paper now explores.

2 An Algorithmic Meta Classification for $\mathcal{H}$-Subgraph-Free Graphs

Many classes of $\mathcal{H}$-subgraph-free graphs are widely studied. For example, the $K_{1,d+1}$-subgraph-free graphs are exactly the graphs of maximum degree d and classes of graphs that exclude a bounded-length path as a subgraph are directly related to classes of graphs of bounded treedepth [42]. Several works provide (almost) complete dichotomy results for specific problems on $\mathcal{H}$-subgraph-free graphs for finite $\mathcal{H}$, notably Independent Set and Max-Cut; see e.g. [2,10, 30,31,36]. Using similar proof ideas as in [30,31,36], Johnson et al. [33] recently showed a broadly applicable algorithmic meta classification for $\mathcal{H}$-subgraph-free graphs.

To state this result, we first need to define the notion of *edge subdivision* in a graph G. Let $\ell \geq 1$ be an integer. The *ℓ-subdivision* of an edge $uv \in E(G)$ replaces uv by a path of length $\ell+1$. Then the *ℓ-subdivision* of G simultaneously ℓ-subdivides each edge of G. For a graph class $\mathcal{G}$, the class $\mathcal{G}^\ell$ consists of all graphs that are the ℓ-subdivision of a graph in $\mathcal{G}$. We say that a graph problem Π is *computationally hard under edge subdivision of subcubic graphs* if for every integer $j \geq 1$, there is an integer $\ell \geq j$ such that if Π is computationally hard on the class $\mathcal{G}$ of subcubic graphs, then Π is computationally on the class $\mathcal{G}^\ell$.

Theorem 2 [33]. *Consider a graph problem Π that is:*
(C1) efficiently solvable on every graph class of bounded treewidth;
(C2) computationally hard on subcubic graphs;
(C3) computationally hard under edge subdivision of subcubic graphs.
Let $\mathcal{S}$ be the set of graphs in which each component is either a path or a subdivision of the claw ($K_{1,3}$). Then for any finite set $\mathcal{H}$ of graphs, on the class of $\mathcal{H}$-subgraph-free graphs, Π is efficiently solvable if $\mathcal{H}$ contains a graph from $\mathcal{S}$ and is computationally hard otherwise.

Call any problem that satisfies (C1), (C2), and (C3) a *C123-problem.* A broad range of problems is C123, including for example Independent Set, Vertex Cover, Steiner Tree, Disjoint Paths, Max-Cut, Treewidth, and many others. Hence, a dichotomy between polynomial-time solvability and NP-completeness on $\mathcal{H}$-subgraph-free graphs for these problems, depending on whether $\mathcal{H}$ contains a graph from the set $\mathcal{S}$, follows directly from this classification. Refer to Johnson et al. [33, Table 1] for proofs and more examples.

Recent works have sought to find further C123 problems, in particular by addressing problems for which one or more of the properties (C1), (C2), (C3) were not yet known. Johnson et al. [35] showed that Multiway Cut is NP-complete on subcubic graphs and Eagling-Vose et al. [21] showed that Graph Homomorphism satisfies (C3). Combined with older results in the literature, this enables a dichotomy result for these problems as in the previous paragraph. Eagling-Vose et al. [21] consider various graph homomorphism variants as well.

Open Problem 1. *Which other graph problems are C123?*

Next, we observe that since the proofs of Theorem 1 and 2 are both purely structural, they enable also other dichotomy results than between polynomial-time solvability and NP-completeness. For example, Johnson et al. [33] show that Theorem 2 combined with known results in the literature implies that DIAMETER is almost linear-time solvable on $\mathcal{H}$-subgraph-free graphs when $\mathcal{H}$ contains a graph from $\mathcal{S}$ and cannot be solved in subquadratic time otherwise, unless the Orthogonal Vectors Conjecture is false. Eagling-Vose et al. [21] show, using Theorem 2, that for a variant of the Quantified Constraint Satisfaction Problem there is a distinction between "efficiently solvable" and "computationally hard", where "efficiently solvable" still means polynomial-time solvability, but "computationally hard" means Π_{2k}^{P}-complete, i.e. hardness in the polynomial hierarchy.

Open Problem 2. *Which other problems are C123, but offer a distinction between "efficiently solvable" and "computationally hard" different from polynomial-time solvability versus NP-completeness?*

For a further and more extensive discussion of the quirks and features of Theorem 2, refer to Johnson et al. [33]. However, here we briefly consider the role of the parameter treewidth in Theorem 2. Johnson et al. [33] noted that the classes of $\mathcal{H}$-subgraph-free graphs for finite sets $\mathcal{H}$ that have bounded treewidth are exactly those for which $\mathcal{H}$ contains a graph of $\mathcal{S}$. One may expect that the use of stronger parameters than treewidth could lead to more expansive meta classifications, as they could cover more classes of $\mathcal{H}$-subgraph-free graphs. However, this seems unlikely. Recall that the *pathwidth* of a graph is the smallest width of any path decomposition of it, which is simply a tree decomposition where the tree T must be a path.

Proposition 1. *Let p be a graph parameter that is bounded whenever the pathwidth of a graph is bounded. Let Π be a C123-problem. If Π is efficiently solvable on every graph class in which p is bounded, then for any finite set $\mathcal{H}$ of graphs, p is bounded on $\mathcal{H}$-subgraph-graphs exactly when $\mathcal{H}$ contains a graph from $\mathcal{S}$, unless the notions of efficiently solvable and computationally hard coincide.*

Proof. Let $\mathcal{H}$ be any finite set of graphs. If $\mathcal{H}$ contains a graph $H \in \mathcal{S}$, then the proof of Theorem 2 (see Johnson et al. [33]) ensures that the pathwidth of any $\mathcal{H}$-subgraph-free graph is bounded. Thus, by assumption, p is bounded and Π is efficiently solvable. Otherwise, if $\mathcal{H}$ does not contain a graph in $\mathcal{S}$, then Π is computationally hard on the class of $\mathcal{H}$-subgraph-free graphs. If p is bounded on the class of $\mathcal{H}$-subgraph-free graphs, then Π is efficiently solvable, and the notions of efficiently solvable and computationally hard coincide. □

As a consequence, if we consider stronger parameters than pathwidth, including treewidth, cliquewidth, and mim-width, then noting that INDEPENDENT SET is polynomial-time solvable on graph classes in which any of these parameters are bounded [4,12,15], we obtain that these parameters are bounded on $\mathcal{H}$-subgraph-free graphs for finite sets $\mathcal{H}$ exactly when $\mathcal{H}$ contains a graph of $\mathcal{S}$, unless P = NP.

Hence, these parameters would not yield a different algorithmic meta classification for $\mathcal{H}$-subgraph-free graphs. A similar result follows for $\mathcal{H}$-minor-free graphs with respect to Theorem 1.

Open Problem 3. *Which graph parameters allow for a different algorithmic meta classification for $\mathcal{H}$-subgraph-free graphs?*

Per Proposition 1, any such parameter would need to be weaker than or incomparable to pathwidth.

3 Beyond the Meta Classification

There exists a wide variety of problems that are beyond the algorithmic meta classification of Theorem 2. Mostly, these problems break just one of the conditions (C1), (C2), or (C3). Hence, a first course of action to study problems to which Theorem 2 does not apply, is to investigate such problems.

3.1 Problems That Only Do Not Satisfy (C1)

A canonical example is the Steiner Forest problem. This asks, given a graph G and a set $T \subseteq V(G) \times V(G)$ of terminal pairs, to find a smallest forest F as a subgraph of G such that, for any pair $(s,t) \in T$, both s and t are in the same tree of F. By results of Bateni, Hajiaghayi, and Marx [6] and Gassner [28], this problem is polynomial-time solvable on graphs of treewidth 2, yet NP-complete on graphs of treewidth 3. Hence, it does not satisfy (C1). On the other hand, Steiner Forest is NP-complete on subcubic graphs and this hardness is maintained under edge subdivision [33], thus satisfying (C2) and (C3). Still, this leaves the complexity open on a broad set of graph classes. However, Bodlaender et al. [11] show the following:

Theorem 3 [11]**.** *For a graph H,* Steiner Forest *on H-subgraph-free graphs is:*

- *polynomial-time solvable if $H \subseteq 2K_{1,3}+P_3+sP_2$, $2P_4+P_3+sP_2$, P_9+sP_2, or $S_{1,1,4}+sP_2$ for any $s \geq 0$;*
- *NP-complete if $H \supseteq iK_{1,3}+jP_4$ where $i+j \geq 3$ or $H \notin \mathcal{S}$.*

Recall that $H_1 + H_2$ for graphs H_1, H_2 is the disjoint union of H_1 and H_2, while sH for a graph H is the disjoint union of s copies of H. Moreover, P_ℓ is the path on ℓ vertices and $S_{x,y,z}$ is the graph formed by subdividing the edges of a $K_{1,3}$ respectively $x-1$, $y-1$, and $z-1$ times.

The NP-completeness results in Theorem 3 follow directly from the hardness on subcubic graphs, that this hardness is maintained under edge subdivision, and from the aforementioned result of Bateni, Hajiaghayi, and Marx [6] and Gassner [28]. The polynomial-time algorithms in Theorem 3 all relate to algorithms for Steiner Forest parameterized by the c-deletion number. For any

$c \geq 1$, the *c-deletion number* of a graph G is the smallest number of vertices that need to be removed from G to obtain a graph with components of at most c vertices (see [5,13,19,20,24,25,29]). Note that $c = 1$ corresponds to the vertex cover number. We have a good understanding of STEINER FOREST in relation to the c-deletion number:

Theorem 4 [11,22,29]. STEINER FOREST *can be solved in polynomial-time on graphs for which the c-deletion number is* 1 *for any* $c \geq 1$ *and on graphs for which the* 2*-deletion number is at most* 2. STEINER FOREST *is NP-complete on graphs for which the* 3*-deletion number is* 2. *Finally,* STEINER FOREST *can be solved in* $2^{O(k \log k)} n^{O(1)}$ *time on graphs for which the* 1*-deletion number is* k *and cannot be solved in* $2^{o(k \log k)} n^{O(1)}$ *time unless the Exponential Time Hypothesis fails.*

As an example for how Theorem 4 enables the proof of Theorem 3, we sketch the case of P_7-subgraph-free graphs. We may assume that the graph is 2-connected, or we can easily create separate instances for each 2-connected component, solve these independently, and combine the solutions in a canonical way. Next, let us assume that we can solve the instance in polynomial time if the input graph G is P_6-subgraph-free. Hence, we can assume that G has a subgraph P isomorphic to P_6. Suppose there is a connected component C of $G - V(P)$ that has more than one vertex. Then no vertex of C can be adjacent to the first two or last two vertices of P, or G would contain a P_7 as a subgraph. Let u, v be the middle two vertices of P. By 2-connectivity, there must be a vertex u' of C adjacent to u and a vertex v' of C adjacent to v. Then there is a P_7 as a subgraph in G, by following P until u, then uu', then a path from u' to v' in C, then $v'v$ and then continuing along P, a contradiction. Hence, G has 1-deletion number at most 6, as witnessed by P. Then Theorem 4 implies that we can solve STEINER FOREST in polynomial time.

Another problem in the same category as STEINER FOREST is SUBGRAPH ISOMORPHISM. Bodlaender et al. [10] gave an almost complete dichotomy of the computational complexity of SUBGRAPH ISOMORPHISM. To obtain a complete dichotomy, it suffices to resolve the following:

Open Problem 4 [10]. *What is the computational complexity of* SUBGRAPH ISOMORPHISM *on* P_5*-subgraph-free graphs and* $2P_5$*-subgraph-free graphs?*

We also note that, from Theorem 3 and the results of Bodlaender et al. [10], one can also observe that the computational complexities of STEINER FOREST and SUBGRAPH ISOMORPHISM differ on $\mathcal{H}$-subgraph-free graphs, as discussed in more detail in Bodlaender et al. [11].

3.2 Problems That Only Do Not Satisfy (C2)

A canonical example is the FEEDBACK VERTEX SET problem. This asks, given a graph G, to find a smallest set $X \subseteq V(G)$ such that the removal of X from G results in a forest. FEEDBACK VERTEX SET is known to be polynomial-time solvable on subcubic graphs [45] and thus does not satisfy (C2). However, it satisfies (C3) trivially. Indeed, any problem that is polynomial-time solvable on subcubic graphs satisfies (C3) trivially. However, FEEDBACK VERTEX SET satisfies it even in a strong sense, as the hardness of the problem is maintained when ℓ-subdividing for any $\ell \geq 1$. Note that FEEDBACK VERTEX SET also satisfies (C1), as it can be solved in polynomial time on graphs of bounded treewidth [3].

Observe that FEEDBACK VERTEX SET becomes NP-complete on graphs of maximum degree 4 [26], and thus on $K_{1,5}$-subgraph-free graphs. Hence, the crux to understanding its complexity is to study graphs that can contain vertices of degree 4, but not in certain settings. Let $S_{w,x,y,z}$ be the graph formed by subdividing the edges of a $K_{1,4}$ respectively $w-1$, $x-1$, $y-1$, and $z-1$ times. Johnson et al. [34] show that FEEDBACK VERTEX SET is NP-complete on $S_{2,2,2,2}$-subgraph-free graphs that have maximum degree 4. At the same time, they show the following lemma that is useful for algorithmic results:

Lemma 1 [34]. *Let $q, r \geq 0$ be integers. Then the class of connected $S_{1,1,q,r}$-subgraph-free graphs that are not subcubic and whose bridges are each incident on a vertex of degree* 1, *has bounded treewidth depending only on q, r.*

Proof (Sketch). The proof actually shows that even the treedepth is bounded. Recall that the *treedepth* of a graph G is the minimum depth of any tree T on vertex set $V(G)$ such that for any $uv \in E(G)$, u is an ancestor of v in T or vice versa. Let G be a graph from the class in the lemma statement. Consider $v_0 \in V(G)$ with at least four neighbors v_1, v_2, v_3, v_4. If G does not have bounded treedepth, then neither does $G' = G - \{v_0, v_1, v_2, v_3, v_4\}$. Thus, G' has a long path [42]. Let z be the middle vertex of this path P. Using Menger's theorem, one can argue that there exist two edge disjoint paths between v_0 and z. Thus, there exists a cycle C that contains z and a vertex v of degree 4 that has exactly two neighbors on the cycle. If C is long, then we readily find a subgraph isomorphic to $S_{1,1,q,r}$ in G, with v as the center of the star. Otherwise, if C is short, one can inspect the pieces of P formed by the intersection of C with P. If there is only one such piece that is long, then this piece must be so long that z cannot be the middle vertex of P. If there are multiple long pieces, then one can agai find a subgraph isomorphic to $S_{1,1,q,r}$ in G (e.g. by finding a cycle with the same properties as C that is long). □

Note that the lemma, combined with the polynomial-time algorithm for subcubic graphs, the polynomial-time algorithm for bounded treewidth graphs, and a straightforward reduction step, immediately gives that FEEDBACK VERTEX SET can be solved in polynomial time on $S_{1,1,q,r}$-subgraph-free graphs for any $q, r \geq 1$.

Summarizing, the following is currently known about the computational complexity of FEEDBACK VERTEX SET on H-subgraph-free graphs:

Theorem 5 [34]. *For a connected graph H,* FEEDBACK VERTEX SET *on H-subgraph-free graphs is:*

- *polynomial-time solvable if $H \in \mathcal{S}$ or $H = S_{1,1,q,r}$ for $q, r \geq 1$;*
- *NP-complete if H contains a cycle, H has more than one vertex of degree at least 3, $H = K_{1,5}$, or $H = S_{2,2,2,2}$.*

If we look to generalize Lemma 1, note that all its assumptions are strictly necessary. For example, without the connectivity assumption, one could have a disjoint union of a $K_{1,4}$ and any path in the class, which has treedepth dependent on the length of the path. Considering $S_{1,p,q,r}$-subgraph-free graphs for $p, q, r \geq 1$ seems challenging, even more so to derive a statement that would be directly helpful to the FEEDBACK VERTEX SET problem.

Open Problem 5 [34]. *What is the computational complexity of* FEEDBACK VERTEX SET *on $S_{1,p,q,r}$-subgraph-free graphs for $p, q, r \geq 2$?*

Other problems in the same category as FEEDBACK VERTEX SET include INDEPENDENT FEEDBACK VERTEX SET, CONNECTED VERTEX COVER, COLORING, and MATCHING CUT. Refer to Johnson et al. [34] and references therein for the current state of the art on these problems. For these problems, Lemma 1 is also helpful. Indeed, what we know of their complexity is similar to Theorem 5, except for the following:

Open Problem 6 [34]. *What is the computational complexity of* CONNECTED VERTEX COVER *and* MATCHING CUT *on $S_{2,2,2,2}$-subgraph-free graphs?*

3.3 Problems That Only Do Not Satisfy (C3)

A canonical example is the HAMILTONIAN CYCLE problem. This asks, given a graph G, to find a cycle that is a subgraph of G and contains all vertices of G. Such a cycle simply ceases to exist even if one 1-subdivides a graph, except if that graph was a cycle [39], and thus HAMILTONIAN CYCLE does not satisfy (C3). However, HAMILTONIAN CYCLE is polynomial-time solvable on graphs of bounded treewidth [4] and NP-complete on subcubic graphs [27], and thus satisfies (C1) and (C2). In fact, it is NP-complete on bipartite subcubic graphs of arbitrary girth [1].

The preceding means that the remaining interesting $\mathcal{H}$-subgraph-free graph classes for finite sets $\mathcal{H}$ for HAMILTONIAN CYCLE are the following:

Open Problem 7 [38,39]. *What is the computational complexity of* HAMILTONIAN CYCLE *on $\mathcal{H}$-subgraph-free graphs when $\mathcal{H}$ contains a subcubic forest with multiple vertices of degree 3?*

While our understanding is far from complete, partial progress on this open problem has been made. In particular, the case when $\mathcal{H}$ contains elements from the set $\{\mathbb{H}_1, \mathbb{H}_2, \ldots\}$ drew recent attention. Here, for an integer $j \geq 1$, $\mathbb{H}_j$ is the graph that is built from a path of length j by attaching two pendant vertices to each of the endpoints of the path (so that the graph looks like an elongated letter "H"). Also here, treewidth can sometimes aid:

Lemma 2 [39]. *For every $\ell \geq 1$, the class of $(\mathbb{H}_\ell, \mathbb{H}_{\ell+1}, \ldots)$-subgraph-free graphs and the class of $(\mathbb{H}_\ell, \mathbb{H}_{2\ell}, \mathbb{H}_{3\ell} \ldots)$-subgraph-free graphs have bounded treewidth (bounded only in ℓ).*

Proof (Sketch). The class of $(\mathbb{H}_\ell, \mathbb{H}_{\ell+1}, \ldots)$-subgraph-free graphs is in fact $\mathbb{H}_\ell$-minor-free. Hence, it has treewidth at most $\ell + 2$, following [7]. For the second part, suppose that the class of $(\mathbb{H}_\ell, \mathbb{H}_{2\ell}, \mathbb{H}_{3\ell} \ldots)$-subgraph-free graphs does not have bounded treewidth. Then a sufficiently large grid appears as a minor of some graph G in the class [44]. There exists a set D of vertices of degree at least 3 in G, one in each part of G that corresponds to the vertices of the central row of this grid minor. An algebraic argument now shows that between some pair of these vertices, some path of length exactly $c\ell$ exists for some integer c. This yields a subgraph of G isomorphic to $\mathbb{H}_{c\ell}$. □

Combined with the known polynomial-time algorithm for graphs of bounded treewidth, this immediately yields polynomial-time algorithms for Hamiltonian Cycle on the classes of graphs of Lemma 2.

Furthermore, Lozin [38] showed that Hamiltonian Cycle can be solved in polynomial time on T'-subgraph-free graphs for any tree T' in the class $\mathcal{T}$ of subcubic trees with exactly two vertices of degree 3 that are adjacent, and on long-H-subgraph-free graphs, where long-H is the 1-subdivision of $\mathbb{H}_1$. Lozin et al. [39] show a polynomial-time algorithm for Hamiltonian Cycle on T-subgraph-free graphs for another specific subcubic tree T. Notably, these results imply polynomial-time algorithms on $\mathbb{H}_1$-, $\mathbb{H}_2$-, and $\mathbb{H}_3$-subgraph-free graphs. However, we do not know the following:

Open Problem 8 [39]. *Is* Hamiltonian Cycle *NP-complete on $\mathbb{H}_j$-subgraph-free graphs for some $j \geq 4$?*

Lozin [38] did show that Hamiltonian Cycle remains NP-complete on Ξ_1- and Ξ_2-subgraph-free graphs for two particular subcubic trees Ξ_1 and Ξ_2.

Summarizing, the following is known about the computational complexity of Hamiltonian Cycle on $\mathcal{H}$-subgraph-free graphs for finite sets $\mathcal{H}$ (but also see the discussion in [38, Theorem 1]):

Theorem 6 [1,4,38,39]. *For a finite set $\mathcal{H}$,* Hamiltonian Cycle *on $\mathcal{H}$-subgraph-free graphs is:*

- *polynomial-time solvable if $\mathcal{H}$ contains the tree T, the long-H graph, a graph in $\mathcal{T}$, or a graph in $\mathcal{S}$;*
- *NP-complete if $\mathcal{H}$ contains only graphs that contain a cycle or that are a forest with a vertex of degree at least 4, or if $\mathcal{H} = \{\Xi_1\}$ or $\mathcal{H} = \{\Xi_2\}$.*

Other problems in the same category as Hamiltonian Cycle include k-Induced Disjoint Paths, C_5-Coloring, and Star 3-Coloring. While Lemma 2 is useful for all these problems, their computational complexity is different. In particular, it seems that the complexity landscape of problems that satisfy (C1) and (C2) but not (C3) is very rich. Refer to Lozin et al. [39] and references therein for the current state of the art on these problems, further discussions, and more open problems.

4 Discussion

As this paper shows, the algorithmic meta classification of Theorem 2 provides an intriguing starting point for the investigation of problems on $\mathcal{H}$-subgraph-free graphs. It immediately gives dichotomy results on the computational complexity of many problems. At the same time, it spotlights problems for which not all conditions (C1), (C2), (C3) hold or are not yet known to hold.

A next step is to find further algorithmic meta classifications for other ways of forbidding substructures. Most interesting in this respect seem $\mathcal{H}$-free graphs. Lozin and Razgon [40] show that for any finite set $\mathcal{H}$ of graphs, the class of $\mathcal{H}$-free graphs has bounded treewidth if and only if $\mathcal{H}$ contains a complete graph, a completely bipartite graph, a graph from $\mathcal{S}$, and a line graph of a graph in $\mathcal{S}$. Hickingbotham [32] shows that the same holds for bounded pathwidth. This immediately yields an algorithmic meta classification (see Johnson et al. [33]). Unfortunately, thusfar the only problem to which we know this classification applies is Weighted Edge Steiner Tree [9].

Open Problem 9 [33]**.** *Which natural problems, other than* Weighted Edge Steiner Tree, *satisfy the conditions of the meta classification for $\mathcal{H}$-free graphs?*

Johnson et al. [33] showed an algorithmic meta classification for $\mathcal{H}$-topological-minor-free graphs. A graph G is *$\mathcal{H}$-topological-minor-free graph* if G contains no graph $H \in \mathcal{H}$ as a topological minor, i.e., no such H can be obtained by removing vertices and edges from G and dissolving vertices in G. Recall that the vertex dissolution operation contracts both edges incident on a vertex of degree 2. The result of Johnson et al. [33] is similar to Theorem 1, except that planar graphs are replaced by planar subcubic graphs. This result has a wide reach as well; see Johnson et al. [33, Table 1] for examples and proofs.

Looking further ahead, one may hope for even more broadly applicable algorithmic meta classifications on $\mathcal{H}$-subgraph-free graphs. A challenge here is that some well-known problems do not behave nicely on $\mathcal{H}$-subgraph-free graphs. For example, the Clique problem is polynomial on $\mathcal{H}$-subgraph-free graphs for any set $\mathcal{H}$ [33] and there even exist problems that are constant time solvable on $\mathcal{H}$-(subgraph-)free graphs for any set $\mathcal{H}$, but are PSPACE-complete in general [41]. Hence, this research area provides exciting avenues for further work.

Acknowledgments. I want to thank my coauthors of the works that motivated this invited talk: Hans Bodlaender, Matthew Johnson, Vadim Lozin, Barnaby Martin, Jelle Oostveen, Sukanya Pandey, Daniel Paulusma, Mark Siggers, and Siani Smith. I also want to thank Daniel Paulusma for helpful discussions on this paper and for suggesting Proposition 1.

References

1. Alekseev, V.E., Boliac, R., Korobitsyn, D.V., Lozin, V.V.: NP-hard graph problems and boundary classes of graphs. Theor. Comput. Sci. **389**, 219–236 (2007)
2. Alekseev, V.E., Korobitsyn, D.V.: Complexity of some problems on hereditary graph classes. Diskret. Mat. **4**, 34–40 (1992)
3. Arnborg, S., Lagergren, J., Seese, D.: Easy problems for tree-decomposable graphs. J. Algor. **12**, 308–340 (1991)
4. Arnborg, S., Proskurowski, A.: Linear time algorithms for NP-hard problems restricted to partial k-trees. Disc. Appl. Math. **23**, 11–24 (1989)
5. Barefoot, C.A., Entringer, R., Swart, H.: Vulnerability in graphs - a comparative survey. J. Comb. Math. Comb. Comput. **1**, 13–22 (1987)
6. Bateni, M., Hajiaghayi, M.T., Marx, D.: Approximation schemes for steiner forest on planar graphs and graphs of bounded treewidth. J. ACM **58**, 21:1–21:37 (2011)
7. Bienstock, D., Robertson, N., Seymour, P.D., Thomas, R.: Quickly excluding a forest. J. Comb. Theory Ser. B **52**, 274–283 (1991)
8. Bodlaender, H.L.: A partial k-arboretum of graphs with bounded treewidth. Theor. Comput. Sci. **209**, 1–45 (1998)
9. Bodlaender, H.L., Brettell, N., Johnson, M., Paesani, G., Paulusma, D., van Leeuwen, E.J.: Steiner trees for hereditary graph classes: a treewidth perspective. Theor. Comput. Sci. **867**, 30–39 (2021)
10. Bodlaender, H.L.: Subgraph Isomorphism on graph classes that exclude a substructure. Algorithmica **82**, 3566–3587 (2020)
11. Bodlaender, H.L., et al.: Complexity framework for forbidden subgraphs IV: the steiner forest problem. In: Proceedings of IWOCA 2024. LNCS, vol. 14764, pp. 206–217. Springer, Heidelberg (2024). DOI: https://doi.org/10.1007/978-3-031-63021-7_16
12. Bui-Xuan, B., Telle, J.A., Vatshelle, M.: Fast dynamic programming for locally checkable vertex subset and vertex partitioning problems. Theor. Comput. Sci. **511**, 66–76 (2013)
13. Bulteau, L., Dabrowski, K.K., Köhler, N., Ordyniak, S., Paulusma, D.: An algorithmic framework for locally constrained homomorphisms. In: Proceedings of WG 2022. LNCS, vol. 13453, pp. 114–128. Springer, Heidelberg (2022). https://doi.org/10.1007/978-3-031-15914-5_9
14. Courcelle, B.: The monadic second-order logic of graphs. I. Recognizable sets of finite graphs. Inf. Comput. **85**, 12–75 (1990)
15. Courcelle, B., Makowsky, J.A., Rotics, U.: Linear time solvable optimization problems on graphs of bounded clique-width. Theory Comput. Syst. **33**, 125–150 (2000)
16. Cygan, M., et al.: Parameterized Algorithms. Springer, Heidelberg (2015)
17. de Ridder, H.N., et al.: Information System on Graph Classes and their Inclusions (ISGCI) (2001–2024). https://www.graphclasses.org/
18. Demaine, E.D., Hajiaghayi, M.: The bidimensionality theory and its algorithmic applications. Comput. J. **51**, 292–302 (2008)
19. Drange, P.G., Dregi, M.S., van't Hof, P.: On the computational complexity of vertex integrity and component order connectivity. Algorithmica **76**, 1181–1202 (2016)
20. Dvorák, P., Eiben, E., Ganian, R., Knop, D., Ordyniak, S.: Solving integer linear programs with a small number of global variables and constraints. Proc. IJCAI **2017**, 607–613 (2017)

21. Eagling-Vose, T., Martin, B., Paulusma, D., Smith, S.: Graph homomorphism, monotone classes and bounded pathwidth. In: Proceedings of CiE 2024. LNCS, vol. 14773, pp. 233–251 (2024). https://doi.org/10.1007/978-3-031-64309-5_19
22. Feldmann, A.E., Lampis, M.: Parameterized algorithms for steiner forest in bounded width graphs. In: Proceedings of ICALP 2024. LIPIcs, vol. 272, pp. 61:1–61:20 (2024)
23. Fomin, F.V., Demaine, E.D., Hajiaghayi, M.T., Thilikos, D.M.: Bidimensionality. In: Encyclopedia of Algorithms, pp. 203–207 (2016)
24. Fujita, S., Furuya, M.: Safe number and integrity of graphs. Disc. Appl. Math. **247**, 398–406 (2018)
25. Fujita, S., MacGillivray, G., Sakuma, T.: Safe set problem on graphs. Disc. Appl. Math. **215**, 106–111 (2016)
26. Garey, M.R., Johnson, D.S.: Computers and Intractability: A Guide to the Theory of NP-Completeness. W. H. Freeman & Co., New York (1979)
27. Garey, M.R., Johnson, D.S., Tarjan, R.E.: The Planar Hamiltonian Circuit problem is NP-complete. SIAM J. Comput. **5**, 704–714 (1976)
28. Gassner, E.: The Steiner Forest problem revisited. J. Disc. Algor. **8**, 154–163 (2010)
29. Gima, T., Hanaka, T., Kiyomi, M., Kobayashi, Y., Otachi, Y.: Exploring the gap between treedepth and vertex cover through vertex integrity. Theor. Comput. Sci. **918**, 60–76 (2022)
30. Golovach, P.A., Paulusma, D.: List coloring in the absence of two subgraphs. Disc. Appl. Math. **166**, 123–130 (2014)
31. Golovach, P.A., Paulusma, D., Ries, B.: Coloring graphs characterized by a forbidden subgraph. Disc. Appl. Math. **180**, 101–110 (2015)
32. Hickingbotham, R.: Induced subgraphs and path decompositions. Electron. J. Comb. **30** (2023)
33. Johnson, M., et al.: Complexity framework for forbidden subgraphs I: the framework. Algorithmica (2025)
34. Johnson, M., Martin, B., Pandey, S., Paulusma, D., Smith, S., van Leeuwen, E.J.: Complexity framework for forbidden subgraphs III: when problems are polynomial on subcubic graphs. In: Proceedings of MFCS 2023. LIPIcs, vol. 272, pp. 57:1–57:15 (2023)
35. Johnson, M., Martin, B., Pandey, S., Paulusma, D., Smith, S., van Leeuwen, E.J.: Edge multiway cut and node multiway cut are NP-hard on subcubic graphs. In: Proceedings of SWAT 2024. LIPIcs, vol. 294, pp. 29:1–29:17 (2024)
36. Kamiński, M.: Max-Cut and containment relations in graphs. Theor. Comput. Sci. **438**, 89–95 (2012)
37. Korpelainen, N., Lozin, V.V., Malyshev, D.S., Tiskin, A.: Boundary properties of graphs for algorithmic graph problems. Theor. Comput. Sci. **412**, 3545–3554 (2011)
38. Lozin, V.: The Hamiltonian cycle problem and monotone classes. In: Proceedings of IWOCA 2024. LNCS, vol. 14764, pp. 460–471. Springer, Heidelberg (2024). https://doi.org/10.1007/978-3-031-63021-7_35
39. Lozin, V., et al.: Complexity framework for forbidden subgraphs II: edge subdivision and the "H"-graphs. In: Proceedings of ISAAC 2024. LNCS (2024)
40. Lozin, V., Razgon, I.: Tree-width dichotomy. Eur. J. Comb. **103**, 103517 (2022)
41. Martin, B., Paulusma, D., Smith, S.: Hard problems that quickly become very easy. Inf. Process. Lett. **174**, 106213 (2022)
42. Nešetřil, J., de Mendez, P.O.: Sparsity - Graphs, Structures, and Algorithms, Algorithms and Combinatorics, vol. 28. Springer, Heidelberg (2012)

43. Robertson, N., Seymour, P.D.: Graph minors. I. Excluding a forest. J. Comb. Theory Ser. B **35**, 39–61 (1983)
44. Robertson, N., Seymour, P.D.: Graph minors. V. Excluding a planar graph. J. Comb. Theory Ser. B **41**, 92–114 (1986)
45. Ueno, S., Kajitani, Y., Gotoh, S.: On the nonseparating independent set problem and feedback set problem for graphs with no vertex degree exceeding three. Disc. Math. **72**, 355–360 (1988)

Contributed Papers

Parameterized Complexity of Feedback Vertex Set with Connectivity Constraints

Ankit Abhinav[1], Satyabrata Jana[2(✉)], Nidhi Purohit[3], Abhishek Sahu[1], and Saket Saurabh[4,5]

[1] National Institute of Science Education and Research, An OCC of Homi Bhabha National Institute, Bhubaneswar 752050, Odisha, India
{ankit.abhinav,abhisheksahu}@niser.ac.in
[2] University of Warwick, Coventry, UK
satyamtma@gmail.com
[3] National University of Singapore, Singapore, Singapore
[4] The Institute of Mathematical Sciences, HBNI, Chennai, India
[5] University of Bergen, Bergen, Norway

Abstract. The Feedback Vertex Set (FVS) problem, together with several of its variants, is arguably one of the most well-studied problems in the field of Parameterized Complexity. Two versions of the problem that have garnered significant interest involve the inclusion of an independence constraint and a connectivity constraint in the solution. This paper introduces generalized versions of both these variants, known as At least-c-FVS and At most-c-FVS, respectively, serving as extensions of Connected FVS and Independent FVS, respectively. The problem At most-c-FVS (resp., At least-c-FVS) is defined as follows: given a graph G and an integer k, the objective is to determine whether there exists a subset $S \subseteq V(G)$, with $|S| \leq k$, such that the subgraph $G - S$ is a forest and each component of $G[S]$ contains at most c (resp., at least c) vertices. We study these problems in the realm of Parameterized Complexity and obtain the following results:

– At most-c-FVS parameterized by k has a kernel of size $\mathcal{O}(k^{3+c})$, and admits an FPT algorithm running in time $2^{\mathcal{O}(k)+c\cdot\log(k^2)} \cdot n^{\mathcal{O}(1)}$.

– At least-c-FVS parameterized by k has no kernel of size $k^{f(c)}$ for any computable function f (unless co-NP $\subseteq$ NP/poly), but admits an FPT algorithm running in time $2^{\mathcal{O}(k)} \cdot n^{\mathcal{O}(1)}$.

Keywords: Feedback vertex set · FPT · Kernelization · Lower Bound

S. Jana—Supported by the Engineering and Physical Sciences Research Council (EPSRC) via the project MULTIPROCESS (grant no. EP/V044621/1)

S. Saurabh—Supported by the European Research Council (ERC) under the European Union's Horizon 2020 research and innovation programme (grant agreement No. 819416); and he also acknowledges the support of Swarnajayanti Fellowship grant DST/SJF/MSA-01/2017-18.

R. Královič and V. Kůrková (Eds.): SOFSEM 2025, LNCS 15538, pp. 23–36, 2025.
https://doi.org/10.1007/978-3-031-82670-2_3

1 Introduction

The problem Feedback Vertex Set (FVS) has substantial significance in the domain of networks and interconnected systems [7,10,14]. When depicted through graph representations, the existence of cycles or loops within these networks introduces complexities and impediments that compromise their functional integrity. The FVS serves as an essential strategy to mitigate these issues, striving to ascertain and eliminate the minimal subset of vertices necessary to break all cycles in the graph, thus enhancing the network's operational efficiency and resilience. Formally, a *feedback vertex set* (shortly, fvs) in a graph is a set of vertices that when deleted results in a forest (an acyclic graph). In the Feedback Vertex Set problem, we are given a graph G and an integer k, and the objective is to check whether there is a feedback vertex set S of size at most k. It is one of the first problems to be shown NP-complete and appears in Karp's list of 21 NP-complete problems [12].

FVS is arguably one of the most studied problems in the realm of parameterized complexity. The first known fixed-parameter tractable (FPT) algorithms for FVS date back to the late 1980s and early 1990s [7]. These algorithms relied on the groundbreaking graph-minor theory of Robertson and Seymour [20,21]. Over time, there have been multiple advancements and refinements, leading to the current best deterministic FPT algorithm for FVS, which runs in $\mathcal{O}^*(3.460^k)$[1] time [10]. The fastest known randomized algorithm for this problem runs in time $\mathcal{O}^*(2.7^k)$ and is given by Li and Nederlof [14]. Recently, a factor $(1+\epsilon)$ approximation algorithm for FVS, which has better running time than the best-known (randomized) FPT algorithm for every $\epsilon \in (0,1)$ is obtained by Jana et al. [11].

In the past decade, various variants of FVS have been extensively studied in undirected graphs, including Connected FVS [5,18], Conflict-free FVS [2], Subset FVS [16], Group FVS [6], Simultaneous FVS [3], and Independent FVS [1,15,17]. Among these variants, two problems that are relevant for our work are Independent Feedback Vertex Set (IFVS) and Connected Feedback Vertex Set (CFVS). CFVS and IFVS impose specific requirements on the desired FVS. In CFVS, we require that $G[S]$ induces a connected subgraph, and in IFVS, we require that $G[S]$ is an independent set. Both problems have been thoroughly explored in the domain of parameterized complexity.

In 2011, Misra et al. [17] introduced IFVS and devised an algorithm with a running time of $\mathcal{O}^*(5^k)$. Furthermore, they designed a polynomial kernel of size $\mathcal{O}(k^3)$ for this variant of the problem. In parameterized complexity, each problem instance comes with a parameter k and the parameterized problem is said to admit a *polynomial kernel* if there is a polynomial-time algorithm (the degree of polynomial is independent of k), called a *kernelization algorithm*, that reduces the input instance down to an instance with size bounded by a polynomial $p(k)$ in k, while preserving the answer. This reduced instance is called a $p(k)$ kernel for the problem [8]. Subsequently, Agrawal et al. presented an improved FPT algorithm with a running time of $\mathcal{O}^*(4.148^k)$ [1]. Currently, the

[1] We use $\mathcal{O}^*$ notation to hide factors polynomial in the input size.

most efficient algorithm known for IFVS is provided by Li and Pilipczuk, which runs in $\mathcal{O}^*(3.619^k)$ time [15]. Misra et al. [18] were the first to study CFVS from the perspective of parameterized complexity. They designed a single-exponential fixed-parameter algorithm, running in $\mathcal{O}^*(46.2^k)$ time along with the additional result, that CFVS does not admit a polynomial kernel unless co-NP $\subseteq$ NP/poly. Later, using the seminal cut and count method, Cygan et al. gave the fastest known randomized algorithm for CFVS, achieving a runtime of $\mathcal{O}^*(4^k)$ [5].

In this paper, we study the generalizations of these two variants, each focusing on different aspects, one capturing independence, and the other emphasizing connectivity. Our research looks into these variants, revealing their intricacies with the following theme.

Independence vs **Connectivity**

Formally, the problems we study are the following. For a graph H, we denote the sizes of a largest and smallest connected components in H as $\ell\texttt{comp}(H)$ and $\texttt{scomp}(H)$, respectively.

At least-c-FVS (At most-c-FVS) **Parameter:** k
Input: An undirected graph G, non-negative integers k and c.
Question: Is there a vertex set $S \subseteq V(G)$ of size at most k such that $G - S$ is a forest and $\texttt{scomp}(G[S]) \geq c$ ($\ell\texttt{comp}(G[S]) \leq c$)?

A takeaway message from our research is that the independence generalization does not make the problem more challenging than Independent FVS, and similarly the connectivity generalization of the problem does not make it easier.

1.1 Our Results and Methods

We obtain the following results for At least-c-FVS and At most-c-FVS.

1. At most-c-FVS parameterized by k has a kernel of size $\mathcal{O}(k^{3+c})$ (Theorem 5), and admits an FPT algorithm running in time $\mathcal{O}^*(k^{2c} \cdot 2^{\mathcal{O}(k)})$ (Theorem 6).
2. At least-c-FVS parameterized by k has no kernel of size $k^{f(c)}$ for any computable function f unless co-NP $\subseteq$ NP/poly (Theorem 7), but admits a fixed-parameter tractable (FPT) algorithm running in time $\mathcal{O}^*(2^{\mathcal{O}(k)})$ (Theorem 8).

In Sect. 2, we design a kernel for At most-c-FVS with size $\mathcal{O}(k^{3+c})$. We follow the outline of the known kernelization algorithm for Feedback Vertex Set of Thomassé [22], but, due to the inherent generality of our problem, our implementation differs in several crucial steps. Similarly to the kernel for Feedback Vertex Set, our first objective is to bound the maximum degree of the graph. This is one of our main technical results. In this step, we show that if the degree of a vertex is more than $2k^3 + 7k^2 + 7k$, then there is an *irrelevant edge* incident to it. This is achieved by doing some structural analysis. It is important to note that we are able to bound the degree of most vertices by a polynomial function that is independent of c. We are unable to bound the degree of all but at most k vertices. However, we are guaranteed to take all these vertices in our solution. Unlike the classical Feedback Vertex Set, we cannot delete these vertices, as this is needed to bound the size of a component in the final solution. Observe that we cannot contract a degree 2 vertex immediately, as this could be part of the solution to adhere to the size constraint on the components. Thus, we also need to come up with a new way to reduce the vertices of degree 2 in the graph. This is the step in which we incur k^c in the size of the kernel. After this, with some standard combinatorial techniques, we are able to obtain the desired kernel.

In Sect. ??, we give an FPT algorithm for At most-c-FVS running in $\mathcal{O}^*(k^{2c} \cdot 2^{\mathcal{O}(k+o(k))})$ time. Our algorithm starts similar to the known FPT algorithm for IFVS (two-way branching) by Misra et al. [17]. It initially uses several non-trivial (six-way) branching strategies to reduce the problem to a variant of the path hitting set problem: Exact Path Hitting. Although a straightforward application of *color coding* can solve this problem in $\mathcal{O}^*(k^k)$ time, we improve the runtime dependency to $\mathcal{O}^*(4^{k+o(k)} \cdot k^{2c})$ by employing the *divide and color* technique [13]. It might be tempting to find an fvs using a known algorithm and then applying dynamic programming algorithm to it. However, the running time of this algorithm ends up having a term of $(f(c))^k$, as we at least need to remember the sizes (c) of the current components, which is asymptotically larger than $2^{\mathcal{O}(k)}$ even for $c = \log k$. But, our algorithm has a running time of the form $2^{\mathcal{O}(k)}$, even for $c = k/\log k$.

In SecT. ??, we establish the mentioned lower bound of $k^{f(c)}$ on the kernel size of At least-c-FVS. We achieve this by demonstrating a *polynomial parameter transformation* (PPT) from a suitable parameterization of the Red Blue Dominating Set (RBDS) problem. In Sect. ??, we present an FPT algorithm for At least-c-FVS, using an algorithm (initially introduced by Guo et al. [9]) to enumerate efficient representations of minimal feedback vertex sets, which ensures that at least one is contained in any optimal solution. Upon guessing such a representation, we employ random vertex coloring, which then transforms the problem into finding a colorful forest (by guessing its structure) where each component has a size at least c. Subsequently, we use the findings of Amini et al. [4] to obtain such a forest and consequently design an FPT algorithm running in time $\mathcal{O}^*(741^k)$. Missing proofs (marked with $\star$) are in the full version of the paper

Notation. We use $[n]$ to denote the set of natural numbers $\{1, 2, \dots, n\}$. The input graphs considered in our paper are finite, undirected, and connected. For a graph G, $V(G)$ and $E(G)$ refer to its vertex and edge sets, respectively. $|G|$ denotes the number of vertices in G. The sizes of a largest component and a smallest component in G are denoted by $\ell\texttt{comp}(G)$ and $\texttt{scomp}(G)$, where $\ell\texttt{comp}(G) = \max\{|V(C)| : C \text{ is a component of } G\}$ and $\texttt{scomp}(G) = \min\{|V(C)| : C \text{ is a component of } G\}$. Note that a n-vertex graph G is connected if and only if $\ell\texttt{comp}(G) = n$. Furthermore, for a vertex v in $V(G)$, we use $\deg_G(v)$ to denote its degree, that is, the number of edges incident in v, in the (multi) graph G. A path in G is said to be *degree-2 path* if all the vertices except the end vertices of the path are of degree 2 in G. For a vertex subset $S \subseteq V(G)$, $G[S]$ and $G - S$ represent the graphs induced on S and $V(G) \setminus S$, respectively. Further for an ease of notation, we use $G - v$ when the set $S = \{v\}$ is singleton. Similarly we use $G - e$ to denote the graph $((V(G), E(G) \setminus \{e\})$. If G' is an induced subgraph of G, $G - G'$ denotes the graph $G - V(G')$. Moreover, for a vertex subset $S \subseteq V(G)$, $N_G(S)$ and $N_G[S]$ denote the open neighborhood and the closed neighborhood of S in G, respectively. That is, $N_G(S) = \{v \mid (u,v) \in E(G), u \in S\} \setminus S$ and $N_G[S] = N_G(S) \cup S$. A connected component is called a cyclic component if it contains a cycle. A vertex set $S \subseteq V(G)$ is a *feedback vertex set* (in short, fvs) of the graph G if $G - S$ has no cycle. S is called at least-c-feedback vertex set (in short, at least-c-fvs) if $\texttt{scomp}(G[S]) \geq c$ and an at most-c-fvs if $\ell\texttt{comp}(G[S]) \leq c$. For a path P, we use $\texttt{int}(P)$ to denote the subpath of P obtained by removing both endpoints of the path, and $V(P)$ to denote the set of all vertices on the path. The $\mathcal{O}^\star()$ notation is used to suppress polynomial factors in running time. For extended preliminaries, we refer to ??.

2 Kernel for At most-c-FVS

In this section, we obtain a kernel of size $\mathcal{O}(k^{3+c})$ for At most-c-FVS. We start by introducing a generalization of At most-c-FVS, Ext at most-c-FVS, which is easier to handle while designing the kernelization algorithm.

> Ext at most-c-FVS **Parameter:** k
> **Input:** An undirected Graph $G = (V, E)$, $T \subseteq V$, a non-negative integer k.
> **Question:** Does G have an fvs $S \supseteq T$ of size at most k such that $\ell\texttt{comp}(G[S]) \leq c$?

We design a polynomial kernel for Ext at most-c-FVS and then using some simple gadgets reduce the problem to At most-c-FVS. The first reduction rule handles sanity checks; its correctness is self-evident.

Reduction Rule 1 (Sanity Check)

1. *If $|T| > k$, then return* No *and stop.*
2. *If $\ell\texttt{comp}(G[T]) > c$, then return* No *and stop.*
3. *If $|T| = k$ and $G - T$ is not forest, then return* No *and stop.*
4. *If $|T| \leq k, \ell\texttt{comp}(G[T]) \leq c$ and $G - T$ is a forest, then return* Yes *and stop.*

Next, we apply the following reduction rule to eliminate self-loops and reduce the multiplicity of any edge to at most 2. The correctness of this rule is self-evident.

Reduction Rule 2 (Boundary Conditions)

1. *Let v be a vertex in graph $G - T$ which has a self-loop. Then add v to T; the resulting instance is $(G, T \cup \{v\}, k)$.*

2. *If (x, y) is an edge in G with multiplicity more than two, then reduce it to two.*

In the next two reduction rules, our goal is to remove vertices of degree at most one and bound the number of vertices in a degree 2 path.

Reduction Rule 3 (Degree 1 rule). *Let $x \in V(G-T)$ be a vertex of degree at most one in $G-T$. We remove x from G and return the instance $(G-x, T, k)$.*

The safety/correctness of Reduction Rule 3 can be easily verified. A *safe* reduction rule refers to one that does not alter the (Yes/No) answer to the input instance.

In the following, we provide a reduction rule that allows us to bound the number of vertices in a degree-2 path by k^{c-1}. Let $\mathcal{P}_{x,y}$ denote a degree-2 path between x and y in $G-T$. Note that if there is a vertex v on the path that has neighbors in more than c components in $G[T]$, then it can not be a part of the desired solution and hence *safely* removed. Note that if a solution contains multiple vertices from a path, removing all but one vertex will still result in a valid solution. Hence, we can always assume that we have a solution $S \supseteq T$ of size at most k such that S includes at most one vertex of any such path. Note that we do not demand that S is minimal in any sense. Therefore, if there exists a pair of vertices that have neighbors in the same set of components, we can *safely* remove one of them by "short-circuiting" their neighbors.

Reduction Rule 4 (Degree 2 rule). *Let x and y be a pair of vertices in $G-T$ with a degree-2 path $\mathcal{P}_{x,y}$ between them in $G-T$ where all the internal vertices are of degree-2 in $G-T$. By $\mathcal{C}_T$, we denote the set of all maximal connected components in $G[T]$. Let u and v be a pair of distinct internal vertices in $\mathcal{P}_{x,y}$ such that $u, v \notin N(x) \cup N(y)$. If $\{C : C \in \mathcal{C}_T, N(u) \cap V(C) \neq \emptyset\} \subseteq \{C : C \in \mathcal{C}_T, N(v) \cap V(C) \neq \emptyset\}$ or $|\{C : C \in \mathcal{C}_T, N(v) \cap V(C) \neq \emptyset\}| \geq c$, then we remove the vertex v from the graph G, add the edge $e = v_1v_2$ where v_1 and v_2 are the neighbors of v in the path $\mathcal{P}_{x,y}$ and return the instance $(G' = G - v + e, T, k)$.*

Lemma 1. *Reduction Rule 4 is safe.*

Proof. (i) In scenarios where $|\{C : C \subseteq \mathcal{C}_T, N(v) \cap C \neq \emptyset\}| \geq c$, the safeness follows from the fact that no solution to (G, T, k) includes the vertex v. This is due to the fact that any solution must contain all vertices of T and $\ell\texttt{comp}(G[T \cup \{v\}] > c$. That is, if $T \subseteq S$ is a solution of size at most k then $v \notin S$. Thus, the same set S continues to be a solution for (G', T, k). The reverse direction follows the same lines.

(ii) Let x and y be a pair of vertices in $G-T$ connected by a degree-2 path $\mathcal{P}_{x,y}$ (between them) in $G-T$. Consider u and v, a pair of distinct internal vertices in $\mathcal{P}_{x,y}$ such that $\{C : C \in \mathcal{C}_T, N(u) \cap V(C) \neq \emptyset\} \subseteq \{C : C \in \mathcal{C}_T, N(v) \cap V(C) \neq \emptyset\}$. Let $G' = G - v + e$.

In the forward direction, let (G, T, k) be a Yes instance and $S \supseteq T$ be a solution of size at most k. We can assume that it includes at most one vertex from the set $\{u, v\}$. In fact, we can assume that it includes at most one vertex from the path $\mathcal{P}_{x,y}$. If S does not include v, then S is also a solution to (G', T, k). Now, let us assume that S includes the vertex v. We will show that $S' = S \setminus \{v\} \cup \{u\}$ is a solution to (G', T, k). Clearly $|S'| \leq k$. Now, let us assume, for contradiction, that S' is not a solution to (G', T, k). Since both vertices u and v belong to the degree-2 path $\mathcal{P}_{x,y}$ in $G - T$, any cycle in $G - T$ passing through u must also pass through v, and vice versa. Therefore, S' is an fvs of size at most k.

Hence, the only reason for it not being a solution to (G', T, k) is $\ell\texttt{comp}(G'[S']) > c$. Since $\ell\texttt{comp}(G'[S']) > c$, there must exist a component C in $G'[S']$ whose size is greater than c and u must belong to this component C. Let $\mathcal{C}_u$ denote all the maximal connected components in $G[T]$ that have a neighbor of u. Similarly, let $\mathcal{C}_v$ denote all the maximal connected components in $G[T]$ that have a neighbor of v. By our choice of S, we know that it contains at most one vertex from $\mathcal{P}_{x,y}$. Thus, for the constructed S', $S' \cap (N_{G'}(u) \setminus T) = \emptyset$., i.e., $S' = S \setminus \{v\} \cup \{u\}$ does not contain any neighbors of u that are outside T. As $\mathcal{C}_u \subseteq \mathcal{C}_v$, the size of the component containing v in $G[S]$ must be greater than c, a contradiction.

In the reverse direction, let $T \subseteq S'$ be a solution of size at most k to (G', T, k). As we can derive G from G' by subdividing precisely one edge (the edge between the neighbors of v), every cycle present in $G - S'$ is also present in $G' - S'$. Therefore, S' must indeed be a solution to (G, T, k). This completes the proof. □

Observation 1 (⋆). *After exhaustive application of Reduction Rule 4, every degree-2 path in $G - T$ contains at most k^{c-1} vertices.*

Next, for each vertex of $G - T$, we bound the number of neighbors of the vertex in T.

Reduction Rule 5. *Let u be a vertex in $G - T$ such that $|N(u) \cap T| \geq c + 1$. Then we remove an edge (arbitrarily choosen) uv where $v \in N(u) \cap T$ and return the instance $(G - e, T, k)$.*

Lemma 2 (⋆). *Reduction Rule 5 is safe.*

Observation 2. *After exhaustive application of Reduction Rules 1 to 5, for any vertex v in $G - T$, $|N(v) \cap T| \leq c$.*

Our algorithm exhaustively applies all these reduction rules. Thus, from now on, we assume that the input graph (G, T, k) has been reduced with respect to the Reduction Rules 1 to 5. In the next reduction rule, we focus on the vertices that are the unique intersection of at least $k + 1$ cycles.

Definition 1 (ℓ-flower). Let v be a vertex in a graph G, and let $\ell \in \mathbb{N}$. An ℓ-flower centering at v is a set of ℓ distinct cycles in G such that each cycle contains v and no two cycles share any vertex other than v. The vertex v is said to be in the center of the flower.

Reduction Rule 6. *Let v be a vertex in the graph $G - T$ that is in the center of a $(k+1)$-flower in G. Then we add v to T and return the instance $(G, T \cup \{v\}, k)$.*

The correctness of this rule essentially follows from the fact that any vertex that is at the center of a $(k+1)$-flower must be present in any fvs of size at most k in the graph G.

Lemma 3 ($\star$). *Reduction Rule 6 is safe.*

We next design a reduction rule that helps us to bound the maximum degree of any vertex of $V(G) \setminus T$ in the graph G. This is the most involved reduction rule of our kernelization algorithm. This essentially amounts to finding an *irrelevant edge* to remove in polynomial time. Toward the design of this reduction rule, we use the following known result.

Theorem 3 ([22] **Corollary 2.1**). *Let v be a vertex of a graph G which is not a loop. If there is no $(k+1)$-flower centering at v, there exists a set of vertices $X \subseteq V(G) \setminus \{v\}$ of size at most $2k$ intersecting every cycle containing v. Moreover, the set X is computable in time polynomial in $|V(G)|$.*

Let $G_T = G - T$ and $v \in V(G_T)$. By applying Theorem 3 on v and G_T, we get a set $X_v \subseteq V(G_T) \setminus \{v\}$ of size at most $2k$ intersecting every cycle containing v in G_T (guaranteed to exist by Theorem 3). Let $C_1, C_2, \ldots, C_p$ be the connected components of $G_T - (X_v \cup \{v\})$. Observe that there is at most one edge from v to each of these components; If there are two or more edges from v to some C_i, then the subgraph induced by $\{v\} \cup C_i$ contains a cycle, a contradiction since X_v does not intersect any such cycle. Also, if there are strictly more than k cyclic components then we can return No, since we get $(k+1)$ vertex disjoint cycles in this case and no fvs of size at most k can intersect all these cycles. Due to Reduction Rule 6, there are at most k vertices in X_v that have multiple edges with v with the maximum multiplicity of any such edge being two from Reduction Rule 2. Additionally, there is at most one edge from v to each remaining vertex in X_v, thus bounding the total number of edges by at most $3k$, that are present between v and X_v. We summarize the above discussion in the following observation.

Observation 4. *Let v be a vertex in $V(G_T)$, $X_v \subseteq V(G_T) \setminus \{v\}$ denote a set of size at most $2k$, of the kind guaranteed to exist by Theorem 3 and let $C_1, C_2, \ldots, C_p$ be the connected components of $G_T - (X_v \cup \{v\})$. Then, there is at most one edge from v to each of these components and at most of k of these components contains a cycle.*

We use $\mathcal{C}_v$ to denote the set of components in $G_T - (X_v \cup \{v\})$ that satisfy the following properties:

- there is exactly one edge from v to each component and
- each component is a tree.

That is, $\mathcal{C}_v$ are the components among $C_1, C_2, \ldots, C_p$ that are not cyclic. Since v has at most one neighbor within each component of $\mathcal{C}_v$ (by Observation 4), we can conclude that $\deg_{G_T}(v) \leq |\mathcal{C}_v| + 3k + k \leq |\mathcal{C}_v| + 4k$. Now in each of these components $C \in \mathcal{C}_v$, there is at least one vertex that is adjacent to some vertex in X_v. This is because if no vertex in C is adjacent to any vertex in X_v, then C contains a vertex of degree one in G_T, a contradiction since our instance is reduced with respect to Reduction Rule 3.

Bounding the Maximum Degree. Our next objective is to reduce the degree of vertices in G_T. Suppose v is a vertex of degree more than $2k^3 + 7k^2 + 7k$ in G_T. Then, $|X_v| \leq 2k$ and $|\mathcal{C}_v| > 2k^3 + 7k^2 + 3k$. Using the notion of *rich* and *poor pair* (as defined below), we are able to identify an edge (by Reduction Rule 7) incident on v that can be *safely* removed, thus reducing its degree.

Definition 2 (Super-Rich, Rich and Poor pair). For a vertex $z \in X_v \cup \{v\}$, we define $\mathcal{C}(z)$ as the set comprising all components in $\mathcal{C}_v$ that contain a neighbor of z. Regarding a pair of vertices x and y in $X_v \cup \{v\}$, we classify the pair $\{x, y\}$ as super-rich if $|\mathcal{C}(x) \cap \mathcal{C}(y)| \geq k+4$ and rich if $|\mathcal{C}(x) \cap \mathcal{C}(y)| = k+3$. Otherwise, we term the pair as poor.

Lemma 4 ($\star$). *Let $e \in E(G_T)$. Then, for each super-rich pair $\{x, y\}$ in G_T, at least one of the vertices in the pair must be included in all fvs of size at most k in the graph $G_T - e$.*

We say that $H^\star$ is a supergraph of H, if H is a subgraph of $H^\star$.

Lemma 5 ($\star$). *Let $e \in E(G_T)$, and let $G^\star$ be a supergraph of $G_T - e$, then for each super-rich pair $\{x, y\}$ in G_T, at least one of the vertices in the pair must be included in all fvs of size at most k in $G^\star$.*

Marking Scheme. We start with the following marking procedure, which marks $\mathcal{O}(k^2)$ components of $\mathcal{C}_v$. Recall that $\mathcal{C}_v$ is a set of some components in $G_T - (X_v \cup \{v\})$.

> **Procedure** `Mark-1`(v). Let C be a component in $\mathcal{C}_v$. We mark the component C if there exists a pair of vertices x and y in $X_v \cup \{v\}$ such that:
> - $\{x, y\}$ is rich or poor.
> - $C \in \mathcal{C}(x) \cap \mathcal{C}(y)$.

Claim 1 ($\star$). *`Mark-1`(v) process marks at most $2k^3 + 7k^2 + 3k$ components of $\mathcal{C}_v$.*

Claim 2 ($\star$). *If v is a vertex of degree more than $2k^3+7k^2+7k$ in G_T, then there always exists an unmarked component in $\mathcal{C}_v$ following the `Mark-1`(v) procedure.*

Reduction Rule 7 (Irrelevant Edge Rule). *Let v be a vertex of degree strictly more than $2k^3 + 7k^2 + 7k$ in G_T and C be an unmarked component (which is a tree) in $\mathcal{C}_v$ after the procedure* `Mark-1`(v). *And, let x be the neighbor (must be unique) of v in C. Then we remove the edge $e = vx$ and return the instance is $(G - e, T, k)$.*

We note that although G_T is employed to identify an irrelevant edge, we apply Reduction Rule 7 to G.

Lemma 6. *Reduction Rule 7 is safe.*

Proof. Let (G, T, k) be an input instance to Reduction Rule 7 and $(G' = G - e, T, k)$ be the corresponding output instance. The forward direction is simple. As $G - e$ is a subgraph of G, any solution of (G, T, k) is indeed a solution to $(G - e, T, k)$.

In the reverse direction, suppose (G', T, k) is a `Yes`-instance, and let $S' \subseteq V(G)$ be a solution. Without loss of generality, we may assume that S' is a minimal in the following sense. That is, there is no subset $T \subseteq S^\star \subsetneq S'$ such that $S^\star$ is a at most-c-fvs. That is, for every vertex $x \in S' \setminus T$, we have a cycle in $G'[V \setminus S' \cup \{x\}]$. We show that S' remains a solution to (G, T, k) with a proof based on the following three cases.

- **Case (i)**: $|S' \cap \{x, v\}| = 0$. If neither of the vertices x and v belongs to S', then the size of the largest component in subgraph $G[S']$ is at most c. We now show that S' is an fvs of G. If not, then $G - S'$ contains a cycle, say $\mathtt{Q}_x$, that must contain the edge xv, otherwise $G' - S'$ also has the same cycle, contradicting that S' is a solution to (G', T, k). Let u denote the vertex in $\mathtt{Q}_x \cap X_v$ that has a neighbor in the component C. Clearly, $u \notin S'$. Since C is unmarked, the pair $\{u, v\}$ must be super-rich in G_T. Since, G' is a supergraph of $G_T - e$, by Lemma 4, we have that one of the vertices of each super-rich pair must be present in any fvs of size at most k in G', contradicting the fact that $S' \cap \{u, v\} = \emptyset$. This implies that S' is a solution to (G, T, k).
- **Case (ii)**: $|S' \cap \{x, v\}| = 1$. When S' contains at most one vertex from $\{x, v\}$, then $G' - S' = G - S'$ and $G[S'] = G'[S']$. Hence S' is a solution to (G, T, k).
- **Case (iii)**: $|S' \cap \{x, v\}| = 2$. Since S' is a minimal fvs in G', $G'[V \setminus S' \cup \{x\}]$ has a cycle, say $\mathtt{Q}_x$ that contains x. As C is a tree, $\mathtt{Q}_x$ contains at least two vertices, say p and q in $V(\mathtt{Q}_x) \cap X_v$ such that $p, q \notin S'$ and $C \in \mathcal{C}(p) \cap \mathcal{C}(q)$. Since C is unmarked, the pair $\{p, q\}$ must be super-rich in G_T. Since, G' is a supergraph of $G_T - e$, by Claim 4, we have that one of the vertices of each super-rich pair must be present in any fvs of size at most k in G', contradicting the fact that $S' \cap \{p, q\} = \emptyset$. Therefore, S' cannot be a minimal solution when $|S' \cap \{x, v\}| = 2$. This implies that either $|S' \cap \{x, v\}| = 1$ or $|S' \cap \{x, v\}| = 0$. Thus, the case $|S' \cap \{x, v\}| = 2$ is not possible and hence, we fall in the first two cases.

This completes the proof. □

Observation 1. *Following the exhaustive application of Reduction Rule 7 the degree of every vertex in $G - T$ is bounded by $2k^3 + 7k^2 + 8k$.*

By exhaustive application of Reduction Rules 1 to 7 on an input instance (G, T, k) of the problem EXT AT MOST-c-FVS, in polynomial time, we either derive a Yes or No answer or arrive at an equivalent instance (G', T', k') where none of the rules is applicable. In the following lemma, we show a size bound of $\mathcal{O}(k^{c+3})$ for (G', T', k'), when the input (G, T, k) is a Yes instance.

Lemma 7. *Let (G, T, k) be an input instance of* EXT AT MOST-c-FVS, *and (G', T', k') be a resulting instance following exhaustive application of Reduction Rules 1 to 7. If (G, T, k) is a* Yes *instance, then G' has $\mathcal{O}(k^{c+3})$ vertices.*

Proof. Let (G, T, k) be a Yes instance of AT MOST-c-FVS. Following exhaustive applications of the Reduction Rules 1-7, the degree of each vertex in $G' - T'$ is at most $k^\star = 2k^3 + 7k^2 + 8k$. Let S be a solution to (G', T', k'), $F = G' - S$ and $S' = S \setminus T'$.

- The number of vertices in F that have a neighbor in S' (denoted by Q) is at most $k^\star k$. In particular, since G' is reduced with respect to the Reduction Rule 3, every leaf in F has at least one neighbor in S', so there are at most $|Q|$ leaves (denoted by L) in F.
- Since, F is a forest, the number of vertices with degree at least three (denoted by P) in F is bounded by the number of leaves in F. So $|P| \leq |L| \leq |Q|$.
- Let N be the set of degree 2 vertices in F, which have no neighbors in S' but *may* have neighbors in T'. As every degree-2 path in $G' - T'$ contains at most k^{c-1} vertices (by Observation 1, $|N| \leq k^{c-1} \cdot (|Q| + |P|) \leq 2k^{c-1}|Q|$. The last inequality follows from the fact that in a forest the number of maximal degree 2 paths are upper bounded by the number of leaves and the number of number of vertices of degree at least 3 (see [19, Lemma 9] for a proof of this fact).

Hence, the number of vertices in G' is bounded by $|S| + |L| + |P| + |N| \leq k + 2|Q| + 2k^{c-1}|Q|$. As $|Q| \leq k^\star k$, $|V(G)| \leq k + 2k^\star k + 2k^c k^\star$. Putting $k^\star = 2k^3 + 7k^2 + 8k$, we obtain $|V(G')| \leq 4k^{c+3} + 14k^{c+2} + 16k^{c+1} + 14k^3 + 16k^2 + k = \mathcal{O}(k^{c+3})$. □

Hence, we obtain the following result.

Lemma 8. EXT AT MOST-c-FVS *admits a kernel of size $\mathcal{O}(k^{c+3})$ vertices.*

Equivalence. Given an instance (G, k) of AT MOST-c-FVS, we create an equivalent instance $(G, T = \{\phi\}, k)$ of EXT AT MOST-c-FVS. And by applying Lemma 8 to $(G, T = \{\phi\}, k)$, we derive a kernel/an equivalent instance (G', T', k') where size $|V(G')| = \mathcal{O}(k^{c+3})$. Subsequently, we transform back (G', T', k') of EXT AT MOST-c-FVS into an equivalent instance (G'', k'') of AT MOST-c-FVS, where $|V(G'')|$ is also bounded by $\mathcal{O}(k^{c+3})$, utilizing Lemma 9, which results in the desired kernel for AT MOST-c-FVS in Theorem 5.

Lemma 9. *Given an instance (G', T', k') of* EXT AT MOST-c-FVS, *an equivalent instance (G'', k'') of* AT MOST-c-FVS, *where $|V(G'')| = \mathcal{O}(k^{c+3})$ can be constructed in polynomial time.*

Proof. Given an instance (G', T', k') of Ext at most-c-FVS, we construct an equivalent instance (G'', k'') At most-c-FVSas follows. For each of the terminal vertices (say $v \in T'$), we add a set of $k+1$ triangles $((v, v_1, v_2), (v, v_3, v_4), \ldots, (v, v_{2k+1}, v_{2k+2}))$ that intersect only at v. Formally, $V(G'') = V(G') \cup \{v_1, v_2, \ldots, v_{2k+2}; v \in T'\}$, and $E(G'') = E(G') \cup \{(v, v_1), (v, v_2), \ldots, (v, v_{2k+2}); v \in T'\} \cup \{(v_{2i-1}, v_{2i}) \colon v \in T', i \in [k+1]\}$. The equivalence of instances (G', T', k') and (G'', k'') is apparent from the fact that any solution of size at most k for either of the instances contains all the vertices of T'. □

This leads to the desired kernel.

Theorem 5. At most-c-FVS *admits a kernel with* $\mathcal{O}(k^{c+3})$ *vertices.*

3 Other Results

In Sect. ??, we obtain the following result (Theorem 6) for At most-c-FVS. Our algorithm for Theorem 6 starts similar to the known FPT algorithm for IFVS (two-way branching) by Misra et al. [17]. But here we use several non-trivial (six-way) branching strategies to reduce the problem to a variant of the path hitting set problem: Exact Path Hitting. Although a straightforward application of *color coding* can solve this problem in $\mathcal{O}^*(k^k)$ time, we improve the runtime dependency to $\mathcal{O}^*(4^{k+o(k)} \cdot k^{2c})$ by employing the *divide and color* technique [13].

Theorem 6 (⋆). At most-c-FVS *can be solved in* $\mathcal{O}^*(k^{2c} \cdot 2^{\mathcal{O}(k)})$ *time.*

In Sect. ??, we establish the following result (Theorem 7), a lower bound on the kernel size of At least-c-FVS. We achieve this by demonstrating a *polynomial parameter transformation* (PPT) from a suitable parameterization of the Red Blue Dominating Set (RBDS) problem.

Theorem 7 (⋆). At least-c-FVS *admits no kernel of size* $k^{f(c)}$ *unless* co-NP $\subseteq$ NP/poly.

In Sect. ??, we present the following result (Theorem 8) for an FPT algorithm for At least-c-FVS. We achieve this by using an algorithm (initially introduced by Guo et al. [9]) to enumerate efficient representations of minimal feedback vertex sets, which ensures that at least one is contained in any optimal solution. Upon guessing such a representation, we employ random vertex coloring, which then transforms the problem into finding a colorful forest (by guessing its structure) where each component has a size at least c. Subsequently, we use the findings of Amini et al. [4] to obtain such a forest and consequently design an FPT algorithm running in time $\mathcal{O}^*(741^k)$.

Theorem 8 (⋆). At least-c-FVS *can be solved in time* $\mathcal{O}^*(2^{\mathcal{O}(k)})$ *time.*

References

1. Agrawal, A., Gupta, S., Saurabh, S., Sharma, R.: Improved algorithms and combinatorial bounds for independent feedback vertex set. In: Guo, J., Hermelin, D. (eds.) IPEC 2016, Aarhus, Denmark, vol. 63 of LIPIcs, pp. 2:1–2:14. Schloss Dagstuhl - Leibniz-Zentrum für Informatik (2016). https://doi.org/10.4230/LIPIcs.IPEC.2016.2
2. Agrawal, A., Jain, P., Kanesh, L., Lokshtanov, D., Saurabh, S.: Conflict free feedback vertex set: a parameterized dichotomy. In: Potapov, I., Spirakis, Worrell, J. (eds.) MFCS 2018, vol. 117 of LIPIcs, pp. 53:1–53:15. Schloss Dagstuhl - Leibniz-Zentrum für Informatik (2018). https://doi.org/10.4230/LIPIcs.MFCS.2018.53
3. Agrawal, A., Lokshtanov, D., Mouawad, A.E., Saurabh, S.: Simultaneous feedback vertex set: a parameterized perspective. ACM Trans. Comput. Theory **10**(4):18:1–18:25 (2018). https://doi.org/10.1145/3265027
4. Amini, O., Fomin, F.V., Saurabh, S.: Counting subgraphs via homomorphisms. SIAM J. Disc. Math. **26**(2), 695–717 (2012). https://doi.org/10.1137/100789403
5. Cygan, M., et al.: Solving connectivity problems parameterized by treewidth in single exponential time. ACM Trans. Algor. **18**(2), 17:1–17:31 (2022). https://doi.org/10.1145/3506707
6. Cygan, M., Pilipczuk, M., Pilipczuk, M.: On group feedback vertex set parameterized by the size of the cutset. Algorithmica **74**(2), 630–642 (2016). https://doi.org/10.1007/s00453-014-9966-5
7. Downey, R.G., Fellows, M.R.: Fixed parameter tractability and completeness. In: Ambos-Spies, K., Homer, S., Schöning, U. (eds.) Complexity Theory: Current Research, Dagstuhl Workshop, 2–8 February 1992, pp. 191–225. Cambridge University Press, Cambridge (1992)
8. Fomin, F.V., Lokshtanov, D., Saurabh, S., Zehavi, M.: Kernelization: Theory of Parameterized Preprocessing. Cambridge University Press, Cambridge (2019)
9. Guo, J., Gramm, J., Hüffner, F., Niedermeier, R., Wernicke, S.: Compression-based fixed-parameter algorithms for feedback vertex set and edge bipartization. J. Comput. Syst. Sci. **72**(8), 1386–1396 (2006). https://doi.org/10.1016/j.jcss.2006.02.001
10. Iwata, Y., Kobayashi, Y.: Improved analysis of highest-degree branching for feedback vertex set. Algorithmica **83**(8), 2503–2520 (2021). https://doi.org/10.1007/s00453-021-00815-w
11. Jana, S., Lokshtanov, D., Mandal, S., Rai, A., Saurabh, S.: Parameterized approximation scheme for feedback vertex set. In: Leroux, J., Lombardy, S., Peleg, D. (eds.) 48th International Symposium on Mathematical Foundations of Computer Science, MFCS 2023, Bordeaux, France, 28 August–1 September 2023, vol. 272 of LIPIcs, pp. 56:1–56:15. Schloss Dagstuhl - Leibniz-Zentrum für Informatik (2023). https://doi.org/10.4230/LIPIcs.MFCS.2023.56
12. Karp, R.M.: Reducibility among combinatorial problems. In: Miller, R.E., Thatcher, J.W. (eds.) Proceedings of a Symposium on the Complexity of Computer Computations, at the IBM Thomas J. Watson Research Center, Yorktown Heights, New York, USA, 20–22 March 1972, The IBM Research Symposia Series, pp. 85–103. Plenum Press, New York (1972). https://doi.org/10.1007/978-1-4684-2001-2_9
13. Kneis, J., Mölle, D., Richter, S., Rossmanith, P.: Divide-and-Color. In: Fomin, F.V. (ed.) WG 2006. LNCS, vol. 4271, pp. 58–67. Springer, Heidelberg (2006). https://doi.org/10.1007/11917496_6

14. Li, J., Nederlof, J.: Detecting feedback vertex sets of size k in O^* $(2.7k)$ time. ACM Trans. Algor. **18**(4), 34:1–34:26 (2022). https://doi.org/10.1145/3504027
15. Li, S., Pilipczuk, M.: An improved FPT algorithm for independent feedback vertex set. Theory Comput. Syst. **64**(8), 1317–1330 (2020). https://doi.org/10.1007/s00224-020-09973-w
16. Lokshtanov, D., Ramanujan, M.S., Saurabh, S.: Linear time parameterized algorithms for subset feedback vertex set. ACM Trans. Algor. **14**(1), 7:1–7:37 (2018). https://doi.org/10.1145/3155299
17. Misra, N., Philip, G., Raman, V., Saurabh, S.: On parameterized independent feedback vertex set. Theor. Comput. Sci. **461**, 65–75 (2012). https://doi.org/10.1016/j.tcs.2012.02.012
18. Misra, N., Philip, G., Raman, V., Saurabh, S., Sikdar, S.: FPT algorithms for connected feedback vertex set. J. Comb. Optim. **24**(2), 131–146 (2012). https://doi.org/10.1007/s10878-011-9394-2
19. Raman, V., Saurabh, S., Subramanian, C.R.: Faster fixed parameter tractable algorithms for finding feedback vertex sets. ACM Trans. Algor. **2**(3), 403–415 (2006). https://doi.org/10.1145/1159892.1159898
20. Robertson, N., Seymour, P.D.: Graph minors. xiii. The disjoint paths problem. J. Comb. Theory B **63**(1), 65–110 (1995). https://doi.org/10.1006/jctb.1995.1006
21. Robertson, N., Seymour, P.D.: Graph minors. XX. Wagner's conjecture. J. Comb. Theory B **92**(2), 325–357 (2004). https://doi.org/10.1016/j.jctb.2004.08.001
22. Thomassé, S: A $4k^2$ kernel for feedback vertex set. ACM Trans. Algor. **6**(2), 32:1–32:8 (2010). https://doi.org/10.1145/1721837.1721848

Online b-Matching with Stochastic Rewards

Susanne Albers and Sebastian Schubert(✉)

Department of Computer Science, Technical University of Munich, Garching, Germany
albers@in.tum.de, sebastian.schubert@tum.de

Abstract. The b-matching problem is an allocation problem where the vertices on the left-hand side of a bipartite graph, referred to as servers, may be matched multiple times. In the setting with stochastic rewards, an assignment between an incoming request and a server turns into a match with a given success probability. Mehta and Panigrahi (FOCS 2012) introduced online bipartite matching with stochastic rewards, where each vertex may be matched once. The framework is equally interesting in graphs with vertex capacities. In Internet advertising, for instance, the advertisers seek successful matches with a large number of users. We develop (tight) upper and lower bounds on the competitive ratio of deterministic and randomized online algorithms, for b-matching with stochastic rewards. Our bounds hold for both offline benchmarks considered in the literature. As in prior work, we first consider vanishing probabilities. We show that no randomized online algorithm can achieve a competitive ratio greater than $1 - 1/e \approx 0.632$, even for identical vanishing probabilities and arbitrary uniform server capacities. Furthermore, we conduct a primal-dual analysis of the deterministic StochasticBalance algorithm. We prove that it achieves a competitive ratio of $1 - 1/e$, as server capacities increase, for arbitrary heterogeneous non-vanishing edge probabilities. This performance guarantee holds in a general setting where servers have individual capacities and for the vertex-weighted problem extension. To the best of our knowledge, this is the first result for StochasticBalance with arbitrary non-vanishing probabilities. We remark that our impossibility result implies in particular that, for the AdWords problem, no online algorithm can be better than $(1 - 1/e)$-competitive in the setting with stochastic rewards.

1 Introduction

Online bipartite matching and the generalized AdWords problem have received tremendous research interest over the last three decades, see e.g. [1–4, 7, 9–11, 13–16, 18]. In this realm Mehta and Panigrahi [15] introduced the online matching problem with stochastic rewards, which focuses on *successful* allocations. Given a bipartite graph, the left-hand side vertices are known in advance. The right-hand side vertices arrive one by one. In each step, an incoming vertex may be assigned to an available neighbor. Then, with a certain probability, the assignment turns

R. Královič and V. Kůrková (Eds.): SOFSEM 2025, LNCS 15538, pp. 37–50, 2025.
https://doi.org/10.1007/978-3-031-82670-2_4

into an actual match. Mehta and Panigrahi motivate their study by the fact that in Internet advertising, an advertiser only pays if a user clicks on the assigned ad (pay-per-click) and click-through rates are known.

Further and recent work on matchings with stochastic rewards includes [6–8, 10,17,19]. Almost all the contributions address the uncapacitated setting where each vertex may be matched once. However, matchings with stochastic rewards are also highly relevant in capacitated environments. In fact, in the AdWords problem, the left-hand side vertices of the bipartite graph represent advertisers with budgets who wish to show ads to a large number of users. Furthermore, the left-hand side vertices could be service providers that offer support in terms of telecommunications, data storage or job processing to a large number of clients. A client with a service request may accept an offer with a certain probability. Alternatively, the left-hand side vertices could be online retailers that sell large quantities of products. A potential customer buys a recommended product with a certain chance.

Problem Definition: In this paper we study online b-matching with stochastic rewards. The b-matching problem is a well-known capacitated allocation framework. In particular, it models interesting special cases of the AdWords problem. Formally, we are given a bipartite graph $G = (S \cup R, E)$. The vertices of S are servers. Each server $s \in S$ has a capacity of b, indicating that may it be matched up to b times. The vertices of R are requests that have to be assigned to the servers. The set S of servers is known in advance. The requests of R arrive online, one by one. Whenever a new request $r \in R$ arrives, its incident edges are revealed. Each such edge $\{s, r\}$ has a success probability $p_{s,r}$ that is revealed as well. An algorithm has to assign the request immediately and irrevocably to an eligible server with remaining capacity, provided that there is one. If an assignment is made, request r accepts it with the probability of the corresponding edge, i.e., the assignment turns into a successful match. The outcome of the random choice is independent of past ones. If r does not accept, it leaves the system and the remaining capacity of the proposed server is unchanged. The goal is to maximize the expected number of successful matches. Throughout this paper we will use the term *assignment* to refer to the decision of an algorithm. The terms *success* or *match* are used for assignments that succeeded.

Benchmarks and Competitiveness: We evaluate the performance of online algorithms using competitive analysis. For online matching with stochastic rewards, addressing the uncapacitated variant with $b = 1$, there has been some discussion in the literature [7,15] on how to quantify the value of an optimal solution. When Mehta and Panigrahi [15] introduced the problem, they compared their algorithms against the offline and non-stochastic optimum. More specifically, they consider the optimal solution to the problem where the entire graph is known in advance and the reward of adding an edge $\{s, r\}$ to the matching is (deterministically) $p_{s,r}$. The reward that can be accrued per server is upper bounded by 1. The problem of maximizing total reward is known as the budgeted allocation problem. Mehta and Panigrahi show that the expected number of matches of any online algorithm for the matching problem with stochastic rewards is upper bounded by the value of a (fractional) optimal solution to the budgeted allocation problem. This fact immediately carries over to the capacitated problem, assuming that the accrued reward per server is upper bounded by its capacity b. Hence the fractional optimum to the budgeted allocation problem is a suitable benchmark for evaluating an online algorithm. In the following, we will refer to it as the *non-stochastic benchmark*.

The other benchmark that is used in the literature is referred to as the *clairvoyant* or simply *stochastic benchmark*. It corresponds to an offline algorithm assigning requests. More specifically, the benchmark is defined as the expected number of matches produced by an optimal algorithm that knows the entire graph including the edge probabilities in advance. The algorithm has to assign requests according to their arrival order and is only informed after an assignment whether it succeeded or failed. Golrezaei et al. [6] showed that, for any graph, the stochastic benchmark is upper bounded by the non-stochastic benchmark. Hence the stochastic benchmark potentially admits better competitive ratios.

Given an input graph G, let $\mathbb{E}[\text{ALG}(G)]$ denote the expected number of matches of an online algorithm ALG for the online b-matching problem with stochastic rewards. Let $\text{OPT}(G)$ and $\text{SOPT}(G)$ be the value of the non-stochastic and stochastic benchmark on G, respectively. Algorithm ALG is c-competitive against the non-stochastic benchmark if $\mathbb{E}[\text{ALG}(G)] \geq c \cdot \text{OPT}(G)$ holds, for all G. Analogously, ALG is c-competitive against the stochastic benchmark if $\mathbb{E}[\text{ALG}(G)] \geq c \cdot \text{SOPT}(G)$, for all G.

1.1 Related Work

First, we review a few results on online matching *without* stochastic rewards, i.e. an assignment is a match (with probability 1). Online bipartite matching was introduced in a seminal paper by Karp et al. [13]. Each vertex may be matched once. Karp et al. showed that the best competitive ratio of deterministic online algorithms is equal to $1/2$. Furthermore, they proposed a randomized RANKING algorithm that achieves an optimal competitiveness of $1-1/e \approx 0.632$, see [2,5,13]. Online b-matching was studied in [12], assuming that all servers have a uniform capacity of b. The best competitive ratio of deterministic online algorithms is equal to $1 - 1/(1+1/b)^b$, which tends to $1 - 1/e$ as $b \to \infty$ [12].

Online matching with stochastic rewards was defined by Mehta and Panigrahi [15]. Again, in the original problem each vertex may be matched only once, i.e. $b = 1$. Mehta and Panigrahi consider equal success probabilities $p_{s,r} = p$, for all edges $\{s, r\} \in E$. All their results are in comparison with OPT. Mehta and Panigrahi present a simple STOCHASTICBALANCE algorithm that assigns incoming requests to an eligible neighbor that is assigned the least amount of requests so far. They show that its competitive ratio is at least 0.567 and at most 0.588 for vanishing probabilities, meaning that $p \to 0$. Moreover, they analyze RANKING and show that it achieves a competitive ratio of at least 0.534, for equal non-vanishing probabilities. Finally, they show that no (randomized) algorithm has a competitive ratio greater than $0.621 < 1 - 1/e$. This upper bound has recently been improved to 0.597 by leveraging reinforcement learning [20].

Huang and Zhang [10] were the first to successfully conduct a primal-dual analysis of the problem. They improve the competitiveness of StochasticBalance against Opt to 0.576 if all edges have equal and vanishing probabilities. Furthermore, they also show that a generalized version of StochasticBalance is at least 0.572-competitive if all edges have unequal vanishing probabilities. Concurrently, Goyal and Udwani [7] analyzed the performance of a randomized PerturbedGreedy algorithm against the stochastic benchmark SOpt, for the more general vertex-weighted problem. They show a competitive ratio of $1 - 1/e$ if the edge probabilities are *decomposable*, i.e., they are the product of two factors, one for each of the two vertices of an edge. For unequal but vanishing probabilities, they give an algorithm that is 0.596-competitive. In the most recent work, Huang et al. [8] improve the competitiveness of Ranking against Opt to 0.572, if all probabilities are equal and vanishing. They further show that StochasticBalance is 0.613-competitive against SOpt in the case of equal and vanishing probabilities. Their result slightly worsens to 0.611 if the probabilities are unequal, but still vanishing.

We remark that the lower bounds, i.e. the competitive ratios of the various algorithms mentioned in the last two paragraphs also hold for general b-matching with stochastic rewards. This follows from a standard vertex-splitting argument: Replace each server of capacity b by b vertices that may be matched only once. On the resulting graph, execute the corresponding algorithm. However, the upper bounds cannot immediately be extended to other values of b. The only prior work that covers online b-matching with stochastic rewards is by Golrezaei et al. [6]. The authors study more general personalized assortments optimization in management science. Their InventoryBalance algorithm achieves a competitive ratio of $1 - 1/e$ against Opt for arbitrary edge probabilities, if $b \to \infty$. They also show that this is tight, by giving an upper bound of $1 - 1/e$ if all edges have probability 1. For $b = 1$, their algorithm reduces to a greedy algorithm that is 1/2-competitive.

We finally mention the Adwords problem [16], which has been studied without stochastic rewards. There is a set of advertisers, each with a daily budget, who wish to link their ads to search keywords and issue respective bids. Queries along with their keywords arrive online and must be allocated to the advertisers. The optimal competitiveness is $1 - 1/e$, under the small-bids assumption [3,16]. The b-matching problem models the basic setting where each advertiser issues bids of value 0 or 1.

1.2 Our Contribution

In this paper we develop tight upper and lower bounds on the competitive ratio of deterministic and randomized online algorithms for b-matching with stochastic rewards. Our bounds hold for both benchmarks Opt and SOpt. Our study is specifically motivated by the following question: Is it possible to beat the barrier of $1 - 1/e$, for competitiveness, in the (easiest) setting of equal and vanishing probabilities when comparing against the stochastic benchmark? Observe that, for $b = 1$ and vanishing probabilities, deterministic online algorithms achieve competitive ratios greater than 1/2, which is the best bound in the framework without stochastic rewards. The question was also raised by Goyal and Udwani [7]. Vanishing probabilities are sensible in various applications. In Internet advertising, a user clicks on an assigned ad with low probability.

In Sect. 2 we develop an upper bound. Here we consider vanishing probabilities, as almost all prior work. We prove that no randomized online algorithm can achieve a competitive ratio greater than $1-1/e$ against the stochastic benchmark SOpt, for online b-matching with stochastic rewards. This holds even for equal vanishing edge probabilities and for all values of b. To the best of our knowledge, this is the first hardness result for the stochastic benchmark in the literature. We immediately obtain the same upper bound against the non-stochastic benchmark Opt because, as stated above, for any graph the value of Opt is at least as large as that of SOpt [6]. In conclusion, surprisingly, it is impossible to break the barrier of $1 - 1/e$, which is a recurring performance guarantee in matching problems. In particular for Adwords, a prominent application of matchings with stochastic rewards, no further improvement is possible.

Technically, in our upper bound construction, we define a family of graphs, for general $n = |S|$ and b. It generalizes graphs used by Mehta and Panigrahi [15] to prove their upper bound of $0.621 < 1-1/e$. For our problem, the mathematical analysis is much more involved. A key problem is to estimate the value of the stochastic benchmark SOpt. An offline algorithm is hindered by the fact that it does not know in advance which edges will be successful and that it has to serve requests in the order of arrival. We resolve this issue by considering a Greedy strategy, for the graph family.

In Sect. 3 we present a constructive, algorithmic result. We analyze a generalization of the deterministic StochasticBalance algorithm, for online b-matching with stochastic rewards. Specifically, we perform a primal-dual analysis, considering the harder benchmark Opt. In order to ease the technical exposition, we first assume that all servers have a uniform capacity of b. We prove that StochasticBalance achieves a competitive ratio of $1 - 1/e$ with respect to Opt, as $b \to \infty$. This bound holds for arbitrary individual, non-vanishing edge probabilities. The performance guarantee immediately carries over to the easier stochastic benchmark SOpt. Finally, we show how to extend the result to more general settings: (1) Each server $s \in S$ has an individual capacity b_s. (2) Each server $s \in S$ has a weight w_s and any successful match incident to s has a weight/value of w_s. The goal is to maximize the expected total weight of successful matches. StochasticBalance remains $(1 - 1/e)$-competitive as the minimum server capacity increases, i.e., $\min_{s \in S} b_s \to \infty$. We remark that, due to the complexity of the primal-dual analysis, we present bounds as b or $\min_{s \in S} b_s$ tend to infinity. Already for $b = 1$, the analysis of StochasticBalance involves solving integral equations using numerical methods [8,10].

Our contribution for StochasticBalance is one of the few existing results that hold for arbitrary non-vanishing probabilities. StochasticBalance was analyzed for identical non-vanishing probabilities, the best ratio of 0.567 being achieved as $p \to 0$ [15]. The randomized algorithms Ranking and Perturbed-Greedy attain competitive ratios of 0.572 and $1-1/e$ for equal or decomposable, non-vanishing probabilities [7,8]. Finally, in comparison, StochasticBalance appears to be a more favourable algorithm than InventoryBalance, for online b-matching with stochastic rewards. It is a simpler, more intuitive and well studied matching algorithm [8,10,15], attaining competitive ratios greater than 1/2, for $b = 1$.

2 Upper Bound of $1 - 1/e$ for Vanishing Probabilities

We give an upper bound on the competitiveness of any (randomized) algorithm for the online b-matching problem with stochastic rewards. Note that for $p = 1$, both benchmarks are identical, and it is known [16] that no algorithm achieves a competitive ratio greater than $1 - 1/e$. However, it was unknown if one could improve upon this in the case of smaller edge probabilities or when using the stochastic benchmark. We will show that this is impossible, resolving the open question raised in [7]. To the best of our knowledge, we develop the first hardness result with respect to the stochastic benchmark. As mentioned before, the stochastic benchmark is easier than the non-stochastic one. Hence we immediately obtain the same upper bound for the non-stochastic benchmark.

We investigate the family of graphs G_n^b with n servers $S = \{s_1, s_2, \ldots, s_n\}$ of capacity b and $n \cdot b/p$ requests. The online side is divided into n rounds, each containing b/p identical requests. All requests of round i, $1 \le i \le n$, are connected to servers $\{s_i, s_{i+1}, \ldots, s_n\}$. The success probability of every edge is $p \to 0$.

Note that the family G_n^b is closely related to other instances used to derive an upper bound on the competitive ratio of online matching problems (cf. hardness results in [13,16]). Most relevant to this paper, Mehta and Panigrahi [15] use the family G_n^1 to show their upper bound of $0.621 < 1 - 1/e$ for $b = 1$ against Opt. They first prove that StochasticBalance is an optimal online algorithm for G_n^1, for all n. In fact, their proof also works for all values of b. Afterward, they compute the exact expected number of successes generated by StochasticBalance with input G_n^1 for $n = 1, 2$ and 3 and compare it against $\text{Opt}(G_n^1) = n$. The smallest competitive ratio of 0.621 is achieved on G_3^1.

Lemma 1 (extension of Lemma 12 in [15]). StochasticBalance *is optimal for input graph G_n^b, for any combination of n and b.*

We will show our upper bound similarly but need new technical ideas. In Sect. 2.1, we analyze the expected number of successful matches of StochasticBalance on G_n^b. This turns out to be much more complex for arbitrary values of n and b. By Lemma 1, this upper bounds the performance of any online algorithm. Afterwards, in Sect. 2.2, we quantify $\text{SOpt}\left(G_n^b\right)$. This will be again more involved than determining $\text{Opt}\left(G_n^b\right)$, which is simply $n \cdot b$.

2.1 Expected Number of Matches by StochasticBalance on G_n^b

We develop an upper bound for the expected number of matches generated by StochasticBalance on G_n^b. For equal edge probabilities, StochasticBalance simply assigns an incoming request r to a neighbor with remaining capacity that has minimum *load*. The load of a server s is defined as the sum of edge probabilities to requests assigned to s. Observe that StochasticBalance is *opportunistic* in the sense that never leaves a request unassigned if it has a neighbor with remaining capacity.

The following lemma focuses on any round of requests. It not only holds true for StochasticBalance but any opportunistic algorithm. Due to space constraints, all proofs of this subsection are only given in the full version of the paper.

Lemma 2. *Let the random variable R_i, $1 \leq i \leq n$, denote the number of matches during the assignment of the requests belonging to round i in G_n^b. Let m be the number of unused capacity from servers in $\{s_i, s_{i+1}, \dots, s_n\}$ at the start of round i. It holds that $R_i \sim \min\{Pois(b), m\}$. More precisely, it holds*

$$\Pr[R_i = k] = \begin{cases} \frac{b^k}{k!e^b} & k < m\,, \\ 1 - \sum_{j=0}^{m-1} \frac{b^j}{j!e^b} & k = m\,. \end{cases}$$

Moreover, $\mathbb{E}[R_i] \leq b$.

With the help of Lemma 2, we can now upper bound the expected number of successes for each server when executing StochasticBalance on G_n^b.

Lemma 3. *Let the random variable S_j denote the total number of matches of server s_j after executing* StochasticBalance *on G_n^b. It holds that*

$$\mathbb{E}[S_j] \leq \min\left\{\sum_{i=1}^{j} \frac{b}{n-i+1}, b\right\} = b \cdot \min\left\{\sum_{i=1}^{j} \frac{1}{n-i+1}, 1\right\}.$$

A reader may recognize the expression $\min\left\{\sum_{i=1}^{j} \frac{1}{n-i+1}, 1\right\}$, cf. [16] and the analysis of the (fractional) Water-Filling algorithm on G_n^1 with $p = 1$.

Lemma 4. *Let the random variable* SBal (G_n^b) *denote the number of matches generated by* StochasticBalance *on G_n^b. It holds that*

$$\frac{\mathbb{E}\left[\text{SBal}\left(G_n^b\right)\right]}{n \cdot b} \leq \frac{\left\lceil\left(1-\frac{1}{e}\right)(n+1)\right\rceil \cdot b}{n \cdot b} \xrightarrow{n\to\infty} 1 - \frac{1}{e}\,.$$

2.2 Stochastic Benchmark on G_n^b

We quantify the stochastic benchmark SOpt on G_n^b. Recall that the stochastic benchmark is defined as the expected number of matches of the best possible algorithm that knows the entire graph, including edge probabilities, in advance. However, said algorithm does not know which edges will be successful a priori and still needs to match the requests according to the arrival order. By definition, the expected number of matches of any algorithm with the restrictions above lower bounds SOPT.

In the following we simply analyze the algorithm GREEDY on G_n^b, as a lower bound for SOPT. GREEDY assigns an incoming request to the server with the smallest index among all servers with remaining capacity. Note that GREEDY needs to know the graph in advance to identify the indices of the servers.

To simplify the analysis, we only analyze the exact expected number of successes generated by GREEDY on G_n^1. This will suffice to show the same upper bound for all values of b. Let $T_{n,m}$ be the number of matches by GREEDY on G_n^1, where only the last m ($\leq n$) servers, i.e. $s_{n-m+1}, \ldots, s_n$, are present. Note that we want to determine $\mathbb{E}\left[T_{n,n}\right]$. We use the recursive nature of G_n^1 together with the law of total expectation to develop a recurrence relation for $\mathbb{E}\left[T_{n,m}\right]$. All proofs of this subsection are given in the full version of this paper.

Lemma 5. *For $m < n$, it holds that*

$$\mathbb{E}\left[T_{n,m}\right] = \sum_{k=0}^{m-1} \frac{1}{k!e} \cdot \left(k + \mathbb{E}\left[T_{n-1,m-k}\right]\right) + \left(1 - \sum_{k=0}^{m-1} \frac{1}{k!e}\right) \cdot m\,. \tag{1}$$

Moreover, for $m = n$, we have

$$\mathbb{E}\left[T_{n,n}\right] = \frac{1}{e} \cdot \mathbb{E}\left[T_{n-1,n-1}\right] + \sum_{k=1}^{n-1} \frac{1}{k!e} \cdot \left(k + \mathbb{E}\left[T_{n-1,n-k}\right]\right) + \left(1 - \sum_{k=0}^{n-1} \frac{1}{k!e}\right) \cdot n\,. \tag{2}$$

The following lemma solves the recurrence relation. The proof is by induction over n.

Lemma 6. *For all $m \leq n$, it holds that*

$$\mathbb{E}\left[T_{n,m}\right] = m - \sum_{k=0}^{m-1} \frac{(n-k)^{m-k-1}}{(m-k-1)!e^{n-k}}\,.$$

Lemma 7. *Let the random variable* GRE$\left(G_n^1\right)$ *denote the number of matches generated by* GREEDY *on G_n^1. It holds that*

$$\frac{\mathbb{E}\left[\mathrm{GRE}\left(G_n^1\right)\right]}{n} = 1 - \frac{1}{n} \sum_{k=1}^{n} \frac{k^{k-1}}{(k-1)!e^k} \xrightarrow{n \to \infty} 1\,.$$

Theorem 1. *No (randomized) algorithm achieves a competitive ratio greater than $1 - 1/e$ against* SOPT *for the online b-matching problem with stochastic rewards, for all b, even for equal and vanishing edge probabilities.*

Note that the upper bound of $1 - 1/e$ is tight for all values of b. Goyal and Udwani [7] show that PERTURBEDGREEDY achieves this competitiveness against SOPT for $b = 1$ and equal edge probabilities. As mentioned before, the result can be generalized to arbitrary values of b using a standard vertex-splitting argument.

Corollary 1. *No (randomized) algorithm achieves a competitive ratio greater than $1 - 1/e$ against* OPT *for the online b-matching problem with stochastic rewards, for all b, even for equal and vanishing edge probabilities.*

3 Competitiveness of STOCHASTICBALANCE for $b \to \infty$

In this section, we analyze the STOCHASTICBALANCE algorithm for large server capacities using the standard primal-dual framework by Devanur et al. [5]. Huang and Zhang [10] were the first to successfully apply the primal-dual framework to online matching with stochastic rewards. They focus on the case $b = 1$ with vanishing probabilities and show that the configuration LP has to be used for the analysis to obtain a non-trivial competitiveness for STOCHASTICBALANCE. However, it turns out that the standard matching LP suffices to show the best possible competitive ratio of $1 - 1/e$ for $b \to \infty$, even for non-vanishing probabilities.

For any $s \in S$, let the *load* l_s of a server s denote the sum of edge probabilities of requests assigned to s. If all edge probabilities are identical, STOCHASTICBALANCE simply assigns incoming requests to a neighbor with remaining capacity that has minimum load. For arbitrary edge probabilities, the algorithm is generalized with the help of a non-decreasing function f, which is determined during the analysis to optimize the competitive ratio.

Algorithm 1: Generalized STOCHASTICBALANCE

while *a new request $r \in R$ arrives* **do**
  Let $N(r)$ denote the set of neighbors of r with remaining capacity;
  if $N(r) = \emptyset$ **then**
    Do not assign r;
  else
    assign r to $\arg\max\{p_{s,r}(1 - f(l_s)) : s \in N(r)\}$ (break ties arbitrarily);
  end
end

In the following, we conduct a primal-dual analysis of STOCHASTICBALANCE. We obtain a competitive ratio of $1 - 1/e$ against the non-stochastic benchmark OPT. The result holds true for arbitrary, non-vanishing edge probabilities, if $b \to \infty$. Since the stochastic benchmark is easier than the non-stochastic one, the same competitive ratio is also achieved against SOPT. Afterward, we outline the changes that are necessary to extend this result to the more general vertex-weighted variant of the problem, where each server s moreover has an individual server capacity b_s.

Theorem 2. StochasticBalance *achieves a competitive ratio of* $1-1/e$ *for the online b-matching problem with stochastic rewards for arbitrary edge probabilities against both benchmarks, if* $b \to \infty$.

Theorem 3. StochasticBalance *achieves a competitive ratio of* $1-1/e$ *for the vertex-weighted online b-matching problem with stochastic rewards for arbitrary edge probabilities and individual server capacities against both benchmarks, if* $b_{\min} := \min_{s \in S} b_s \to \infty$.

In the remainder of this section, we sketch how to prove Theorem 2. Due to space constraints, most of the lemmas and technical details are omitted here. The complete proofs are given in the full version of this paper.

First, consider the standard (relaxed) primal and dual LP of online b-matching with stochastic rewards given below. Here, the primal variable $m(s,r)$ for each edge indicates the *probability* that $e = \{s,r\} \in E$ is chosen by the algorithm (irrespective of if the assignment succeeded or not). Note that these probabilities are generally different from 0 or 1, even for deterministic algorithms, as the randomness of assignments succeeding can influence the decisions of the algorithm. In the primal LP, the first set of constraints ensures that the server capacities are observed in expectation. The second set of constraints ensures that each request is assigned to at most one server. Note that the primal program also corresponds to the budgeted allocation problem (cf. the discussion on benchmarks in the introduction). As mentioned before, one can show that the expected number of matches of any online algorithm is upper bounded by the optimal solution to the primal program, which is exactly the non-stochastic benchmark.

$$
\begin{aligned}
\textbf{P:} \max \quad & \sum_{\{s,r\} \in E} p_{s,r} \cdot m(s,r) \\
\text{s.t.} \quad & \sum_{r:\{s,r\} \in E} p_{s,r} \cdot m(s,r) \le b, \quad (\forall s \in S) \\
& \sum_{s:\{s,r\} \in E} m(s,r) \le 1, \quad (\forall r \in R) \\
& m(s,r) \ge 0, \quad (\forall \{s,r\} \in E)\,.
\end{aligned}
$$

$$
\begin{aligned}
\textbf{D:} \min b \cdot \sum_{s \in S} x(s) + \sum_{r \in R} y(r) \\
\text{s.t. } p_{s,r} \cdot x(s) + y(r) \ge p_{s,r}, \quad (\forall \{s,r\} \in E) \\
x(s),\, y(r) \ge 0, \quad (\forall s \in S, \forall r \in R)\,.
\end{aligned}
$$

Now, we can explain the primal-dual framework. We construct a solution to the primal program by setting the primal variables according to StochasticBalance. The value of the primal solution is thus the expected number of matches generated by StochasticBalance. In parallel, we create a feasible solution to the dual program with the value of the primal solution multiplied by a constant $1/c$. It then follows by weak duality that StochasticBalance is c competitive against the optimal solution to the primal program, i.e. the non-stochastic benchmark.

More specifically, depending on the realization of edge successes and failures, we maintain a (random) assignment to the variables $\hat{m}(s,r)$, $\hat{x}(s)$ and $\hat{y}(r)$. Let the random variables P and D denote the value of the random primal and dual solutions, respectively. We denote a change in the values P and D by ΔP and ΔD, respectively. Moreover, we define $m(s,r) := \mathbb{E}[\hat{m}(s,r)]$, $x(s) := \mathbb{E}[\hat{x}(s)]$ and $y(r) := \mathbb{E}[\hat{y}(r)]$, where expectation is taken over the randomness of edges succeeding or failing. Initially, all variables are 0. Whenever StochasticBalance assigns a request r to a server s, we set $\hat{m}(s,r)$ to 1. Moreover, we increase $\hat{x}(s)$ by and set $\hat{y}(r)$ to

$$\Delta\hat{x}(s) = p_{s,r} \cdot \frac{f(l_s)}{b \cdot c} \quad \text{and} \quad \hat{y}(r) = p_{s,r} \cdot \frac{1 - f(l_s)}{c},$$

respectively. Here, l_s denotes the load of server s *before* the assignment. Note that it always holds that $\Delta P/c = p_{s,r}/c = \Delta D$. Thus, summing over all assignments of the algorithm and taking expectation yields $\mathbb{E}[P]/c = \mathbb{E}[D]$. This is exactly the desired relation between the values of the primal and dual solutions to show c-competitiveness. All that is left to show now is that f and c can be chosen in a way that guarantees the feasibility of the dual solution.

Determining the optimal combination of f and c for $b = 1$ involves solving integral equations using numeric methods, even for vanishing probabilities [8,10]. While these integral equations can be adapted and solved numerically for larger values of b, we are interested the best possible competitive ratio of StochasticBalance as $b \to \infty$[1]. For this, we use the simple combination of $c = 1 - 1/e$ and

$$f(x) = \begin{cases} e^{\frac{x}{b}-1} & x \leq b, \\ 1 & x > b. \end{cases}$$

We argue that this yields a feasible solution in the limit $b \to \infty$. By the choice of f, we always have $\hat{x}(s) \geq 0$ and $\hat{y}(r) \geq 0$. It immediately follows that $x(s) \geq 0$ and $y(r) \geq 0$. Hence we only need to show that our dual solution satisfies the first set of constraints.

The following lemma allows us to lower bound $\hat{x}(s)$ by only considering the load of a server s. This will simplify the analysis later on, since we then do not need to consider the exact set of edges assigned to s anymore.

Lemma 8. *For any load l_s of s, where $\{p_1, p_2, \ldots, p_j\}$ with $\sum_{i=1}^{j} p_i = l_s$ are the individual probabilities of the edges assigned to s, it holds that*

$$\hat{x}(s) = \frac{1}{b \cdot c} \cdot \sum_{i=0}^{j-1} f\left(\sum_{l=1}^{i} p_l\right) \cdot p_{i+1} \geq \frac{1}{b \cdot c} \int_{-1}^{l_s - 1} f(x)\, \mathrm{d}x\,.$$

[1] As for other capacitated online matching problems, the competitive ratio is expected to improve for growing values of b.

Let E_s denote the set of edges incident to s. We will focus on a single server s and fix the randomness of all other servers. More precisely, we fix the outcome, i.e. success or failure, of every edge in $E \setminus E_s$. Formally, we define the binary vector Z that has one entry $Z(e)$ for every edge $e \in E$. $Z(e) = 1$ then means that the edge e succeeds when it is chosen by the algorithm. We use Z_{-s} and Z_s to denote the parts of Z that contain all the entries for edges $E \setminus E_s$ and E_s, respectively.

The major step in the analysis of StochasticBalance is to establish the following statement, which will be Lemma 14 in the full proof.

Lemma 14. *For any realization Z_{-s} of all edges in $E \setminus E_s$, it holds for any edge $\{s, r\} \in E$*

$$\mathbb{E}_{Z_s}\left[p_{s,r} \cdot \hat{x}(s) + \hat{y}(r) \mid Z_{-s}\right] \geq p_{s,r} \cdot (1 - \varepsilon),$$

with $\varepsilon \to 0$ as $b \to \infty$.

Taking expectation over Z_{-s} will then show the feasibility of the dual solution for large server capacities. For the proof of Lemma 14, we need in turn two lemmas. To simplify the notation, we define $p_j := p_{s,r_j}$, $l_s^j := \sum_{i=0}^{j} p_i$ and

$$I_j := \frac{1}{b} \cdot \int_{-1}^{l_s^j - 1} f(x)\, \mathrm{d}x,$$

for all $j \in [k]$. Given the random outcomes Z_{-s} of all edges in $E \setminus E_s$, let $R_s = \{r_1, r_2, \ldots, r_k\}$ denote the set of requests assigned to s ordered by arrival times if $Z_s = \mathbf{0}$. By analyzing the assignment of StochasticBalance and characterizing the requests that are matched with server s, we show the following (cf. Lemmas 11 and 12).

For any realization of the random outcomes Z_{-s} of all edges in $E \setminus E_s$ and the corresponding set of requests R_s, it holds that

$$\mathbb{E}_{Z_s}\left[\hat{x}(s) \mid Z_{-s}\right] = \frac{1}{c} \cdot \left(\sum_{j=0}^{k-1} (I_{j+1} - I_j) \cdot (1 - \Pr\left[L_s \leq j\right]) \right),$$

where we define $I_0 := 0$. Moreover, consider any edge $\{s, r\} \in E$. For any realization of the random outcomes Z_{-s} of all edges in $E \setminus E_s$ and the corresponding set of requests R_s, it holds that

$$\mathbb{E}_{Z_s}\left[\hat{y}(r) \mid Z_{-s}\right] \geq (1 - \Pr\left[L_s \leq k\right]) \cdot \frac{p_{s,r}}{c} \cdot \left(1 - f\left(l_s^k\right)\right).$$

Proof of Theorem 2. All that remains to show is that our constructed dual solution is feasible if $b \to \infty$. Consider any edge $\{s, r\} \in E$. Recall that $x(s) := \mathbb{E}[\hat{x}(s)]$ and $y(r) := \mathbb{E}[\hat{y}(r)]$. Taking expectation over Z_{-s} on both sides of Lemma 14 yields

$$p_{s,r} \cdot x(s) + y(r) \geq p_{s,r} \cdot (1 - \varepsilon),$$

by the tower property of conditional expectation. Hence our constructed dual solution is almost feasible. Recall that we constructed the solutions such that $\mathbb{E}[\text{SBAL}] = \mathbb{E}[P] = c \cdot \mathbb{E}[D]$. Moreover, we can obtain a feasible dual solution by dividing all dual variables by $(1 - \varepsilon)$. The resulting dual solution has value $\mathbb{E}[D'] = \mathbb{E}[D]/(1 - \varepsilon)$. Therefore, we have

$$\mathbb{E}[\text{SBAL}] = \left(1 - \frac{1}{e}\right) \cdot \mathbb{E}[D] = \left(1 - \frac{1}{e}\right) \cdot (1 - \varepsilon) \cdot \mathbb{E}[D'] \geq \left(1 - \frac{1}{e}\right) \cdot (1 - \varepsilon) \cdot \text{OPT},$$

where the inequality follows from weak duality. This implies a competitiveness of

$$\left(1 - \frac{1}{e}\right) \cdot (1 - \varepsilon) \xrightarrow{b \to \infty} \left(1 - \frac{1}{e}\right).$$

Since SOPT $\leq$ OPT, we immediately obtain the same competitiveness against the stochastic benchmark. □

References

1. Aggarwal, G., Goel, G., Karande, C., Mehta, A.: Online vertex-weighted bipartite matching and single-bid budgeted allocations. In: Proceedings of 22nd Annual ACM-SIAM Symposium on Discrete Algorithms (SODA), pp. 1253–1264. SIAM (2011)
2. Birnbaum, B., Mathieu, C.: On-line bipartite matching made simple. SIGACT News **39**(1), 80–87 (2008)
3. Buchbinder, N., Jain, K., Naor, J.S.: Online primal-dual algorithms for maximizing ad-auctions revenue. In: Arge, L., Hoffmann, M., Welzl, E. (eds.) ESA 2007. LNCS, vol. 4698, pp. 253–264. Springer, Heidelberg (2007). https://doi.org/10.1007/978-3-540-75520-3_24
4. Buchbinder, N., Naor, J., Wajc, D.: Lossless online rounding for online bipartite matching (despite its impossibility). In: Proceedings of 34th ACM-SIAM Symposium on Discrete Algorithms (SODA), pp. 2030–2068. SIAM (2023). https://doi.org/10.1137/1.9781611977554.ch78
5. Devanur, N., Jain, K., Kleinberg, R.: Randomized primal-dual analysis of RANKING for online bipartite matching. In: Proceedings of 24th Annual ACM-SIAM Symposium on Discrete Algorithms (SODA), pp. 101–107 (2013)
6. Golrezaei, N., Nazerzadeh, H., Rusmevichientong, P.: Real-time optimization of personalized assortments. Manag. Sci. **60**(6), 1532–1551 (2014). https://doi.org/10.1287/mnsc.2014.1939
7. Goyal, V., Udwani, R.: Online matching with stochastic rewards: optimal competitive ratio via path-based formulation. Oper. Res. **71**(2), 563–580 (2023). https://doi.org/10.1287/opre.2022.2345
8. Huang, Z., Jiang, H., Shen, A., Song, J., Wu, Z., Zhang, Q.: Online matching with stochastic rewards: advanced analyses using configuration linear programs. In: Proceedings of 19th International Conference on Web and Internet Economics (WINE). Lecture Notes in Computer Science, vol. 14413, pp. 384–401. Springer, Heidelberg (2023). https://doi.org/10.1007/978-3-031-48974-7_22

9. Huang, Z., Tröbst, T.: Applications of online matching. In: Echenique, F., Immorlica, N., Vazirani, V. (eds.) Online and Matching-Based Market Design, pp. 109–129. Cambridge University Press (2023)
10. Huang, Z., Zhang, Q.: Online primal dual meets online matching with stochastic rewards: configuration LP to the rescue. In: Proceedings of 52nd Annual ACM SIGACT Symposium on Theory of Computing (STOC), pp. 1153–1164 (2020)
11. Huang, Z., Zhang, Q., Zhang, Y.: Adwords in a panorama. In: Proceedings of 61st IEEE Annual Symposium on Foundations of Computer Science (FOCS), pp. 1416–1426 (2020)
12. Kalyanasundaram, B., Pruhs, K.: An optimal deterministic algorithm for online b-matching. Theor. Comput. Sci. **233**(1–2), 319–325 (2000)
13. Karp, R., Vazirani, U., Vazirani, V.: An optimal algorithm for on-line bipartite matching. In: Proceedings of 22nd Annual ACM Symposium on Theory of Computing (STOC), pp. 352–358 (1990)
14. Mehta, A.: Online matching and ad allocation. Found. Trends Theor. Comput. Sci. **8**(4), 265–368 (2013). https://doi.org/10.1561/0400000057
15. Mehta, A., Panigrahi, D.: Online matching with stochastic rewards. In: Proceedings of 53rd Annual IEEE Symposium on Foundations of Computer Science (FOCS), pp. 728–737 (2012)
16. Mehta, A., Saberi, A., Vazirani, U., Vazirani, V.: Adwords and generalized online matching. J. ACM **54**(5), 22 (2007)
17. Mehta, A., Waggoner, B., Zadimoghaddam, M.: Online stochastic matching with unequal probabilities. In: Proceedings of 26th Annual ACM-SIAM Symposium on Discrete Algorithms (SODA), pp. 1388–1404. SIAM (2015). https://doi.org/10.1137/1.9781611973730.92
18. Udwani, R.: Adwords with unknown budgets and beyond. In: Proceedings of 24th ACM Conference on Economics and Computation (RC), p. 1128. ACM (2023). https://doi.org/10.1145/3580507.3597724
19. Udwani, R.: When stochastic rewards reduce to deterministic rewards in online bipartite matching. In: Proceedings of 6th Symposium on Simplicity in Algorithms (SOSA), pp. 321–330. SIAM (2024). https://doi.org/10.1137/1.9781611977936.29,
20. Zhang, Q., Shen, A., Zhang, B., Jiang, H., Du, B.: Online matching with stochastic rewards: provable better bound via adversarial reinforcement learning. In: Forty-First International Conference on Machine Learning, ICML 2024, Vienna, Austria, 21–27 July 2024. OpenReview.net (2024). https://openreview.net/forum?id=TujtZgdRxB

Shortest Longest-Path Graph Orientations for Trees

Yuichi Asahiro[1(✉)], Jesper Jansson[2], Avraham A. Melkman[3], Eiji Miyano[4], Hirotaka Ono[5], Quan Xue[6], Yoshichika Yano[7], and Shay Zakov[8]

[1] Kyushu Sangyo University, Fukuoka, Japan
asahiro@is.kyusan-u.ac.jp
[2] Kyoto University, Kyoto, Japan
jj@i.kyoto-u.ac.jp
[3] Ben-Gurion University of the Negev, Be'er Sheva, Israel
melkmana@gmail.com
[4] Kyushu Institute of Technology, Iizuka, Japan
miyano@ai.kyutech.ac.jp
[5] Nagoya University, Nagoya, Japan
ono@nagoya-u.jp
[6] The University of Hong Kong, Hong Kong, China
quan.xue@connect.polyu.hk
[7] The University of Tokyo, Tokyo, Japan
yyano-ut@g.ecc.u-tokyo.ac.jp
[8] Ruppin Academic Center, Kfar Monash, Israel
Zakov.Shay@ruppin365.net

Abstract. Graph orientation transforms an undirected graph into a directed graph by assigning a direction to each edge. Among the many different optimization problems related to graph orientations, we focus here on the Shortest Longest-Path Orientation problem (SLPO) which is a generalization of the well-known Minimum Graph Coloring problem. The input to SLPO is an edge-bi-weighted undirected graph in which every edge has two (possibly different and not necessarily positive) lengths associated with its two directions. The goal is to find an orientation of the input graph that minimizes the length of the longest simple directed path. Recently, polynomial-time algorithms for simple graph structures such as paths, cycles, stars, and trees were proposed, and a new polynomial-time inapproximability result was also established. This paper presents (i) an $O(n^2 \log n)$-time algorithm for trees, which is a significant improvement over the previously fastest algorithm whose time complexity was $\Omega(n^{14})$ and (ii) polynomial-time algorithms for trees and spiders that run even faster than (i) as long as every edge weight is an integer and the total weight of the edges is sub-exponential.

Keywords: Graph orientation · Longest path · Tree · Spider

1 Introduction

An *orientation* of an undirected graph is an assignment of a direction to each of its edges. Many interesting optimization problems involving graph orientation

R. Královič and V. Kůrková (Eds.): SOFSEM 2025, LNCS 15538, pp. 51–64, 2025.
https://doi.org/10.1007/978-3-031-82670-2_5

have been extensively researched, e.g., to find an orientation where the total arc-connectivity is maximized [12], an orientation minimizing the maximum outdegree [2,4–6,16], and an orientation maximizing the number of source-target vertex pairs from a given set that become connected by directed source-to-target paths [9,14]. Certain graph orientation problems are equivalent to well-known classic problems. For example, Minimum Vertex Cover (or Maximum Independent Set) is equivalent to orienting an undirected graph such that the number of vertices with outdegree at least one is minimized (or with outdegree zero is maximized) [1]. A natural generalization of Minimum Vertex Cover (or Maximum Independent set) where the outdegree threshold is raised from one (or zero) to any positive integer W was studied in [1].

A graph orientation problem called Unweighted Shortest-Longest-Path Orientation (USLPO) can be viewed as a generalization of Minimum Graph Coloring. The objective of USLPO is to find an orientation of an undirected, unweighted graph that minimizes the length of a longest simple directed path. For any undirected graph G, let $H(G)$ and $\chi(G)$ denote the length of a longest simple directed path in an optimal solution to USLPO for G and the chromatic number of G, respectively. It is known that $H(G) + 1 = \chi(G)$ [10,11,15,17], even when the output orientation must be acyclic [8]. This equality immediately implies the intractability of USLPO: USLPO is NP-hard since Minimum Graph Coloring is NP-hard [13]. Moreover, USLPO cannot be approximated within a ratio of $(3/2 - \varepsilon)$ for any constant $\varepsilon > 0$ in polynomial time unless P=NP even if restricted to 4-regular planar graphs, since it is NP-hard to determine if a 4-regular planar graph G satisfies $\chi(G) \leq 3$ [7].

This paper considers a generalization of USLPO named Shortest Longest-Path Orientation (SLPO) that takes an *edge-bi-weighted* graph as input, in which every edge $\{u, v\}$ has two (potentially different and not necessarily positive) weights $w(u, v)$ and $w(v, u)$ representing the lengths of its two possible directions (u, v) and (v, u). The goal of SLPO is to find an orientation minimizing the length of a longest directed path in the resulting directed graph. If some edge lengths (weights) are negative, then a longest directed path is not necessarily a maximal directed path. Hence, we consider two variants SLPO_m and SLPO_s, in which the longest directed path is taken, respectively, among maximal simple directed paths *only* and among *all* simple directed paths. As stated above, USLPO is NP-hard which immediately implies that SLPO_m and SLPO_s are also NP-hard for general graphs. In fact, SLPO_m and SLPO_s are NP-hard even for subcubic planar graphs [3], where a graph is subcubic if the degree of every vertex is at most three. On the positive side, polynomial-time algorithms for simple graph structures (more precisely, trees, paths, cycles, and stars), are known [3].

This paper further investigates the computational complexity of SLPO_m and SLPO_s. The contributions are as follows.

1. An $O(n \log \Delta)$-time algorithm which determines if there is an orientation with cost at most a fixed value B for SLPO_m on trees, where n is the number of vertices and Δ is the maximum degree of a vertex. This leads to an $O(n^2 \log n)$-time algorithm for SLPO_m on trees, which is a significant improvement over

Table 1. Summary of the results from [3] (top) and this paper (bottom), where n is the number of vertices, Δ is the maximum degree of a vertex, and $Z = \sum_{\{u,v\}} \max\{|w(u,v)|, |w(v,u)|\}$. The time complexities indicated by "*" need the assumption that every edge weight is an integer.

Graph class	SLPO_m	SLPO_s	Reference/Theorems
Path	$O(n \log n)$	$O(n)$	[3]
Cycle	$O(n^2 \log n)$	$O(n)$	[3]
Star	$O(n \log n)$	$O(n \log n)$	[3]
Tree	$\Omega(n^{14})$	$\Omega(n^{14})$	[3]
Subcubic Planar	NP-hard	NP-hard	[3]
Tree	$O(n^2 \log n)$	$O(n^2 \log n)$	Theorems 1 and 2
	$O(n \log \Delta \log Z)^*$	$O(n \log \Delta \log Z)^*$	
Spider	—	$O((n + \Delta \log \Delta) \log Z)^*$	Theorem 3

the previously fastest algorithm that runs in polynomial time but has a time complexity of $\Omega(n^{14})$ [3]; and also an $O(n \log \Delta \log Z)$-time algorithm for SLPO_m on trees under the assumption that all edge weights are integers, where $Z = \sum_{\{u,v\}} \max\{|w(u,v)|, |w(v,u)|\}$. The latter runs faster than the former when Z is sub-exponential. Utilizing these algorithms for SLPO_m on trees, we can solve SLPO_s on trees in the same time complexity.
2. Even faster algorithm for SLPO_s on spiders (a subclass of trees), also under the assumption that all edge weights are integers and Z is sub-exponential.

Table 1 summarizes the previously known and the new results in this paper.

The organization of the paper is as follows. The problems SLPO_s and SLPO_m are defined formally in Sect. 2. Sections 3 and 4 respectively describe our new algorithms for trees and spiders. Concluding remarks are given in Sect. 5. Due to space limitations, many details and all proofs are omitted.

2 Preliminaries

We shall use the following definitions and terminology from Sect. 2 in [3]. Let $G = (V, E)$ be an undirected graph. The vertex set and the edge set of G are denoted by $V(G)$ and $E(G)$, respectively. For a vertex v, its unweighted degree is denoted by $deg(v)$, and Δ denotes the maximum unweighted degree of all vertices. Replacing each undirected edge $\{u, v\} \in E$ by either the directed edge (*di-edge* for short) (u, v) or the di-edge (v, u) gives a directed graph (*di-graph* for short). The resulting di-graph $\widetilde{G}$ is called *an orientation* of G. The vertex set and the di-edge set of $\widetilde{G}$ are respectively denoted by $V(\widetilde{G})$ $(= V(G))$ and $E(\widetilde{G})$. Let $\mathcal{O}(G)$ denote the set of all orientations of G. We sometimes regard an orientation as a set of di-edges: If a di-edge (u, v) is included in $E(\widetilde{G})$, we write $(u, v) \in \widetilde{G}$. The unweighted indegree and the unweighted outdegree of a vertex v in $\widetilde{G}$ are denoted by $deg^-_{\widetilde{G}}(v)$ and $deg^+_{\widetilde{G}}(v)$, respectively.

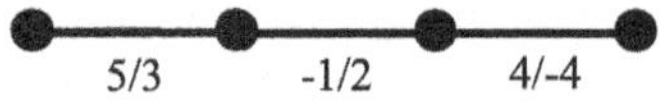

(a) The w_l/w_r below each edge specifies the weights w_l and w_r of that edge's left and right directions.

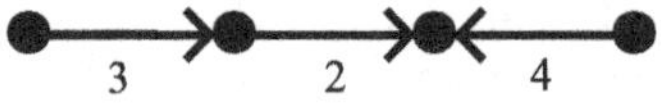

(b) An orientation $\widetilde{G}$ with $h_s(\widetilde{G}) = 5$ and $h_m(\widetilde{G}) = 5$.

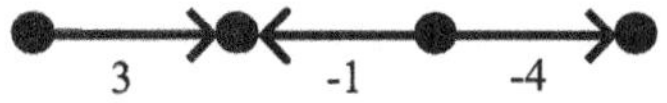

(c) An orientation $\widetilde{G}$ with $h_s(\widetilde{G}) = 3$ and $h_m(\widetilde{G}) = 3$.

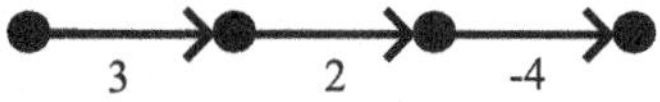

(d) An orientation $\widetilde{G}$ with $h_s(\widetilde{G}) = 5$ and $h_m(\widetilde{G}) = 1$.

Fig. 1. (a) An edge-bi-weighted graph G, (b) an example orientation of G, (c) an optimal orientation of G under cost function H_s (here, $H_s(G) = 3$), and (d) an optimal orientation of G under cost function H_m (here, $H_m(G) = 1$).

An *edge-bi-weighted graph* $G = (V, E, w)$ is an undirected graph $G = (V, E)$ in which every edge $\{u, v\} \in E$ has a pair of weights $w(u, v)$ and $w(v, u)$ associated with the two directions (u, v) and (v, u), respectively. The weights $w(u, v)$ and $w(v, u)$ are possibly nonpositive. We define $Z = \sum_{\{u,v\}\in E} \max\{|w(u, v)|, |w(v, u)|\}$. A *directed path* (*di-path* for short) in a digraph $\widetilde{G}$ is a sequence $\langle v_1, v_2, \ldots, v_q\rangle$ of vertices such that for $k \in \{1, 2, \ldots, q-1\}$, the graph $\widetilde{G}$ contains the di-edge (v_k, v_{k+1}). We say that this di-path *starts from* v_1, *ends at* v_q, and *passes* v_i for $2 \leq i \leq q - 1$. The *length* of a di-path $\overrightarrow{P} = \langle v_1, v_2, \ldots, v_q\rangle$ in an orientation of an edge-bi-weighted graph is: $W(\overrightarrow{P}) = \sum_{k=1}^{q-1} w(v_k, v_{k+1})$. The di-path $\overrightarrow{P}$ is *simple* if all vertices in $\overrightarrow{P}$ are distinct. Also, $\overrightarrow{P}$ is *maximal* if it is not contained in any di-path with more vertices. If some edge weights are negative, then a longest di-path is not necessarily maximal. Hence, two alternative cost functions for an orientation are defined; see Fig. 1 for an example that shows the difference between them.

Definition 1 [3]. *Define the following two cost measures for an orientation $\widetilde{G}$ of an edge-bi-weighted graph G:*

$$h_s(\widetilde{G}) = \max\{W(\overrightarrow{P}) \mid \overrightarrow{P} \textit{ is a simple di-path in } \widetilde{G}\} \textit{ and}$$
$$h_m(\widetilde{G}) = \max\{W(\overrightarrow{P}) \mid \overrightarrow{P} \textit{ is a maximal simple di-path in } \widetilde{G}\}.$$

The corresponding two cost functions for orienting G are:

$$H_s(G) = \min\{h_s(\widetilde{G}) \mid \widetilde{G} \in \mathcal{O}(G)\} \textit{ and } H_m(G) = \min\{h_m(\widetilde{G}) \mid \widetilde{G} \in \mathcal{O}(G)\}.$$

Note that a di-path including only one vertex is a simple di-path with zero length. Hence $h_s(\widetilde{G}) \geq 0$ holds for any $\widetilde{G}$, and thus $H_s(G) \geq 0$ holds for any G.

The two problem variants that we consider in this article are the following:

> The Shortest Longest-Path Orientation Problem, variants SLPO_s & SLPO_m:
> **Input:** An undirected, edge-bi-weighted graph G.
> **Output:** An orientation $\widetilde{G}$ of G such that $h_s(\widetilde{G}) = H_s(G)$ (for SLPO_s) or $h_m(\widetilde{G}) = H_m(G)$ (for SLPO_m).

For an edge-bi-weighted graph G and a fixed value B, an orientation $\widetilde{G}$ of G is *B-feasible* if $h_x(\widetilde{G}) \leq B$ for SLPO_x ($x \in \{s, m\}$). Our basic strategy for solving $\text{SLPO}_s/\text{SLPO}_m$ is to design an algorithm that answers the following question:

Question 1. Does G have a B-feasible orientation?

3 Trees

In this section, we first design a dynamic programming algorithm to answer Question 1 for cost function H_m, and then utilize it to solve both of SLPO_m and SLPO_s. First, we introduce several definitions only used for trees in Sect. 3.1 Then, Sects. 3.2 and 3.3 describe three procedures that are to be applied locally to subtrees of the input tree in order to transform it into a star. Section 3.4 presents an algorithm that uses these procedures in order to answer Question 1 for H_m. Finally, Sect. 3.5 describes the main algorithm for SLPO_m and how to apply it to also solve SLPO_s restricted to trees.

3.1 Preliminaries for Trees

Let $T = (V, E)$ be an undirected edge-bi-weighted tree with root r. Assume that $V = \{v_1, v_2, \ldots, v_{n-1}, v_n\}$ and $v_n = r$ without loss of generality. The parent of a vertex v_i in T is denoted by $p(v_i)$ if it exists. For a vertex v_i, I_i denotes the set of indices of children of v_i in T. The subtree rooted at v_i in T is denoted by T_i.

Consider an orientation $\widetilde{T}$ of T. Let $\widetilde{T}_i$ be the directed subtree rooted at v_i in $\widetilde{T}$. For a di-path $\overrightarrow{P}$ from v_i to v_j (note that it is unique) in a directed tree $\widetilde{T}$ of T, the length of $\overrightarrow{P}$ is defined as the total weight of di-edges in $\overrightarrow{P}$, and is denoted by $W(i, j)$. On the other hand, if there is no path from v_i to v_j, $W(i, j) = \infty$. Let $V_{\uparrow}^i$ be the set $\{v_j \mid \widetilde{T}_i$ contains an upward di-path from v_j to $v_i\}$, i.e., the set of all vertices that can reach v_i by following an upward path in $\widetilde{T}_i$. Similarly, $V_{\downarrow}^i$ is defined as the set $\{v_j \mid \widetilde{T}_i$ contains an downward di-path from v_i to $v_j\}$. We use the convention that a path may have zero edges, so $v_i \in V_{\uparrow}^i$ and $v_i \in V_{\downarrow}^i$, by which $V_{\uparrow}^i \neq \emptyset$, $V_{\downarrow}^i \neq \emptyset$, and $W(i, i) = 0$. Note that $V_{\uparrow}^i \cup V_{\downarrow}^i = V(T_i)$ does not necessarily hold. Next, let $W_{\uparrow}^i$ for SLPO_m be the length of a longest di-path from vertices in $V_{\uparrow}^i$ to v_i, and analogously for $W_{\downarrow}^i$. Formally,

$$W_{\uparrow}^i = \max\left\{W(j, i) \mid v_j \in V_{\uparrow}^i,\ deg_{\widetilde{T}_i}^-(v_j) = 0\right\} \text{ and}$$

$$W_{\downarrow}^i = \max\left\{W(i, j) \mid v_j \in V_{\downarrow}^i,\ deg_{\widetilde{T}_i}^+(v_j) = 0\right\}.$$

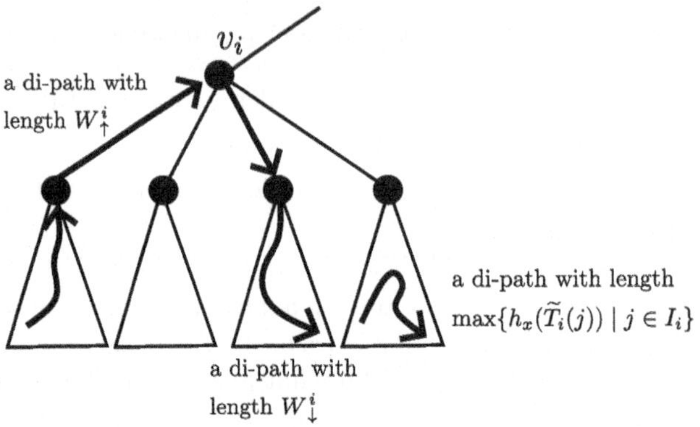

Fig. 2. Illustration of Observation 1. We consider three di-paths which may be (a part of) a (maximal) longest di-path.

For SLPO_s, $W^i_\uparrow$ and $W^i_\downarrow$ are defined to take all simple di-paths into account:

$$W^i_\uparrow = \max\left\{W(j,i) \mid v_j \in V^i_\uparrow\right\} \text{ and } W^i_\downarrow = \max\left\{W(i,j) \mid v_j \in V^i_\downarrow\right\}.$$

For $\widetilde{T}_i$ and a child v_j of v_i, i.e., $j \in I_i$, $\widetilde{T}_i(j)$ represents the directed subtree rooted at v_j in $\widetilde{T}_i$. If we are given an orientation $\widetilde{T}$ of T, then $\widetilde{T}_j$ equals $\widetilde{T}_i(j)$. However, we sometimes need to define or construct a part of an orientation of the whole graph before the other parts. For this purpose, we define $\widetilde{T}_i(j)$. The next observation gives the maximum length of a (maximal) simple di-path in $\widetilde{T}_i$, where $\max\{h_x(\widetilde{T}_i(j)) \mid j \in I_i\}$ for $x \in \{m, s\}$ represents the maximum length of a (maximal) simple di-path in $\widetilde{T}_i$, which does not pass v_i. Figure 2 illustrates this observation.

Observation 1. *For a directed subtree $\widetilde{T}_i$, it holds that*

$$h_m(\widetilde{T}_i) = \max\left\{W^i_\uparrow + W^i_\downarrow,\ \max\left\{h_m(\widetilde{T}_i(j)) \mid j \in I_i\right\}\right\} \text{ and}$$
$$h_s(\widetilde{T}_i) = \max\left\{W^i_\uparrow,\ W^i_\downarrow,\ W^i_\uparrow + W^i_\downarrow,\ \max\left\{h_s(\widetilde{T}_i(j)) \mid j \in I_i\right\}\right\}.$$

In the above, we assumed that the root vertex $r = v_n$. However the choice of the root vertex r is not important for the following reason. Based on Observation 1, for a directed tree $\widetilde{T}$ with root $r = v_n$, it holds that $h_m(\widetilde{T}) = h_m(\widetilde{T_n})$ and $h_s(\widetilde{T}) = h_s(\widetilde{T_n})$. Pick a child v_i of $r(= v_n)$, where $i \neq n$. When computing $h_m(\widetilde{T_n})$ or $h_s(\widetilde{T_n})$, a di-edge between r and v_i are in consideration as a part of computing $W^n_\uparrow$ and $W^n_\downarrow$. Before this, di-edges between v_i and children of v_i were taken into account during computation of $W^i_\uparrow$ and $W^i_\downarrow$. Thus, letting v_i as a root of $\widetilde{T}$ instead of r, just changes the order of the computation of $W_\uparrow$'s and $W_\downarrow$'s. Thus, we can choose an arbitrary vertex as a root of a tree.

The $\Omega(n^{14})$-time algorithm in [3] was a dynamic programming algorithm that implemented Observation 1 directly. To obtain a fast algorithm, we need a more

Algorithm 1: OrientStarDown$_B(T_i)$

Input: an edge-bi-weighted star (subgraph) T_i of T, where T_i is rooted at v_i, the vertex set of T_i is $\{v_i, u_1, u_2, \ldots, u_d\}$, edge set or T_i is $\{\{v_i, u_1\}, \{v_i, u_2\}, \ldots, \{v_i, u_d\}\}$, and $u_1, u_2, \ldots, u_d$ are all leaves of T

Output: an orientation of T_i and its cost

Sort $w(v_i, u_j)$'s in the non-decreasing order. Without loss of generality, we assume $w(v_i, u_1) \leq w(v_i, u_2) \leq \cdots \leq w(v_i, u_d)$;

For each $j \in \{1, 2, \ldots, d-1\}$, $c_j \leftarrow \max\{w(u_k, v_i) \mid j+1 \leq k \leq d\}$ and let $\bar{j}$ be an index such that $w(u_{\bar{j}}, v) = c_j$;

Find the smallest $j \in \{1, 2, \ldots, d-1\}$ such that $c_j + w(v_i, u_j) \leq B$ holds. If there exists such j, then let $\widetilde{T}_i$ be an orientation of T_i in which each edge $\{v_i, u_k\}$ is directed toward u_k for $1 \leq k \leq j$, and toward v_i for $j+1 \leq k \leq d$, and $C \leftarrow c_j + w(v_i, u_j)$. Otherwise, $C \leftarrow \infty$.

Let $\widetilde{T}_i^{up}$ (or $\widetilde{T}_i^{down}$) be an orientation of T_i in which every edge $\{v_i, u_k\}$ is directed toward v_i (or toward u_k). If $\max\{w(u_k, v_i) \mid 1 \leq k \leq d\} \leq B$, then $C^{up} \leftarrow 0$, otherwise $C^{up} \leftarrow \infty$. Also, $C^{down} \leftarrow w(v_i, u_d)$.

$C_{\min} \leftarrow \min\{C, C^{up}, C^{down}\}$. If $C_{\min} = C$, C^{up}, or C^{down}, then output $(\widetilde{T}_i, C)$, $(\widetilde{T}_i^{up}, C^{up})$, or $(\widetilde{T}_i^{down}, C^{down})$, respectively;

refined approach. To this end, we propose a new technique for answering Question 1 efficiently by using graph transformations and another type of dynamic programming. Our strategy is to minimize $W_\uparrow^i$ and $W_\downarrow^i$ for every vertex v_i under the condition that $\max\{h_x(\widetilde{T}_i(j)) \mid j \in I_i)\} \leq B$ for $x \in \{m, s\}$ and a fixed value B in a bottom-up manner by maintaining two values $L_\uparrow^i$ and $L_\downarrow^i$ for each vertex v_i, defined as follows:

$$L_\uparrow^i = \min\{W_\uparrow^i \mid \widetilde{T}_i \text{ is an orientation of } T_i,\ \max\{h_x(\widetilde{T}_i(j)) \mid j \in I_i\} \leq B\} \text{ and}$$
$$L_\downarrow^i = \min\{W_\downarrow^i \mid \widetilde{T}_i \text{ is an orientation of } T_i,\ \max\{h_x(\widetilde{T}_i(j)) \mid j \in I_i\} \leq B\},$$

where $x = m$ for SLPO$_m$ and $x = s$ for SLPO$_s$. For a leaf v_i, $L_\uparrow^i = L_\downarrow^i = 0$.

Algorithms to compute $L_\uparrow^i$ and $L_\downarrow^i$ of a subtree in the input tree for SLPO$_m$ are given in Sect. 3.2. In Sect. 3.3, we transform a tree into another one based on these algorithms. Based on Observation 1, an algorithm to answer Question 1 for SLPO$_m$ is given in Sect. 3.4. Finally, Sect. 3.5 describes the whole algorithm for SLPO$_m$ and how to apply it to SLPO$_s$.

3.2 Procedures for Stars

Let T be a tree with root r. Suppose that children of a vertex v_i are all leaves (note that there must exist such a vertex). Let $u_1, \ldots, u_d$ for $d \geq 1$ be the children of v_i. The subtree T_i rooted at v_i is a star, whose vertex set is $\{v_i, u_1, u_2, \ldots, u_d\}$. In this subsection, we propose an algorithm for a star, which will be used as a subroutine. For a fixed B, an algorithm OrientStarDown$_B$ (listed in Algorithm 1)

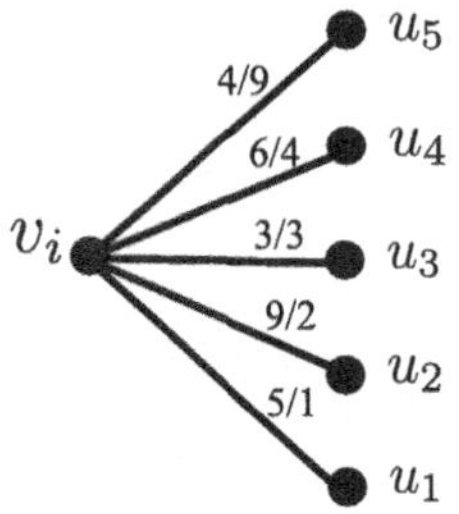

(a) A star T_i.

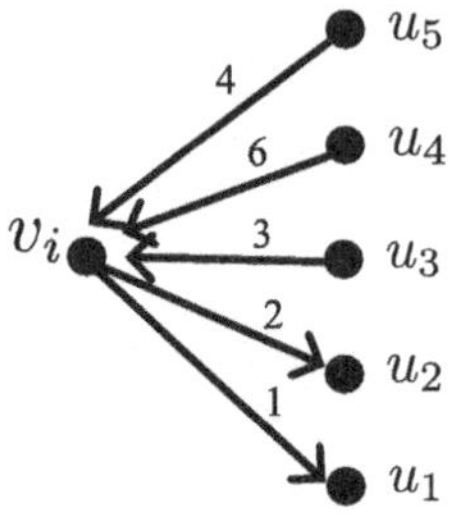

(b) Computing $c_2 = w(u_4, v_i) = 6$ and hence $\overline{2} = 4$ for T_i.

Fig. 3. An example of OrientStarDown$_B$ for a star T_i in (a) and $B = 8$. (b) is the obtained orientation $\widetilde{T}_i$ of T_i.

outputs an orientation of T_i such that the length of the maximal di-path (di-edge) starting from v is minimum and every other maximal di-path ending at v or passing v has length at most B, if such an orientation exists.[1]

Figure 3(b) illustrates an example of the output of OrientStarDown$_B$ for the star T_i in Fig. 3(a) and $B = 8$. In Step 2, we compute c_2 by directing the edges $\{v_i, u_1\}$ and $\{v_i, u_2\}$ as (v_i, u_1) and (v_i, u_2), respectively, and directing other edges toward v_i. Then, we obtain $c_2 = \max\{w(u_k, v_i) \mid 3 \leq k \leq 5\} = w(u_4, v_i)$ and hence we set $\overline{2} = 4$, where $\overline{2}$ indicates a di-edge having maximum weight among di-edges (u_k, v_i)'s for $2 < k$. Then, we see that the longest maximal di-path in this orientation is $\langle u_{\overline{2}}, v_i, u_2 \rangle = \langle u_4, v_i, u_2 \rangle$ with length 8 and thus $C = 8$ in Fig. 3(b). As for Step 4, $C^{up} = \infty$ since $w(u_2, v_i) = 9 > B(= 8)$ and also $C^{down} = w(v_i, u_5) = 9$. Step 5 may output $\widetilde{T}_i^{down}$ (or $\widetilde{T}_i^{up}$) even if $C^{down} > B$ (or $C^{up} > B$), when $C > C^{down}$ (or $C > C^{up}$). The reason is that $\widetilde{T}_i^{down}$ (or $\widetilde{T}_i^{up}$) may minimize the length of a longest maximal di-path passing v_i in the whole orientation of T due to the possible existence of an edge with negative weight. (This is not the case in the example in Fig. 3.) Finally, the orientation in Fig. 3(b) is the output of OrientStarDown$_B(T_i)$ for $B = 8$, since $C < C^{down}$.

Let $\widetilde{T}_i^D$ be the orientation obtained by OrientStarDown$_B$ for T_i. Define an index i_D as follows: If $C_{\min} = C^{up}$, then let $i_D = 0$ and let $w(v, u_{i_D}) = w(v, u_0) = 0$ (though there is no vertex u_0). If $C_{\min} = C^{down}$, then let $i_D = d$ (and hence $w(v, u_{i_D}) = w(v, u_d)$). Otherwise if $C_{\min} = C$, then let i_D be the index j found in Step 3 of OrientStarDown$_B$. The following lemma guarantees that OrientStar-Down$_B$ outputs an intended orientation which gives $L_\downarrow^i (= h_m(\widetilde{T}_i^D))$.

Lemma 1. *In $\widetilde{T}_i^D$, the length of the longest di-edge starting from v_i is the minimum among all orientations of $\widetilde{T}_i$ such that every maximal di-path ending at v_i or passing v_i has length at most B. Moreover, the orientation $\widetilde{T}_i^D$ of T_i can be obtained in $O(d \log d)$ time.*

[1] This algorithm differs from the known ones for stars in [3].

Algorithm 2: DeleteLeaf(T, v_i, $L^i_\downarrow$, $L^i_\uparrow$)

Input: an edge-bi-weighted rooted tree T, a vertex v_i whose children are all leaves in T, and two values $L^i_\downarrow$ and $L^i_\uparrow$

Output: an edge-bi-weighted tree T'

Let the children of v_i in T be $u_1, u_2, \ldots, u_d$ $(d \geq 1)$;

Remove the vertices $u_1, u_2, \ldots, u_d$ and the edges $\{v_i, u_1\}, \{v_i, u_2\}, \ldots, \{v_i, u_d\}$ from T;

Update the weights of the edge $\{p(v_i), v_i\}$ as $w(p(v_i), v_i) \leftarrow w(p(v_i), v_i) + L^i_\downarrow$ and $w(v_i, p(v_i)) \leftarrow w(v_i, p(v_i)) + L^i_\uparrow$;

Output the resulting T (as T');

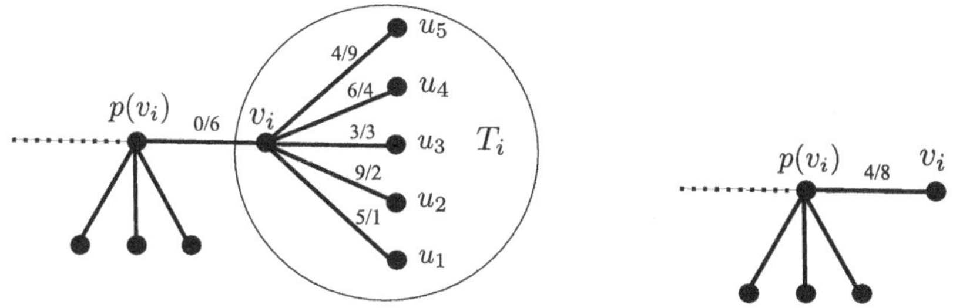

(a) A tree T and T_i (in Fig. 3), where $i_D = 2$ and $i_U = 5$.

(b) Another tree T' constructed from T by DeleteLeaf.

Fig. 4. An example of DeleteLeaf for $B = 8$.

We define an algorithm OrientStarUp$_B$ in the same way as OrientStarDown$_B$, but where the order of v_i and u_k is swapped when weights of edges are handled and where the directions of edges determined in Steps 3 and 4 are the opposite. (We omit the description of OrientStarUp$_B$, since it is very similar to OrientStar-Down$_B$.) Then, Let $\widetilde{T}^U_i$ be the orientation with cost $L^i_\uparrow (= h_m(\widetilde{T}^U_i))$, which is obtained by applying OrientStarUp$_B$ to T_i. The index i_U is defined similarly to the index i_D.

Lemma 2. *In $\widetilde{T}^U_i$, the length of the longest di-edge ending at v_i is the minimum among all orientations of $\widetilde{T}_i$ such that every maximal di-path starting from v_i or passing v_i has length at most B. Moreover, the orientation $\widetilde{T}^U_i$ of T_i can be obtained in $O(d \log d)$ time.*

3.3 Transformation

Let T be an edge-bi-weighted tree, which is not a star. Taking $L^i_\downarrow$ and $L^i_\uparrow$ computed by the procedures in the previous subsection as input, the algorithm DeleteLeaf (listed in Algorithm 2) constructs another tree T' that has fewer vertices than T. The resulting T' may be a star.

Figure 4 illustrates an example of DeleteLeaf for $B = 8$. In Fig. 4(a), the part T_i is the same as in Fig. 3, where $i_D = 2$ with $L^2_\downarrow = 2$ and $i_U = 5$ with $L^2_\uparrow =$

4. Thus, DeleteLeaf constructs another tree in Fig. 4(b), in which $w(p(v_i), v_i))$ is updated to $w(p(v_i), v_i) + (v_i, u_2) = 8$ and also $w(v_i, p(v_i))$ is updated to $w(v_i, p(v_i)) + w(u_5, v_i) = 4$.

Procedure DeleteLeaf only modifies T around v_i and its neighboring vertices, so its running time can be bounded as follows.

Lemma 3. *Given an edge-bi-weighted rooted tree T and a vertex v_i (with $L^i_{\downarrow}$ and $L^i_{\uparrow}$) as input, DeleteLeaf runs in $O(d)$ time, where $d = deg(v_i)$.*

The following lemma relates the optimal cost of T' to that of T. This leads to an algorithm OrientTree$_B$ in the next subsection.

Lemma 4. *$H_m(T) \leq B$ if and only if $H_m(T') \leq B$.*

3.4 An Algorithm to Answer Question 1

In this subsection, we describe the algorithm OrientTree$_B$ which answers Question 1 based on the results in the previous subsections. The basic strategy of OrientTree$_B$ is as follows. First we construct a sorted list S of vertices in the input tree T. This list indicates the order of vertices to apply OrientStarDown$_B$, OrientStarUp$_B$, and DeleteLeaf. By DeleteLeaf, T is transformed to another tree T'. Then we remove the first element of S for T', where the first element of the resulting new list indicates the next target to apply OrientStarDown$_B$, OrientStarUp$_B$, and DeleteLeaf. In this way, we iteratively apply these three procedures and then obtain a star. Finally, we apply BestOrientStar$_m$ in [3] to compute an optimal orientation of the obtained star.

We start with the construction of the list S of vertices in T. First choose a vertex arbitrarily as the root r. By using the breadth first search starting from r, we construct a sorted list S which contains all non-leaf vertices in T, and if $i \leq j$, the distance between $S[i]$ and r is not smaller than that between $S[j]$ and r, where a tie is broken arbitrarily. Suppose that we apply DeleteLeaf to the first vertex $S[1]$ of S and its children in T, and then obtain a new tree T', One can see that children of $S[2]$ must be leaves in T', even if $S[1]$ is a child of $S[2]$ in T, since DeleteLeaf transforms a star formed by $S[1]$ and its children in T to a vertex. Hence we can apply DeleteLeaf to $S[2]$ for T' as the next step. Since the construction of S is done mainly by the breadth first search, we have:

Lemma 5. *The construction of the list S is done in $O(n)$ time.*

The algorithm OrientTree$_B$ is listed in Algorithm 3. It uses an algorithm named BestOrientStar$_m$ from [3], which solves SLPO$_m$ on stars exactly. The following lemma guarantees the correctness of OrientTree$_B$.

Lemma 6. *OrientTree$_B$ answers Question 1 in $O(n \log \Delta)$ time.*

3.5 Time Complexity and Cost Function H_8

The whole algorithm for cost function H_m is listed in Algorithm 4. Here it outputs the value $H_m(T)$ only. However, it is easy to output an orientation with cost $H_m(T)$ by modifying OrientTree$_B$ so that it also outputs an obtained orientation when answering "Yes".

Algorithm 3: OrientTree$_B(T, S)$

Input: an edge-bi-weighted tree T which is not a star, and a list S of vertices
Output: "Yes" or "No"

while *T is not a star* **do**
  Suppose the first element $S[1]$ of S is v_i;
  $(\widetilde{T}_i^D, L_\downarrow^i) \leftarrow$ OrientStarDown$_B(T_i)$ and $(\widetilde{T}_i^U, L_\uparrow^i) \leftarrow$ OrientStarUp$_B(T_i)$;
  $T \leftarrow$ DeleteLeaf$(T, v_i, L_\downarrow^i, L_\uparrow^i)$, and delete $S[1]$ from S;
Apply BestOrientStar$_m$ in [3] to T. If the obtained orientation has cost at most B, output "Yes", otherwise, output "No";

Algorithm 4: OrientTree$_m(T)$

Input: an edge-bi-weighted tree T which is not a star
Output: $H_m(T)$

Construct the list S of T;
Find the minimum B such that OrientTree$_B(T, S)$ returns "Yes";
Output B;

Algorithm 5: AddLeaf(G)

Input: an edge-bi-weighted graph G
Output: an edge-bi-weighted graph G'

For each vertex v_i of degree at least two, add two vertices x_i and y_i to G;
To G, add an edge $\{v_i, x_i\}$ with weights $w(v_i, x_i) = \infty$ and $w(x_i, v_i) = 0$, and then add an edge $\{v_i, y_i\}$ with weights $w(v_i, y_i) = 0$ and $w(y_i, v_i) = \infty$;
return G (as G');

Theorem 1. *If the input is a tree, then SLPO$_m$ can be solved in $O(n^2 \log n)$ time. Moreover, if every edge weight of the input tree is an integer, SLPO$_m$ can be solved in $O(n \log \Delta \log Z)$ time.*

To solve SLPO$_s$, we reduce SLPO$_s$ to SLPO$_m$. To this end, we introduce an algorithm AddLeaf (listed in Algorithm 5). This algorithm transforms an edge-bi-weighted graph G to another edge-bi-weighted graph G' in linear time by adding leaves to G. The resulting graph G' does not always maintain the structural properties of G. For example, if G is a tree then so is G', but if G is a cycle then G' is not. However, the next lemma shows that the optimal cost of G for cost function H_s transfers to G'.

Lemma 7. $H_s(G) = H_m(G')$

The important point here is that AddLeaf transforms a tree to another tree. Thus, Theorem 1 and Lemma 7 immediately yield the following theorem.

Theorem 2. *If the input is a tree, then SLPO$_s$ can be solved in $O(n^2 \log n)$ time. Moreover, if every edge weight of the input tree is an integer, SLPO$_s$ can be solved in $O(n \log \Delta \log Z)$ time.*

4 Spiders with Cost Function H_s

This section presents an algorithm that solves SLPO$_s$ on spiders under the assumption that all edge weights are integers. It runs faster than the algorithm for trees when Z is sub-exponential. A *spider* is a tree with exactly one vertex r of degree greater than 2, referred to as the *root vertex*. Let G be a spider which has root vertex r with degree $\Delta \geq 3$ and undirected paths $P_i = (v_{i,1}, v_{i,2}, \ldots, v_{i,n_i}, r)$ of length n_i for $1 \leq i \leq \Delta$. We call each P_i a *leg*. For simplicity, we let $v_{i,n_i+1} = r$.

Here we briefly sketch the main idea of the algorithm. For a leg P_i with $n_i + 1$ vertices, the optimal cost $H_s(P_i)$ under H_s can be obtained in $O(n_i)$ time by the algorithm BestOrientPath$_s$ in [3]. However, just combining optimal orientations of $P_1, \ldots, P_\Delta$ may not give an optimal orientation for G. Instead, we proceed as follows.

1. For each leg P_i, we obtain an orientation such that $\{v_{i,n_i}, r\}$ is directed toward r (or toward v_{i,n_i}) and any maximal di-path not including (v_{i,n_i}, r) (or (r, v_{i,n_i})) inside P_i has length at most B, utilizing BestOrientPath$_s$.
2. Based on these two orientations for each leg, we construct a star G' from an input spider G, which preserves the lengths of longest di-paths, ending at, starting from, and passing r in G.
3. From an orientation of G' obtained by the algorithm BestOrientStar$_s$ in [3], we construct an orientation of G.

The above procedure answers Question 1 for a fixed B in $O(n + \Delta \log \Delta)$ time. Then, by utilizing the above procedure in a binary search manner on B having $O(Z)$ candidates, we obtain the main result of this section.

Theorem 3. *Suppose that the input is a spider in which every edge weight is an integer. Then, SLPO$_s$ can be solved in $O((n + \Delta \log \Delta) \log Z)$ time.*

5 Concluding Remarks

In this paper, we presented efficient algorithms for SLPO$_m$ and SLPO$_s$ on trees and spiders. Some open questions are:

- Are faster algorithms possible? For example, linear-time algorithms to solve SLPO$_m$ for paths, cycles, and stars are plausible targets.

- Can we remove the assumption that all edge weights are integers and Z is sub-exponential, imposed to obtain faster algorithms for trees and spiders?
- Is it possible to design polynomial-time algorithms for other graph classes, such as unicyclic graphs, cactus graphs, and graphs with bounded treewidth?
- Is there a polynomial-time approximation algorithm for subcubic planar graphs?
- Is there any graph class, for which the (in)tractability of SLPO_s differs from SLPO_m, e.g., SLPO_m is polynomial-time solvable while SLPO_s is NP-hard?

Acknowledgments. This work was supported by JSPS KAKENHI Grant Numbers JP22K11915 and JP24K02902.

References

1. Asahiro, Y., Jansson, J., Miyano, E., Ono, H.: Graph orientations optimizing the number of light or heavy vertices. J. Graph Algor. Appl. **19**(1), 441–465 (2015)
2. Asahiro, Y., Jansson, J., Miyano, E., Ono, H., Zenmyo, K.: Approximation algorithms for the graph orientation minimizing the maximum weighted outdegree. J. Comb. Optim. **22**(1), 78–96 (2011)
3. Asahiro, Y., et al.: Shortest longest-path graph orientations. In: Proceedings of the 29th International Computing and Combinatorics Conference (COCOON 2023). LNCS, vol. 14422 , pp. 141–154. Springer, Heidelberg (2023). https://doi.org/10.1007/978-3-031-49190-0_10
4. Asahiro, Y., Miyano, E., Ono, H., Zenmyo, K.: Graph orientation algorithms to minimize the maximum outdegree. Int. J. Found. Comput. Sci. **18**(2), 197–215 (2007)
5. Borradaile, G., Iglesias, J., Migler, T., Ochoa, A., Wilfong, G., Zhang, L.: Egalitarian graph orientations. J. Graph Algor. Appl. **21**(4), 687–708 (2017)
6. Chrobak, M., Eppstein, D.: Planar orientations with low out-degree and compaction of adjacency matrices. Theor. Comput. Sci. **86**(2), 243–266 (1991)
7. Dailey, D.P.: Uniqueness of colorability and colorability of planar 4-regular graphs are NP-complete. Disc. Math. **30**(3), 289–293 (1980)
8. Deming, R.W.: Acyclic orientations of a graph and chromatic and independence numbers. J. Comb. Theory Ser. B **26**(1), 101–110 (1979)
9. Elberfeld, M., et al.: On the approximability of reachability-preserving network orientations. Internet Math. **7**(4), 209–232 (2011)
10. Gallai, T.: On directed graphs and circuits. In: Theory of Graphs (Proceedings of the Colloquium held at Tihany 1966), pp. 115–118. Akadémiai Kiadó (1968)
11. Hasse, M.: Zur algebraischen Begründung der Graphentheorie. I. Mathematische Nachrichten **28**(5–6), 275–290 (1965)
12. Hörsch, F.: On orientations maximizing total arc-connectivity. Theor. Comput. Sci. **978**, 114176 (2023)
13. Karp, R.M.: Reducibility among combinatorial problems. In: Proceedings of Complexity of Computer Computations. The IBM Research Symposia Series, pp. 85–103. Plenum Press (1972)

14. Medvedovsky, A., Bafna, V., Zwick, U., Sharan, R.: An algorithm for orienting graphs based on cause-effect pairs and its applications to orienting protein networks. In: Crandall, K.A., Lagergren, J. (eds.) WABI 2008. LNCS, vol. 5251, pp. 222–232. Springer, Heidelberg (2008). https://doi.org/10.1007/978-3-540-87361-7_19
15. Roy, B.: Nombre chromatique et plus longs chemins d'un graphe. Revue française d'informatique et de recherche opérationnelle **1**(5), 129–132 (1967)
16. Venkateswaran, V.: Minimizing maximum indegree. Disc. Appl. Math. **143**(1–3), 374–378 (2004)
17. Vitaver, L.M.: Determination of minimal coloring of vertices of a graph by means of Boolean powers of the incidence matrix. In: Proceedings of the USSR Academy of Sciences, vol. 147, pp. 758–759. Nauka (1967). (in Russian)

Parameterized Complexity of Generalizations of Edge Dominating Set

Shubhada Aute[1(✉)], Fahad Panolan[2], Souvik Saha[3], Saket Saurabh[3,4], and Anannya Upasana[3]

[1] Department of Computer Science and Engineering, IIT Hyderabad, Kandi, India
cs21resch11001@iith.ac.in
[2] School of Computer Science, University of Leeds, Leeds, UK
f.panolan@leeds.ac.uk
[3] The Institute of Mathematical Sciences, HBNI, Chennai, India
{souviks,saket,anannyaupas}@imsc.res.in
[4] University of Bergen, Bergen, Norway

Abstract. The objective of this article is to propose two natural generalizations of covering edges by edges (Edge Dominating Set) and study these problems from the multivariate lens. The first is simply considering Edge Dominating Set on hypergraphs, called Hyperedge Dominating Set. Given a hypergraph $\mathcal{H} = (\mathcal{U}, \mathcal{F})$, a set $F \subseteq \mathcal{F}$ is called a *hyperedge dominating set* if all hyperedges intersect with at least one hyperedge $e \in F$. The objective of the Hyperedge Dominating Set problem is to determine whether a hyperedge dominating set of size at most k exists. We find it quite surprising that such generalization is missing from the literature. The second extension we consider is the t-Path Edge Dominating Set problem. In this problem, the input consists of a graph G and an integer k, and the goal is to find a set $\mathcal{P}$ of at most k paths, each of length at most t, such that for every edge in G, at least one of its endpoints belongs to the vertex set $V(P)$ for some $P \in \mathcal{P}$. We show the following results and add to the literature on Edge Dominating Set.

- Hyperedge Dominating Set is FPT parameterized by $k+d$, where d is the maximum size of a hyperedge in the input hypergraph.
- A kernel of size $\mathcal{O}(k^d)$ can be obtained for the Hyperedge Dominating Set problem, where d is the maximum size of a hyperedge in the input hypergraph.
- The problem of finding a Hyperedge Dominating Set is computationally difficult; specifically it is W[2]-hard when parameterized by k. This hardness result holds even when each vertex is contained in at most 2 hyperedges and the intersection between any two hyperedges is at most 1.
- t-Path Edge Dominating Set is FPT when parameterized by $k+t$. Additionally, it has a kernel of size $\mathcal{O}(k^3t^3)$.

Keywords: Edge Dominating Set · FPT · Hypergraph · Kernel

R. Královič and V. Kůrková (Eds.): SOFSEM 2025, LNCS 15538, pp. 65–79, 2025.
https://doi.org/10.1007/978-3-031-82670-2_6

1 Introduction

Covering things by things is ubiquitous in theoretical computer science. In most cases, these can be abstracted as either the classical Set Cover problem or the Hitting Set problem. In these problems, we are given a hypergraph $(\mathcal{U}, \mathcal{F})$, here $\mathcal{U} = \{u_1, \ldots, u_n\}$ is a universe (set of vertices), and $\mathcal{F} = \{F_1, \ldots, F_m\}$ is a family of subsets over $\mathcal{U}$ also called hyperedges, and a positive integer k. In the Set Cover problem, the goal is to find a subfamily $\mathcal{F}' \subseteq \mathcal{F}$ of size at most k, the union of which contains all the elements of $\mathcal{U}$. In the Hitting Set problem, we want to find a subset $\mathcal{U}' \subseteq \mathcal{U}$ of size at most k that has a non-empty intersection with each element of $\mathcal{F}$. These problems, together with their numerous variants and generalizations, have been some of the most explored research directions in the field of parameterized complexity. Motivated by these studies, in this paper, we consider two generalizations of classical Edge Dominating Set and study them from the Parameterized Complexity perspective [3,9,14,19,20,22].

In the Edge Dominating Set (EDS) problem, we are given a graph G and an integer k, and the task is to find a set of k edges that dominate all the edges in G. We say that an edge $e_1 = (a_1, b_1)$ dominates (or covers) another edge $e_2 = (a_2, b_2)$ if $\{a_1, b_1\} \cap \{a_2, b_2\} \neq \emptyset$. Note that the edge dominating set is a dominating set in the line graph. This problem is known to be NP-Complete even when restricted to planar or bipartite graphs with maximum degree 3 [32]. It is also known to admit a factor 2 approximation algorithm in polynomial time [13]. Fernau was the first to consider the problem from the perspective of parameterized complexity and, by employing enumeration-based techniques, obtained an FPT algorithm [10]. After a series of improvements, the current best algorithm was given by Iwaide and Nagamochi which runs in time $\mathcal{O}^\star(2.2351^k)$ [18][1]. The problem also admits a polynomial kernel with $\mathcal{O}(k^2)$ vertices and $\mathcal{O}(k^3)$ edges [30]. In fact, people have also tried to improve constants, and the current best-known kernel has $\max\{\frac{k^2}{2} + \frac{7k}{2}, 6k\}$ vertices and $\frac{8k^3}{27} + \mathcal{O}(k^2)$ edges [15].

Escoffier et al. even considered this problem in the world of FPT approximation and designed an FPT algorithm that is faster than the best known FPT exact algorithm and has a ratio better than 2. In fact, they give an FPT-approximation scheme [8]. Finally, we would also like to mention that several exact algorithms have been made for the problem [11,27,28,31]. This demonstrates the extensive research on Edge Dominating Set through algorithmic approaches designed to address its NP-hardness.

1.1 Edge Dominating Set on Hypergraphs

We first define the notion of domination by a vertex or an edge in a graph. A *vertex dominates* itself, all its neighbors, and all the edges that are incident with it. Similarly, an *edge dominates* its two endpoints, and all the edges incident with either of its endpoints. These problems can be broadly classified into four

[1] The $\mathcal{O}^\star$ notation hides polynomial factors.

Dominatee / Dominator	Vertex	Edge / Hyperedge
Vertex	DOMINATING SET	VERTEX COVER HITTING SET
Edge / Hyperedge	EDGE COVER SET COVER	EDGE DOMINATING SET HYPEREDGE DOMINATING SET*

Fig. 1. An overview of domination problems in graphs and their counterparts in hypergraphs. Problem marked with $*$ has not been previously studied.

Table 1. The figure shows the FPT status with respect to the parameter k (solution size) of the various domination problems in graphs and hypergraphs.

Dominatee	Dominator	Problem	FPT Status
In Graphs			
Vertices	Vertices	DOMINATING SET	W[2]-hard [6]
Vertices	Edges	EDGE COVER	P [25]
Edges	Edges	EDGE DOMINATING SET	FPT [10,18,30]
Edges	Vertices	VERTEX COVER	FPT [4,17]
In Hypergraphs (d-Hypergraphs)			
Vertices	Vertices	DOMINATING SET	W[2]-hard [6]
Vertices	Hyperedges	SET COVER (d-SET COVER)	W[2]-hard [5] (FPT [5])
Hyperedges	Vertices	HITTING SET (d-HITTING SET)	W[2]-hard [7] (FPT [2,24])
Hyperedges	Hyperedges	HYPEREDGE DOMINATING SET (d-HYPEREDGE DOMINATING SET)	W[2]-hard (FPT)

categories based on whether the dominator and the dominatee is a vertex set or an edge set. Each type gives rise to some classical well-studied problems in graph algorithms. A brief overview of the classification is given in Fig. 1. In particular, a set of vertices dominating all the vertices is the DOMINATING SET problem, a set of vertices dominating all the edges is the VERTEX COVER problem, a set of edges dominating all the vertices is the EDGE COVER problem, and a set of edges dominating all the edges is the EDGE DOMINATING SET problem. Our first generalization is obtained by considering these basic problems from graphs to hypergraphs.

By considering a natural generalization of these problems on hypergraphs (d-hypergraphs), we get another set of well-studied problems. A d-hypergraph is a hypergraph where each hyperedge has size at most d, called a d-hyperedge. In the case where we want to dominate all the edges (d-hyperedges) by a set of vertices, we get the HITTING SET (d-HITTING SET) problem. Also, when all the vertices need to be covered by a set of hyperedges (d-hyperedges), we get the SET COVER (d-SET COVER) problem. The status of each of these problems with respect to parameterized complexity is mentioned in Table 1. The domination of vertices by vertices remains an equivalent problem in terms of complexity in hypergraphs. Since DOMINATING SET in graphs is W[2]-hard parameterized by k, VERTEX DOMINATING SET in hypergraphs is also W[2]-hard parameterized by k. However, the case where all the hyperedges need to be dominated by a set of hyperedges has not been studied. We fill this gap in our knowledge by studying the HYPEREDGE DOMINATING SET (HEDS) problem through the lens of parameterized complexity. In particular, we prove the following.

Theorem 1 ($\star^2$). *There is an algorithm of running time $\mathcal{O}(d^{dk} \cdot 2^{dk}|\mathcal{F}|)$ for d-HYPEREDGE DOMINATING SET.*

Our algorithm is inspired by the techniques for EDGE DOMINATING SET problem studied in these papers [10,18,30].

Theorem 2. *d-HYPEREDGE DOMINATING SET admits a kernel of size $\mathcal{O}((dk)^{d^2})$.*

Theorem 2 is proved using the concept of representative sets [12]. Given a hypergraph H defined on $\mathcal{U} = \{u_1, \ldots, u_n\}$ and $\mathcal{F} = \{F_1, \ldots, F_m\}$, we can define a dual hypergraph H^* on $\mathcal{U}^*$ and $\mathcal{F}^*$ as follows. For each $F \in \mathcal{F}$, create a vertex $x_F \in \mathcal{U}^*$ and for each $u \in \mathcal{U}$, create a subset of $\mathcal{U}^*$ corresponding to the hyperedges in $\mathcal{F}$ that contain u and add this subset to $\mathcal{F}^*$. Notice that a set of vertices S of a hypergraph H is a vertex dominating set in H if and only if the set of hyperedges corresponding to S is a hyperedge dominating set in the dual hypergraph H^*. Specifically, $S \subseteq \mathcal{U}$ is a vertex dominating set of size at most k in a hypergraph H where the degree of every vertex is bounded by d if and only if the set of at most k d-hyperedges corresponding to S is a hyperedge dominating set in the dual d-hypergraph H^*. Thus, our FPT algorithm and kernelization results for d-HYPEREDGE DOMINATING SET (d-HEDS) also extend to VERTEX DOMINATING SET in hypergraphs where the degree of every vertex is bounded by d. Notice that even though the degree of every vertex in H^* is at most d, the size of the hyperedges in H^* is not bounded. So, a simple branching algorithm may not work.

However, when we allow the size of the hyperedges to be unbounded, the problem becomes W[2]-hard. In fact, we show that the problem HYPEREDGE DOMINATING SET is hard even for a specific case where the frequency of every element is bounded, i.e., every vertex has bounded degree and any pair of sets in the family has a bounded intersection.

Theorem 3 ($\star$). HYPEREDGE DOMINATING SET *is* W[2]-hard *parameterized by k even when the intersection of any two edges is bounded by* 1 *and the degree of any vertex is bounded by* 2.

We show that the above theorem implies that HYPEREDGE DOMINATING SET is W[2]-hard even when the hypergraph is $K_{2,2}$-free. Here $K_{2,2}$-free means that there are no two vertices present in two hyperedges. This result is in contrast to the fact that DOMINATING SET is FPT in $K_{i,j}$-free graphs, that is, graphs that do not contain the complete bipartite graph with i and j vertices as subgraph [26].

[2] Proofs of results marked with $\star$ are omitted due to paucity of space.

1.2 Covering Edges by Paths: t-PATH EDGE DOMINATING SET

Our next generalization is motivated by the following analogy: VERTEX COVER and EDGE DOMINATING SET can be viewed as *domination* of edges by paths of length 1 and 2, respectively. Here, we consider a vertex as a path of length 1 and an edge as a path of length 2. We extend this to study *domination* of edges by paths of length t. In this problem, given a graph G and an integer k as input, the goal is to find at most k paths, each with length at most t, which dominate all edges in the graph. In simple words, deleting vertices appearing on these paths results in an independent set. In this paper, we design an FPT algorithm and a kernel for the t-PATH EDGE DOMINATING SET (t-PATH EDS) problem. We give a randomized FPT algorithm parameterized by $k + t$ using the technique of color coding, and then we derandomize it by utilizing an (n, k)-perfect hash family [5].

Theorem 4 *Given an instance $\mathcal{I} = (G = (V, E), k)$ of t-PATH EDGE DOMINATING SET, there is a deterministic FPT algorithm that runs in time $(8e)^{kt}4^t2^{\mathcal{O}(log^2 kt)}n^{\mathcal{O}(1)}$ and finds at most k paths, each of length at most t, that dominate all the edges in E.*

One might ponder why we do not study this problem parameterized by k or t alone? We give a reduction from an NP-hard problem s-t HAMILTONIAN PATH to our problem. In the s-t HAMILTONIAN PATH problem, we are given a graph G, two vertices $s, t \in V(G)$ and the task is to check if a path exists that goes through every vertex exactly once and starts from s and ends at t. The reduction is as follows. Given G, we subdivide every edge. We add paths $P_1 = (s_1, s_2, s)$ and $P_2 = (t, t_1, t_2)$ to the graph with subdivided edges. Let the modified graph be G'. Then, $(G', 1, 2n + 1)$ is an equivalent instance of t-PATH EDGE DOMINATING SET. It is easy to see that G has a s-t hamiltonian path if and only if G' has a path on $2n+1$ vertices that dominates all its edges. An FPT algorithm for t-PATH EDGE DOMINATING SET parameterized by k alone would imply a polynomial time algorithm for the s-t HAMILTONIAN PATH, implying P = NP.

Similarly, an FPT algorithm parameterized by t alone would imply a polynomial time algorithm for VERTEX COVER, which is our problem corresponding to $t = 1$ and is known to be NP-hard, thus, implying P = NP.

Theorem 5 (⋆). *t-PATH EDGE DOMINATING SET admits a kernel of size $\mathcal{O}(k^3t^3)$.*

We would like to remark that the best known kernel for VERTEX COVER (corresponding to $t = 1$) has size $\mathcal{O}(k)$ and for EDGE DOMINATING SET (corresponding to $t = 2$) has size $\mathcal{O}(k^2)$ [15,21,29].

2 Preliminaries

We use $\mathbb{N}$ to denote the set of positive integers. For a graph G, we denote its vertex set and edge set as $V(G)$ and $E(G)$, respectively. A hypergraph $\mathcal{H}$ on a universe $\mathcal{U} = \{u_1, \dots, u_n\}$ is a family $\mathcal{F}$ of subsets of $\mathcal{U}$. For a set $W \subseteq \mathcal{U}$, $\mathcal{F} - W$ denotes all the sets in $\mathcal{F}$ that do not contain any element from W. The frequency or equivalently the degree of a vertex v in a hypergraph is the number of hyperedges which contains v. A path P in a graph G is a sequence of distinct vertices $v_1, \dots, v_\ell$, with $\ell > 1$, such that $(v_i, v_{i+1}) \in E(G)$, for each $i \in [\ell - 1]$. We say that a path is of length t if the path contains t distinct vertices. For a path P, let $V_P = \{v : v \in V(P)\}$ be the set of vertices contained in the path P. For a set of paths $\mathcal{P} = \{P_1, \dots, P_\ell\}$, let $V_\mathcal{P} = \{v : v \in \bigcup_{P \in \mathcal{P}} V(P)\}$ be the set of vertices contained in the paths in $\mathcal{P}$. Let $\mathcal{P}$ be a set of paths that dominate all the edges in a graph, then we denote the set of vertices in $V_\mathcal{P}$ as *used.* For a vertex v, $N(v) = \{u : (u, v) \in E\}$ denotes the set of neighbors of v. For a set of vertices X, $N(X) = (\bigcup_{v \in X} N(v)) \setminus X$. For a subset of vertices S, $G[S]$ denotes the graph induced on S, i.e., the graph on the vertex set S and the set of edges present between any two vertices of S.

For two sets A and B, $A \setminus B$ denotes the set of elements in A, but not in B. For an integer i, we denote the set $\{1, \dots, i\}$ by $[i]$. For integers i and j, we denote the set $\{i, \dots, j\}$ by $[i, j]$. For sets A and B, by $A \uplus B$, we denote the disjoint union of the sets.

3 Kernel for *d*-Hyperedge Dominating Set

In this section, we will develop a kernel for d-Hyperedge Dominating Set using representative sets. We begin with the definition of representative sets.

Definition 1 ([12]). *(q-**Representative Family**) Given a family $\mathcal{A}$ over a universe $\mathcal{U}$, a subfamily $\mathcal{A}' \subseteq \mathcal{A}$ is said to q-represent $\mathcal{A}$ if for every set $B \subseteq \mathcal{U}$ of size q such that there is an $A \in \mathcal{A}$ and $A \cap B = \emptyset$, there is a set $A' \in \mathcal{A}'$ such that $A' \cap B = \emptyset$. If $\mathcal{A}' \subseteq \mathcal{A}$ is q-representative for $\mathcal{A}$ we write $\mathcal{A}' \subseteq^q_{rep} \mathcal{A}$.*

Theorem 6 ([12]). *Given a family $\mathcal{A}$ of sets of size p over a universe, and $q \in \mathbb{N}$, a q-representative family $\hat{\mathcal{A}} \subseteq \mathcal{A}$ for $\mathcal{A}$ with at most $\binom{p+q}{p}$ sets can be computed in time $\mathcal{O}(|\mathcal{A}|(\binom{p+q}{p}p^\omega + \binom{p+q}{p}^{\omega-1})$. Here, $\omega < 2.373$ is the matrix multiplication exponent.*

Theorem 2. *d-Hyperedge Dominating Set admits a kernel of size $\mathcal{O}((dk)^{d^2})$.*

Proof. In the d-HEDS problem, we have a universe $\mathcal{U}$ and family $\mathcal{F}$ of sets of cardinality at most d and an integer k as input. We compute a q-representative set, with $q = dk$, of $\mathcal{F}$ using the algorithm in Theorem 6. Let $\hat{\mathcal{F}}$ be the q-representative set of $\mathcal{F}$. We know $|\hat{\mathcal{F}}| \leq \binom{dk+d}{d} = \mathcal{O}((dk)^d)$ and we can find $\hat{\mathcal{F}}$ in time $\mathcal{O}(|\mathcal{F}|((dk)^d d^\omega + (dk)^{d\omega - d})) = \mathcal{O}(|\mathcal{F}|(dk)^{d\omega})$. Note that $|\mathcal{F}| \geq \binom{dk+d}{d}$, otherwise, we can take $\hat{\mathcal{F}} = \mathcal{F}$. Now, let $\hat{\mathcal{U}}$ be the union of elements in the sets in $\hat{\mathcal{F}}$. Then, $|\hat{\mathcal{U}}| \leq d \cdot \binom{dk+d}{d}$. For all possible subsets U of $\hat{\mathcal{U}}$, of size at most d, we add a set from $\mathcal{F}$ to $\hat{\mathcal{F}}$, if it exists, which contains the elements of U. The number of

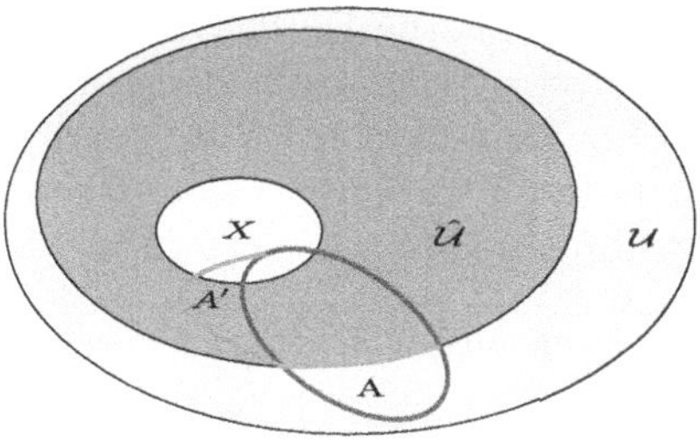

Fig. 2. An illustration of the proof of Theorem 2.

sets added to $\hat{\mathcal{F}}$ is at most $\binom{|\hat{\mathcal{U}}|}{d}+\binom{|\hat{\mathcal{U}}|}{d-1}+\ldots+\binom{|\hat{\mathcal{U}}|}{1} \le \binom{|\hat{\mathcal{U}}|+d-1}{d} \le \binom{d\binom{dk+d}{d}+d-1}{d}$. Note that the size of $\hat{\mathcal{F}}$ now is at most $\binom{dk+d}{d}+\binom{d\binom{dk+d}{d}+d-1}{d}=\mathcal{O}((dk)^{d^2})$. Note that we also assume $|\mathcal{F}| \ge \mathcal{O}((dk)^{d^2})$, otherwise we can return the original instance itself. We claim that $(\hat{\mathcal{U}}, \hat{\mathcal{F}})$ is our desired kernel. The total time taken is $\mathcal{O}(|\mathcal{F}|((dk)^{d^2}))$. Our construction is polynomial in the size of the input instance.

Let $\hat{S}$ be a solution for the instance $(\hat{\mathcal{U}}, \hat{\mathcal{F}})$ and $\mathcal{U}_{\hat{S}} = \bigcup_{F\in\hat{S}} F$. Then, $\mathcal{U}_{\hat{S}}$ hits all the sets in $\hat{\mathcal{F}}$. We claim that $\mathcal{U}_{\hat{S}}$ hits all the sets in $\mathcal{F}$. Suppose not. Then let A be a set in $\mathcal{F}$ that is not hit by $\mathcal{U}_{\hat{S}}$. Since $\hat{\mathcal{F}}$ contains a q-representative set, there must exist some set A' in $\hat{\mathcal{F}}$ which is disjoint from $\mathcal{U}_{\hat{S}}$. But $\mathcal{U}_{\hat{S}}$ hits all the sets in $\hat{\mathcal{F}}$, which is a contradiction.

Let S be a solution for the instance $(\mathcal{U}, \mathcal{F})$ and $\mathcal{U}_S = \bigcup_{F\in S} F$. We know that S dominates all the sets in $\mathcal{F}$. We claim that there exists a subfamily S' in $\hat{\mathcal{F}}$ of cardinality at most $|S|$ that dominates all the sets in $\hat{\mathcal{F}}$. For a set $A \in S$, if $A \in \hat{\mathcal{F}}$ then we pick A in S'. Otherwise, we have two cases. If $\hat{\mathcal{U}} \cap A \neq \emptyset$, then we choose a set A' that is added for the subset $\hat{\mathcal{U}} \cap A$ in the construction of $\hat{F}$. If $\hat{\mathcal{U}} \cap A = \emptyset$, we don't add any set corresponding to A. This completes the construction of S' and notice that $|S'| \le |S|$. Next we need to prove that S' is a solution to $(\hat{\mathcal{U}}, \hat{\mathcal{F}})$. Fix an arbitrary set $X \in \hat{\mathcal{F}}$. We know that X is dominated by S. Let $A \in S$ be a set that dominates X. That is, $X \cap A \neq \emptyset$. If $A \in \hat{\mathcal{F}}$, then $A \in S'$ (by construction of S'), and hence X is dominated by S'. Suppose this is not the case. Notice that $\emptyset \neq X \cap A \subseteq A \cap \hat{\mathcal{U}}$. We have added a set A' to S' corresponding to A, where $A' \cap \hat{\mathcal{U}} = A \cap \hat{\mathcal{U}}$. This implies that A' dominates X. See Fig. 2 for an illustration. □

4 FPT Algorithm for t-PATH EDGE DOMINATING SET

In this section, we present an algorithm for solving t-PATH EDGE DOMINATING SET (t-PATH EDS). Given a graph $G = (V, E)$ and an integer k, the objective of this problem is to find at most k paths, each with a length of at most t, that cover all the edges in G. Recall that a path is considered to have a length of t if it consists of t vertices. A path P is said to cover (dominate) an edge $e = (a, b)$ if the set of vertices V_P intersects with $\{a, b\}$. It is important to note that when

$t = 2$, the problem is the same as the EDGE DOMINATING SET problem. The paths in the solution may have overlapping vertices or edges.

A randomized algorithm for t-PATH EDS is described in Algorithm 3 and Theorem 7 states the main result. However, before we state our main result, we list a few lemmas that will be used to prove the correctness of our algorithm.

Lemma 1 ([16]). *There is an algorithm to find a vertex cover of size k (if it exists) in a graph G, in time $\mathcal{O}^*(1.25284^k)$.*

Our next lemma uses the notion of a *colorful path*. We first define what we mean by a colorful path in a graph and mention a known result that states how to find one, if it exists.

Definition 2 (*Colorful Path*). *For a graph $G = (V, E)$, a positive integer k and a coloring $f : V \rightarrow [k]$ of the vertices of G using k colors, we call P a colorful path in G if all the vertices in P get distinct colors. That is, for any two distinct vertices $u, v \in V_P$, $f(u) \neq f(v)$.*

Lemma 2 ([1,5]). *Let G be a graph, k be a positive integer, and $f : V \rightarrow [k]$ be a coloring of V using k colors. There exists a deterministic algorithm that checks in time $2^k n^{\mathcal{O}(1)}$ whether G contains a colorful path on k vertices and, if this is the case, returns one such path.*

For a graph G, a set of colors Col, and a coloring $f : V \rightarrow [|Col|]$, we define a subroutine called $\mathsf{Path}(Col)$ which is TRUE if invoking Lemma 2 on the instance $(G, |Col|, f)$ returns a colorful path that uses all the colors in Col. For implementation details, please refer to [5].

4.1 COLORFUL t-PATH COVER and t-PATH COVER

We describe a problem called COLORFUL t-PATH COVER that takes a graph G, a positive integer k, a vertex subset $V' \subseteq V$, and a coloring $f : V' \rightarrow [q]$ of V' using q colors, and checks if there exist at most k colorful paths in $G[V']$, each of length at most t, say $\mathcal{P}$, such that $\{f(v) : v \in V_{\mathcal{P}}\} = [q]$. The paths need not be vertex disjoint which implies that $q \leq kt$. In essence, we want to find at most k colorful paths, each of length at most t, that use all the colors. We define it formally below.

COLORFUL t-PATH COVER
Input: $\mathcal{I} = (G, k, V', q, f)$ where $G = (V, E)$ is a graph, $k \in \mathbb{N}$, $V' \subseteq V$, a set of q colors, and $f : V' \rightarrow [q]$ is a coloring of V' using q colors.
Parameter: $k + t$.
Output: A set $\mathcal{P}$ such that $|\mathcal{P}| \leq k$, each $P \in \mathcal{P}$ is a colorful path of length at most t and $\{f(v) : v \in V_{\mathcal{P}}\} = [q]$.

We can solve an instance $\mathcal{I} = (G, k, V', q, f)$ of COLORFUL t-PATH COVER by invoking Algorithm 1 that uses Lemma 3 as a subroutine.

Algorithm 1. An FPT algorithm for COLORFUL t-PATH COVER

Input: $(\mathcal{I} = (G = (V, E), k, V', \mathcal{C}ol, f))$
Parameter: $k + t$
Output: YES if there exist at most k paths in $G[V']$, each of length at most t, that together use all the colors in $\mathcal{C}ol$, else NO.

```
for i = 1 to k do
    Invoke Lemma 3 on (Col, i).
    If the output is YES, then return YES.
end for
return NO.
```

Lemma 3 (⋆). *Let $G = (V, E)$ be a graph, $V' \subseteq V$, $\mathcal{C}ol$ be a set of colors, k be a positive integer, and $f : V' \to [|\mathcal{C}ol|]$ be a function. Then, for a subset of colors $col \subseteq \mathcal{C}ol$ and some $i \in \mathbb{Z}^+$,* Cover(col, i) *is* TRUE *if there exist i paths, each of them colorful and of length at most t, in $G[V']$, say $\mathcal{P}$, such that $\{f(v) : v \in V_{\mathcal{P}}\} = col$. It is* FALSE *otherwise.* Cover(col, i) *can be computed using the following recursive formula.*

$$\mathsf{Cover}(col, i) = \bigvee_{\substack{C \subseteq col \text{ where } |C| \le t, \\ col' = (col \setminus C) \cup W \text{ where } W \subset C, \text{ and} \\ \mathsf{Path}(C) = \mathsf{TRUE}}} \mathsf{Cover}(col', i-1)$$

with

$$\mathsf{Cover}(col, 1) = \begin{cases} \mathsf{Path}(col) & \text{if } |col| \le t \\ \mathsf{FALSE} & \text{otherwise} \end{cases}$$

Moreover, Cover$(\mathcal{C}ol, k)$ *can be computed in time* $2^{2(kt+t)} \cdot k^2 t \cdot n^{\mathcal{O}(1)}$.

The running time of Algorithm 1 is $2^{2(kt+t)} \cdot \mathcal{O}(k^3 t) \cdot n^{\mathcal{O}(1)}$, since Lemma 3 is invoked at most k times. Its correctness follows from the correctness of the above lemma.

We now describe another problem called t-PATH COVER that takes a graph G, an integer k, a partition $V_1 \uplus V_2 \uplus V_3$ of V, and checks if there exist at most k paths in $G[V_1 \cup V_2]$, each of length at most t, that cover all the vertices in V_1 but do not use any vertex from V_3. The paths need not be vertex disjoint. We define it formally below.

t-PATH COVER
Input: An instance $\mathcal{I} = (G, k, V_1, V_2, V_3)$ where $G = (V, E)$ is a graph, k is a positive integer, and $V_1 \uplus V_2 \uplus V_3$ is a partition of V.
Parameter: $k + t$
Output: A set $\mathcal{P}$ such that $|\mathcal{P}| \le k$, each $P \in \mathcal{P}$ is a path of length at most t in $G[V_1 \cup V_2]$, $V_1 \subseteq V_{\mathcal{P}}$ and $V_3 \cap V_{\mathcal{P}} = \emptyset$.

Algorithm 2. An FPT algorithm for t-PATH COVER

Input: $(\mathcal{I} = (G = (V, E), k, V_1, V_2, V_3))$
Parameter: $k + t$
Output: YES if there exist at most k paths in $G[V_1 \cup V_2]$, each of length at most t, that cover all the vertices in V_1 and do not contain any vertex from V_3, else NO.

1: Color the vertices of V_1 such that every vertex gets a distinct color. Let $\widetilde{\chi} : V_1 \rightarrow [kt - |V_1| + 1, kt]$ be a coloring where vertices of V_1 get distinct colors.
2: **for** $r = 0$ to $kt - |V_1|$ **do**
3: Color the vertices in V_2 uniformly at random using r colors. Let $\chi : V_2 \rightarrow [r]$ be a coloring of V_2 uniformly at random. (Note that when $r = 0$, we do not color the vertices of V_2 and proceed with just the coloring on V_1.)
4: Let $Col = \mathsf{Range}(\chi) \cup \mathsf{Range}(\widetilde{\chi})$.
5: Define a coloring function $f : (V_1 \cup V_2) \rightarrow [|Col|]$ where $f(v) = \widetilde{\chi}(v)$ if $v \in V_1$ and $f(v) = \chi(v)$ if $v \in V_2$.
6: Invoke Algorithm 1 on the instance $(G, k, V_1 \cup V_2, Col, f)$.
7: If the algorithm outputs YES, then **return** YES.
8: **end for**
9: **return** NO instance.

Lemma 4 ($\star$). *Given an instance $\mathcal{I} = (G, k, V_1, V_2, V_3)$ of t-PATH COVER, there is a randomized algorithm that, given a* YES *instance, returns* YES *with probability at least $\left(1 - \frac{1}{e^c}\right)$ for some constant $c > 1$, and given a* NO *instance, always returns* NO. *The algorithm runs in time $2^{2(kt+t)} \cdot e^{kt} \cdot k^3 t^2 \cdot n^{\mathcal{O}(1)}$.*

A description of the algorithm is given in Algorithm 2.

4.2 Algorithm for t-PATH EDS

Our randomized FPT algorithm is described in Algorithm 3.

Theorem 7. *Given an instance $\mathcal{I} = (G = (V, E), k)$ of t-PATH EDS, there is a randomized* FPT *algorithm that runs in time $(8e)^{kt} 4^t k^4 t^2 n^{O(1)}$ and finds at most k paths, each of length at most t, that dominate all edges in E. Given a* YES *instance, it returns a solution with probability at least $\left(1 - \frac{1}{e^c}\right)$, for a constant $c > 1$. It returns* NO *when given a* NO *instance.*

Proof. We first compute a vertex cover A of size at most kt using Lemma 1. Recall that we call a vertex *used* if it occurs in one of the k solution paths. We guess a subset $A_{\mathsf{used}} \subseteq A$ that consists of used vertices. Note that $|A_{\mathsf{used}}|$ should be at most kt as the total number of used vertices cannot exceed kt. Then, $A_{\mathsf{left}} = A \setminus A_{\mathsf{used}}$ is the remaining part of the vertex cover A, and $I = V \setminus A$ is an independent set. Since vertices in A_{left} are not used, the vertices in $N_{\mathsf{left}} = N(A_{\mathsf{left}}) \cap I$ must be used to dominate the edges going across A_{left} and N_{left}. In case of a YES instance, $|N_{\mathsf{left}}| \leq kt$. Moreover, for a YES instance, $G[A_{\mathsf{left}}]$ must be an independent set because any edge completely contained in $G[A_{\mathsf{left}}]$ can only be dominated by vertices in A_{left}.

Algorithm 3. An FPT algorithm for t-PATH EDGE DOMINATING SET

Input: $(\mathcal{I} = (G = (V, E), k))$
Parameter: $k + t$
Output: YES if there exists a set $\mathcal{P}$ such that $|\mathcal{P}| \leq k$, each $p \in \mathcal{P}$ is a path of length at most t and the paths in $\mathcal{P}$ dominate every edge in E, else NO.

1: Find a vertex cover A of size at most kt using Lemma 1.
2: If a vertex cover of size kt doesn't exist, then **return** NO instance.
3: Let $I = V \setminus A$.
4: **for** each subset $A_{\text{used}} \subseteq A$ of size at most kt **do**
5: Let $A_{\text{left}} = A \setminus A_{\text{used}}$, $N_{\text{left}} = N(A_{\text{left}}) \cap I$ and $B = I \setminus N_{\text{left}}$.
6: Let $Z = A_{\text{used}} \cup N_{\text{left}}$.
7: If $G[A_{\text{left}}]$ is not an independent set or $|Z| > kt$, go to Step 4 and proceed with the next guess of A_{used}.
8: Invoke Algorithm 2 on the instance $(G, k, Z, B, A_{\text{left}})$.
9: If the subroutine returns YES, **return** YES and terminate.
10: Else, go to Step 4 and proceed with the next guess of A_{used}.
11: **end for**
12: **return** NO instance.

Given this structure of a graph $(A_{\text{used}}, A_{\text{left}}, N_{\text{left}}, B)$, where $B = I \setminus N_{\text{left}}$, solving t-PATH EDS on
(G, k) reduces to solving t-PATH COVER on $(G, k, Z, B, A_{\text{left}})$, where $Z = A_{\text{used}} \cup N_{\text{left}}$. The correctness holds due to Claim 8 listed below.

Claim 8. *A solution to the instance (G, k, Z, B, A_{left}) of t-PATH COVER, where $Z = A_{used} \cup N_{left}$, is also a solution to the instance (G, k) of t-PATH EDS.*

Proof. Let $\mathcal{P}_{\text{cov}}$ be the solution to the instance $(G, k, Z, B, A_{\text{left}})$ of t-PATH COVER. Then, $\mathcal{P}_{\text{cov}}$ consists of at most k paths, each of length at most t, that cover all the vertices in Z and do not contain any vertex from A_{left}. Edges within $G[A_{\text{used}}]$, edges going across A_{used} and A_{left}, edges going across A_{used} and B all contain vertices of A_{used} and hence, have a non-empty intersection with the paths in $\mathcal{P}_{\text{cov}}$. Similarly, edges going across A_{left} and N_{left} contain the vertices of N_{left} and edges going across A_{used} and N_{left} have used vertices as both their endpoints. Thus, these kinds of edges also have a non-empty intersection with the paths in $\mathcal{P}_{\text{cov}}$. As every edge in the graph has a non-empty intersection with the paths in $\mathcal{P}_{\text{cov}}$, the paths in $\mathcal{P}_{\text{cov}}$ dominate every edge in the graph, and hence, the claim holds. □

Since Algorithm 2 is invoked as a subroutine, we have that Algorithm 3 also returns YES with probability at least $\left(1 - \frac{1}{e^c}\right)$. If the given instance is a NO instance, Algorithm 2 always returns NO, and hence, Algorithm 3 also returns NO.

Running Time: Computing a vertex cover A of size at most kt takes time $\mathcal{O}^*(1.25284^{kt})$. There are at most 2^{kt} choices of A_{used} and for each choice of A_{used}, Steps 5 and 7 take polynomial time and invoking Algorithm 2 takes

$2^{2(kt+t)} \cdot e^{kt} \cdot \mathcal{O}(k^4t^2) \cdot n^{\mathcal{O}(1)}$ time. So, the total time taken is $\mathcal{O}(1.25284^{kt}) \cdot n^{\mathcal{O}(1)} + 2^{2(kt+t)} \cdot (2e)^{kt} \cdot \mathcal{O}(k^4t^2) \cdot n^{\mathcal{O}(1)}$. Thus, the algorithm runs in time $(8e)^{kt}4^tk^4t^2n^{\mathcal{O}(1)}$. □

Remark 1. Notice that, the way the algorithm is presented, it returns YES with probability at least $(1 - \frac{1}{e^c})$ if the given instance is a YES instance and NO when the given instance is a NO instance. We can tweak Lemma 3 to return a set of colorful paths using the standard technique of backlinks. The necessary modifications in all the algorithms will give us a set of solution paths.

We can derandomize the algorithm using an (n, k)-perfect hash family. An (n, k)-perfect hash family $\mathcal{F}$ is a family of functions from $[n]$ to k such that for every set $S \subseteq [n]$ of size k, there exists a function $f \in \mathcal{F}$ that *splits S evenly*. That is, for every $1 \leq j, j' \leq k$, $|f^{-1}(j) \cap S|$ and $|f^{-1}(j') \cap S|$ differ by at most 1. For any $n, k \geq 1$, one can construct an (n, k)-perfect hash family of size $e^k k^{\mathcal{O}(logk)} logn$ in time $e^k k^{\mathcal{O}(logk)} nlogn$ [5,23].

The deterministic algorithm does the following. It finds a vertex cover A of size at most kt, guesses a subset A_{used} of used vertices from A and partitions the vertices of the graph into $Z \uplus B \uplus A_{\text{left}}$, as described in Algorithm 3. If $|Z| \leq kt$, then for each $r \in [0, kt - |A_{\text{used}}|]$, it constructs an $(|Z|, |A_{\text{used}}| + r)$-perfect hash family $\mathcal{F}_r$ and for each $f \in \mathcal{F}_r$, invokes Algorithm 1 on $(G, k, Z, [|A_{\text{used}}| + r], f)$. The properties of an (n, k)-perfect hash family ensure that if there exists a set $\mathcal{P}$ of at most k paths, each of length at most t, in $G[Z \cup B]$ covering all the vertices in Z, then there is a function $f \in \mathcal{F}_r$ that is injective on $V_{\mathcal{P}}$ and consequently, Algorithm 1 finds a set of at most k colorful paths with the required properties for the coloring f. A vertex cover A of size at most kt can be computed in $\mathcal{O}(1.25284^{kt}) \cdot n^{\mathcal{O}(1)}$ time. There are 2^{kt} choices for A_{used}. For each choice of A_{used}, partitioning the vertices into $Z \uplus B \uplus A_{\text{left}}$ takes polynomial time and for each $r \in [kt - |A_{\text{used}}|]$, constructing a $(|Z|, |A_{\text{used}}| + r)$-perfect hash family $\mathcal{F}_r$ takes $e^{kt}(kt)^{\mathcal{O}(logkt)} ktlog(kt)$ time and invoking Algorithm 1 for each $f \in \mathcal{F}_r$ takes $e^{kt}(kt)^{\mathcal{O}(logkt)} log(kt) \cdot 2^{2(kt+t)}\mathcal{O}(k^3t)n^{\mathcal{O}(1)}$. So, the total time taken is $(8e)^{kt}4^t \cdot 2^{\mathcal{O}(log^2kt)}n^{\mathcal{O}(1)}$.

Theorem 4. *Given an instance $\mathcal{I} = (G = (V, E), k)$ of t-PATH EDGE DOMINATING SET, there is a deterministic* FPT *algorithm that runs in time $(8e)^{kt}4^t2^{\mathcal{O}(log^2kt)}n^{\mathcal{O}(1)}$ and finds at most k paths, each of length at most t, that dominate all the edges in E.*

5 Conclusion

In this paper, we study two generalizations of EDS from the perspective of parameterized complexity. In particular, we study the EDS problem in hypergraphs, called d-HEDS. We study another extension of EDS where we want to dominate the edges by paths of length t. We give FPT algorithms and polynomial kernels for both problems. We also demonstrate the hardness of the HYPEREDGE DOMINATING SET problem when each element has bounded frequency and any pair of sets in the family intersect in at most one element.

References

1. Alon, N., Yuster, R., Zwick, U.: Color-coding: a new method for finding simple paths, cycles and other small subgraphs within large graphs. In: Leighton, F.T., Goodrich, M.T. (eds.) Proceedings of the Twenty-Sixth Annual ACM Symposium on Theory of Computing, 23–25 May 1994, Montréal, Québec, Canada, pp. 326–335. ACM (1994). https://doi.org/10.1145/195058.195179
2. Bevern, R.: Towards optimal and expressive kernelization for d-hitting set. In: Gudmundsson, J., Mestre, J., Viglas, T. (eds.) COCOON 2012. LNCS, vol. 7434, pp. 121–132. Springer, Heidelberg (2012). https://doi.org/10.1007/978-3-642-32241-9_11
3. Björklund, A., Husfeldt, T., Koivisto, M.: Set partitioning via inclusion-exclusion. SIAM J. Comput. **39**(2), 546–563 (2009). https://doi.org/10.1137/070683933
4. Chen, J., Kanj, I.A., Xia, G.: Improved parameterized upper bounds for vertex cover. In: Královič, R., Urzyczyn, P. (eds.) MFCS 2006. LNCS, vol. 4162, pp. 238–249. Springer, Heidelberg (2006). https://doi.org/10.1007/11821069_21
5. Cygan, M., et al.: Parameterized Algorithms. Springer, Heidelberg (2015). https://doi.org/10.1007/978-3-319-21275-3
6. Downey, R.G., Fellows, M.R.: Fixed-parameter tractability and completeness II: on completeness for W[1]. Theor. Comput. Sci. **141**(1&2), 109–131 (1995). https://doi.org/10.1016/0304-3975(94)00097-3
7. Downey, R.G., Fellows, M.R.: Parameterized Complexity. Monographs in Computer Science. Springer, Heidelberg (1999). https://doi.org/10.1007/978-1-4612-0515-9
8. Escoffier, B., Monnot, J., Paschos, V.T., Xiao, M.: New results on polynomial inapproximability and fixed parameter approximability of EDGE DOMINATING SET. In: Thilikos, D.M., Woeginger, G.J. (eds.) IPEC 2012. LNCS, vol. 7535, pp. 25–36. Springer, Heidelberg (2012). https://doi.org/10.1007/978-3-642-33293-7_5
9. Feige, U.: A threshold of ln n for approximating set cover. J. ACM **45**(4), 634–652 (1998). https://doi.org/10.1145/285055.285059
10. Fernau, H.: EDGE DOMINATING SET: efficient enumeration-based exact algorithms. In: Bodlaender, H.L., Langston, M.A. (eds.) IWPEC 2006. LNCS, vol. 4169, pp. 142–153. Springer, Heidelberg (2006). https://doi.org/10.1007/11847250_13
11. Fomin, F.V., Gaspers, S., Saurabh, S., Stepanov, A.A.: On two techniques of combining branching and treewidth. Algorithmica **54**(2), 181–207 (2009). https://doi.org/10.1007/s00453-007-9133-3
12. Fomin, F.V., Lokshtanov, D., Panolan, F., Saurabh, S.: Efficient computation of representative families with applications in parameterized and exact algorithms. J. ACM **63**(4), 29:1–29:60 (2016). https://doi.org/10.1145/2886094
13. Fujito, T., Nagamochi, H.: A 2-approximation algorithm for the minimum weight edge dominating set problem. Discret. Appl. Math. **118**(3), 199–207 (2002). https://doi.org/10.1016/S0166-218X(00)00383-8
14. Garey, M.R., Johnson, D.S.: Computers and Intractability: A Guide to the Theory of NP-Completeness. W. H. Freeman, New York (1979)
15. Hagerup, T.: Kernels for edge dominating set: simpler or smaller. In: Rovan, B., Sassone, V., Widmayer, P. (eds.) MFCS 2012. LNCS, vol. 7464, pp. 491–502. Springer, Heidelberg (2012). https://doi.org/10.1007/978-3-642-32589-2_44
16. Harris, D.G., Narayanaswamy, N.S.: A faster algorithm for vertex cover parameterized by solution size. CoRR abs/2205.08022 (2022). https://doi.org/10.48550/arXiv.2205.08022

17. Harris, D.G., Narayanaswamy, N.S.: A faster algorithm for vertex cover parameterized by solution size. In: Beyersdorff, O., Kanté, M.M., Kupferman, O., Lokshtanov, D. (eds.) 41st International Symposium on Theoretical Aspects of Computer Science, STACS 2024, 12–14 March 2024, Clermont-Ferrand, France. LIPIcs, vol. 289, pp. 40:1–40:18. Schloss Dagstuhl - Leibniz-Zentrum für Informatik (2024). https://doi.org/10.4230/LIPIcs.STACS.2024.40
18. Iwaide, K., Nagamochi, H.: An improved algorithm for parameterized edge dominating set problem. In: Rahman, M.S., Tomita, E. (eds.) WALCOM 2015. LNCS, vol. 8973, pp. 234–245. Springer, Cham (2015). https://doi.org/10.1007/978-3-319-15612-5_21
19. Johnson, D.S.: Approximation algorithms for combinatorial problems. In: Aho, A.V., et al. (eds.) Proceedings of the 5th Annual ACM Symposium on Theory of Computing, 30 April–2 May 1973, Austin, Texas, USA, pp. 38–49. ACM (1973). https://doi.org/10.1145/800125.804034
20. Karp, R.M.: Reducibility among combinatorial problems. In: Miller, R.E., Thatcher, J.W. (eds.) Proceedings of a symposium on the Complexity of Computer Computations, held 20–22 March 1972, at the IBM Thomas J. Watson Research Center, Yorktown Heights, New York, USA, pp. 85–103. The IBM Research Symposia Series, Plenum Press, New York (1972). https://doi.org/10.1007/978-1-4684-2001-2_9
21. Lampis, M.: A kernel of order 2 k-c log k for vertex cover. Inf. Process. Lett. **111**(23–24), 1089–1091 (2011). https://doi.org/10.1016/j.ipl.2011.09.003
22. Lin, B., Ren, X., Sun, Y., Wang, X.: Constant approximating parameterized k-setcover is w[2]-hard. CoRR abs/2202.04377 (2022). https://arxiv.org/abs/2202.04377
23. Naor, M., Schulman, L.J., Srinivasan, A.: Splitters and near-optimal derandomization. In: 36th Annual Symposium on Foundations of Computer Science, Milwaukee, Wisconsin, USA, 23–25 October 1995, pp. 182–191. IEEE Computer Society (1995). https://doi.org/10.1109/SFCS.1995.492475
24. Niedermeier, R., Rossmanith, P.: An efficient fixed-parameter algorithm for 3-hitting set. J. Discret. Algorithms **1**(1), 89–102 (2003). https://doi.org/10.1016/S1570-8667(03)00009-1
25. Norman, R.Z., Rabin, M.O.: An algorithm for a minimum cover of a graph (1959). https://api.semanticscholar.org/CorpusID:120383003
26. Philip, G., Raman, V., Sikdar, S.: Polynomial kernels for dominating set in graphs of bounded degeneracy and beyond. ACM Trans. Algorithms **9**(1), 11:1–11:23 (2012). https://doi.org/10.1145/2390176.2390187
27. Raman, V., Saurabh, S., Sikdar, S.: Efficient exact algorithms through enumerating maximal independent sets and other techniques. Theory Comput. Syst. **41**(3), 563–587 (2007). https://doi.org/10.1007/s00224-007-1334-2
28. van Rooij, J.M.M., Bodlaender, H.L.: Exact algorithms for edge domination. In: Grohe, M., Niedermeier, R. (eds.) IWPEC 2008. LNCS, vol. 5018, pp. 214–225. Springer, Heidelberg (2008). https://doi.org/10.1007/978-3-540-79723-4_20
29. Soleimanfallah, A., Yeo, A.: A kernel of order 2k-c for vertex cover. Discret. Math. **311**(10–11), 892–895 (2011). https://doi.org/10.1016/j.disc.2011.02.014
30. Xiao, M., Kloks, T., Poon, S.-H.: New parameterized algorithms for the edge dominating set problem. In: Murlak, F., Sankowski, P. (eds.) MFCS 2011. LNCS, vol. 6907, pp. 604–615. Springer, Heidelberg (2011). https://doi.org/10.1007/978-3-642-22993-0_54

31. Xiao, M., Nagamochi, H.: A refined exact algorithm for edge dominating set. In: Agrawal, M., Cooper, S.B., Li, A. (eds.) TAMC 2012. LNCS, vol. 7287, pp. 360–372. Springer, Heidelberg (2012). https://doi.org/10.1007/978-3-642-29952-0_36
32. Yannakakis, M., Gavril, F.: Edge dominating sets in graphs. SIAM J. Appl. Math. **38**(3), 364–372 (1980). https://doi.org/10.1137/0138030

Beyond Image-Text Matching: Verb Understanding in Multimodal Transformers Using Guided Masking

Ivana Beňová[1,2(✉)], Jana Košecká[3], Michal Gregor[2], Martin Tamajka[2], Marcel Veselý[2], and Marián Šimko[2]

[1] Faculty of Information Technology, Brno University of Technology, Brno, Czech Republic

[2] Kempelen Institute of Intelligent Technologies, Bratislava, Slovakia
{ivana.benova,michal.gregor,martin.tamajka,marcel.vesely, marian.simko}@kinit.sk

[3] George Mason University, Fairfax, VA, USA
kosecka@gmu.edu

Abstract. Probing methods are widely used to evaluate the multimodal representations of vision-language models (VLMs), with dominant approaches relying on zero-shot performance in image-text matching tasks. These methods typically assess models on curated datasets focusing on linguistic aspects such as counting, relations, or attributes. This work uses a complementary probing strategy called **guided masking**. This approach selectively masks different modalities and evaluates the model's ability to predict the masked word. We specifically focus on probing verbs, as their comprehension is crucial for understanding actions and relationships in images, and it presents a more challenging task than subjects, objects, or attributes comprehension. Our analysis targets VLMs that use region-of-interest (ROI) features obtained from object detectors as input tokens. Our experiments demonstrate that selected models can accurately predict the correct verb, challenging previous conclusions based on image-text matching methods, which suggested VLMs fail in situations requiring verb understanding. The code for experiments will be available https://github.com/ivana-13/guided_masking.

Keywords: multimodal models · probing · understanding · verb phrases · foundational models · image-text matching · guided masking

1 Introduction

Recent advances in multimodal transformers have fused vision and language using attention and self-supervised learning. Initially developed for NLP [8], these models have evolved into vision-language models (VLMs) like

---

The original version of the chapter has been revised. The author names, text part and table alignment in Tables 2 and 3 has been corrected. A correction to this chapter can be found at https://doi.org/10.1007/978-3-031-82670-2_26

R. Královič and V. Kůrková (Eds.): SOFSEM 2025, LNCS 15538, pp. 80–93, 2025.
https://doi.org/10.1007/978-3-031-82670-2_7

LXMERT [27], ALBEF [14], CLIP [22], BLIP [13], or FLAVA [26]. Pre-trained on large, noisy image-caption pairs datasets, they use tasks such as masked language modeling, image-text matching, or contrastive learning to capture correlations between image and text tokens.

Several probing methods have been introduced to understand multimodal models' capabilities. A common approach is image-text matching in a zero-shot setting, where models are tested on carefully curated datasets. Image-caption pairs are used to test the model by classifying the pairs as matching or not, while the captions have controlled edits. Studies using this approach have evaluated object recognition [25], verb comprehension [11], or word order [28].

Despite its wide use, image-text matching has limitations [2,12,30,32,33]. It relies on holistic representations, making it sensitive to minor variations (changing "trail" to "pathway" in the caption). Furthermore, the random selection of negative captions during pre-training may introduce bias, making it easier to identify non-matching pairs during training.

To address these issues, we propose to use a complementary method, guided masking, that selectively ablates specific elements in the caption and evaluates the model's ability to predict the masked word. This approach offers a focused analysis of verb comprehension, identified as challenging for VLMs by [11].

While recent models like LLAVA [18] show remarkable performance, they are not always suitable for tasks requiring interpretability, control, or resource-efficient experimentation. Additionally, finding datasets these models haven't have yet to see during pre-training remains challenging. In contrast, the models we chose to study, use region-of-interest (ROI) features, are valuable for controlled, foundational research, and allow detailed evaluation of visual and textual inputs.

The main contributions of our work are:

- Proposing **guided masking** for a detailed evaluation of VLMs across linguistic aspects such as subject, verb, attributes, counting, and spatial relations.
- Quantitative analyzes of verb understanding in VLMs using the SVO-Probes [11] and V-COCO [10] datasets, showing models correctly predicted verbs in over 75% of cases.
- Conducting sensitivity analysis through visual token ablation to study how visual inputs affect predictions. By focusing on multimodal models using regions of interest (ROI) of the image obtained with object detectors as inputs, we show in a controlled setting that visual information impacts verb prediction.

2 Related Work

This section covers foundational vision-language models, probing methods, and VLMs' specific work on verb understanding.

Foundational Vision-Language Transformers. Vision-language transformers fuse visual and textual inputs using region-of-interest (ROI) features from pre-trained object detectors or patches. Notable models using ROI features include ViLBERT [20], LXMERT [27], VisualBERT [15], and UNITER [7], with pretraining tasks such as masked language modeling, masked region modeling, or image-text matching. Advanced techniques, like composition-aware hard negative mining [32], have enhanced fine-grained multimodal understanding.

Vision-language transformer models integrate visual and textual inputs by jointly training models on both modalities. Commonly, vision inputs are tokenized (region-of-interest (ROI) features obtained via object detectors or later as patches), and textual inputs are processed through language transformers. Datasets such as Conceptual Captions (CC) [24] are used to pretrain these models. The typical pretraining objectives include masked language modeling (MLM), masked region modeling (MRM), image-text matching (ITM), and cross-modal contrastive learning (CMCL). Advanced techniques, like composition-aware hard negative mining [32] or vision-language replaced token detection [33], have been introduced to enhance fine-grained multimodal understanding.

The pretraining of fusion encoders with image-text matching is performed on holistic image-text pair representation. For example, LXMERT [27], UNITER [7], and VisualBERT [15] derive representations using the final hidden state of the `[CLS]` token, while ViLBERT [20] employs element-wise multiplication of holistic visual `[IMG]` and language `[CLS]` tokens. Dual-encoder models like CLIP [22], FLAVA [26], and BLIP [13], which use contrastive learning and have patch-based visual input, perform well on downstream tasks but are more challenging to probe due to their architecture. Tools like gScoreCAM [6] have been used to explain model behavior.

Probing Methods for Model Understanding. Probing methods to investigate models' capabilities have been developed. Early works like VALUE [4] focused on explaining individual layers, heads, or fusion techniques, but fine-grained probing tasks emerged later. These methods rely on specially curated datasets with foiled captions and often use image-text matching evaluation in a zero-shot setting. A foil caption was created for every image for verb understanding [11]. In this caption, only the part representing the studied linguistic aspect was changed; in this case, the verb (see Fig. 1 for an example). Other studies have focused on counting [21], spatial relationships [17], word order [28], and color, size, position, and adversarial captions [23].

Building on these results, other researchers have explored similar tasks in different settings. For example, [31] introduced the SeeTrue benchmark for visual entailment. citeyuksekgonul2022and has investigated the exploration of word order, attributes, and relations, employing an image retrieval approach on holistic text and image representations. The authors in [19] used a text prompt with masks and a set of possible words to study the understanding of spatial commonsense in different models. The model's task is to calculate the probability of each possible word filling the masked position.

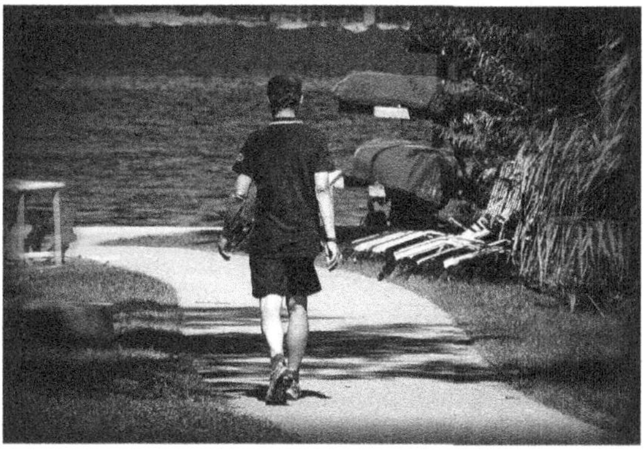

Fig. 1. Image from the SVO-Probes dataset [11]. The dataset includes image-caption pairs where sentences match the image (positive example) or contain mismatches in subject, verb, or object (negative example). These pairs are used to probe models via zero-shot image-text matching. Positive caption: *A person walking on a trail.* Verb-negative caption: *A person runs on the trail.*

Verb Understanding in VLMs. One of the introduced probing tasks is image-text matching. Hendricks et al. [11] found that VLMs struggle with verbs compared to subjects and objects in image-text matching task, performing near chance (60% on binary classification) when understanding actions was required, compared to 67% when understanding the subject and 73% when understanding object was required. This is the reason why we focus on verb understanding.

Ablation Studies. Ablation studies further probe multimodal models. Frank et al. [9] evaluated how models like ViLBERT and LXMERT predict masked language tokens (class of masked bounding box), given ablated inputs in the other modality. They revealed a strong dependence of predicting masked phrases on missing visual input, but not the other way around when there is missing language input. For evaluation, they used the value of cross-entropy loss for masked language modeling and Kullback-Leibler divergence loss for masked region modeling, specifically the increase or no effect in the value of loss in the presence of ablation.

Our work is closest to [9,19], but instead of masking general phrases, we focus on the analysis of verbs as in [11]. We study the model's ability to predict the correct word in the masked caption with or without ablated visual input from the whole vocabulary.

3 Probing with Masking

Guided Masking. To better understand the effect of local caption changes on cross-modal representations, we propose the following probing technique:

$$\boldsymbol{P} = f_{lang}(I, C'), \tag{1}$$

where I is input image, $C = \{w_1, w_2, \ldots w_i, \ldots w_n\}$ is a matching caption containing n words w, $C' = \{w_1, w_2, \ldots [MASK]_i, \ldots w_n\}$ is created by masking a word w_i we want to probe, f_{lang} is representing the pre-trained language head of the multimodal model which predicts masked word(s), and $\boldsymbol{P}$ is probability distribution across all the tokens in the vocabulary.

This technique presents a compelling advantage as it avoids creating new foiled caption datasets and goes beyond binary image-text matching. Our approach delves into exploring the model's most probable token predictions, thereby offering potential alternatives considered by the model. It provides deeper insights into examining local connections between vision and language tokens. We focus our study on verb understanding, meaning all masked words w_i are verbs. However, this approach can be used to study understanding of different aspects. We compare model performance with and without visual input to assess the role of visual grounding. This nuanced evaluation underscores the method's efficacy in providing a more comprehensive and insightful analysis.

3.1 Guided Masking Evaluation

To avoid false negatives during evaluation, we suggest following a robust evaluation approach. By doing that, verbs in varying grammatical forms and even synonyms, to some extent, are considered.

- **Lemmatization:** Grammatical forms are handled by lemmatizing original and predicted words using *nltk* Lemmatizer. For example, "sitting" and "sits" are considered equivalent.
- **Synonyms:** We compare the top n predictions ($n = 5$) with the original verb. For instance, if the original verb is "jogging" and the model predicts "running," they have the same meaning but are not the same. The top 5 words are close in meaning, and probabilities beyond this drop significantly.
- **Semantic Variation:** In cases where different but valid words (e.g., "laying" vs. "resting") are predicted, comparing the top 5 lemmatized predictions can capture some semantic flexibility, although not always perfectly.

Evaluation of Cross-Modal Grounding. We evaluate verb cross-modal grounding by testing model performance with visual ablation. [1] observed that alignment between language and visual tokens results in high transformer attention. We assessed this relationship through ablation. If the performance diminishes upon removing visual inputs, it suggests that the model has acquired alignments between phrases and objects.

We employed vision ablation to evaluate verb grounding in visual inputs of subjects engaged in activities. The process of visual ablation is following:

1. Determine the caption's subject through Trankit's part-of-speech tagging [29]. For example, in caption *A person walking on a trail.*, the word *person* is a subject.

2. Obtain all labels predicted for each visual token in the image by Faster R-CNN. For example, *man, shirt, shoe, path, head, tree,...*
3. Using the WordNet graph (approximating semantic similarity), find which label of visual tokens is closest to the subject from the caption. For example, label *man* is most similar to word *person*.
4. Mask the features of the bounding box with the selected label, together with all features whose bounding boxes intercede with the image's subject bounding box.

4 Experimental Results

We used the ViLBERT, LXMERT, UNITER, and VisualBERT models, implemented in VOLTA [3], to probe verb understanding. Experiments were conducted on the SVO-Probes dataset to compare our results using guided masking with results in [11].

SVO-Probes Dataset. The SVO-Probes dataset contains captions created from ⟨subject, verb, object⟩ triplets, where the verb should be visually recognizable. For each caption, one image forms a positive image-caption pair, and another forms a subject-negative, verb-negative, or object-negative pair, as shown in Fig. 2.

Fig. 2. Images from the SVO-Probes dataset [11]. Caption: *"A girl sitting on grass."* The image on the left creates a positive pair, while the image on the right creates a negative pair.

This dataset was created with the help of Amazon Mechanical Turk annotators. The SVO triplets were collected from the Conceptual Captions (CC) dataset, while images were downloaded from the web to prevent overlap with the CC dataset.

Image-Text Matching Baseline. In the first row of Table 1, we can see the results reported in [11] for probing verb understanding with a base multimodal transformer (MMT) that closely replicates the ViLBERT architecture. The results represent the accuracy of the average prediction of positive and negative pairs. The difference in accuracy we measured (ITM for ViLBERT) could have been influenced by various factors, including the distinct implementation of ViLBERT and the slightly lower number of samples we could obtain (some links had broken when we downloaded the data). Since the average accuracy for image-text matching is around 46%, the conclusion about understanding verbs was negative.

Additionally, we computed the results for image-text matching while performing vision ablation on subjects and by masking the entire image. The outcomes of these experiments are in Table 1 and reveal that ablating the vision causes the models to predict the negative label more frequently. Strikingly, this leads to an improvement in average accuracy due to the imbalanced dataset. This suggests that the image-text matching evaluation method needs to be improved.

Table 1. Performance on probing for verb understanding with image-text matching using SVO-Probes dataset averaged over all (Average), positive (Positive), and negative (Negative) image-caption pairs. The results in the first row (MMT) were published in [11], and we obtained the results in the rest of the table using VOLTA's model implementation.

	Average	Positive	Negative
#Examples	34k	11k	23k
MMT	60.8%	93.8%	27.8%
Image-Text Matching			
#Examples	33893	11571	22322
ViLBERT	46.3%	95.4%	20.9%
LXMERT	45.0%	**95.5%**	18.8%
UNITER	**47.2%**	94.5%	**22.7%**
VisualBERT	47.1%	94.4%	22.6%
Vision ablation			
ViLBERT	**56.5%**	63.4%	**53.0%**
LXMERT	47.5%	86.5%	27.2%
UNITER	44.2%	**95.6%**	17.5%
VisualBERT	46.3%	91.6%	22.9%
Masking whole image			
ViLBERT	**65.2%**	8.3%	**94.7%**
LXMERT	55.3%	45.5%	60.4%
UNITER	43.1%	**85.2%**	21.2%
VisualBERT	58.3%	40.3%	67.7%

Guided Masking. We evaluated the models' ability to predict masked verbs in captions using the guided masking technique, shown in Table 2. Only positive image-caption pairs from the SVO-Probes dataset were used because by masking the verb in the verb-negative image-caption pair, we would obtain the same caption as by masking verb in the positive image-caption. All models reached almost 75% top-5 accuracy, suggesting stronger verb understanding than previously thought in [11]. Vision ablation (masking the visual token of the subject performing activity) led to a 2.7% accuracy drop, while whole-image masking caused a 12% drop. It is essential to state that the masking of visual tokens assumes that the Faster R-CNN prediction of the subject was correct. This, however, is not always the case, causing errors in masking the tokens.

To better understand to what extent the observed token predictions are due tolanguage priors vs the result of multimodal pretraining, we compare the VLMs with BERT [8]. BERT's top 5 accuracy was only 36.1%. Since the VLM's are initialized with BERT, comparing the results of complete image ablation and BERT can also suggest how over-fitting BERT on CC captions boosts Performance. This baseline clarifies why masking the whole image decreases Performance by only 11. The language model and its fine-tuned versions in image-language models are adept at predicting verbs in many instances. A more comprehensive evaluation beyond the top 5 predictions (considering caption semantic variety) could yield even more substantial improvements with added visual input.

Table 2. Probing on positive image-caption pairs of the SVO-Probes dataset for verb understanding with guided masking probing technique. BERT - using the guided masking technique with BERT. Of all 11,571 samples, 44 (0.4%) were not evaluated.

	Top 5			
	ViLBERT	UNITER	LXMERT	VisualBERT
Guided masking	73.9%	74.4%	74.6%	74.3%
Vision ablation	71.5%	72.2%	71.6%	72.2%
Masking whole image	62.9%	62.5%	59.6%	59.9%
BERT	36.1%			

V-COCO Dataset. The V-COCO dataset, a subset of MS-COCO [16], was used to test the ability to associate objects in the scene with the semantic roles of the action. Since the full captions were not part of V-COCO, for this experiment, we used the captions from the MS-COCO dataset for training, validation, and testing. Some captions did not align with the actions in V-COCO (see Fig. 3).

For example, in caption *'Player hitting a ball with a baseball bat"*, *player"* is the agent of the action, *bat"* is the instrument, and *ball"* is the object. This leads to the realization that there are different types of activities, depending on the number of visual tokens affecting them. Grounding of activities such as *sitting"*, *standing"*, and *running"* is affected only by a single image token containing the entity. However, if activities such as *playing guitar"*, *kicking football"*, and *hitting the ball with baseball bat"* are grounded, they should be connected with multiple image tokens.

Fig. 3. An example of two images in the MS-COCO dataset. The action assigned to an image in the V-COCO dataset is not guaranteed to be contained in the MS-COCO caption used with our guided masking technique. The image on the left exemplifies where the action and the masked verb are identical. The action names associated with the image in the V-COCO dataset are *"hold"*, *"stand"*, *"walk"*, *"look"*, and *"carry"*. The description in MS-COCO is *"A man walks with his surfboard on the sand."* The masked verb is *"walks"*. The image on the right exemplifies where the action and the verb differ. The action names associated with the image in the V-COCO dataset are *"hold"*, *"sit"*, and *"drink"*. The description in MS-COCO is *"An older person with a child, both eating donuts."* The masked verb is *"eating"*.

The results of the masked language modeling probing technique on the V-COCO dataset can be seen in Table 3. The accuracy of predicting the correct verb in the caption is over 80%. The vision ablation leads to a 11% performance decrease, further supporting the claim of grounding the verb token in image tokens. Compared with SVO-Probes, the accuracy of BERT's only predictions is higher. Human-generated captions in MS-COCO contain context, and their vocabulary is not restricted in the same way as in SVO-Probes. This nature of captions influenced BERT's results.

Table 3. Probing the V-COCO dataset with captions from MS-COCO for verb understanding with guided masking probing technique. BERT - using the guided masking technique with BERT. Out of all 10345 samples, three (0.03%) were not evaluated.

	Top 5			
	ViLBERT	UNITER	LXMERT	VisualBERT
Guided masking	81.1%	81.3%	80.5%	80.2%
Vision ablation	79.8%	79.5%	79.2%	78.5%
Masking whole image	72.5%	76.3%	73.5%	74.8%
BERT	58.5%			

4.1 Image-Text Matching and Explainability

To compare the image-text matching probing method with guided masking, we used the explainability tool from [5], which generates relevancy maps based on the model's attention layers. We focused on the LXMERT architecture and analyzed how relevancy maps highlight the positive relevancy between image and text inputs and the output prediction.

Positive Example. The SVO-Probes image in Fig. 4 (left) is paired with the verb-negative caption *"A woman lies on a beach."*, which LXMERT correctly identifies as a mismatch with a 97.83% probability. The relevancy map for the caption (see Fig. 5) shows that the model focused on the verb *"lies"* when making a prediction, and that is the key to finding the mismatch. Similarly, in the image (see Fig. 4 right), the woman's region of interest is most relevant.

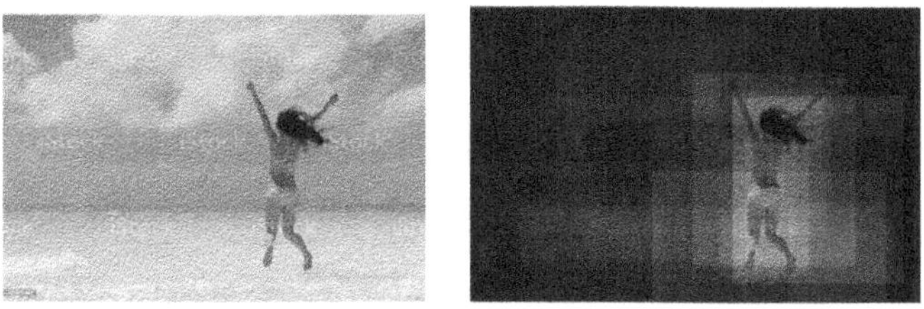

Fig. 4. An SVO-Probes image studied when paired with the negative verb caption *"A woman lies on a beach."* The image is shown without and with the visualization of the relevancy map on the left and the right, respectively. The woman's region of interest (ROI) is the most relevant for the model's prediction that the pair is not a match.

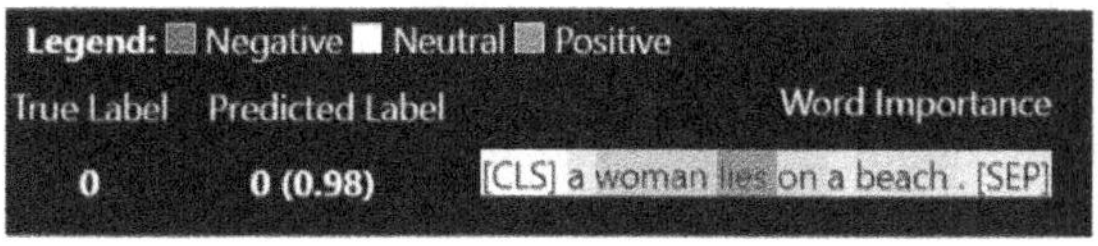

Fig. 5. The visualization of the relevancy map on the input caption *"A woman lies on a beach."* with Fig. 4 for LXMERT's image-text matching correct prediction that this caption and image do not match.

Negative Example. For the caption *"The person ran on the trail."*, LXMERT again correctly identifies the mismatch for image in Fig. 6 (left) with a 98.72% probability. However, as shown in Fig. 7, the relevancy map focuses on the words *"person"* and *"trail"* rather than the verb *"ran"*, suggesting that the model may not fully capture the verb mismatch that drives the negative prediction. This suggests that the model's prediction during image-text matching (correct or incorrect) does not necessarily have to be because of not understanding the verb. Models can have a higher relevance to other parts of text or image than the verb or activity (in this case, the model focuses on the word "trail" because the path on the image may not look like a traditional trail). Therefore, we can not fully rely on the results of an image-text matching task alone to evaluate verb understanding in VLMs, and doing the explainability analysis for all samples would be time-consuming and computationally exhausting.

Fig. 6. An SVO-Probes image studied when paired with the negative verb caption *"The person ran on the trail."* The image is shown without and with the visualization of the relevancy map on the left and the right, respectively. The region of interest (ROI) of the man is the most relevant to the classification.

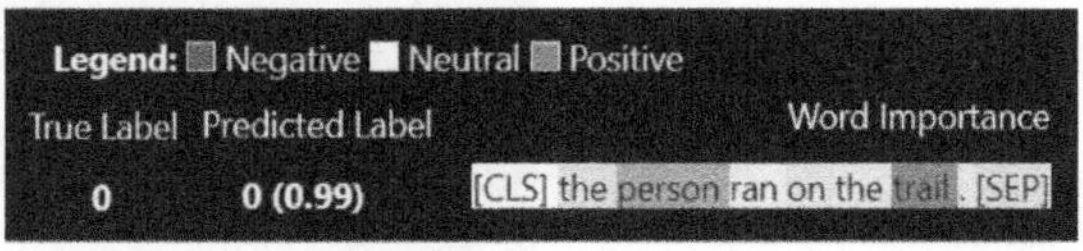

Fig. 7. Visualization of the relevancy map on the input caption *"The person ran on the trail."* with Fig. 6 for LXMERT's image-text matching correct prediction that this caption and image do not match.

When we applied guided masking with the caption *"A person walking on a trail"* and masked the verb *"walking"*, LXMERT predicted the correct verb with 66% probability, further showing that guided masking directly probes verb understanding.

This experiment highlights the need for explainability tools in image-text matching to ensure models focus on verbs and activities as intended. Due to the manual nature of relevancy analysis, it is impractical for large datasets, emphasizing the limitations of image-text matching probing and the value of guided masking for deeper insight. In contrast, guided masking offers insights through its five considered predictions.

5 Ethical Policy

This section examines the prospective benefits and potential hazards associated with this paper. Although the introduced probing technique contributes to the advancement of interpretable deep learning models, it is crucial to acknowledge the limited scope of this study, which is centered solely around English image-caption datasets characterized by North American and Western European biases. It is essential to recognize that the quality of datasets significantly influences the outcomes and the implications that can be generalized for other models adopting the guided masking probing technique.

6 Computing Infrastructure and Budget

All results were calculated on a local Linux server (Ubuntu 20.04.3 LTS) with 4 NVIDIA RTX 3090 GPUs, AMD Ryzen Threadripper 3970X 32-core CPU, and 128 GB DDR4. From these resources, we used 1 GPU. Replicating all experiments with guided masking, vision ablation, and comparison to BERT on all three datasets would take approximately 8 GPU days. Additionally, roughly 15 GPU days were spent on other experiments that are not reported in this paper.

7 Conclusion

While multimodal vision-language transformers have demonstrated impressive performance on downstream tasks, assessing their fine-grained understanding remains challenging due to the complexity of both the models and tasks involved.

This paper suggests using an alternative approach called guided masking for probing and evaluating specific aspects of multimodal transformers. Unlike traditional methods that rely on image-text matching, guided masking systematically ablates either vision or language modalities to evaluate the model's ability to predict masked tokens. This method offers several advantages: It eliminates the need for custom datasets with foiled captions and aligns more closely with standard pretraining masked language modeling (MLM). Additionally, the guided ablation of visual tokens sheds light on how vision and language models ground meaning, allowing direct comparisons with language-only models like BERT.

Our study applied this method to multimodal transformers that process ROI features from object detector on the vision side, evaluating their ability to understand verbs. However, guided masking is flexible and can be extended to probe other linguistic aspects such as subjects, objects or attributes. Furthermore, it can be adapted to multimodal transformers that use ViT patch features, such as ALBEF, VLMo, or X-VLM, by employing appropriate methods for visual ablation. Overall, guided masking is a versatile tool that can be applied to any model with masked language modeling as a pretraining objective.

The second essential contribution of this paper is a quantitative analysis of verb comprehension in four pre-trained VLMs evaluated on the SVO-Probes and V-COCO datasets. Our results indicate that these models correctly predicted verbs in over 75% cases on the SVO-Probes dataset and over 80% on V-COCO. These findings suggest that multimodal models possess a stronger verb understanding than previously recognized.

In conclusion, guided masking provides a more nuanced and accessible framework for probing multimodal transformers, paving the way for deeper insights into their linguistic and visual grounding capabilities.

Acknowledgments. This research was partially supported by *DisAI - Improving scientific excellence and creativity in combating disinformation with artificial intelligence and language technologies*, a project funded by Horizon Europe under GA No. 101079164 and by the *MIMEDIS*, a project funded by the Slovak Research and Development Agency under GA No. APVV-21-0114.

References

1. Aflalo, E., et al.: Vl-interpret: an interactive visualization tool for interpreting vision-language transformers. In: Proceedings of the IEEE/CVF Conference on Computer Vision and Pattern Recognition, pp. 21406–21415 (2022)
2. Bi, J., et al.: Vl-match: enhancing vision-language pretraining with token-level and instance-level matching. In: Proceedings of the IEEE/CVF International Conference on Computer Vision, pp. 2584–2593 (2023)
3. Bugliarello, E., Cotterell, R., Okazaki, N., Elliott, D.: Multimodal pretraining unmasked: a meta-analysis and a unified framework of vision-and-language BERTs. Trans. Assoc. Comput. Linguist. **9**, 978–994 (2021)
4. Cao, J., Gan, Z., Cheng, Yu., Yu, L., Chen, Y.-C., Liu, J.: Behind the scene: revealing the secrets of pre-trained vision-and-language models. In: Vedaldi, A., Bischof, H., Brox, T., Frahm, J.-M. (eds.) ECCV 2020. LNCS, vol. 12351, pp. 565–580. Springer, Cham (2020). https://doi.org/10.1007/978-3-030-58539-6_34
5. Chefer, H., Gur, S., Wolf, L.: Generic attention-model explainability for interpreting bi-modal and encoder-decoder transformers. In: Proceedings of the IEEE/CVF International Conference on Computer Vision, pp. 397–406 (2021)
6. Chen, P., Li, Q., Biaz, S., Bui, T., Nguyen, A.: gscorecam: What objects is clip looking at? In: Proceedings of the Asian Conference on Computer Vision, pp. 1959–1975 (2022)
7. Chen, Y.C., et a.: Uniter: learning universal image-text representations (2019)
8. Devlin, J., Chang, M.W., Lee, K., Toutanova, K.: BERT: pre-training of deep bidirectional transformers for language understanding. arXiv preprint arXiv:1810.04805 (2018)
9. Frank, S., Bugliarello, E., Elliott, D.: Vision-and-language or vision-for-language? On cross-modal influence in multimodal transformers. arXiv preprint arXiv:2109.04448 (2021)
10. Gupta, S., Malik, J.: Visual semantic role labeling. arXiv preprint arXiv:1505.04474 (2015)
11. Hendricks, L.A., Nematzadeh, A.: Probing image-language transformers for verb understanding. arXiv preprint arXiv:2106.09141 (2021)
12. Herzig, R., et al.: Incorporating structured representations into pretrained vision & language models using scene graphs. arXiv preprint arXiv:2305.06343 (2023)
13. Li, J., Li, D., Xiong, C., Hoi, S.: Blip: bootstrapping language-image pre-training for unified vision-language understanding and generation. In: International Conference on Machine Learning, pp. 12888–12900. PMLR (2022)
14. Li, J., Selvaraju, R., Gotmare, A., Joty, S., Xiong, C., Hoi, S.C.H.: Align before fuse: vision and language representation learning with momentum distillation. In: Advances in Neural Information Processing Systems, vol. 34, pp. 9694–9705 (2021)
15. Li, L.H., Yatskar, M., Yin, D., Hsieh, C.J., Chang, K.W.: VisualbERT: a simple and performant baseline for vision and language. arXiv preprint arXiv:1908.03557 (2019)
16. Lin, T.-Y., et al.: Microsoft COCO: common objects in context. In: Fleet, D., Pajdla, T., Schiele, B., Tuytelaars, T. (eds.) ECCV 2014. LNCS, vol. 8693, pp. 740–755. Springer, Cham (2014). https://doi.org/10.1007/978-3-319-10602-1_48
17. Liu, F., Emerson, G., Collier, N.: Visual spatial reasoning. arXiv preprint arXiv:2205.00363 (2022)
18. Liu, H., Li, C., Wu, Q., Lee, Y.J.: Visual instruction tuning (2023)

19. Liu, X., Yin, D., Feng, Y., Zhao, D.: Things not written in text: exploring spatial commonsense from visual signals. arXiv preprint arXiv:2203.08075 (2022)
20. Lu, J., Batra, D., Parikh, D., Lee, S.: VilBERT: pretraining task-agnostic visiolinguistic representations for vision-and-language tasks. In: Advances in Neural Information Processing Systems, vol. 32 (2019)
21. Parcalabescu, L., Gatt, A., Frank, A., Calixto, I.: Seeing past words: testing the cross-modal capabilities of pretrained v&l models on counting tasks. arXiv preprint arXiv:2012.12352 (2020)
22. Radford, A., et al.: Learning transferable visual models from natural language supervision. In: International Conference on Machine Learning, pp. 8748–8763. PMLR (2021)
23. Salin, E., Farah, B., Ayache, S., Favre, B.: Are vision-language transformers learning multimodal representations? A probing perspective. In: AAAI 2022 (2022)
24. Sharma, P., Ding, N., Goodman, S., Soricut, R.: Conceptual captions: a cleaned, hypernymed, image alt-text dataset for automatic image captioning. In: Proceedings of the 56th Annual Meeting of the Association for Computational Linguistics (Volume 1: Long Papers), pp. 2556–2565 (2018)
25. Shekhar, R., et al.: Foil it! Find one mismatch between image and language caption. arXiv preprint arXiv:1705.01359 (2017)
26. Singh, A., et al.: Flava: a foundational language and vision alignment model. In: Proceedings of the IEEE/CVF Conference on Computer Vision and Pattern Recognition, pp. 15638–15650 (2022)
27. Tan, H., Bansal, M.: Lxmert: learning cross-modality encoder representations from transformers. arXiv preprint arXiv:1908.07490 (2019)
28. Thrush, T., et al.: Winoground: probing vision and language models for visio-linguistic compositionality. In: Proceedings of the IEEE/CVF Conference on Computer Vision and Pattern Recognition, pp. 5238–5248 (2022)
29. Van Nguyen, M., Lai, V.D., Veyseh, A.P.B., Nguyen, T.H.: Trankit: a light-weight transformer-based toolkit for multilingual natural language processing. arXiv preprint arXiv:2101.03289 (2021)
30. Yang, Z., Kafle, K., Dernoncourt, F., Ordonez, V.: Improving visual grounding by encouraging consistent gradient-based explanations. In: Proceedings of the IEEE/CVF Conference on Computer Vision and Pattern Recognition, pp. 19165–19174 (2023)
31. Yarom, M., et al.: What you see is what you read? Improving text-image alignment evaluation. arXiv preprint arXiv:2305.10400 (2023)
32. Yuksekgonul, M., Bianchi, F., Kalluri, P., Jurafsky, D., Zou, J.: When and why vision-language models behave like bags-of-words, and what to do about it? arXiv e-prints, pp. arXiv–2210 (2022)
33. Zeng, Y., Zhang, X., Li, H.: Multi-grained vision language pre-training: aligning texts with visual concepts. arXiv preprint arXiv:2111.08276 (2021)

On the Complexity of Minimum Membership Dominating Set

D. Karthika[1(✉)], R. Muthucumaraswamy[1], Matthias Bentert[2], Sriram Bhyravarapu[3], Saket Saurabh[2,3], and Sanjay Seetharaman[3]

[1] Department of Mathematics, Sri Venkateswara College of Engineering, Sriperumbudur, Kanchipuram 602117, India
{2022pm0001,msamy}@svce.ac.in

[2] University of Bergen, Bergen, Norway
matthias.bentert@uib.no

[3] The Institute of Mathematical Sciences, HBNI, Chennai, India
{sriramb,saket,sanjays}@imsc.res.in

Abstract. In this paper, we study a variant of Domination called Minimum Membership Dominating Set, in short MMDS. The input to the problem is a graph G and an integer k (which is the membership parameter). The goal is to compute a set $S \subseteq V(G)$ such that for each $v \in V(G)$, $1 \leq |N[v] \cap S| \leq k$. Notice that there is no requirement on the size of S. We extend the study on this problem from the parameterized complexity perspective. The following are the results of this paper.

- Agrawal et al. (Algorithmica 2023) showed that MMDS is W[1]-hard parameterized by pathwidth of the input graph. They asked as open questions if MMDS is FPT when parameterized by maximum degree, distance to bounded degree graphs or maximum number of leaves in a spanning tree.
  We show that MMDS is NP-hard even on graphs with maximum degree three. This answers the first two questions in negative. We consider the parameter distance to disjoint paths that generalizes the maximum number of leaves in a spanning tree and show that the problem is FPT.
- Recently Sangam et al. (2024) showed that MMDS is FPT parameterized by the combined parameters distance to cluster and membership. However the running time of the algorithm is triple exponential. We design a single exponential FPT algorithm parameterized by distance to cluster alone. We also obtain an FPT algorithm parameterized by distance to co-cluster.
- We show that MMDS can be solved in time $k^{5\mathsf{cw}} n^{O(1)}$ where cw is the cliquewidth of the input graph. This implies a polynomial time algorithm for graphs of bounded cliquewidth. In particular distance hereditary graphs that have cliquewidth at most three, thereby resolving an open question asked by Sangam et al. in the above paper.

R. Královič and V. Kůrková (Eds.): SOFSEM 2025, LNCS 15538, pp. 94–107, 2025.
https://doi.org/10.1007/978-3-031-82670-2_8

1 Introduction

A set of vertices $D \subseteq V(G)$ in a graph $G = (V, E)$ is said to be a *dominating set*, if every vertex v in G is either in D or at least one neighbor of v is in D. Given a graph G and an integer k the DOMINATING SET problem asks if there exists a dominating set of size at most k. Agrawal et al. [1] studied the variant of DOMINATING SET called MINIMUM MEMBERSHIP DOMINATING SET (MMDS, in short). In MMDS, there is an additional constraint that every vertex has a bounded number (which is an integer that is part of the input) of closed neighbors that are in a dominating set. For a vertex $v \in V(G)$ and subset $S \subseteq V(G)$, the *membership* of v in S is defined as $|N[v] \cap S|$, denoted by $M(v, S)$. We also say that v is dominated $M(v, S)$ many times by S. Note that there is no constraint on the size of the dominating set.

MINIMUM MEMBERSHIP DOMINATING SET (MMDS)

Input: A graph $G = (V, E)$ and an integer k.
Question: Does there exist a dominating set S such that $\max_{v \in V(G)} M(v, S) \leq k$?

Kuhn et al. [2] introduce the "membership" variant for the SET COVER problem called MINIMUM MEMBERSHIP SET COVER (MMSC, in short). They described how the MINIMUM MEMBERSHIP SET COVER problem can be formulated as a linear program and provide a $\mathcal{O}(\ln n)$-approximation algorithm for this problem. Dom et al. [3] investigated how natural generalizations and variations of MMSC problem behave in terms of the consecutive ones property: a restriction in the input sets under which some set covering problems become polynomial-time solvable. They [3] established polynomial-time solvability, NP-completeness, and approximability results for various cases. A related problem that is "dual" to MMSC is the problem MINIMUM MEMBERSHIP HITTING SET (MMHS, in short) studied in [4,5].

Another variant of DOMINATING SET is the problem called PERFECT CODE. A subset C of $V(G)$ is said to be a *perfect code* in G if C is an independent set and every vertex in $V(G) \setminus C$ has exactly one neighbor in C. In the PERFECT CODE problem, the goal is to check if there exists a perfect code in G. The existence of a perfect code in graphs has been studied extensively, see for instance [6–11].

The generalized domination problem, also called $[\sigma, \rho]$-DOMINATING SET, is a problem that generalizes many other domination problems. Here, the problem is additionally specified by two sets of non-negative integers σ and ρ. The objective is to decide if there exists $D \subseteq V(G)$ such that for all $v \in D$, there exists $i \in \sigma$ such that $|N(v) \cap D| = i$ and for all $v \in V(G) \setminus D$, there exists $j \in \rho$ such that $|N(v) \cap D| = j$. The problem was introduced and studied by Telle [12,13]. Since MMDS is the special case where $\sigma = \{0, 1, \ldots, k-1\}$ and $\rho = \{1, 2, \ldots, k\}$, algorithms for $[\sigma, \rho]$-DOMINATING SET which depends on the values in σ and ρ, also apply to this problem. In particular, we have that given a graph G along with its tree decomposition of width tw and an integer k, we can solve MMDS in time $(2k+1)^{\mathsf{tw}} n^{\mathcal{O}(1)}$ [14,15].

Chellali et al. [16] introduced the problem called $[j,k]$-Dominating Set. A $[j,k]$-dominating set is a subset $S \subseteq V(G)$ such that for every vertex $v \in V(G) \setminus S$, $j \leq |N(v) \cap S| \leq k$ for non-negative integers j and k; that is, every vertex $v \in V(G) \setminus S$ is adjacent to at least j but not more than k vertices in S. They focused on small values j and k, and relate the concept of $[j,k]$-dominating set to a host of other concepts in domination theory, including perfect domination, efficient domination, nearly perfect sets, 2-packings, and k-dependent sets. For a graph G, a set $D \subseteq V(G)$ is called a $[1,j]$-dominating set if every vertex in $V(G) \setminus D$ has at least one and at most j neighbors in D. A set $D \subseteq V(G)$ is called a $[1,j]$-total dominating set if every vertex in $V(G)$ has at least one and at most j neighbors in D. In the $[1,j]$-(Total) Dominating Set, we are given a graph G and an integer k and the objective is to test whether there exists a $[1,j]$-(total) dominating set of size at most k. Recently, Meybodi et al. [17] studied the problems $[1,j]$-Dominating Set and $[1,j]$-Total Dominating Set from the perspective of parameterized complexity. Notice that these problems require a membership constraint only on the open neighborhood of vertices.

Recently Sangam et al. [18] proved that MMDS is NP-complete on bipartite graphs when the maximum degree of the graph is equal to $k+2, k \geq 5$. They investigated the parameterized complexity of the problem for the parameter twin cover and the combined parameters distance to cluster, membership. They proved that MMDS is FPT parameterized by combined parameters distance to cluster and membership by obtaining a triple exponential time algorithm. Further they obtained a linear-time algorithm for trees.

Our Results: We study the problem from the viewpoint of parameterized complexity. MMDS is W[1]-hard parameterized by pathwidth of the input graph [1]. Agrawal et al. [1] asked as open questions if MMDS is FPT when parameterized by maximum degree, distance to bounded degree graphs or maximum number of leaves in a spanning tree. We show that MMDS is NP-hard even on graphs with maximum degree three[1]. Thus the problem is para-NP-hard parameterized by maximum degree. This also implies that the problem is W[1]-hard parameterized by distance to bounded degree graphs. This result is presented in Sect. 3. We consider the parameter distance to disjoint paths that generalizes the maximum number of leaves in a spanning tree and show that the problem is FPT. This result is presented in Sect. 5.

We consider the graphs with bounded distance to cluster and bounded distance to co-cluster and obtain efficient algorithms for these parameters. We design a single exponential FPT algorithm parameterized by distance to cluster alone improving a triple exponential algorithm from [18]. We also give an FPT algorithm parameterized by distance to co-cluster. These results are presented in Sects. 6 and 7 respectively.

[1] We would like to note that the result in [11] subsumes our result. We were unaware of this at the time of submission and we thank the anonymous reviewers for bringing this to our notice.

The problem is known to be W[1]-hard parameterized by tree-width of the graph [1]. Since cliquewidth is a generalization of tree-width, the problem is W[1]-hard parameterized by cliquewidth. We design an FPT algorithm parameterized by the combined parameters cliquewidth and membership and this is presented in Sect. 4. As a corollary we obtain a polynomial time algorithm for the problem for distance hereditary graphs resolving an open question from [18].

2 Preliminaries

We assume that our graphs are simple and undirected. Given a graph $G = (V, E)$, n represents the number of vertices, and m represents the number of edges. For a subset $S \subseteq V$, by $G[S]$ we mean the subgraph of G induced by S. For every vertex $v \in V$, by $N(v)$ we mean open neighborhood of v and by $N[v]$ we mean closed neighborhood of v. We denote the set $\{1, 2, \ldots, r\}$ with the set $[r]$ and the set $\{0, 1, \ldots, r\}$ with the set $[0, r]$.

Given a graph $G = (V, E)$, a *distance to co-cluster modulator* is a set $D \subseteq V(G)$ such that $G - D$ is a partition of sets $I_1, I_2, \ldots, I_\ell$ such that for all $i \in [\ell]$, I_i is an independent set and for each pair of sets I_j and $I_{j'}$, where $j \neq j'$, each vertex in I_j is adjacent to every vertex in $I_{j'}$. A *disjoint union of paths modulator* is a set $D \subseteq V(G)$ such that $G - D$ is a collection of a vertex disjoint paths. A *cluster modulator* is a set $D \subseteq V(G)$ such that $G - D$ is a *cluster* graph, a disjoint union of cliques. Given a graph G and an integer t, the modulators for distance to disjoint union of paths [19] and distance to cluster [20] of size at most t can be found in $4^t n^{O(1)}$ and $1.9102^t n^{\mathcal{O}(1)}$ respectively. Thus we assume that the modulators for each of the above parameters are given as input in the respective sections. The proofs of the results marked ($\star$) are presented in the full version of the paper.

Definition 1 (Clique-width [21]). *Let $w \in \mathbb{N}$. A w-expression Φ defines a graph G_Φ where each vertex receives a label from $[w]$. The graph consisting of a solitary vertex v with label i has the w-expression $v(i)$. Graphs that contain two or more vertices are defined inductively using the three operations described below. Let $G_{\Phi'}$ and $G_{\Phi''}$ be graphs given by the w-expressions Φ' and Φ'' respectively.*

1. *Disjoint union: The graph G_Φ which is the disjoint union of $G_{\Phi'}$ and $G_{\Phi''}$ is given by the w-expression $\Phi = \Phi' \oplus \Phi''$.*
2. *Relabel: Let the graph G_Φ be $G_{\Phi'}$ where each vertex labeled i in $G_{\Phi'}$ is relabeled with the label j. The graph G_Φ is given by the w-expression $\Phi = \rho_{i \to j}(\Phi')$.*
3. *Join: Let the graph G_Φ be obtained from $G_{\Phi'}$ by adding edges between all the vertex pairs (u, v), where u has label i and v has label j. The graph G_Φ is given by the w-expression $\Phi = \eta_{i,j}(\Phi')$.*

The clique-width *of a graph G denoted by* cw(G) *is the minimum number w such that there is a w-expression Φ that defines G.*

3 para-NP-hard Parameterized by Maximum Degree

We show that MMDS is NP-hard even when the graph has maximum degree three and $k = 1$. We establish the hardness by a reduction from the variant of 3-SAT in which each clause is of length exactly 3 and each variable occurs exactly twice unnegated and twice negated. This variant was shown to be NP-complete by Darmann and Döcker in [22].

Theorem 1. MMDS *is* NP*-hard even on graphs with maximum degree three.*

Proof. Given a set S, we can verify if S is a membership dominating set of G in polynomial time. Hence MMDS is in NP. Let ϕ be a 3-CNF boolean formula with n variables $X = \{x_1, x_2, ...x_n\}$ and m clauses $\mathcal{C} = \{C_1, C_2, ..., C_m\}$ such that each clause is of length exactly three, and each variable appears twice unnegated and twice negated. We would like to note that the reduction is an extension of the reduction given by Cull and Nelson in [23] that established the hardness of PERFECT CODE. We construct the graph G_ϕ as follows. For each variable x_i, $1 \leq i \leq n$, we have a variable gadget H_i and for each clause C_j, $1 \leq j \leq m$, we have a clause gadget R_j. These gadgets are illustrated in Fig. 1.

The set of vertices in the graph is given by $V(G_\phi) = V(H_i) \cup V(R_j), 1 \leq i \leq n$ and $1 \leq j \leq m$. The set of edges in the graph is defined as follows:

- For each i, if x_i appears in the clause C_j then we add the edge $x_i\widehat{x_i}$ where $\widehat{x_i} \in V(R_j)$.
- For each i, if $\bar{x}_i$ appears in the clause C_j then we add the edge $\bar{x}_i\widehat{\bar{x}_i}$ where $\widehat{\bar{x}_i} \in V(R_j)$.

The vertices in the gadgets H_i and R_j for all i, j have degree at most three. Since each literal appears in exactly 2 clauses, it is easy to see that the maximum degree of G_ϕ is three. Also, the construction of G_ϕ can be done in polynomial time. In the following lemma, we show that there are exactly two ways in which an MMDS intersects with a variable gadget.

Lemma 1 ($\star$). *If S is a membership dominating set with $k = 1$ then in each H_i either $x_i \in S$ or $\bar{x}_i \in S$ but not both.*

Lemma 2 ($\star$). *Consider a gadget R_j corresponding to the clause $C_j = (x_i \vee \bar{x}_j \vee x_k)$. Let S be a membership dominating set of G_ϕ with $k = 1$. If $x_i \notin S$ (resp. $\bar{x}_j$, x_k) then $\widehat{x}_i \notin S$ (resp. $\widehat{\bar{x}}_j \notin S$, $\widehat{x}_k \notin S$).*

We show that ϕ is satisfiable iff $(G_\phi, 1)$ is a YES-instance of MMDS.

Lemma 3. *ϕ is satisfiable if and only if $(G_\phi, 1)$ is a* YES*-instance of* MMDS.

Proof. Let $\{a_1, a_2, \ldots, a_n\}$ be a satisfying assignment of ϕ where a_i corresponds to the truth value of the variable x_i. We construct a set $S \subseteq V(G_\phi)$ as follows. For each $i \in [n]$ if a_i is true then include $x_i \in S$, otherwise include $\bar{x}_i \in S$. For the former case (resp. latter case), we add the red colored vertices in Fig. 2 on the left (resp. right) to S. Consider a clause C_q that contains the literals say

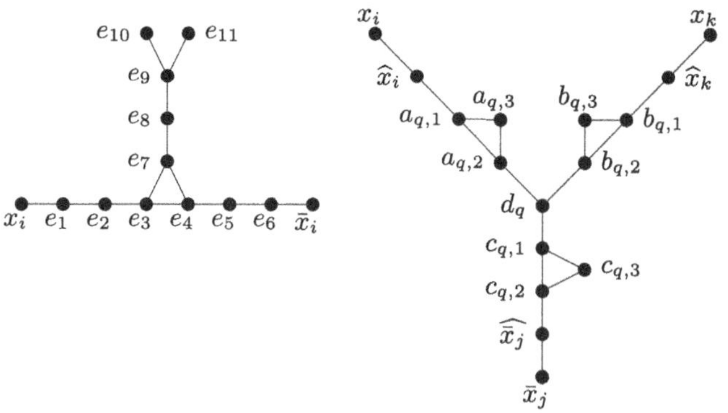

Fig. 1. Illustration of the variable gadget H_i (in the left) for the variable x_i, $i \in [n]$ and clause gadget R_q (on the right) for the clause $C_q = (x_i \vee \bar{x}_j \vee x_k)$, $q \in [m]$. The vertices x_i, $\bar{x}_j$ and x_k are not part of the clause gadget R_q.

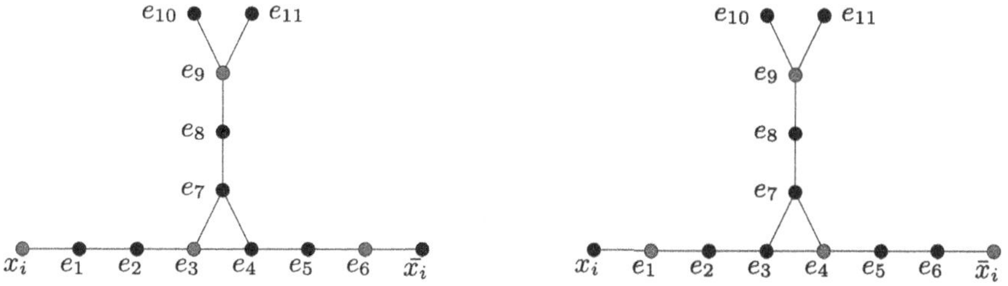

Fig. 2. The figure on the left (resp. right) illustrates the solution vertices (colored red) when a_i is true (resp. a_i is false) in the satisfying assignment. (Color figure online)

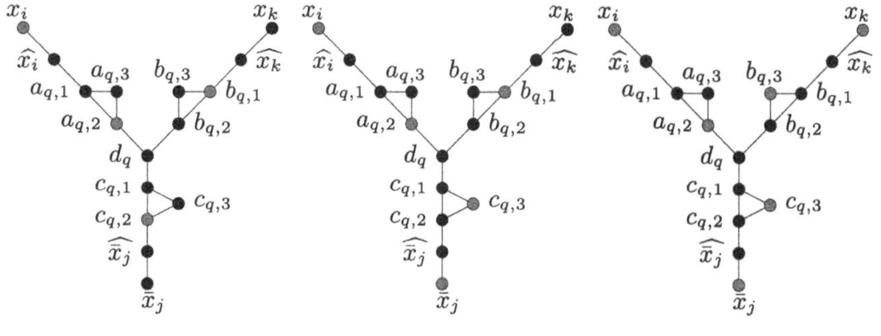

Fig. 3. Illustration of the solution vertices (colored in red) from the clause gadget R_q corresponding to clause $C_q = (x_i \vee \bar{x}_j \vee x_k)$ for three cases: exactly one of the literal is set to true say x_i, two of the literals are set to true say x_i, $\bar{x}_j$ and all three of them set to true respectively. (Color figure online)

$x_i, \bar{x}_j, x_k$. Since there is a satisfying assignment for ϕ, at least one of the literals in each clause is true. Depending on the literals that are set to true in C_q, we add the red colored vertices from R_q into S (that are uniquely determined based on the true literals). A few cases are illustrated in Fig. 3. It is easy to see that each $v \in V(G_\phi)$ we have $|N[v] \cap S| = 1$ and thus S is a membership dominating set with $k = 1$.

Conversely, if G_ϕ has a dominating set S with membership value $k = 1$ then we show that there is a satisfying assignment for ϕ. From Lemma 1, it is the case that either $x_i \in S$ or $\bar{x}_i \in S$ but not both. For every variable x_i in the variable gadget H_i, if $x_i \in S$ then we set a_i to be true. Otherwise we know that $\bar{x}_i \in S$ and thus we set a_i as false. We show that $\{a_1, a_2, ...a_n\}$ is a satisfying assignment of ϕ.

Consider a clause $C_q = (x_i \vee \bar{x}_j \vee x_k)$. Suppose for a contradiction that C_q is not satisfied. Then from Lemma 2, we have that $\widehat{x}_i$, $\widehat{\bar{x}}_j$ and $\widehat{x}_k$ are not in S. Thus to dominate these vertices we need $a_{q,1}$, $b_{q,1}$ and $c_{q,2}$ to be in S. This further implies that the vertices $a_{q,2}, a_{q,3}, b_{q,3}, b_{q,2}, c_{q,3}$ and $c_{q,1}$ are not in S. To dominate d_q we need $d_q \in S$ which contradicts the membership for the vertices $a_{q,2}$, $b_{q,2}$ and $c_{q,1}$. Thus C_q is satisfied. □

This completes the proof of Theorem 1. □

As a consequence of Theorem 1 we obtain the following corollary.

Corollary 1. MMDS *is* para-NP-*hard parameterized by maximum degree.*

4 Cliquewidth

In this section, we show that MMDS is fixed parameter tractable when parameterized by the combined parameters cliquewidth and membership.

Theorem 2. *Given a* cw*-expression of the input graph G and an integer k,* MMDS *can be solved in $2^{2\mathsf{cw}}(k+2)^{4\mathsf{cw}}n^{\mathcal{O}(1)}$, where* cw *is the cliquewidth of G.*

Let $G = (V, E)$ be the input graph and $\mathsf{cw} \in \mathbb{N}$ be an integer. It is known [24] that one can compute a $(2^{3\mathsf{cw}+2} - 1)$-expression (also known as "clique expression") of G in time $f(\mathsf{cw}) \cdot n^{\mathcal{O}(1)}$, if G has cliquewidth at most cw. Thus we can assume that we are given with a w-expression of G. We also assume that the w-expression is *irredundant* where no edge of G is introduced twice in the expression. Given a w-expression of G, one can find a irredundant w-expression of G in polynomial time [21]. Thus the input instance is (G, w, k).

We first provide an overview of the algorithm. Let Ψ be an irredundant w-expression of G and k be an integer that indicates the membership value. At any sub-expression Φ of Ψ, we maintain three pieces of information at each label $q \in [w]$: n_q denoting the number of dominating set vertices assigned the label q, m_q denoting the maximum of the membership values over all the vertices assigned the label q, and f_q denoting the existence of a non-dominated vertex assigned the label q. The value n_q helps when we perform a join operation (say $\eta_{q,p}$) to update the membership values of the vertices assigned the label p. Therefore we observe that it is sufficient to limit the value of n_q to $k+1$. The value m_q on the other hand is limited to at most k, otherwise we say that the current subexpression Φ is a NO-instance of (G, w, k). We then use a dynamic programming based algorithm at each subexpression of Φ.

Proof of Theorem 2. Let (G, w, k) be the input instance, Ψ be an irredundant w-expression of G and $k \in \mathbb{N}$. We now describe the dynamic programming based algorithm over the w-expression of G. For each subexpression Φ of Ψ and a subset $S \subseteq V(G_\Phi)$, we have a boolean table entry $T[\Phi; \mathcal{N}, \mathcal{M}, \mathcal{F}]$ where

$$\mathcal{F} = (f_1, f_2, \dots f_w) \in \{0,1\}^w,$$

$$\mathcal{N} = (n_1, n_2, \dots n_w) \in \{0, 1, \dots, k+1\}^w, \text{ and}$$

$$\mathcal{M} = (m_1, m_2, \dots, m_w) \in \{0, 1, \dots, k+1\}^w.$$

We now explain the variables. The variables f_q represents the existence of a vertex v assigned the label q such that $N[v] \cap S = \emptyset$. That is, $f_q = 1$ if $N[v] \cap S = \emptyset$, and 0 otherwise. Let $\widehat{n}_q$, where $q \in [w]$, represent the number of vertices assigned the label $q \in [w]$ belonging to the set S. That is, $\widehat{n}_q = |\{v \in S \mid v \text{ is assigned the label } q\}|$. The variable $n_q = \min\{\widehat{n}_q, k+1\}$. Similarly let $\widehat{m}_q$, where $q \in [w]$, represent the maximum of the memberships of all vertices assigned the label $q \in [w]$. That is, $\widehat{m}_q = \max\{|N[v] \cap S| \mid v \text{ is assigned the label } q\}$. The variable $m_q = \min\{\widehat{m}_q, k+1\}$. We say a set S *satisfies* the variables $\mathcal{N}$, $\mathcal{M}$ and $\mathcal{F}$ if it achieves the variable values n_q, m_q and f_q, for each $q \in [w]$. The number of entries at a subexpression Φ is at most $2^w(k+2)^{2w}$.

Definition of the Table Entry. For each subexpression Φ of Ψ, we set the entry $T[\Phi; \mathcal{N}, \mathcal{M}, \mathcal{F}]$ to be TRUE if and only if there exists a set $S \subseteq V(G_\Phi)$ that satisfies $\mathcal{N}$, $\mathcal{M}$ and $\mathcal{F}$. Otherwise, we set the entry to FALSE. We now give the computation of $T[\Phi; \mathcal{N}, \mathcal{M}, \mathcal{F}]$ at each type of operation.

Introduce Node: $\Phi = v(i)$
We set the two entries to be TRUE corresponding to whether $v \in S$ or $v \notin S$. For the former case, we set the entry to be TRUE where $n_i = 1$, $m_i = 1$, $n_j = 0$ and $m_j = 0$, for all $j \in [w] \setminus \{i\}$ and $f_q = 0$ for all $q \in [w]$. For the latter case, we set the entry to be TRUE where $n_q = 0$ and $m_q = 0$ for all $q \in [w]$. Also $f_i = 1$ and $f_q = 0$ for all $q \in [w] \setminus \{i\}$. All other entries are set to FALSE.

Disjoint union: $\Phi = \Phi' \oplus \Phi''$
The graph G_Φ results in a disjoint union of two graphs $G_{\Phi'}$ and $G_{\Phi''}$. We set the entry $T[\Phi; \mathcal{N}, \mathcal{M}, \mathcal{F}]$ to be TRUE if and only if there exists two entries $T[\Phi'; \mathcal{N}', \mathcal{M}', \mathcal{F}']$ and $T[\Phi''; \mathcal{N}'', \mathcal{M}'', \mathcal{F}'']$ such that both of them are TRUE and, $n_q = \min\{n'_q + n''_q, k+1\}$ and $m_q = \max\{m'_q, m''_q\}$ for all $q \in [w]$. Also $f_q = \max\{f'_q, f''_q\}$ for all $q \in [w]$.
For each TRUE entry at Φ', we go over each TRUE entry at Φ'' and set the corresponding entry at Φ to TRUE. All the other entries are set to FALSE. Thus to compute the values of the entries at Φ, we need to process $2^{2w}(k+2)^{4w}$ pairs of entries and the above conditions can be checked in $\mathcal{O}(w)$ times.

Relabel node: $\Phi = \rho_{i \to j}(\Phi')$
Notice that the graph G_Φ is obtained as a resultant of changing the label of each vertex of label i to j. We set the entry $T[\Phi; \mathcal{N}, \mathcal{M}, \mathcal{F}]$ to be TRUE if and

only if there exists an entry $T[\Phi';\mathcal{N}',\mathcal{M}',\mathcal{F}']$ that is TRUE and, $n_q = n'_q$ and $m_q = m'_q$ for all $q \in [w] \setminus \{i,j\}$, $n_i = 0$, $m_i = 0$, $n_j = \min\{n'_j + n'_i, k+1\}$ and $m_q = \max\{m'_i, m'_j\}$. Also $f_i = 0$, $f_j = \max\{f'_i, f'_j\}$ and $f_q = f'_q$ for all $q \in [w] \setminus \{i,j\}$.
We go over all entries at Φ' that are TRUE and set the corresponding entry in Φ to be TRUE. All the other entries are set to FALSE. Thus to compute the values of the entries at Φ, we need to process $2^w(k+2)^{2w}$ entries and the above conditions can be checked in $\mathcal{O}(w)$ times.

Join node: $\Phi = \eta_{i,j}(\Phi')$
The graph G_ϕ is obtained by adding edges between each vertex of label i and each vertex of label j in $G_{\Phi'}$. We set the entry $T[\Phi;\mathcal{N},\mathcal{M},\mathcal{F}]$ to be TRUE if and only if there exists an entry $T[\Phi';\mathcal{N}',\mathcal{M}',\mathcal{F}']$ that is TRUE and the following conditions hold:
- $n_q = n'_q$ for all $q \in [w]$,
- $f_q = f'_q$ and $m_q = m'_q$ for all $q \in [w] \setminus \{i,j\}$,
- $m_i = \min\{m'_i + n'_j, k+1\}$ and $m_j = \min\{m'_j + n'_i, k+1\}$,
- $f_i = 1$ if $f'_i = 1$ and $n'_j = 0$, and $f_i = 0$ otherwise. Similarly, $f_j = 1$ if $f'_j = 1$ and $n'_i = 0$, and $f_j = 0$ otherwise.

We go over all entries at Φ' that are TRUE and set the corresponding entry in Φ to be TRUE. All the other entries are set to FALSE. Thus to compute the values of the entries at Φ, we need to process $2^w(k+2)^{2w}$ entries and the above conditions can be checked in $\mathcal{O}(w)$ times.

To decide if there exists a membership dominating set of size at most k we check the entries at Ψ. If there exists an entry $T[\Psi;\mathcal{N},\mathcal{M},\mathcal{F}]$ such that it is TRUE and

- $f_q = 0$ for all $q \in [w]$ indicating that every vertex in G is dominated at least once, and
- $m_q \leq k$ for all $q \in [w]$ indicating that the membership value of each vertex is at most k,

then we decide that there exists a membership dominating set of size at most k. The correctness of the above algorithm follows from the description of the algorithm. Since the number of subexpressions is $n^{\mathcal{O}(1)}$, the running time of the algorithm is $2^{2w}(k+2)^{4w}n^{\mathcal{O}(1)}$. □

Since distance hereditary graphs have cliquewidth at most three and finding a 3-expression of a distance hereditary graph can be done in linear time [25], from Theorem 2, we have the following result.

Corollary 2. MMDS *is polynomial time solvable on distance hereditary graphs.*

5 Distance to Disjoint Paths

In this section, we show that MMDS is FPT parameterized by distance to disjoint paths. We first show that the interesting case is when the membership of the graph is bounded in terms of the modulator size.

Lemma 4. *Let D be a disjoint union of paths modulator of G. If $k \geq |D| + 3$, then (G, k) is a* YES*-instance of* MMDS.

Proof. We construct a membership dominating set $S \subseteq V(G)$ such that for each vertex $v \in V(G)$, we have $|N[v] \cap S| \leq |D| + 3$. Initially we set $S = D$. For each vertex $w \in G - D$ such that $N(w) \cap D = \emptyset$, we add the vertex w to S. We now show that S is our desired set. From the above construction, for each $v \in D$ none of its neighbors in $V(G) \setminus D$ belong to the set S. That is, $N(v) \cap S \cap (V(G) \setminus D) = \emptyset$. Therefore $|N[v] \cap S| \leq |D|$.

Since $G - D$ is a disjoint union of paths, the degree of each vertex in $G - D$ is at most two. A vertex v from $V(G) \setminus D$ is added to S if it does not have a neighbor in D. Therefore $|N[v] \cap S| \leq 3$. For the remaining vertices $v \in V(G) \setminus S$, at most two of its neighbors outside D belong to S and thus $|N[v] \cap S| \leq |D| + 2$. The upper bound of $|D| + 3$ holds even when G itself is a disjoint union of paths, in which case $|D| = 0$. □

Theorem 3. MMDS *is* FPT *parameterized by distance to disjoint union of paths.*

Proof. If $k \geq |D| + 3$, then G has a membership dominating set from Lemma 4. Thus, the interesting case is when $k < |D| + 3$. Given a graph G and a tree-decomposition of G of width $\mathcal{O}(\mathsf{tw})$, there is an algorithm [14] for MMDS running in time $(2k + 1)^{\mathsf{tw}} n^{\mathcal{O}(1)}$. Note that a tree-decomposition of width $\mathcal{O}(\mathsf{tw})$ can be computed in time $\mathsf{FPT}(\mathsf{tw})$ [26]. Since $\mathsf{tw}(G) \leq |D| + 1$ and $k < |D| + 3$, we have the desired result. □

6 Distance to Cluster

In this section, we show that MMDS is FPT parameterized by distance to cluster. We first show that the interesting case is when the membership of the graph is bounded in terms of the modulator size.

Lemma 5 (⋆). *Let $A \subseteq V(G)$ be a cluster modulator of G. If $k \geq |A| + 1$, then (G, k) is a* YES*-instance of* MMDS.

Theorem 4. MMDS *can be solved in time $2^{\mathcal{O}(d \log d)} n^{\mathcal{O}(1)}$, where d is the size a cluster modulator of the input graph.*

Proof. Let $A = \{w_1, \ldots, w_d\}$ be a cluster modulator. Let $\{C_1, C_2, \ldots, C_\ell\}$ be the set of cliques in $G - A$ and $|C_j| = n_j$, for all $j \in [\ell]$. We order the vertices of $G - A$ as $v_{1,1}, \ldots, v_{1,n_1}, \ldots, v_{\ell,1}, \ldots, v_{\ell,n_\ell}$, where $v_{i,j}$ represents the jth vertex in the clique C_i. Let $S \subseteq V(G)$ be a membership dominating set that corresponds to the solution. We guess the set of vertices $I \subseteq A$ that belong to S. That is, we guess $I = A \cap S$. We use a dynamic programming based algorithm over the vertices of the cliques in the above order.

From Lemma 5, we have that the membership of G is at most $|A|$. We now define the table entries. For each $(x_1, \ldots, x_d) \in [0, k]^d$, $q \in [\ell]$, $j \in [0, n_q]$ and $s \in [0, k]$, we define an entry $T[q, j, (x_1, \ldots, x_d), s]$ that computes the minimum size of a set $S' \subseteq \{v_{1,1}, \ldots, v_{q,j}\}$ such that

- each vertex w_i is dominated exactly x_i times in $S' \cup I$. i.e., $|N[w_i] \cap (S' \cup I)| = x_i$,
- each vertex v in $C_1, C_2, ..., C_{q-1}$ (where $q > 1$) is dominated at least once and at most k times, i.e., $1 \leq |N[v] \cap (S' \cup I)| \leq k$, and
- exactly s vertices are picked in clique C_q, i.e., $|S' \cap C_q| = |S' \cap \{v_{q,1}, \ldots, v_{q,j}\}| = s$.

For the base case we set,

$$T[1, 0, (x_1, \ldots, x_d), s] = \begin{cases} 0, & \text{if} s = 0, x_i = |N[w_i] \cap I| \text{ for each } i, \\ \infty & \text{otherwise.} \end{cases}$$

Lemma 6. *The recurrence defined for $T[1, 0, (x_1, \ldots, x_d), s]$ is correct.*

The proof of the above lemma is trivial as there are no vertices to be picked into our solution.

Recursive Step: If $s = 0$, then we set $T[q, j, (x_1, \ldots, x_d), 0] = T[q, j-1, (x_1, \ldots, x_d), 0]$. When $s > 0$, we compute as follows. We first discuss the recursive step when $j > 0$. For each $i \in [d]$, let $x'_i = x_i - 1$ if there is an edge between $v_{q,j}$ and w_i, otherwise let $x'_i = x_i$. Then, we have

$$T[q, j, (x_1, \ldots, x_d), s] = \min\{T[q, j-1, (x'_1, \ldots, x'_d), s-1] + 1, \\ T[q, j-1, (x_1, \ldots, x_d), s]\}.$$

Lemma 7 (⋆). *The recurrence defined for $T[q, j, (x_1, \ldots, x_d), s]$ is correct.*

Next, when $j = 0$ and $s \neq 0$ we have $T[q, j, (x_1, \ldots, x_d), s] = \infty$. Let $r = k - \max_{j \in [n_{q-1}]} |N[v_{q-1,j}] \cap I|$. If all vertices in C_{q-1} are adjacent to I, then we set

$$T[q, 0, (x_1, \ldots, x_d), 0] = \min_{s' \in [0, k - \max_{j \in [n_{q-1}]} |N[v_{q-1,j}] \cap I|]} T[q-1, n_{q-1}, (x_1, \ldots, x_d), s']$$

$$T[q, 0, (x_1, \ldots, x_d), 0] = \min_{s' \in [0,r]} T[q-1, n_{q-1}, (x_1, \ldots, x_d), s'].$$

Otherwise, there is a vertex in clique C_{q-1} that is not adjacent to I. Then we set

$$T[q, 0, (x_1, \ldots, x_d), 0] = \min_{s' \in [k - \max_{j \in [n_{q-1}]} |N[v_{q-1,j}] \cap I|]} T[q-1, n_{q-1}, (x_1, \ldots, x_d), s']$$

$$T[q, 0, (x_1, \ldots, x_d), 0] = \min_{s' \in [r]} T[q-1, n_{q-1}, (x_1, \ldots, x_d), s'].$$

This ensures that all vertices in the clique C_{q-1} are dominated at least once and at most k times by the corresponding solution.

Solution: For a set I of solution vertices from A, we ask if there exists a set S from $G - A$ such that each vertex $v \in V(G)$ has $1 \leq |N[v] \cap (S \cup I)| \leq k$. Towards this, we ask if there exists an entry $T[\ell, n_\ell, (x_1, \ldots, x_d), s]$ such that the value is not ∞ and the following conditions hold:

- the tuple $(x_1, \ldots, x_d)$ belongs to $[k]^d$,
- $s \leq k - \max_{j \in n_\ell}(|N[v_{\ell,j}] \cap I|)$,
- $s \geq 1$ if there is a vertex $v_{\ell,j}$ that is not adjacent to any vertex in I.

If there exists no such entry then we look for another set $I' \subseteq A$ such that $I' = A \cap S$. If for each of the sets, we do not obtain such an entry we say that the membership of the graph G is greater than k and (G, k) is a NO-instance.

Running Time. We consider all the subsets I of A and run the above procedure. Since there are $(k+1)^d n$ table entries and each entry can be computed in polynomial time, the total running time is $(k+1)^d 2^d n^{\mathcal{O}(1)}$. By Lemma 5, the total running time is $2^{\mathcal{O}(d \log d)} n^{\mathcal{O}(1)}$. □

7 Distance to Co-Cluster

In this section, we show that MMDS is FPT parameterized by distance to co-cluster. We first show that the interesting case is when the membership of the graph is bounded in terms of the modulator size.

Lemma 8. *Let $D \subseteq V(G)$ be a distance to co-cluster modulator of G. If $k \geq |D| + 2$, then (G, k) is a* YES-*instance of* MMDS.

Proof. Let $I_1, I_2, \ldots, I_\ell$ be the maximal independent sets in $G - D$. If $\ell = 1$ then D is our membership dominating set. Indeed, since the graph is connected, each vertex in I_1 is dominated by a neighbor in D and thus the claim holds. Suppose that $\ell \geq 2$. We construct a membership dominating set $S \subseteq V(G)$ such that for each vertex $v \in V(G)$, we have $|N[v] \cap S| \leq |D| + 2$. Initially we set $S = D$. We arbitrarily pick two vertices say $v \in I_j$ and $w \in I_{j'}$ for some $j \neq j'$, and add them to S. Since $G - D$ is a co-cluster graph and the vertices v and w are from different sets, each vertex of $G - D$ is dominated. Since $|S| \leq |D| + 2$, the claim holds. □

Theorem 5. MMDS *is* FPT *when parameterized by distance to co-cluster.*

Proof. (Proof Sketch). Let $G = (V, E)$ be a graph and $D \subseteq V(G)$ of size t be such that $G - D$ is a co-cluster graph. Let $I_1, I_2, \ldots, I_\ell$ be the independent sets in $G - D$ such that each vertex of I_i is adjacent to every vertex of I_j for all $i \neq j$. Let S be a hypothetical solution.

- Let the set of vertices $D' \subseteq D$ be such that $D' = S \cap D$. Notice that the vertices in $N[D']$ are dominated. Let $Z = D \setminus N[D']$.

- Let $g : Z \to 2^D$ be the function that guesses the "type" of a vertex $w \in V(G) \setminus D$ that dominates $v \in Z$. More precisely, for each $A \subseteq D$, the type $T_A = \{v \in V(G) \setminus D \mid N(v) \cap D = A\}$. The type represents the equivalence class or the neighborhood of v with respect to D.
- We guess whether the vertices from $S \cap (V(G) \setminus D)$ come from the same independent set or at least two independent sets.
- For the former case, guess the independent set I_i and we arbitrarily pick a vertex from each type in I_i given by the function g. Notice that every other vertex in I_j is dominated, where $i \neq j$. For each vertex $w \in I_i$ that is not dominated, we pick w to the set S.
- For the latter case, there exists at least two sets I_i and I_j, where $i \neq j$, such that $I_i \cap S \neq \emptyset$ and $I_j \cap S \neq \emptyset$. Similar to the algorithm in Theorem 4, we go over the vertices in the independent sets and store the number of vertices required tô satisfy the memberships of vertices of D and the memberships of the processed vertices in $V(G) \setminus D$. □

References

1. Agrawal, A., Choudhary, P., Narayanaswamy, N.S., Nisha, K.K., Ramamoorthi, V.: Parameterized complexity of minimum membership dominating set. Algorithmica **85**(11), 3430–3452 (2023)
2. Kuhn, F., von Rickenbach, P., Wattenhofer, R., Welzl, E., Zollinger, A.: Interference in cellular networks: the minimum membership set cover problem. In: Wang, L. (ed.) COCOON 2005. LNCS, vol. 3595, pp. 188–198. Springer, Heidelberg (2005). https://doi.org/10.1007/11533719_21
3. Michael Dom, Jiong Guo, Rolf Niedermeier, and Sebastian Wernicke. Minimum membership set covering and the consecutive ones property. In *Algorithm Theory - SWAT 2006, 10th ScandinavianWorkshop on Algorithm Theory, Proceedings*, volume 4059 of *Lecture Notes in Computer Science*, pages 339–350. Springer, 2006
4. Mitchell, J.S.B., Pandit, S.: Minimum membership covering and hitting. Theor. Comput. Sci. **876**, 1–11 (2021)
5. Narayanaswamy, N.S., Dhannya, S.M., Ramya, C.: Minimum membership hitting sets of axis parallel segments. In: Wang, L., Zhu, D. (eds.) COCOON 2018. LNCS, vol. 10976, pp. 638–649. Springer, Cham (2018). https://doi.org/10.1007/978-3-319-94776-1_53
6. Biggs, N.: Perfect codes in graphs. J. Comb. Theory Ser. B **15**(3), 289–296 (1973)
7. Kratochvíl, J.: Perfect codes over graphs. J. Comb. Theory Ser. B **40**(2), 224–228 (1986)
8. Huang, H., Xia, B., Zhou, S.: Perfect codes in Cayley graphs. SIAM J. Discret. Math. **32**(1), 548–559 (2018)
9. Kratochvíl, J.: Perfect codes in graphs and their powers. Ph.D. thesis, Ph.D. dissertation (in Czech), Charles University, Prague (1987)
10. Cesati, M.: Perfect code is W[1]-complete. Inf. Process. Lett. **81**(3), 163–168 (2002)
11. Fellows, M.R., Hoover, M.N.: Perfect domination. Australas. J. Comb. **3**, 141–150 (1991)
12. Telle, J.A.: Complexity of domination-type problems in graphs. Nord. J. Comput. **1**(1), 157–171 (1994)

13. Telle, J.A.: Vertex partitioning problems: characterization, complexity and algorithms on partial k-trees. Ph.D. thesis, University of Oregon (1994)
14. Rooij, J.M.M.: Fast algorithms for join operations on tree decompositions. In: Fomin, F.V., Kratsch, S., van Leeuwen, E.J. (eds.) Treewidth, Kernels, and Algorithms. LNCS, vol. 12160, pp. 262–297. Springer, Cham (2020). https://doi.org/10.1007/978-3-030-42071-0_18
15. Focke, J., et al.: Tight complexity bounds for counting generalized dominating sets in bounded-treewidth graphs. In: Proceedings of the 2023 ACM-SIAM Symposium on Discrete Algorithms, SODA 2023, pp. 3664–3683. SIAM (2023)
16. Chellali, M., Haynes, T.W., Hedetniemi, S.T., McRae, A.A.: [1, 2]-sets in graphs. Discret. Appl. Math. **161**(18), 2885–2893 (2013)
17. Meybodi, M.A., Fomin, F.V., Mouawad, A.E., Panolan, F.: On the parameterized complexity of $[1, j]$-domination problems. Theor. Comput. Sci. **804**, 207–218 (2020)
18. Reddy, S.B., Kare, A.S.: Algorithms for minimum membership dominating set problem (2024). https://arxiv.org/abs/2408.00797
19. Kucera, M., Suchý, O.: Minimum eccentricity shortest path problem with respect to structural parameters. Algorithmica **85**(3), 762–782 (2023)
20. Boral, A., Cygan, M., Kociumaka, T., Pilipczuk, M.: A fast branching algorithm for cluster vertex deletion. Theory Comput. Syst. **58**(2), 357–376 (2016)
21. Courcelle, B., Olariu, S.: Upper bounds to the clique width of graphs. Discret. Appl. Math. **101**(1–3), 77–114 (2000)
22. Darmann, A., Döcker, J.: On simplified NP-complete variants of monotone3-sat. Discret. Appl. Math. **292**, 45–58 (2021)
23. Cull, P., Nelson, I.: Error-correcting codes on the towers of Hanoi graphs. Discret. Math. **208–209**, 157–175 (1999)
24. Oum, S., Seymour, P.: Approximating clique-width and branch-width. J. Comb. Theory Ser. B **96**(4), 514–528 (2006)
25. Golumbic, M.C., Rotics, U.: On the clique-width of some perfect graph classes. Int. J. Found. Comput. Sci. **11**(3), 423–443 (2000)
26. Cygan, M., et al.: Parameterized Algorithms. Springer, Cham (2015). https://doi.org/10.1007/978-3-319-21275-3

On the Structural Parameterized Complexity of Defective Coloring

Sriram Bhyravarapu[1(✉)], Pankaj Kumar[2], and Saket Saurabh[1,3]

[1] Institute of Mathematical Science, HBNI, Chennai, India
{sriramb,saket}@imsc.res.in

[2] School of Computer Science, University of Birmingham, Birmingham, UK

[3] University of Bergen, Bergen, Norway

Abstract. In this paper, we consider the problem DEFECTIVE COLORING. Given a graph G and two positive integers k and Δ^*, the objective is to determine whether it is possible to obtain a coloring (not necessarily proper) of the vertices of G using at most k colors such that each vertex in a color class c has at most Δ^* neighbors in the same color class. DEFECTIVE COLORING is a generalization of GRAPH COLORING with $\Delta^* = 0$. The optimization variant of this problem, which aims to find the minimum number of colors k, is known to be NP-hard even for split graphs and cographs.

Belmonte, Lampis, and Mitsou (SIDMA 2020) showed that DEFECTIVE COLORING is W[1]-hard when parameterized by tree-width, path-width, tree-depth, or feedback vertex set. The problem is W[1]-hard parameterized by modular-width or clique-width as DEFECTIVE COLORING is NP-hard on cographs. They asked as an open question whether DEFECTIVE COLORING is fixed-parameter tractable (FPT) when parameterized by modular-width, clique-width or neighborhood diversity combined with either k or Δ^*. In an effort to address the question concerning modular-width, this study investigates the parameters neighborhood diversity and twin-cover, which are special cases of modular-width. We show that DEFECTIVE COLORING is FPT when parameterized by twin-cover, distance to disjoint paths, or the combined parameters neighborhood diversity and k. The latter result implies an FPT algorithm for complete-d-partite graphs, a subclass of cographs, parameterized by d. This provides a partial response to an open question raised in the above paper. We present an algorithm for graphs with bounded distance to d-degree and as a corollary we obtain an FPTalgorithm parameterized by distance to disjoint paths. Furthermore, the study also presents a 1-additive approximation algorithm for split graphs.

Keywords: defective coloring · parameterized complexity · twin-cover · neighborhood diversity · split graphs · bounded degree graphs

R. Královič and V. Kůrková (Eds.): SOFSEM 2025, LNCS 15538, pp. 108–121, 2025.
https://doi.org/10.1007/978-3-031-82670-2_9

1 Introduction

We study the problem known as DEFECTIVE COLORING. Given a graph G and two positive integers k and Δ^*, we say that G has a *defective coloring* if there exists a coloring of $V(G)$ using at most k colors such that each color class induces a graph with maximum degree at most Δ^*. The problem DEFECTIVE COLORING takes as input a graph G and two integers k and Δ^*, and the task is to decide whether there exists a defective coloring of G. DEFECTIVE COLORING is a well-studied problem that originated with the work of Andrews and Jacobson [16]; Cowen et al. [8]. DEFECTIVE COLORING is a generalization of GRAPH COLORING that corresponds to the case $\Delta^* = 0$. Therefore, all the hardness results of GRAPH COLORING follow.

DEFECTIVE COLORING finds its application in modeling communication networks, with colors representing the assignment of frequencies and Δ^* denoting the amount of tolerance in interference. In practice, communication networks are designed to tolerate interference at frequencies, ensuring a reliable and efficient transmission of information. Naturally, the parameter Δ^* in DEFECTIVE COLORING is effective in measuring tolerance in communication networks. Inspired by these applications, the complexity of DEFECTIVE COLORING is investigated in topological structures, including unit-disk graphs and grid graphs [2–4,7,15]. In the Very Large-Scale Integration (VLSI) design, DEFECTIVE COLORING is used to minimize conflicts, captured by the parameter Δ^*, by assigning colors to different components or circuits on a chip. This helps to reduce the overall complexity and improve the efficiency of the integrated circuits. The flexibility of DEFECTIVE COLORING in addressing challenges related to interference, resource allocation, and connectivity spans various domains, including applications in Social Network Analysis, Parallel Processing, and Resource Allocation in Cloud Computing to name a few.

Cowen, Cowen, and Woodall [8] obtained values of k and Δ^* such that every graph in the class of planar graphs and its subclass outerplanar graphs has a defective coloring with parameters k and Δ^*. DEFECTIVE COLORING has been studied from the parameterized complexity perspective. Belmonte et al. [5] considered the structural parameters such as tree-width, path-width, tree-depth, feedback vertex set number and vertex cover number resulting in several algorithmic and hardness findings. It was shown that DEFECTIVE COLORING is W[1]-hard when parameterized by tree-width, path-width or tree-depth when $k \geq 2$. Surprisingly, for the feedback vertex set number, they showed that it is W[1]-hard when $k = 2$ and FPT for all $k \geq 3$. Further, they show that there is no algorithm that solves DEFECTIVE COLORING in $n^{o(\mathsf{tw})}$ or $n^{o(\mathsf{pw})}$ under ETH, where tw and pw represents the tree-width and the path-width parameters respectively. In the context of parameterized approximation, they demonstrate the existence of an algorithm that for any given k and error $\epsilon > 0$, running in time $(\mathsf{tw}/\epsilon)^{O(\mathsf{tw})}n^{O(1)}$ and approximating the value of Δ^* within a factor of $(1+\epsilon)$. They additionally presented an algorithm running in time $\mathsf{tw}^{O(\mathsf{tw})}$, providing a 2-approximation for the minimum value of k. They also show the near-optimality of this algorithm by proving that, under standard assumptions, no FPT algorithm can achieve a better than 3/2-approximation to k, even when an additional constant additive error is permitted. Defective Coloring is solvable in time $\mathsf{vc}^{O(\mathsf{vc})}n^{O(1)}$ where vc is the vertex cover number of the graph [5]. Moreover, it was shown [20] that there is no $\mathsf{vc}^{o(\mathsf{vc})}n^{O(1)}$ algorithm.

It was posed as an open question in [5] if Defective Coloring is FPT when parameterized by modular-width, clique-width or neighborhood diversity. We try to improve the understanding of the problem with some of these parameters. It was shown in [6] that the optimization version of Defective Coloring is NP-hard on split graphs when $k \geq 2$ or $\Delta^* \geq 1$, and complete d-partite graphs (a subclass of cographs). Since cographs have modular-width at most two, an FPT algorithm parameterized by clique-width or modular-width parameters is ruled out. Also, since the problem is W[1]-hard parameterized by tree-width, it is interesting to study the problem on dense graphs, for instance graphs of bounded distance to cluster number (the minimum number of vertices to delete such that the resultant graphs is a disjoint union of cliques). The parameter modular-width generalizes the parameters neighborhood diversity and twin-cover number in the sense that graphs of bounded neighborhood diversity or twin-cover number have bounded modular-width. In a similar fashion, distance to cluster number generalizes twin-cover number. In this paper, we study the problem with respect to the parameters neighborhood diversity and twin-cover number.

The main result of the paper is an FPT algorithm parameterized by the twin-cover number stated below, the proof of which is presented in Sect. 5.

Theorem 1. Defective Coloring *is FPT parameterized by twin-cover number.*

We consider graphs with bounded distance to d-degree (see definition 1) and obtain an algorithm in terms of the modulator size, k, d and the tree-width of the input graph. As a corollary we obtain an FPT algorithm parameterized by distance to disjoint union of paths, which settles the complexity of this graph class.

Theorem 2. *Given a tree-decomposition of G of width* tw *and a modulator to distance to d-degree of G of size t,* Defective Coloring *can be solved in time* $(k(t+d))^{\mathsf{tw}}n^{O(1)}$.

We then give a simple ILP based FPT algorithm when parameterized by the combined parameters neighborhood diversity and the number of colors. This partially answers an open question posed in [5]. As a corollary, we obtain an FPT algorithm for complete d-partite graphs parameterized by d.

Theorem 3 (⋆). Defective Coloring *is FPT parameterized by neighborhood diversity and the number of colors.*

The optimization version of Defective Coloring takes as input a graph G and an integer Δ^*, the objective is to obtain the minimum number of colors k for which a defective coloring exists. Combinatorial bounds, exact algorithms and approximation algorithms on various graph classes have been extensively studied, see for instance [1,5,6,9,12,14,18,20,21].

Since the problem is NP-hard on split graphs, we investigate the approximation algorithm for split graphs and present a 1-additive approximation algorithm in Sect. 3.

Theorem 4. *There is a 1-additive approximation algorithm for* Defective Coloring *on split graphs.*

Open Questions: Is Defective Coloring FPT parameterized by modular-width, neighborhood diversity or distance to cluster number? Another direction of study could be to study approximation algorithms for some graph classes like chordal graphs, distance hereditary graphs, etc.

2 Preliminaries

We consider graphs that are undirected, connected, finite and simple. If G is disconnected, we apply each of our algorithms independently on each connected component of G. For a graph G, we denote the vertex and edge sets of G by $V(G)$ and $E(G)$ respectively. We denote an edge between u and v by uv. We use n to denote the number of vertices of G. For a set $X \subseteq V$, the graph $G[X]$ denotes the induced subgraph of G on the vertex set X. For any subset $X \subseteq V(G)$, the graph $G - X$ represents the graph induced by the vertices in $V(G) \setminus X$. For an edge $e = uv$, the graph $G - e$ represents the graph induced by the vertex set $V(G)$ without the edge uv. The open neighborhood of a vertex v, denoted by $N(v)$, is the set of vertices adjacent to v and the closed neighborhood of v is denoted by $N[v] = N(v) \cup \{v\}$. For a set $A \subseteq V(G)$, the set $N_G(A) = \bigcup_{v \in A} N(v)$ represents the neighbors of A in the graph G. We use $[k]$ to denote the set $\{1, 2, \ldots, k\}$.

Two vertices $u, v \in V(G)$ are called *twins* if $N(u) \setminus \{v\} = N(v) \setminus \{u\}$. Moreover, if $uv \in E(G)$ then u and v are called *true twins*. For a coloring $f : V(G) \to [k]$, we say a vertex v *satisfies or does not violate color-degree property* if the number of neighbors of v assigned the color $f(v)$ is at most Δ^*. Formally, $|\{u\colon u \in N(v) \text{ and } f(v) = f(u)\}| \leq \Delta^*$.

In parameterized complexity, the running time of an algorithm is measured as a function of input and the input parameter. We say a parameterized problem is *fixed-parameter tractable* (FPT) with respect to a parameter k, if there exists an algorithm that runs in time $f(k)|I|^{O(1)}$ where f is a computable function independent of the input size $|I|$ and k is a parameter associated with the input instance. For more details, the reader is deferred to [10].

A graph $G = (C, I)$ is a *split graph* if $V(G)$ can be partitioned into a clique C and an independent set I. We state two theorems which will be useful in the paper.

Theorem 5 [20]**.** Defective Coloring *can be solved in time* $(k(\Delta^* + 1))^{\mathsf{tw}} n^{O(1)}$ *on any graph* G *on* n *vertices if a tree decomposition of width* tw *of* G *is supplied with the input.*

We use the following result on Integer Linear Programming (ILP) as a subroutine in our algorithm.

Theorem 6 [11,17,19]. Ilp-Feasibility *problem can be solved using* $O(p^{2.5p+o(p)} \cdot L)$ *arithmetic operations and space polynomial in* L*, where* L *is the number of bits in the input and* p *is the number of variables.*

3 Split Graphs

In this section, we present a 1-additive approximation algorithm for split graphs.

Proof (Proof of Theorem 4*).* Let $G = (C, I)$ be a split graph. To prove the theorem, it is sufficient to show that $\chi_{\Delta^*} \in \{\lceil\frac{|C|}{\Delta^*+1}\rceil, \lceil\frac{|C|}{\Delta^*+1}\rceil+1\}$. Any defective coloring of G requires $\lceil\frac{|C|}{\Delta^*+1}\rceil$ many colors. Otherwise, there exists a color class that contains at least $\Delta^* + 2$ vertices from C and thus violates the color-degree property for these vertices.

We now show that $\lceil\frac{|C|}{\Delta^*+1}\rceil + 1$ many colors are sufficient in any defective coloring of G. Let $r = \lceil\frac{|C|}{\Delta^*+1}\rceil$ and $S_1, S_2, \ldots, S_{r+1}$ be the color classes where vertices in S_i are assigned the color i. For each $i \geq 1$, we arbitrarily assign $\Delta^* + 1$ vertices of C except possibly the color class S_r, which is assigned the remaining vertices of C. Notice that the number of vertices of C in S_r is at most $\Delta^* + 1$. We assign all vertices of I to the color class S_{r+1}. It is easy to see that the color classes induce a defective coloring. Thus $\chi_{\Delta^*} \in \{\lceil\frac{|C|}{\Delta^*+1}\rceil, \lceil\frac{|C|}{\Delta^*+1}\rceil + 1\}$. This completes the proof of Theorem 4. □

It was shown in [6], that Defective Coloring is NP-hard on split graphs. From Theorem 4, we get that given a graph G and an integer Δ^*, it is NP-hard to decide whether $\chi_{\Delta^*} = \lceil\frac{|C|}{\Delta^*+1}\rceil$ or $\chi_{\Delta^*} = \lceil\frac{|C|}{\Delta^*+1}\rceil + 1$. Formally,

Corollary 1. *Given a split graph* $G = (C, I)$ *and an integer* Δ^**, it is* NP*-hard to decide whether* $\chi_{\Delta^*} = \lceil\frac{|C|}{\Delta^*+1}\rceil$ *or* $\chi_{\Delta^*} = \lceil\frac{|C|}{\Delta^*+1}\rceil + 1$.

4 Distance to Bounded Degree Graphs

In this section, we present an algorithm for graphs of bounded distance to d-degree. As a corollary we obtain an FPT algorithm parameterized by distance to disjoint union of paths.

Definition 1 (Distance to d-degree graphs). *Given a graph* G*, the minimum size of a set* $S \subseteq V(G)$ *such that* $G[V \setminus S]$ *is a graph of maximum degree* d *is called the* distance to d-degree *of the graph* G.

Proof (Proof of Theorem 2*)c* Let (G, k, Δ^*, t, d) be an instance of Defective Coloring where $S \subseteq V(G)$ is a set of vertices of size at most t such that $G - S$ is a subgraph of G with maximum degree d.

We divide the algorithm into two cases. When $\Delta^* \leq t + d$, from Theorem 5 we have an algorithm running in time $(k(d+t))^{\mathrm{tw}} n^{O(1)}$. Now we discuss the case when $\Delta^* > t + d$. We look at the following reduction rule.

Reduction Rule 1. *Let u and v be two adjacent vertices of $G - S$. Then if follows (G, k, Δ^*, t, d) is a* YES*-instance of* DEFECTIVE COLORING *if and only if $(G - uv, k, \Delta^*, t, d)$ is a* YES*-instance of* DEFECTIVE COLORING.

Proof. The forward direction holds as $G - uv$ is a subgraph of G. Let f be a defective coloring for the instance $(G - uv, k, \Delta^*, t, d)$. We show that f is also a defective coloring for (G, k, Δ^*, t, d). Notice that $N_G(w) = N_{G-uv}(w)$ for all $w \in V(G) \setminus \{u, v\}$. Thus w satisfies the color-degree property in G. The degree of the vertices u and v in $G - uv$ is at most $t + d - 1$ and hence in G is at most $t + d$. Since we are in the case assumption that $\Delta^* > t + d$, irrespective of whether $f(u) = f(v)$ or $f(u) \neq f(v)$, both u and v satisfy the color-degree property in G. □

We repeatedly apply Reduction Rule 1 on the instance (G, k, Δ^*, t, d). This rule is executed at most n times. When the reduction rule is no longer applicable, then $G[V \setminus S]$ is an edge-less graph with S being a vertex cover of G. Since there is a $\mathsf{vc}^{O(\mathsf{vc})} n^{O(1)}$ algorithm for the defective coloring problem [5] where vc is the vertex cover number, we obtain an algorithm running in time $t^{O(t)} n^{O(1)}$. □

Corollary 2 (⋆). DEFECTIVE COLORING *is FPT parameterized by distance to disjoint union of paths.*

5 Twin-Cover

In this section, we prove Theorem 1 by presenting our FPT algorithm for DEFECTIVE COLORING, with twin-cover number as parameter.

Definition 2 (Twin-cover). *A* twin-cover *of a graph G is a set of vertices $S \subseteq V(G)$ such that $G[V \setminus S]$ is a disjoint union of cliques and each pair of vertices in a clique are true twins in G. The* twin-cover number *of G is the size of the smallest set $S \subseteq V(G)$ such that S is a twin-cover.*

Ganian (Theorem 3.4, [13]) showed that a twin-cover of size t can be found in time $1.2738^t n^{O(1)}$. Thus we will assume that a twin-cover of the input graph is also given as input. Let G be a graph and $S \subseteq V(G)$ be a twin-cover of size t. Let (G, k, Δ^*, t) be an instance of DEFECTIVE COLORING. We start with the following observation.

Observation 7. *If $G - S$ has a clique of size at least $(\Delta^* + 1)k + 1$ then (G, k, Δ^*, t) is a* NO*-instance of* DEFECTIVE COLORING.

Proof. Suppose that $G - S$ has a clique C of size $(\Delta^* + 1)k + 1$. Then by pigeonhole principle, in any defective coloring of G using k colors, there is a color that is assigned to at least $\Delta^* + 2$ vertices in C violating the color-degree property of these vertices. □

Thus, we may assume that the maximum size of a clique in $G[V \setminus S]$ is at most $(\Delta^* + 1)k$.

5.1 Overview of the Proof of Theorem 1

Let (G, k, Δ^*, t) be an instance of DEFECTIVE COLORING and $S \subseteq V(G)$ be a twin-cover of size t. We guess the number of colors, say $\ell \leq \min\{k, t\}$, that are used for the vertices in S in a defective coloring of G. Without loss of generality, let these colors be $\{1, 2, \ldots, \ell\}$. Then we do some pre-processing that assigns colors to the vertices of $G - S$. In particular for each clique C, we assign (i) the colors from $[k] \setminus [\ell]$ to the vertices of C, and (ii) the colors from $[\ell]$ that are not assigned to any of the neighbors of C in S.

We partition the cliques in $G - S$ into "types" based on its neighborhood in S. Thus there are at most $2^t - 1$ types and since S is a twin-cover all the cliques in a type have the same neighborhood in S. Now the objective is to assign colors to each clique C from the colors that are assigned to its neighbors in S. Since the coloring of S is fixed, we obtain an upper bound on the number of vertices in C that are assigned the color j. Using this information, we guess the number of vertices from each type that are assigned a color satisfying the color-degree property for each vertex in S. We accomplish this by constructing an instance of ILP, which we call ILP-1 (see Subsect. 5.5). Once ILP-1 returns an assignment, we have to assign colors to the vertices of the cliques across all the types. For this we consider the notion of a "bad clique" whose sizes are big and carefully assign colors to the vertices of these bad cliques. Then we extend the coloring to the remaining cliques maintaining the invariant that the colored vertices do not violate the color-degree property at any stage. This step is accomplished by constructing another instance of ILP called ILP-2 (see Subsect. 5.6).

5.2 Pre-processing

Let $f : S \to [\ell]$ be a coloring of the vertices in S using colors from $\{1, 2, \ldots, \ell\}$, where $\ell \leq \min\{k, t\}$. We assign the colors from $[k] \setminus [\ell]$ to the vertices of $G - S$. Consider a clique C in $G[V \setminus S]$. We denote the set of uncolored vertices of C by d_C. Note that initially $d_C = C$. We assign colors to the vertices of C by invoking Coloring-Scheme$(C, d_C, [k] \setminus [\ell])$ of Algorithm 1.

Algorithm 1: Coloring-Scheme(C, d_C, Y)

Input: a clique C, a set d_C of uncolored vertices of C and a color set Y
Output: a partial coloring of C using the colors from Y

```
1 for each j ∈ Y do
2     arbitrarily pick a set X ⊆ d_C of min{|d_C|, (Δ* + 1)} many vertices of
      C and assign the color j to each of them
3     d_C = d_C \ X
4 return (a partial coloring of C)
```

The provided coloring scheme with $Y = [k] \setminus [l]$ is indeed safe, as none of the colors in $[k] \setminus [\ell]$ are assigned to the vertices in S. Additionally, for each vertex $v \in C$, the color-degree property is not violated, as we assign a maximum of $\Delta^* + 1$ vertices, a color from $[k] \setminus [\ell]$. All the cliques in $G[V \setminus S]$ are colored (partially) using the colors from $[k] \setminus [\ell]$. At the end of this coloring scheme, at most $(k - \ell)(\Delta^* + 1)$ vertices from each clique are assigned colors from $[k] \setminus [\ell]$. The remaining uncolored vertices (if any) of each clique have to be assigned colors from $[\ell]$.

We construct an auxiliary graph G' from G by removing the set of colored vertices from $G[V \setminus S]$. That is, $V(G') = V(G) \setminus \{v \in V(G) \colon v$ is assigned a color from $[k] \setminus [\ell]\}$ and $E(G') = \{vw \in E(G) \colon v, w \in V(G')\}$. Recall that each vertex $u \in V(G') \cap S$ is assigned the color $f(u)$. From now on, we work with the graph G' with the objective to assign colors to the vertices of $G' - S$ from $[\ell]$. This brings us to the following lemma, the proof of which is trivial and follows from the above discussion.

Lemma 1. *Let $f : S \to [\ell]$ be a coloring of the vertices in S. Then (G, k, Δ^*, t) is a* YES*-instance of* Defective Coloring *if and only if (G', ℓ, Δ^*, t) is a* YES*-instance of* Defective Coloring.

Let (G, ℓ, Δ^*, t) be the resultant instance of Defective Coloring obtained after the application of Lemma 1. We partition the cliques of $G - S$ into *types* based on its neighborhood in S. For a nonempty subset $A \subseteq S$, we define a type $T_A = \{C : C$ is a clique in $G - S$ with $N_G(C) = A\}$. Notice that the number of types is at most $2^t - 1$. Throughout the section we consider A to be non-empty, otherwise G is disconnected.

For each T_A, where $A \subseteq S$, we partially color the vertices of the cliques in T_A using the colors from $[\ell] \setminus \bigcup_{u \in A} \{f(u)\}$. Let $M_A = \bigcup_{u \in A} \{f(u)\}$ denote the set of colors that are assigned to the vertices in A. Recall that each vertex in a clique of T_A is adjacent to every vertex in A. For each clique $C \in T_A$, we invoke Coloring-Scheme$(C, d_C, [\ell] \setminus M_A)$ of Algorithm 1. It is easy to see that the above coloring process is safe in the sense that it does not violate the color-degree property of any vertex.

Now, we construct another auxiliary graph G'' from G (similar to the construction of G') by removing the set of vertices from each T_A that are assigned the colors from $[\ell] \setminus M_A$. We work with G'' with the objective to assign colors from M_A to the vertices of the cliques in T_A. This brings us to the following lemma, the proof of which is trivial and follows from the above discussion.

Lemma 2. *Let $f : S \to [\ell]$ be a coloring of the vertices in S. Then (G, ℓ, Δ^*, t) is a* YES*-instance of* Defective Coloring *if and only if (G'', ℓ, Δ^*, t) is a* YES*-instance of* Defective Coloring *where vertices in each type T_A are assigned colors from M_A, for each $A \subseteq S$.*

5.3 Case Division

Let (G, ℓ, Δ^*, t) be the resultant instance of Defective Coloring obtained after the application of Lemma 2. The question now is whether there exists a defective coloring for the graph G where the vertices from each clique of T_A are assigned the colors from M_A? We answer this question based on the following cases: Case 1. $\Delta^* \leq 2t$ and Case 2. $\Delta^* > 2t$.

Case 1. $\Delta^* \leq 2t$. Consider a clique C of $G[V \setminus S]$. From Observation 7, we have that $|C| \leq \ell(\Delta^* + 1) \leq t(\Delta^* + 1)$. By the case assumption we get that $|C| \leq t(2t + 1)$. Since $|S| = t$ and $|C| \leq (2t + 1)t$, the tree-width of G is at most $2t^2 + 2t$. Together with $\ell \leq t$, $\Delta^* \leq 2t$ and $\mathsf{tw}(G) \leq 2t^2 + 2t$, from Theorem 5 we get that DEFECTIVE COLORING can be solved in time $t^{O(t^2)} n^{O(1)}$. Notice that a tree-decomposition of G of width $2t$ can be obtained in linear time.
Case 2. $\Delta^* > 2t$. Consider a type T_A, where $A \subseteq S$. For each color $j \in M_A$, let p_j^A denote the number of vertices in A that are assigned the color j. This brings us to the following observations.

Observation 8. *In any defective coloring of G, any clique $C \in T_A$ has at most $\Delta^* - p_j^A + 1$ vertices assigned the color j.*

Proof. Suppose that there are at least $\Delta^* - p_j^A + 2$ many vertices assigned the color j in C. Any vertex v in C assigned the color j has at most p_j^A neighbors in A of color j. Thus v has at least $\Delta^* + 1$ neighbors assigned the color j violating the color-degree property. □

Observation 9. *Consider a type T_A. Let b_i be the number of vertices over the cliques in T_A that are assigned the color $i \in M_A$ in a defective coloring of G. Further, let $b_i \leq \Delta^* - p_i^A + 1$ for all i. Then we can construct a coloring of the vertices of T_A in linear time.*

Proof. The coloring procedure is as follows. For each $i \in M_A$, arbitrarily pick b_i many uncolored vertices of T_A and assign the color i. Since each vertex of T_A should be assigned a color from M_A, and given b_i for each i, it is the case that the number of vertices of T_A is equal to $\sum_i b_i$. By the definition of b_i, it is easy to see that the color-degree property of each vertex of T_A is satisfied. □

Before we look at the algorithm, we look at some definitions.

5.4 Definitions

From Theorem 8 and Theorem 9, we get that any clique $C \in T_A$ has bounded number of vertices assigned the color j and if each of the colors in M_A are assigned to bounded number of vertices in T_A then it is easy to obtain a defective coloring of T_A. Thus it makes sense to treat the colors that are assigned to at most $\Delta^* - p_j^A + 1$ vertices and at least $\Delta^* - p_j^A + 2$ vertices differently. For the latter case, we will guess the number of vertices which is a number from $\{\Delta^* - p_j^A + 2, \ldots, \Delta^*\}$ for each color j. This motivates us to define the following.

Definition 3 (Choice function). *Let $A \subseteq S$ be a non-empty subset such that T_A is non-empty. For each color $j \in M_A$, we define a choice function $h_A : \{j\} \rightarrow \{\perp, \Delta^* - p_j^A + 2, \Delta^* - p_j^A + 3, \ldots, \Delta^*\}$ that maps the color j to $\Delta^* - p_j^A + i$, where $2 \leq i \leq p_j^A$, if the number of vertices assigned the color j over all the cliques of T_A is $\Delta^* - p_j^A + i$, and to $\perp$ otherwise.*

The number of choice functions for a color $j \in M_A$ is $p_j^A \leq t$. Since the number of colors in M_A is at most t, the number of choice functions over all colors is at most t^t.

Definition 4 (bad-type and bad-vertex). *We say a type T_A, $A \subseteq S$, is a* bad-type *if there exists a color $j \in M_A$ such that $h_A(j) \in \{\Delta^* - p_j^A + 2, \Delta^* - p_j^A + 3, \ldots, \Delta^*\}$. We say the vertices $f^{-1}(j) \cap A$ are* bad *with respect to T_A.*

Definition 5 (Choice-pair set). *Let $\mathcal{H} = \{(A, h_A) \mid A \subseteq S\}$ be a* choice-pair set *that represents the set of pairs (A, h_A) for each $A \subseteq S$. For each $(A, h_A) \in \mathcal{H}$, let $X_A = \{j \in M_A \mid h_A(j) \neq \perp\}$. We say $\mathcal{H}$* respects *S if the following holds:*

1. *for each $(A, h_A) \in \mathcal{H}$, h_A is the choice function for each color in M_A,*
2. *for each $(A, h_A) \in \mathcal{H}$, the number of vertices over all the cliques of T_A assigned the colors from $M_A \setminus X_A$ is equal to $\sum_{C \in T_A} |d_C| - \sum_{j \in X_A} h_A(j)$, and*
3. *each vertex $v \in S$ is adjacent to at most Δ^* vertices that are assigned the color $f(v)$.*

Intuition and Implication of a Choice-Pair Set Respecting. S: Let $\mathcal{H}$ be a choice-pair set that respects S. Then, for each $j \in M_A$, where $A \subseteq S$, the number of vertices over all the cliques of T_A assigned the color j is given by the choice function $h_A(j)$ for all cases expect when it is at most $\Delta^* - p_j^A + 1$. If we are able to compute the number for this case maintaining the color-degree property for every vertex in S, we will try to extend it to a coloring for $V \setminus S$. To compute this exact number of vertices assigned a color from $M_A \setminus X_A$ for each T_A, we use Integer Linear Programming (ILP).

5.5 ILP-1

For each choice-pair set $\mathcal{H}$ respecting S, we construct an instance of ILP as follows. For each color $j \in M_A$ and $A \subseteq S$, we use the variable $n_{A,j}$ to denote the number of vertices of T_A that are assigned the color j. As $|S| \leq t$, the number of variables is at most $t(2^t - 1)$. For each $v \in S$, we denote $m_v = |\{u : u \in N_S(v) \text{ and } f(u) = f(v)\}|$ to be the number of neighbors of v in S that are assigned the color $f(v)$. Let $\widehat{d}_A = \sum_{C \in T_A} |d_C|$ denote the total number of uncolored vertices over all the cliques of T_A. We now describe the constraints.

C1. Consider a type T_A, where $A \subseteq S$. For each $j \in M_A$,
  (a) If $h_A(j) = \perp$, then $n_{A,j} \leq \Delta^* - p_j^A + 1$.
  (b) Else, $n_{A,j} = h_A(j)$.

C2. For each type T_A, where $A \subseteq S$, $\sum_{j \in M_A} n_{A,j} = \widehat{d}_A$.

C3. For each vertex $v \in S$,

$$\sum_{A: v \in A} n_{A,f(v)} \leq \Delta^* - m_v$$

Lemma 3 (⋆). *$\mathcal{H}$ respects S if and only if there is a feasible assignment of ILP-1.*

Any Solution Returned by ILP-1 is Good Enough: For a fixed coloring f of S and a choice-pair set $\mathcal{H}$, notice that any two feasible assignments returned by ILP-1 only differ in the number of vertices assigned the color j for T_A, where $j \in M_A$, when $h_A(j) = \perp$. Otherwise, when $h_A(j) \in \{\Delta^* - p_j^A + 2, \Delta^* - p_j^A + 3, \ldots, \Delta^*\}$, we ensure that $n_{A,j} = h_A(j)$. For the former case, the values of $n_{A,j}$ does not hurt the color-degree property for any vertex in T_A and thus it is sufficient to work with an assignment returned by ILP-1.

5.6 ILP 2

Using the assignment returned by ILP-1, each type T_A that is not a bad-type can be colored from Observation 9. Now, we only consider bad-types.

Claim 1. Let C be a clique in a bad-type T_A such that $|C| \leq \min_{j \in M_A} \Delta^* - p_j^A + 1$. Then any coloring of $V \setminus S$ restricted to C obtained using the feasible assignment returned by ILP-1 satisfies the color-degree property of vertices in C.

The proof of the above claim is easy to see and hence we omit the details. We now define the notion of a "bad-clique", a clique that contains large number of uncolored vertices. We show that the number of such bad-types and bad-cliques are bounded. Subsequently, we formulate another instance of ILP to devise a coloring for these bad-cliques satisfying the feasible assignment returned ILP-1. This, in turn, establishes the existence of a choice-pair set that respects S.

Definition 6 (bad-clique). *Let T_A be a bad-type, where $A \subseteq S$. We call a clique $C \in T_A$ as a* bad-clique *if $|d_C| \geq \min_{j \in M_A} \{\Delta^* - p_j^A + 2\}$.*

Claim 2. Each vertex in S is bad with respect to at most one bad-type.

Proof. Suppose that $v \in S$ is bad with respect to two bad-types, say T_A and $T_{A'}$ where $A, A' \subseteq S$ are non-empty and $A \neq A'$. Then, $v \in A \cap A'$. Let $f(v) = j$. Then from Definition 4, there are at least $\Delta^* - p_j^A + 2$ vertices in T_A and at least $\Delta^* - p_j^{A'} + 2$ vertices in T'_A that are assigned the color j. This implies that there are at least $2\Delta^* - (p_j^A + p_j^{A'}) + 4$ vertices assigned the color j in the neighborhood of v. Since $A \neq A'$ and $v \in A \cap A'$, we have $p_j^A + p_j^{A'} < 2t$. Additionally, by the case assumption $\Delta^* > 2t$, there are at least $\Delta^* + 5 > \Delta^* + 1$ neighbors of v assigned color j violating the color-degree property for v. □

Claim 3. The number of bad-cliques in any bad-type is at most $2t - 1$.

Proof. Let T_A be a bad-type. Recall that the number of colors used for coloring the vertices of cliques across all the types is at most t. The total number of uncolored vertices over all the cliques in T_A is at most $t\Delta^*$. Otherwise, by pigeonhole principle in any coloring (using t colors) of the vertices of cliques, there is a vertex in S for which the color-degree property is violated. Let $c \in M_A$ be a color such that $\Delta^* - p_c^A + 2$ is minimum. The number of bad-cliques in T_A is at most $\frac{t\Delta^*}{\Delta^*-p_c^A+2} < \frac{t\Delta^*}{\Delta^*-p_c^A} = \frac{t\Delta^*-t\cdot p_c^A}{\Delta^*-p_c^A} + \frac{t\cdot p_c^A}{\Delta^*-p_c^A} = \frac{t(\Delta^*-p_c^A)}{\Delta^*-p_c^A} + \frac{t\cdot p_c^A}{\Delta^*-p_c^A} = t + \frac{t\cdot p_c^A}{\Delta^*-p_c^A}$. Since $p_c^A \le t$ and by case assumption $\Delta^* > 2t$, we obtain the desired bound. □

For the rest of the proof, we consider only the bad-cliques of T_A which are at most $2t-1$ from Claim 3. From Claim 2, we get that there are at most t bad-types. Thus the number of bad-cliques across all the types is at most $2t^2 - t$. Let $\mathcal{C} = \{C_1, C_2, \ldots, C_r\}$ be the set of bad-cliques across the bad-types, where $r \le 2t^2 - t$. We show that given a feasible assignment of variables returned by ILP 1, a defective coloring on the vertex set $S \cup \bigcup_{C \in \mathcal{C}} V(C)$ can be extended to obtain a defective coloring for G.

ILP Feasibility Instance 2 (ILP-2): We consider only bad-cliques in this step. Using a feasible assignment of the variables returned by ILP-1, we construct another ILP-feasibility instance. Consider a bad-type T_A, where $A \subseteq S$. For a color $j \in M_A$ and a bad-clique $C \in T_A$, let the variable $q_{C,j}$ denote the number of vertices from d_C that are assigned the color j. The number of variables is at most $2t^3$. We now describe the constraints for a bad-type T_A.

D1. For each bad-clique $C \in T_A$, and $j \in M_A$,

$$0 \le q_{C,j} \le \Delta^* - p_j^A + 1$$

D2(a). If T_A is a bad-type and $T_A \setminus \mathcal{C} = \emptyset$, then for each $j \in M_A$

$$\sum_{C \in \mathcal{C} \cap T_A} q_{C,j} = n_{A,j}$$

D2(b). If T_A is a bad-type and $T_A \setminus \mathcal{C} \neq \emptyset$, then for each $j \in M_A$.

$$\sum_{C \in \mathcal{C} \cap T_A} q_{C,j} \le n_{A,j}$$

D3. For each $C \in \mathcal{C}$, $\sum_{j \in M_A} q_{C,j} = |d_C|$.

Lemma 4 (⋆). *There is a feasible assignment of ILP-2 if and only if there is a defective coloring of the graph satisfying the feasible assignment returned by ILP-1.*

This completes the description and proof of correctness of algorithm in Theorem 1.

Acknowledgement. We are grateful to the reviewers of SWAT whose suggestions helped us improve the result in Sect. 4 and the presentation of the algorithm in Theorem 1. We are also grateful to the reviewers of SOFSEM for their helpful comments.

References

1. Achuthan, N., Achuthan, N.R., Simanihuruk, M.: On minimal triangle-free graphs with prescribed k-defective chromatic number. Disc. Math. **311**(13), 1119–1127 (2011). https://doi.org/10.1016/j.disc.2010.08.013
2. Araújo, J., Bermond, J., Giroire, F., Havet, F., Mazauric, D., Modrzejewski, R.: Weighted improper colouring. J. Disc. Algor. **16**, 53–66 (2012). https://doi.org/10.1016/j.jda.2012.07.001
3. Archetti, C., Bianchessi, N., Hertz, A., Colombet, A., Gagnon, F.: Directed weighted improper coloring for cellular channel allocation. Disc. Appl. Math. **182**, 46–60 (2015). https://doi.org/10.1016/j.dam.2013.11.018
4. Bang-Jensen, J., Halldórsson, M.M.: Vertex coloring edge-weighted digraphs. Inf. Process. Lett. **115**(10), 791–796 (2015). https://doi.org/10.1016/j.ipl.2015.05.007
5. Belmonte, R., Lampis, M., Mitsou, V.: Parameterized (approximate) defective coloring. SIAM J. Disc. Math. **34**(2), 1084–1106 (2020). https://doi.org/10.1137/18M1223666
6. Belmonte, R., Lampis, M., Mitsou, V.: Defective coloring on classes of perfect graphs. Disc. Math. Theor. Comput. Sci. **24**(1) (2022). https://doi.org/10.46298/dmtcs.4926. https://dmtcs.episciences.org/8918
7. Bermond, J., Havet, F., Huc, F., Sales, C.L.: Improper coloring of weighted grid and hexagonal graphs. Disc. Math. Algor. Appl. **2**(3), 395–412 (2010). https://doi.org/10.1142/S1793830910000747
8. Cowen, L.J., Cowen, R., Woodall, D.R.: Defective colorings of graphs in surfaces: partitions into subgraphs of bounded valency. J. Graph Theory **10**(2), 187–195 (1986). https://doi.org/10.1002/jgt.3190100207
9. Cumberbatch, J., Lauri, J., Mitillos, C.: Exact defective colorings of graphs. CoRR arxiv:2109.05255 (2021)
10. Cygan, M., et al.: Parameterized Algorithms, vol. 4. Springer, Heidelberg (2015)
11. Frank, A., Tardos, É.: An application of simultaneous diophantine approximation in combinatorial optimization. Comb. **7**(1), 49–65 (1987). https://doi.org/10.1007/BF02579200
12. Frick, M., Henning, M.A.: Extremal results on defective colorings of graphs. Disc. Math. **126**(1–3), 151–158 (1994). https://doi.org/10.1016/0012-365X(94)90260-7
13. Ganian, R.: Twin-cover: beyond vertex cover in parameterized algorithmics. In: Marx, D., Rossmanith, P. (eds.) IPEC 2011. LNCS, vol. 7112, pp. 259–271. Springer, Heidelberg (2012). https://doi.org/10.1007/978-3-642-28050-4_21
14. Gimbel, J.G., Hartman, C.: Subcolorings and the subchromatic number of a graph. Disc. Math. **272**(2–3), 139–154 (2003). https://doi.org/10.1016/S0012-365X(03)00177-8
15. Gudmundsson, B.A., Magnússon, T.K., Sæmundsson, B.O.: Bounds and fixed-parameter algorithms for weighted improper coloring. In: Crescenzi, P., Loreti, M. (eds.) Proceedings of the 16th Italian Conference on Theoretical Computer Science, ICTCS 2015, Firenze, Italy, 9–11 September 2015. Electronic Notes in Theoretical Computer Science, vol. 322, pp. 181–195. Elsevier (2015). https://doi.org/10.1016/j.entcs.2016.03.013
16. Andrews, J.A., Jacobson, M.S.: On a generalization of chromatic number. Congressus Numerantium **47**, 33–48 (1985)
17. Lenstra Jr., H.W.: Integer programming with a fixed number of variables. Math. Oper. Res. **8**(4), 538–548 (1983). https://doi.org/10.1287/moor.8.4.538

18. Kang, R.J., McDiarmid, C.: The t-improper chromatic number of random graphs. Comb. Probab. Comput. **19**(1), 87–98 (2010). https://doi.org/10.1017/S0963548309990216
19. Kannan, R.: Minkowski's convex body theorem and integer programming. Math. Oper. Res. **12**(3), 415–440 (1987). https://doi.org/10.1287/moor.12.3.415
20. Lampis, M., Vasilakis, M.: Structural parameterizations for two bounded degree problems revisited. In: Gørtz, I.L., Farach-Colton, M., Puglisi, S.J., Herman, G. (eds.) 31st Annual European Symposium on Algorithms, ESA 2023, Amsterdam, The Netherlands, 4–6 September 2023. LIPIcs, vol. 274, pp. 77:1–77:16. Schloss Dagstuhl - Leibniz-Zentrum für Informatik (2023). https://doi.org/10.4230/LIPICS.ESA.2023.77
21. de Mendez, P.O., Oum, S., Wood, D.R.: Defective colouring of graphs excluding a subgraph or minor. Comb. **39**(2), 377–410 (2019). https://doi.org/10.1007/S00493-018-3733-1

Dynamic Range Minimum Queries on the Ultra-wide Word RAM

Philip Bille(✉), Inge Li Gørtz, Máximo Pérez López, and Tord Stordalen

Technical University of Denmark, DTU Compute, Kgs. Lyngby, Denmark
{phbi,inge,mpelo,tjost}@dtu.dk

Abstract. We consider the dynamic range minimum problem on the ultra-wide word RAM model of computation. This model extends the classic w-bit word RAM model with special ultrawords of length w^2 bits that support standard arithmetic and boolean operation and scattered memory access operations that can access w (non-contiguous) locations in memory. The ultra-wide word RAM model captures (and idealizes) modern vector processor architectures.

Our main result is a linear space data structure that supports range minimum queries and updates in $O(\log \log \log n)$ time. This exponentially improves the time of existing techniques. Our result is based on a simple reduction to prefix minimum computations on sequences $O(\log n)$ words combined with a new parallel, recursive implementation of these.

Keywords: Ultra-wide word RAM model · Range minimum queries · Prefix minimum

1 Introduction

Supporting *range minimum queries* (RMQ) on arrays is a well-studied, classic data structure problem, see e.g., [1–4,9–12,14,16,17,19,23,25,27,28]. This paper considers the *dymamic RMQ problem* defined as follows: maintain an array $A[0,\ldots,n-1]$ of w-bit integers and subject to the following operations.

- $\mathsf{rmq}(i,j)$: return a smallest integer in the subarray $A[i..j]$.
- $\mathsf{update}(i,\alpha)$: set $A[i] \leftarrow \alpha$.

On most models of computation, the complexity of the dynamic RMQ problem is well-understood [2,9,11,23,25]. For instance, on the word RAM, a tight $\Theta(\log n/\log\log n)$ time bound on the operations is known [11]. Hence, a natural question is whether practical models of computation capturing modern hardware advances will allow us to improve this bound significantly. One such model is the *ultra-wide word RAM model* (UWRAM) introduced by Farzan et al. [15]. The UWRAM extends the word RAM model with special *ultrawords*

Partially supported by the Danish Research Council via grant DFF-8021-002498.

R. Královič and V. Kůrková (Eds.): SOFSEM 2025, LNCS 15538, pp. 122–135, 2025.
https://doi.org/10.1007/978-3-031-82670-2_10

of w^2 bits. The model supports standard boolean and arithmetic operations on ultrawords and *scattered* memory operations that access w words in parallel. The UWRAM model captures (and idealizes) modern vector processing architectures [13,22,24,26]. We present the details of the UWRAM model of computation in Sect. 2. By extending recent techniques for the UWRAM model [5], we can immediately solve the dynamic RMQ problem using $O(\log\log n)$ time per operation.

1.1 Results and Techniques

Our main result is an exponential improvement of the $O(\log\log n)$ time bound. More precisely, we show the following bound:

Theorem 1. *Given an array A of n w-bit integers, we can construct an $O(n)$ space data structure in $O(n)$ time on the UWRAM that supports* rmq *and* update *in $O(\log\log\log n)$ time.*

Technically, our solution is based on a simple linear space and logarithmic time folklore solution, which we call the *range minimum tree* (see Sect. 3 for a detailed description). The range minimum tree is a balanced binary tree over the input array and supports operations in $O(\log n)$ time by sequentially traversing the tree. On the UWRAM, we show how to efficiently compute the access patterns of the operations on the range minimum tree in parallel using scattered memory access operations and prefix minimum computation on ultrawords. More precisely, we show that given any algorithm for a prefix minimum computation on a sequence of words of length $\ell = O(w)$ stored in a constant number of ultrawords that uses time $t(\ell)$ implies a linear space solution for dynamic RMQ that supports both operations in $O(t(\log n))$ time. If we implement a standard parallel prefix computation algorithm [20] (see also the survey by Blelloch [8]) using the UWRAM techniques in Bille et al. [5] we immediately obtain an algorithm that uses $O(\log \ell)$ time. Our main technical contribution is a new, exponentially faster prefix minimum algorithm that achieves $O(\log\log \ell)$ time. The key idea is a constant time prefix minimum algorithm for "short" sequences of $O(\sqrt{w})$ words that takes advantage of parallel computations on multiple copies of the sequence packed into a constant number of ultrawords (note that a constant number of ultrawords can store $O(\sqrt{w})$ copies of a sequence of $O(\sqrt{w})$ words). We implement the idea recursively and in parallel on each recursion level to obtain a fast solution for general sequences of words of length $O(w)$. Each recursion step uses constant time, and the depth is $O(\log\log \ell)$, leading to the $O(\log\log \ell)$ time bound.

1.2 Outline

The paper is organized as follows. In Sects. 2 and 3 we review the UWRAM model of computation and the range minimum tree. In Sect. 4.1, we present the UWRAM implementation of the range minimum tree that leads to the reduction to prefix minimum computation on word sequences. Finally, in Sect. 5, we present our fast prefix minimum algorithm.

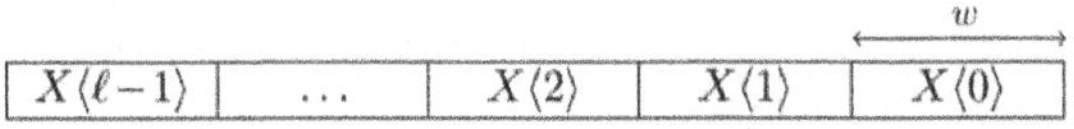

Fig. 1. The layout of a word sequence X.

2 The Ultra-wide Word RAM Model

The *word RAM* model of computation [18] consists of an unbounded memory of w-bit words and a standard instruction set including arithmetic, boolean, and bitwise operations (denoted '&,' '|,' and '∼' for *and*, *or* and *not*) and shifts (denoted '≫' and '≪') such as those available in standard programming languages (e.g., C). We assume that we can store a pointer to the input in a single word and hence $w \geq \log n$, where n is the input size. The time complexity of a word RAM algorithm is the number of instructions, and the space is the number of words the algorithm stores.

The *ultra-wide word RAM* (UWRAM) model of computation [15] extends the word RAM model with special *ultrawords* of w^2 bits. As in [15], we distinguish between the *restricted UWRAM* that supports a minimal set of instructions on ultrawords consisting of addition, subtraction, shifts, and bitwise boolean operations, and the *multiplication UWRAM* that additionally supports multiplications. We extend the notation for bitwise operations and shifts to ultrawords. The UWRAM (restricted and multiplication) also supports contiguous and scattered memory access operations, as described below. The time complexity is the number of instructions (on standard words or ultrawords), and the space complexity is the number of words used by the algorithms, where each ultraword is counted as w words. The UWRAM model captures (and idealizes) modern vector processing architectures [13,22,24,26]. See Farzan et al. [15] for a detailed discussion of the applicability of the UWRAM model.

2.1 Instructions and Componentwise Operations

Recall that ultrawords consist of w^2 bits. We often use ultrawords to store and manipulate small sequences of $O(w)$ words. A *word sequence* X of length ℓ is a sequence of ℓ words (also called the components of X). We number the words from right to left starting from 0 and use the notation $X\langle i\rangle$ to denote the ith word in X (see Fig. 1).

We define common operations on word sequences that we will use later. Let X and Y be word sequences of length ℓ. The *componentwise addition* of X and Y is the word sequence Z such that $Z\langle i\rangle = X\langle i\rangle + Y\langle i\rangle$. The *componentwise comparison* of X and Y is the word sequence Z such that $Z\langle i\rangle = 1$ if $X\langle i\rangle < Y\langle i\rangle$ and 0 otherwise. Given another word sequence I of length ℓ, where each word is either 0 and 1 (we will call this a *binary word sequence*), the *componentwise extract* of X wrt. I is the word sequence Z such that $Z\langle i\rangle = X\langle i\rangle$ if $I\langle i\rangle = 1$ and $Z\langle i\rangle = 0$ otherwise. We can also *concatenate* X and Y, denoted $X \cdot Y$ producing the length 2ℓ word sequence $X\langle \ell-1\rangle \cdots X\langle 0\rangle Y\langle \ell-1\rangle \cdots Y\langle 0\rangle$ or *split* X at any point k into word sequences $X\langle \ell - 1\rangle \cdots X\langle k+1\rangle$ and $X\langle k\rangle \cdots X\langle 0\rangle$. Note that using two split operations, we can compute any contiguous subsequence $X\langle i\rangle \cdots X\langle j\rangle$ of X. All these operations can be implemented in constant time for word sequences of length $O(w)$ on the restricted UWRAM using standard-word level parallelism techniques [5,6,18]. Note that we can manipulate word sequences with length $\leq cw$, for some constant $c > 1$, by storing them in c ultrawords and simulating operations on them in constant time.

The UWRAM also supports a *compress* operation that takes a binary word sequence I of length ℓ and constructs the bitstring of the ℓ bits of I. The inverse *spread* operation takes a bitstring of length ℓ and constructs the corresponding binary word sequence of length ℓ. This is the UWRAM model that we will use throughout the rest of the paper. Note that these operations are widely supported directly in modern vector processing architectures.

Given a word sequence of X of length ℓ, we define the *prefix minimum* of X, denoted $\mathsf{pmin}(X)$, to be the word sequence P of length ℓ such that $P\langle i\rangle = \mathsf{min}(X\langle i\rangle, \ldots, X\langle 0\rangle)$. We also define the *minimum* of X, denoted $\mathsf{min}(X)$, as the smallest entry among all entries in X. Note that we can use a prefix minimum algorithm to compute the minimum. The prefix minimum operation is central in our solutions, and as discussed, we will show how to implement it in $O(\log\log \ell)$ time.

Some of our operations require precomputed constant word sequences, which we assume are available (e.g., computed at "compile-time"). If not, we can compute those needed for Theorem 1 in $\log^{O(1)} n$ time, which is negligible.

2.2 Memory Access

The UWRAM supports standard memory access operations that read or write a single word or a sequence of w contiguous words. More interestingly, the UWRAM also supports *scattered* access operations that access w memory locations (not necessarily contiguous) in parallel. Given a word sequence A containing w memory addresses, a *scattered read* loads the contents of the addresses into a word sequence X, such that $X\langle i\rangle$ contains the contents of memory location $A\langle i\rangle$. Given word sequences X and A of length $O(w)$ a *scattered write* sets the contents of memory location $A\langle i\rangle$ to be $X\langle i\rangle$. We can implement the following *shuffle* operations in constant time using scattered read and write. Given two word sequences A and X of length ℓ, a *shuffled read* computes the word sequence Y, such that $Y\langle i\rangle = X\langle A\langle i\rangle\rangle$. Given word sequences X and A of length ℓ, a *shuffled write* computes the word sequence Y, such that $Y\langle A\langle i\rangle\rangle = X\langle i\rangle$. Scattered memory accesses capture the memory model used in IBM's *Cell* architecture [13]. They also appear (e.g., `vpgatherdd`) in Intel's AVX vector extension [24]. Scattered memory access operations were also proposed by Larsen and Pagh [21] in the context of the I/O model of computation. Note that while the addresses for scattered writes must be distinct, we can read simultaneously from the same address.

3 Range Minimum Tree

Let A be an array of n w-bit integers and assume for simplicity that n is a power of two. The *range minimum tree* T is the perfectly balanced rooted binary tree over A such that the ith leaf corresponds to the ith entry in A. We associate each node v in T with a *weight*, denoted $\mathsf{weight}(v)$. If v is a leaf, the weight is the value represented by the corresponding entry of A, and if v is an internal node, the weight is the minimum of the weights of the descendant leaves. Note that T has height $h = O(\log n)$ and uses $O(n)$ space. See Fig. 2.

For nodes $i, j \in T$, let $\mathsf{lca}(i, j)$ denote the lowest common ancestor of i and j. To answer an $\mathsf{rmq}(i, j)$ query, we traverse the path from i to $\mathsf{lca}(i, j)$ and from j to $\mathsf{lca}(i, j)$ and return the minimum weight of i, j, and of the nodes hanging off to the right and left of these paths (except for the children of $\mathsf{lca}(i, j)$), respectively (see Fig. 2). To perform an $\mathsf{update}(i, \alpha)$, we traverse the path from i to the root. At leaf i, we set the weight to be α, and at each internal node v, we set the weight to be the minimum of the weights of the two children of v. If we do not modify the weight of a node at some node in the traversal, we may stop since no weights need to be updated on the remaining path. See Fig. 3. Both operations traverse paths of $O(\log n)$ length and use constant time at each node. Hence, both operations use $O(\log n)$ time.

We introduce the following *node sequences* to implement the range minimum tree on the UWRAM efficiently. Let i be an index in A corresponding to the ith leaf in T, and let p be the path of nodes from i to the root in T. The *path sequence* is the sequence of nodes on the path p. Define the *left sequence* to be the sequence of nodes that are hanging off to the left of p, i.e., a node v is in the left sequence if it is the left child of a node on p and is not on p itself. Similarly, define the *right sequence* and the *off-path sequence* to be the sequence of nodes to the right of p and the sequence of nodes to the left or right of p, respectively. All node sequences are ordered from i to the root in order of increasing height. See Figs. 2 and 3.

We can use the node sequences to describe the traversed nodes during the rmq and *update* operations on the range minimum tree. Consider an $\mathsf{rmq}(i, j)$ with $u = \mathsf{lca}(i, j)$ of depth d. Then $\mathsf{rmq}(i, j)$ is the minimum of the weights of the leaves i and j and of the nodes of depth $> d + 1$ on the right sequence of i and the left sequence of j. Next, consider an $\mathsf{update}(i, \alpha)$ and let $i^p_0, \ldots, i^p_{h-1}$ and $i^o_0, \ldots, i^o_{h-2}$ be the path and off-path sequence, respectively, for i. Let $\mathsf{weight}(\cdot)$ and $\mathsf{weight}'(\cdot)$ denote the weight of nodes in T before and after the update. Recall that only the nodes on the path sequence may change. We have that $\mathsf{weight}(i^p_0) = \alpha$ and $\mathsf{weight}'(i^p_j) = \min(\mathsf{weight}'(i^p_{j-1}, \mathsf{weight}(i^o_{j-1})))$ for $0 < j < h$. If we unfold the recursion, it follows that

$$\mathsf{weight}'(i^p_j) = \min(\mathsf{weight}(i^o_{j-1}), \ldots, \mathsf{weight}(i^o_0), \alpha) \qquad \text{for} 0 \leq j < h. \tag{1}$$

In other words, the new weights of the nodes on the path sequence are the prefix minimums of the sequence $\mathsf{weight}(i^o_{h-2}), \ldots, \mathsf{weight}(i^o_0), \alpha$.

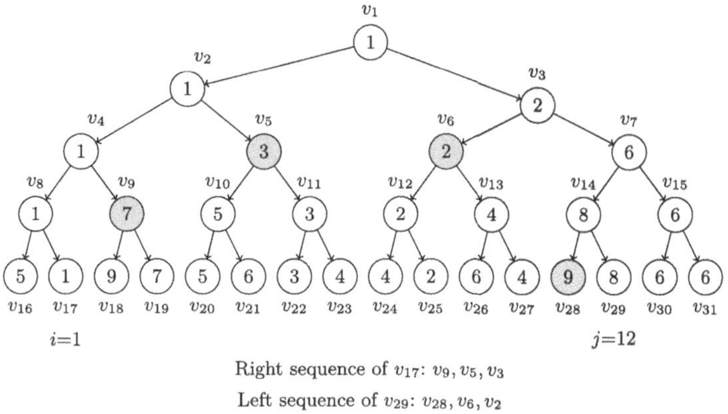

Fig. 2. An example array A, with its range minimum tree. For a query $\mathsf{rmq}(1, 12)$, we illustrate with grey circles the right sequence vertices of $i = 1$ and the left sequence vertices of $j = 12$ of depth greater than $d + 1 = 1$.

4 From Range Minimum Queries to Prefix Minimum on the UWRAM

In this section, we show that any UWRAM data structure that supports prefix minimum computations on word sequences of length $O(\log n)$ implies a UWRAM solution for the range minimum query problem.

Theorem 2. *Let A be an array of n w-bit integers, and let $t(\ell)$ be the time to compute* pmin *on a word sequence of length at most $\ell = O(w)$. Then, we can construct an $O(n)$ space data structure in $O(n)$ time on the UWRAM that supports* rmq *and* update *in $O(1 + t(\log n))$ time.*

In the next section, we will show how to compute pmin on a word sequence of length $\ell = O(w)$ in $O(\log \log \ell)$ time, implying the main result of Theorem 1.

4.1 Range Minimum Tree on the UWRAM

We first show a simple direct implementation of the range minimum tree that achieves the rmq and update time bounds of Theorem 2 but with $O(n \log n)$ space and preprocessing time.

Data Structure. Our data structure consists of the input array A, the range minimum tree T over A, and a data structure that supports lowest common ancestor queries on T. This data structure can be implemented in linear space and preprocessing time to support lca queries in constant time [1,3,19]. Furthermore, for each index i in A, we store the path sequence, the left path sequence, the right path sequence, and the off-path sequence. The sequences are stored as sequences of pointers to the nodes, and together with the left and right sequences, we also store the sequence of depths of the nodes in the sequence.

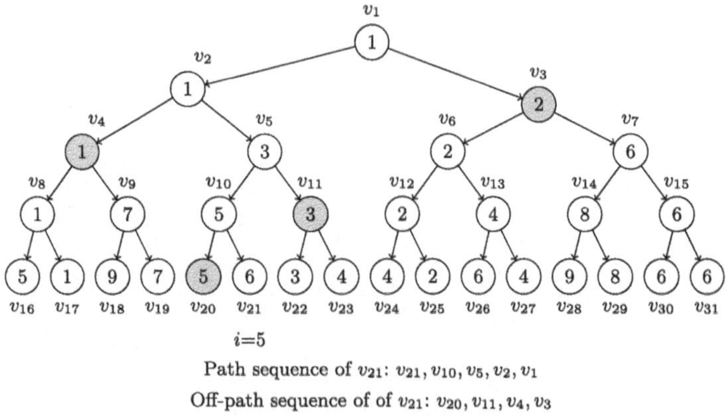

Fig. 3. Illustration of an $\mathsf{update}(5, \alpha)$ query. We draw with grey circles the off-path sequence vertices. Note how the value of v_{10} should be $\min(v_{20}, \alpha)$, the value of v_5 should be $\min(v_{10}, v_{11}) = \min(v_{11}, v_{20}, \alpha)$, and the value of v_2 should be $\min(v_4, v_5) = \min(v_4, v_{11}, v_{20}, \alpha)$, and so on. These values are the prefix minimum of the grey vertices and α.

The array, the range minimum tree, and the lowest common ancestor data structure use $O(n)$ space. Each of the $O(n)$ sequences uses $O(\log n)$ space. In total, we use $O(n \log n)$ space and preprocessing time.

Range Minimum Queries. To answer a $\mathsf{rmq}(i, j)$ query, compute $u = \mathsf{lca}(i, j)$ and the depth d of u. Read the left and right path sequences of i and j into word sequences denoted I^r and J^l, and their corresponding depth sequences into word sequences denoted DI^r and DJ^l. Then, construct masks MI^r and MJ^l containing 1s in the positions in DI^r and DJ^l that are greater than $d+1$ and use these to extract the prefixes $\hat{I}^r$ and $\hat{J}^l$ of I^r and J^l, respectively, that contains the nodes that have depth greater than $d + 1$. Finally, compute the weights $W\hat{I}^r$ and $W\hat{J}^l$ of the nodes in $\hat{I}^r$ and $\hat{J}^l$ using two scattered reads, and return $\mathsf{min}(W\hat{I}^r \cdot W\hat{J}^l)$. The masks MI^r and MJ^l can be computed using a scattered read and a comparison operation. Thus, all operations take constant time, except min, which uses $O(t(\log n))$ time. In total, we use $O(1 + t(\log n))$ time.

Updates. To implement $\mathsf{update}(i, \alpha)$, we first set $\mathsf{weight}(i) = \alpha$. We read the off-path sequence into a word sequence, denoted I^o, and then compute the corresponding sequence of weights WI^o using a scattered read. We then compute $P = \mathsf{pmin}(WI^o \cdot \alpha)$ and perform a scattered write with I^p and P. All operations use constant time, except pmin, which uses $O(t(\log n))$ time. In total, we use $O(1 + t(\log n))$.

In summary, the UWRAM range minimum tree uses $O(n \log n)$ space and preprocessing time and supports rmq and update in $O(1 + t(\log n))$ time.

4.2 Reducing Space

We now improve the space and preprocessing time to $O(n)$ by using a single level of indirection.

Data Structure. We partition A into *blocks* $B_0, \ldots, B_{n/\log n-1}$ each of $\log n$ consecutive entries. We store the minimum of each block in a *block array* B of length $n/\log n$ and construct the UWRAM range minimum data structure from Sect. 4.1 on B. This uses $O(n/\log(n/\log n)) = O(n)$ space and preprocessing time.

Range Minimum Queries. To answer an $\mathsf{rmq}(i,j)$ query there are two cases:

Case 1: i and j are within the same block. Let B_k, where $k = \lfloor i/\log n \rfloor$ be the block containing i and j and compute the corresponding local indices $i' = i \mod \log n$ and $j' = j \mod \log n$ in B_k. We read B_k and return $\mathsf{min}(B_k\langle i'\rangle, \cdots B_k\langle j'\rangle)$.

Case 2: i and j are in different blocks. Let $B_l, B_{l+1}, \ldots, B_r$ be the blocks covering the range from i to j and let i' and j' be the local indices in B_l and B_r. We decompose the range into three parts. We compute the minimum in the leftmost block as $l_{\min} = \mathsf{min}(B_l\langle i'\rangle \cdots, B_l\langle \log n - 1\rangle)$ and the rightmost block as $r_{\min} = \mathsf{min}(B_r\langle \log n - 1\rangle \cdots B_r\langle j'\rangle)$. We then compute the minimum $m_{\min}$ of the middle blocks using the range minimum tree. Finally, we return $\mathsf{min}(l_{\min}, c_{\min}, r_{\min})$.

Updates. Consider an $\mathsf{update}(i, \alpha)$ operation. Let B_k, where $k = \lfloor i/\log n \rfloor$ be the block containing i, and let $i' = i \mod \log n$ be the local index in B_k. We set $B_k\langle i'\rangle = \alpha$. We then read B_k and compute $b_{\min} = \mathsf{min}(B_k)$. If $b_{\min}$ differs, we update T with the new value.

Both operations use constant time except for (prefix) minimum computations and operations on the range minimum tree that take $O(1 + t(\log(n/\log n))) = O(t(\log n))$ time. Hence, the total time is $O(1 + t(\log n))$. In summary, we have shown Theorem 2.

5 Computing Prefix Minimum on Word Sequences

We now show how to efficiently compute the prefix minimum on word sequences of length $\ell = O(w)$ in $O(\log\log \ell)$ time. We first show how to do so in constant time for word sequences of length $O(\sqrt{w})$. We then show how to implement this algorithm in parallel and then recursively leading to the result.

Our algorithm often partitions a word sequence into multiple equal-length sequences to work on them in parallel. We define a b-way word sequence to be a word sequence $X = X_{s-1} \cdots X_0$ where each subsequence X_i is a *block* of length b. Thus, the total length of X is sb. We use $X\langle i,j\rangle$ to denote entry j in block i, that is, $X\langle i,j\rangle = X_i\langle j\rangle$.

$$
\begin{array}{rl}
X = & \langle\ 9\ 2\ 5\ 3\ \rangle \\
\widehat{X} = & \langle\ 9\ 2\ 5\ 3 \quad 9\ 2\ 5\ 3 \quad 9\ 2\ 5\ 3 \quad 9\ 2\ 5\ 3\ \rangle \\
\widetilde{X} = & \langle\ 9\ 9\ 9\ 9 \quad 2\ 2\ 2\ 2 \quad 5\ 5\ 5\ 5 \quad 3\ 3\ 3\ 3\ \rangle \\
C = & \langle\ 1\ 0\ 0\ 0 \quad 1\ 1\ 1\ 1 \quad 1\ 0\ 1\ 0 \quad 1\ 0\ 1\ 1\ \rangle \\
D = & \langle\ 0\ 0\ 0\ 0 \quad 1\ 1\ 1\ 1 \quad 0\ 0\ 0\ 0 \quad 0\ 0\ 1\ 1\ \rangle \\
M = & \langle\ 1\ 0\ 0\ 0 \quad 1\ 1\ 0\ 0 \quad 1\ 1\ 1\ 0 \quad 1\ 1\ 1\ 1\ \rangle \\
E = & \langle\ 0\ 0\ 0\ 0 \quad 1\ 1\ 0\ 0 \quad 0\ 0\ 0\ 0 \quad 0\ 0\ 1\ 1\ \rangle \\
E' = & \langle\ 0\ 0\ 0\ 0 \quad 3\ 2\ 0\ 0 \quad 0\ 0\ 0\ 0 \quad 0\ 0\ 1\ 0\ \rangle \\
E'' = & \langle\ 19\ 18\ 17\ 16 \quad 0\ 0\ 13\ 12 \quad 11\ 10\ 9\ 8 \quad 7\ 6\ 0\ 0\ \rangle \\
P = & \langle\ 19\ 18\ 17\ 16 \quad 3\ 2\ 13\ 12 \quad 11\ 10\ 9\ 8 \quad 7\ 6\ 1\ 0\ \rangle \\
Y = & \langle\ 2\ 2\ 3\ 3\ \rangle
\end{array}
$$

Fig. 4. Our prefix minimum algorithm.

5.1 Prefix Minimum on Small Word Sequences

We now show how to compute the prefix minimum on a word sequence X of length $\ell = O(\sqrt{w})$. For simplicity, we first assume all entries in X are distinct. Our algorithm proceeds as follows. See also the example in Fig. 4.

Step 1: Compare All Pairs of Words in X. We construct a b-way word sequence C containing the results of all pairwise comparisons of words in X. To do so, we first construct the word sequences:

$$\widehat{X} = \underbrace{X \cdot X \cdots X}_{\ell}$$

$$\widetilde{X} = \underbrace{X\langle \ell-1\rangle \cdots X\langle \ell-1\rangle}_{\ell} \cdot \underbrace{X\langle \ell-2\rangle \cdots X\langle \ell-2\rangle}_{\ell} \cdots \underbrace{X\langle 0\rangle \cdots X\langle 0\rangle}_{\ell}$$

We compute these using shuffled read operations on X with the constant word sequences

$$\widehat{A} = \underbrace{\langle \ell-1, \cdots 0\rangle \cdots \langle \ell-1, \cdots 0\rangle}_{\ell}$$

$$\widetilde{A} = \underbrace{\langle \ell-1, \ldots \ell-1\rangle}_{\ell} \cdot \underbrace{\langle \ell-2, \ldots \ell-2\rangle}_{\ell} \cdots \underbrace{\langle 0, \ldots 0\rangle}_{\ell}.$$

Note that both $\widehat{X}$ and $\widetilde{X}$ have length $\ell^2 = O(w)$. We then do a componentwise comparison of $\widehat{X}$ and $\widetilde{X}$. This produces a word sequence C of length ℓ^2, which viewed as an ℓ-way word sequence is defined by:

$$C\langle i,j\rangle = \begin{cases} 1 & \text{if} X\langle i\rangle \leq X\langle j\rangle \\ 0 & \text{otherwise} \end{cases} \tag{2}$$

Thus, the ith block in C stores the comparison of $X\langle i\rangle$ with all other words in X.

Step 2: Compute Prefix Minima. We construct an ℓ-way word sequence E that contains the positions of the prefix minima. First, compute the ℓ-way word sequence D such that

$$D\langle i,j\rangle = \begin{cases} 1 & \text{if} C\langle i,j\rangle = C\langle i,j-1\rangle = \cdots C\langle i,0\rangle = 1 \\ 0 & \text{otherwise} \end{cases} \tag{3}$$

Thus, the block D_i in D takes the rightmost 0 in C_i and "smears" it to the left. In the full version, we show how to compute this operation, which we call *left clear*, in constant time [7]. We then mask out all entries in D where $i > j$ using an ℓ-way mask $M = M_{\ell-1} \cdots M_0$, where $M\langle i,j\rangle = 1$ if $i \leq j$ and 0 otherwise. By (2) we have that $D\langle i,j\rangle = 1$ iff $X\langle i\rangle \leq X\langle j\rangle, X\langle j-1\rangle, \ldots, X\langle 0\rangle$. Hence, $E\langle i,j\rangle = 1$ iff $X\langle i\rangle \leq X\langle j\rangle, X\langle j-1\rangle, \ldots, X\langle 0\rangle$ and $i \leq j$, i.e., $X\langle i\rangle$ is the prefix minimum of $X\langle j-1\rangle, \ldots, X\langle 0\rangle$.

Step 3: Extract Prefix Minimum. We now extract the prefix minima entries indicated by D and then compact them into a word sequence of length ℓ. The key observation is that $E\langle i,j\rangle = 1$ implies that $X\langle i\rangle$ should appear in position j in the final prefix minimum. We use the following constant word sequences:

$$P' = \underbrace{\langle \ell-1, \cdots 0\rangle \cdots \langle \ell-1, \cdots 0\rangle}_{\ell}$$

$$P'' = \langle \ell^2+\ell-1, \ell^2+\ell-2, \ldots \ell\rangle$$

We do a component-wise extraction of the positions in P' wrt. E and of P'' wrt. $\overline{E}$, where $\overline{E}$ is the negation of E, and then $\mid$ the resulting sequences E' and E'' together to get a word sequence P. We then do a shuffled write on $\widetilde{X}$ with P and clear out all but the ℓ rightmost entries. The resulting word sequence Y of length ℓ contains a $X\langle i\rangle$ in position j iff $E\langle i,j\rangle = 1$ as desired. Since we assumed all input words are distinct and since P'' only contains numbers greater than $\ell - 1$, there is no write conflict.

All of the above word sequences have length at most $\ell^2 = O(w)$, and thus, each step takes constant time. Recall that we assumed all entries in X were distinct. If not, we may have a write conflict at the end of step 3. To fix this, we can always double the length of the input word sequence X and represent each entry using two words consisting of the entry and the position in the sequence, thus breaking ties. We can then simulate the above algorithms with constant factor slowdown. In summary, we have shown the following result.

Lemma 1. *Given a word sequence X of length $\ell = O(\sqrt{w})$, we can compute the prefix minimum of X in $O(1)$ time.*

As a corollary, we implement the above operation to compute prefix minima of blocks in a b-way word sequence X. More precisely, let $X = X_{s-1} \cdots X_0$ be a b-way word sequence and define the *b-way prefix minimum* of X to be

$$\mathsf{pmin}^b(X) = \mathsf{pmin}(X_{s-1}) \cdot \mathsf{pmin}(X_{s-2}) \cdots \mathsf{pmin}(X_0)\,.$$

To compute $\mathsf{pmin}^b(X)$, where X has length $\ell = O(w/b)$, consider each block as a word sequence of size b to which we apply the above algorithm in parallel as follows. In Step 1, we create two word sequences $\widehat{X}'$ and $\widetilde{X}'$ that are the concatenation of all the word sequences $\widehat{X}$ and $\widetilde{X}$ of size b^2 corresponding to each block, and we apply all the operations outlined above in parallel for each subsequence of b^2 words. Since the length of X is $\ell = O(w/b)$, the number of blocks is $O(w/b^2)$ and the length of the word sequences $\widehat{X}$ and $\widetilde{X}$ is $O(w)$. We proceed similarly for the other steps.

Since the length of the word sequences we work with is $O(w)$ we can do all the operations in constant time. Note that, if $b > \sqrt{w}$, the condition $\ell = O(w/b)$ implies that $\ell = O(\sqrt{w})$, and also that we have a constant number of blocks to process, which we can do with Lemma 1 instead. We have the following result:

Corollary 1. *Given a b-way word sequence X of length $O(w/b)$ we can compute the b-way prefix minimum of X in constant time.*

5.2 Prefix Minima on General Word Sequences

We now show how to recursively apply Corollary 1 to compute the prefix minima on a word sequence X of length $\ell = O(w)$. Given the b-way prefix minimum of X, we show how to compute the b^2-way prefix minimum of X in constant time. We then show how to apply this recursively to obtain our $O(\log\log \ell)$ algorithm. Let $X^b = X^b_{s-1} \cdots X^b_0$ be the b-way prefix minimum of X. Our algorithm proceeds as follows (see Fig. 5).

Step 1: Compute Prefix Minima of Block Minima. We compute the word sequence

$$B^b = \mathsf{pmin}^b(X^b_{s-1}\langle b-1\rangle \cdots X^b_0\langle b-1\rangle)$$

containing the prefix minimum of the minimum (leftmost) entries of the blocks in X^b. To do so, we first shuffle the leftmost entries of the blocks into a word sequence Y and then compute $B^b = \mathsf{pmin}^b(Y)$ using Corollary 1 (note that Y has length $O(w/b)$ as required).

Step 2: Propagate Prefix Minima of Block Minima. We construct a word sequence E that, for each block of X^b, has the minimum of the previous blocks in each sequence of b blocks. First, we compute the word sequences

$$\widehat{B}^b = \underbrace{B^b\langle s-1\rangle \cdots B^b\langle s-1\rangle}_{b} \cdots \underbrace{B\langle 0\rangle \cdots B\langle 0\rangle}_{b},$$

$$M = \underbrace{0 \cdots 0}_{b^2-b} \cdot \underbrace{1^w \cdots 1^w}_{b} \cdots \underbrace{0 \cdots 0}_{b^2-b} \cdot \underbrace{1^w \cdots 1^w}_{b},$$

where $\widehat{B}^b$ contains b copies of every entry in B^b, and M contains a repeated pattern of b words of 1s followed by $b^2 - b$ 0s. Both word sequences have length

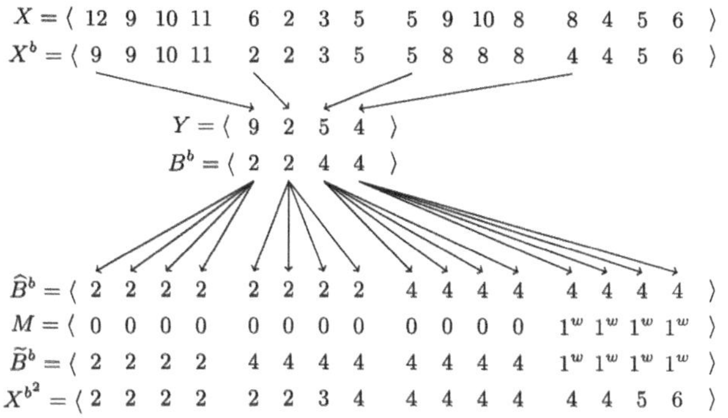

X = ⟨ 12 9 10 11 6 2 3 5 5 9 10 8 8 4 5 6 ⟩
X^b = ⟨ 9 9 10 11 2 2 3 5 5 8 8 8 4 4 5 6 ⟩
Y = ⟨ 9 2 5 4 ⟩
B^b = ⟨ 2 2 4 4 ⟩
B̂^b = ⟨ 2 2 2 2 2 2 2 2 4 4 4 4 4 4 4 4 ⟩
M = ⟨ 0 0 0 0 0 0 0 0 0 0 0 0 1^w 1^w 1^w 1^w ⟩
B̃^b = ⟨ 2 2 2 2 4 4 4 4 4 4 4 4 1^w 1^w 1^w 1^w ⟩
X^{b²} = ⟨ 2 2 2 2 2 2 3 4 4 4 4 4 4 4 5 6 ⟩

Fig. 5. Illustration of Step 2 of Sect. 5.2: computing the 16-way prefix minimum of a word sequence X, given the 4-way prefix minimum of X in X^b.

ℓ, and we compute them using shuffled read operations. After this, we shift left $\widehat{B}^b$ by b words, and we calculate the logical '|' of $\widehat{B}^b$ and M in a word sequence $\widetilde{B}^b$. Finally, we compute the componentwise minimum of $\widetilde{B}^b$ and X^b to produce X^{b^2}.

To see why this is correct, consider a sequence of b blocks of the b^2-way prefix minimum of X, denoted $X^{b^2}_{ib}, X^{b^2}_{ib+1}, \ldots, X^{b^2}_{ib+b-1}$ for any $0 \leq i < \lceil \ell/b^2 \rceil$. We can obtain the value of any $X^{b^2}_{ib+j}$ by the componentwise minimum of X^b_{ib+j}, and a block that contains b copies of the minimum of $X_{ib+j-1}, \ldots X_{ib}$. This minimum is computed in the $(j-1)$th word of B^b_i, so we copy that word b times in $\widetilde{B}^b$, and we shift it left by b words to compute the componentwise minimum with X^b_{ib+j}. Note that for $j = 0$, the value of $X^{b^2}_{ib}$ is the same as the value of X^b_{ib}, therefore for these blocks, we instead compare to the maximum word available (1^w) in order not to change their value. See Fig 5 for a visualization.

Each operation uses constant time, and hence we have the following result.

Lemma 2. *Let X be a word sequence of length $\ell = O(w/b)$. Given a b-way prefix minimum of X, we can compute a b^2-way prefix minimum of X in constant time.*

It now follows that we can compute the prefix minimum of a word of length ℓ by applying Lemma 2 for double exponentially increasing values of b over $O(\log \log \ell)$ rounds. Hence, we have the following result.

Theorem 3. *Given a word sequence X of length $\ell = O(w)$, we can compute the prefix minimum of X in $O(\log \log \log \ell)$ time.*

Plugging in Theorem 3 into our reduction of Theorem 2, we have shown the main result of Theorem 1.

References

1. Alstrup, S., Gavoille, C., Kaplan, H., Rauhe, T.: Nearest common ancestors: a survey and a new algorithm for a distributed environment. Theory Comput. Syst. **37**, 441–456 (2004)
2. Alstrup, S., Husfeldt, T., Rauhe, T.: Marked ancestor problems. In: Proceedings of the 39th Annual IEEE Symposium on Foundations of Computer Science, pp. 534–543 (1998)
3. Bender, M.A., Farach-Colton, M.: The LCA problem revisited. In: Proceedings of the 4th Latin American Symposium on Theoretical Informatics, pp. 88–94 (2000)
4. Berkman, O., Breslauer, D., Galil, Z., Schieber, B., Vishkin, U.: Highly parallelizable problems. In: Proceedings of 21st STOC, pp. 309–319 (1989)
5. Bille, P., Gørtz, I.L., Skjoldjensen, F.R.: Partial sums on the ultra-wide word RAM. Theor. Comput. Sci. **905**, 99–105 (2022)
6. Bille, P., Gørtz, I.L., Stordalen, T.: Predecessor on the ultra-wide word RAM. Algorithmica **86**(5), 1578–1599 (2024)
7. Bille, P., Gørtz, I.L., Stordalen, T., Pérez-López. M.: Dynamic range minimum queries on the ultra-wide word RAM. arXiv:2411.16281 (2024)
8. Blelloch, G.E.: Prefix sums and their applications. In: Synthesis of Parallel Algorithms (1990)
9. Brodal, G.S., Chaudhuri, S., Radhakrishnan, J.: The randomized complexity of maintaining the minimum. In: Proceedings of 5th SWAT, pp. 4–15 (1996)
10. Brodal, G.S., Davoodi, P., Lewenstein, M., Raman, R., Satti, S.R.: Two dimensional range minimum queries and fibonacci lattices. Theor. Comput. Sci. **638**, 33–43 (2016)
11. Brodal, G.S., Davoodi, P., Srinivasa Rao, S.: Path minima queries in dynamic weighted trees. In: Proceedings of 12th WADS, pp. 290–301 (2011)
12. Chazelle, B., Rosenberg, B.: Computing partial sums in multidimensional arrays. In: Proceedings of 5th SOCG, pp. 131–139 (1989)
13. Chen, T., Raghavan, R., Dale, J.N., Iwata, E.: Cell broadband engine architecture and its first implementation–a performance view. IBM J. Res. Dev. **51**(5), 559–572 (2007)
14. Demaine, E.D., Landau, G.M., Weimann, O.: On cartesian trees and range minimum queries. Algorithmica **68**, 610–625 (2014)
15. Farzan, A., López-Ortiz, A., Nicholson, P.K., Salinger, A.: Algorithms in the ultra-wide word model. In: Proceedings of 12th TAMC, pp. 335–346 (2015)
16. Fischer, J., Heun, V.: Space-efficient preprocessing schemes for range minimum queries on static arrays. SIAM J. Comput. **40**(2), 465–492 (2011)
17. Gabow, H.N., Bentley, J.L., Tarjan, R.E.: Scaling and related techniques for geometry problems. In: Proceedings of 16th STOC, pp. 135–143 (1984)
18. Hagerup, T.: Sorting and searching on the word ram. In: Proceedings of 15th STACS, pp. 366–398 (1998)
19. Harel, D., Tarjan, R.E.: Fast algorithms for finding nearest common ancestors. SIAM J. Comput. **13**(2), 338–355 (1984)
20. Ladner, R.E., Fischer, M.J.: Parallel prefix computation. J. ACM **27**(4), 831–838 (1980)
21. Larsen, K.G., Pagh, R.: I/o-efficient data structures for colored range and prefix reporting. In: Proceedings of 23rd SODA, pp. 583–592 (2012)
22. Lindholm, E., Nickolls, J., Oberman, S., Montrym, J.: NVIDIA tesla: a unified graphics and computing architecture. IEEE Micro **28**(2), 39–55 (2008)

23. Pătraşcu, M., Demaine, E.D.: Logarithmic lower bounds in the cell-probe model. SIAM J. Comput. **35**(4), 932–963 (2006)
24. Reinders, J.: AVX-512 instructions. Intel Corporation (2013)
25. Sleator, D., Tarjan, R.E.: A data structure for dynamic trees. J. Comput. Syst. Sci. **26**(3), 362–391 (1983)
26. Stephens, N., et al.: The ARM scalable vector extension. IEEE Micro **37**(2), 26–39 (2017)
27. Yao, A.C.: On the complexity of maintaining partial sums. SIAM J. Comput. **14**(2), 277–288 (1985)
28. Yuan, H., Atallah, M.J.: Data structures for range minimum queries in multidimensional arrays. In: Proceedings of 21st SODA, pp. 150–160 (2010)

Fast Practical Compression of Deterministic Finite Automata

Philip Bille(✉), Inge Li Gørtz, and Max Rishøj Pedersen

Technical University of Denmark, Kongens Lyngby, Denmark
{phbi,inge}@dtu.dk

Abstract. We revisit the popular *delayed deterministic finite automaton* (D^2FA) compression algorithm introduced by Kumar et al. [SIGCOMM 2006] for compressing deterministic finite automata (DFAs) used in intrusion detection systems. This compression scheme exploits similarities in the outgoing sets of transitions among states to achieve strong compression while maintaining high throughput for matching.

Unfortunately, the D^2FA algorithm and later variants of it require at least quadratic compression time since they compare all pairs of states to compute an optimal compression. This is too slow and, in some cases, even infeasible for collections of regular expression in modern intrusion detection systems that produce DFAs of millions of states.

Our main result is a simple, general framework for constructing D^2FA based on locality-sensitive hashing that constructs an approximation of the optimal D^2FA in near-linear time. We apply our approach to the original D^2FA compression algorithm and two important variants, and we experimentally evaluate our algorithms on DFAs from widely used modern intrusion detection systems. Overall, our new algorithms compress up to an order of magnitude faster than existing solutions with either no or little loss of compression size. Consequently, our algorithms are significantly more scalable and can handle larger collections of regular expressions than previous solutions.

Keywords: Finite Automata · Algorithms · Data Compression

1 Introduction

Signature-based deep packet inspection is a key component of modern intrusion detection and prevention systems. The basic idea is to maintain a collection of regular expressions, called *signatures*, that correspond to malicious content, violations of security policies, etc., and then match the collection of signatures against the input. In the typical scenario of high-throughput network traffic, the matching must be fast enough to process the input at the network's speed.

A natural approach to solve this is to construct a *deterministic finite automaton* (DFA) of the collection of signatures and then simulate it on the input. The

Partially supported by the Independent Research Fund Denmark (DFF-9131-00069B and 10.46540/3105-00302B).

R. Královič and V. Kůrková (Eds.): SOFSEM 2025, LNCS 15538, pp. 136–150, 2025.
https://doi.org/10.1007/978-3-031-82670-2_11

DFA matches each character of the input with a single constant time state transition and requires only a single memory access. Unfortunately, DFAs for the collections of signatures used in modern intrusion detection systems are prohibitively large and not feasible for practical implementation [6,51]. To overcome this space issue while still maintaining fast matching, significant work has been done on compressing DFAs [3–5,7–9,15,20–22,25,29–32,34,35,37,42,45,48].

In this paper, we revisit the elegant and powerful *delayed DFA* compression technique introduced by Kumar et al. [29] and applied in many subsequent solutions [7,9,22,30–35,37]. The basic observation is that many states in the DFAs for real-world regular expression collections have similar sets of outgoing transitions, and we can take advantage of this to reduce the space significantly. Specifically, if two states s and s' share many such transitions, we can replace these in s with a special *default transition* to the other state s'.

Kumar et al. [29] proposed an algorithm to compute an optimal set of default transitions to compress any DFA. The key idea is to compute the similarity of all pairs of states, i.e., the number of shared outgoing transitions. We store these in a complete graph, called the *space reduction graph* (SRG), on the states with edges weighted by similarity. Finally, we compute a maximum spanning tree on the SRG and use each edge in the tree as a default transition leading to a compressed version of the original DFA called the *delayed deterministic finite automaton* (D^2FA).

To implement matching, we traverse the D^2FA similar to the Aho-Corasick algorithm for multi-string matching [2]. When we are at a state s and want to match a character α, we first inspect the outgoing transitions at s for a match of α. If we find a match, we continue to that state and process the next character in the input, and if not, we follow the default transition to state s' and repeat the process from s' with character α.

Compared to a standard DFA solution, Kumar et al. [29] showed that the D^2FA dramatically reduces space by more than 90% on real-world collections of regular expressions while still achieving fast matching performance.

Processing a character D^2FA during matching may require following multiple default transitions, thus incurring (as the name suggests) a *delay*. Minimizing the delay is essential in high-throughput applications, and two important variants of D^2FAs that address this problem have been proposed. Kumar et al. [29] gave a modified D^2FA construction that, given an integer parameter L, limits the maximum path of default transitions to L. We refer to this as the *longest delay* variant of the problem. Alternatively, Becchi and Crowley [7,9] gave a modified D^2FA construction that limits the maximum total number of default transitions traversed on any input string S by $|S|$. We refer to this as the *matching delay* variant.

The main bottleneck in the above algorithms is computing the similarity of all pairs of states to construct the SRG. If the input DFA contains n states, this requires at least $\Omega(n^2)$ time and space. This is too slow and, in some cases, even infeasible for collections of regular expression in modern intrusion detection systems that produce DFAs of millions of states.

Contributions. We present a simple, general framework for fast compression of DFAs with default transitions based on locality-sensitive hashing. We apply our approach to general D^2FA compression and the longest delay and matching delay variant and experimentally evaluate our algorithms on collections of regular expressions used in the popular Snort [43], Zeek [38], and Suricata [1] intrusion detection systems (see also www.snort.org, zeek.org, and suricata.io). Overall, we obtain new algorithms that compress up to an order of magnitude faster than existing solutions with either no or little loss of compression size. Consequently, our algorithms are significantly more scalable and can handle larger collections of regular expressions than previous solutions.

Technically, our main idea is to use locality-sensitive hashing to identify approximately similar states quickly. We then add edges between these states weighted by their similarity, producing the *sparse space reduction graph* (SSRG). We then compute the maximum spanning tree on the SSRG and then the D^2FA. The SSRG contains significantly fewer edges, leading to a significant improvement in compression time. While the SSRG approximates the SRG by discarding edges, which may lead to worse compression size, we observe that this loss is negligible experimentally. For the longest delay variant, the previous solution by Kumar et al. [29] also uses a costly heuristic to maintain a bounded diameter maximum spanning forest. If we directly apply our sparsification technique, this heuristic dominates the running time, and we do not experimentally observe a significant speed-up in compression time. Instead, we develop an efficient alternative heuristic that first constructs a maximum spanning tree and then cuts edges to achieve the desired maximum spanning forest with bounded diameter. We show that combining our sparsification with the new heuristic leads to improvements in compression time similar to our other variants with little or no loss of compression size.

Related Work. Substantial work has been done on compression DFAs. For an overview, see surveys [41,50]. A popular approach is compressing the set of transitions [3–5,7,9,20,21,29–32,34,35,37,42,45,48]. This approach includes the popular D^2FA algorithm we focus on in this paper. Another approach is to compress the alphabet to reduce the size of the transition table [8,9,15,25,47]. The main idea is that, if some sets of characters (almost) always cause the same transitions throughout the DFA, they can be replaced by a single character [8,9,15,25]. Alternatively, we can also compress the alphabet by replacing infrequent characters with sequences of frequent characters [47]. Finally, we can also compress the set of states as proposed by Becchi and Cambadi [5]. They showed how states with similar sets of outgoing transitions could be merged into one, thus compressing the set of states.

Using locality-sensitive hashing for fast compression of collections of sets has been used widely in many other contexts [11,17,18,26,28,36,39,46,49]. Our work naturally extends this work to fast DFA compression.

2 Preliminaries

Deterministic Finite Automata. A *deterministic finite automaton* (DFA) is a 5-tuple $D = (Q, \Sigma, \delta, q_0, A)$ where Q is a set of states, Σ is an alphabet, $\delta : Q \times \Sigma \rightarrow Q$ is a transition function, $q_0 \in Q$ is the initial state and $A \subseteq Q$ is a set of accepting states. We let $n = |Q|$ denote the number of states. A DFA can be thought of as a *labeled directed graph* where Q is the set of nodes and each transition $\delta(u, c) = v$ is a labeled, directed edge, denoted $(u, v)_c$. See Fig. 1. For simplicity, we assume every state has exactly one labeled transition for each character in the alphabet, i.e., δ is *total*, as in previous work.

Given a string S and a path p in D we say that p matches S if the concatenation of the labels of p equals S. A path that starts in q_0 and ends in A is *accepting* and D *accepts* a string S if there exists an accepting path that matches S. The *language* of D is the set of strings it accepts.

Locality-Sensitive Hashing. A family of hash functions is *locality-sensitive*, for some similarity measure, if the probability of two objects hashing to the same value is *high* (lower-bounded for some parameter) when they are *similar* (similarity above some threshold) and, conversely, *low* when they are *dissimilar* (see e.g. [23] for formal details). There are different families of locality-sensitive hash functions for different distance or similarity measures, with some of the most popular being *simhash* [16], *MinHash* [14] and *sdhash* [44]. For example, the MinHash of a set is the minimum element according to a uniformly random permutation. The probability that two sets A and B hash to the same value is precisely their Jaccard similarity $(|A \cap B|)/(|A \cup B|)$.

3 Delayed Deterministic Finite Automata

A *delayed deterministic finite automaton* (D^2FA) [29] is a deterministic finite automaton that is augmented with unlabeled *default transitions*. Formally a D^2FA is a 6-tuple $D^2 = (Q, \Sigma, \delta, q_0, A, F)$. As for DFAs, Q is the set of states, Σ is the alphabet, δ is the transition functions, $q_0 \in Q$ is the initial state, and $A \subseteq Q$ is the set of accepting states. The final component $F : Q \rightarrow Q$ is the *default transition function*. Viewed as a graph, default transitions are ϵ-labeled directed edges, where ϵ is the empty string, and each state has at most one outgoing default transition. See Fig. 1.

To transition from a state u according to a character c, we follow a c-labeled transition if it exists or otherwise follow the default transition:

$$\delta(u, c) = \begin{cases} v & \text{if } (u, v)_c \text{ is a edge.} \\ \delta(F(u), c) & \text{otherwise} \end{cases}$$

Note that for δ to be well-defined, reaching a state from u with a c-labeled transition must always be possible. This implies that any cycle of default transitions must have an outgoing c-labeled transition for any character c. To transition from

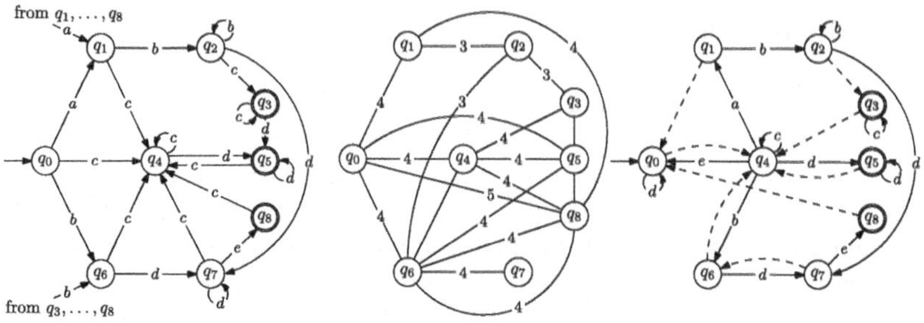

Fig. 1. Example from [29]. DFA D for regular expression `.*((ab+c+)|(cd+)|(bd+e))`. Edges to q_0 are omitted (left). Space reduction graph for D with edges annotated with similarity. Edges with similarity less than 4 omitted, except those connecting q_2 to avoid disconnecting the graph (middle). D^{2}FA equivalent to D. All transitions are shown, default transitions are dashed (right).

a state u according a string $S = c_1 \ldots c_m$ we recursively transition according to each character:

$$\delta(u, c_1c_2 \ldots c_m) = \delta(\delta(u, c_1), c_2 \ldots c_m).$$

Given a character c and a path $p = u_1, \ldots, u_k$ we say p matches c if $\delta(u_1, c) = u_k$ and all but the last transition is default, i.e., $F(u_i) = u_{i+1}$ for $1 \leq i < k$. Note that the concatenation of the labels of p equals c. Given a string $S = c_1 \ldots c_m$ and a path p we say p matches S if p is the concatenation of paths matching the individual characters $c_1, \ldots, c_m$. Note that the concatenation of the labels of p is S. We define acceptance as before. Two D^{2}FAs are equivalent if they have the same language, and two transitions are equivalent if they have the same destination and label.

Given a DFA, we can compress it by replacing sets of equivalent transitions with single default transitions to obtain an equivalent D^{2}FA with fewer total transitions. We define the *similarity* of two states u and v, denoted $sim(u, v)$, to be their number of equivalent transitions, that is, $sim(u, v) = |\{c \in \Sigma \mid \delta(u, c) = \delta(v, c)\}|$. See Fig. 1. Inserting a default transition (u, v) and removing the equivalent transitions from u does not affect the language but saves $sim(u, v) - 1$ transitions. Each transition we can remove without affecting the language we say is *redundant*.

Following a default transition does not consume an input character, which introduces a *delay* when matching. We define the *longest delay* of D^2 to be maximum number of default transitions in any path matching a single character, i.e., the longest delay is the maximum number of default transitions in D^2 to match any single character. Given a string S, we define the *matching delay* of S in D^2 to be the number of default transitions in the path starting in q_0 and matching S.

4 Compressing DFAs with Default Transitions

We now review the algorithm of Kumar et al. [29] that compress a DFA D into an equivalent D^{2}FA. Let D^2 be an initially empty D^{2}FA with the same set of states as D. We proceed as follows.

Step 1: Space Reduction Graph Construct a complete, undirected graph on the states of D, and to each edge (u, v) assign weight $sim(u, v)$. This is the *space reduction graph* (SRG). See Fig. 1.

Step 2: Maximum Spanning Tree Build a maximum spanning tree over the SRG. Root the spanning tree in a central node, i.e., a node of minimal radius, and direct all edges towards the root to obtain a directed spanning tree.

Step 3: Transitions For each edge (u, v) in the tree insert the default transition (u, v) into D^2. Copy every labeled transition from D into D^2 that is not redundant in D^2.

Since the SRG weighs the edges by similarity, computing the maximum spanning tree maximizes the overall compression.

Step 1 uses $O(n^2|\Sigma|)$ to compare all states and construct the SRG. Step 2 constructs the maximum spanning tree with Kruskal's algorithm [27] in $O(n^2 \log n)$ time, and step 3 takes $O(n|\Sigma|)$ time. In total, the running time is $O(n^2 \log n + n^2|\Sigma|)$[1].

5 Fast Compression

We now show how to speed up the DFA compression algorithm using locality-sensitive hashing to sparsify the SRG construction in step 1.

Let D be the input DFA and let r and k be two constant, positive integer parameters. We initialize an undirected graph $G = (Q, E)$ where Q is the set of states in D and $E = \{(q_0, u) \mid u \neq q_0\}$, i.e., we have a edge between the initial state and all other states. We then add edges to the graph in r rounds, where each round proceeds as follows:

First, pick k unique random characters $c_1, \ldots, c_k \in \Sigma$. Then, for each state $v \in Q$ we construct the sequence of k states $V = \delta(v, c_1), \ldots, \delta(v, c_k)$. We hash V into a single hash value $h(v)$ using a standard hashing scheme of Black et al. [12]. We insert v into a table with key $h(v)$. For each unique hash value h_i, consider the set of states C_i that hash to h_i. For each state $u \in C_i$, we pick another state $v \in C_i$ uniformly at random and insert (u, v) into E if it does not already exist.

After r rounds, the algorithm terminates, and we assign weights to each edge of G equal to the similarity of the endpoint states. The resulting graph G is the *sparse space reduction graph* (SSRG).

The above scheme is inspired by Har-Peled et al. [23] hashing bitstrings w.r.t. Hamming distance. Their solution samples positions from the input bitstrings.

[1] The running time is not explicitly stated in the paper, but follows from the description.

In our solution, the sampled positions correspond to the sampled characters from Σ.

Hashing a state takes $O(k)$ time, and sampling an edge takes constant time. Thus, a round takes $O(kn)$ time and we use $O(rkn)$ time for all rounds. Each round inserts at most n edges and hence the SSRG G has $O(rn)$ edges at the end. We compute the similarity between two states in $O(|\Sigma|)$ time, and hence the final algorithm uses $O(rkn + rn|\Sigma|) = O(n|\Sigma|)$ time.

We plug in the modified Step 1 in the algorithm from Sect. 4. Step 1 takes $O(n|\Sigma|)$ time and since G has $O(n)$ edges, Step 2 takes $O(n \log n)$ time. Step 3 takes $O(n|\Sigma|)$ time as before. In total, we use $O(n \log n + n|\Sigma|)$ time.

6 Compression with Bounded Longest Delay

We now consider the bounded longest delay variant of D^2FA, that is, given a DFA D and an integer parameter L, construct a D^2FA D^2 equivalent to D, such that the longest delay of D^2 is at most L. Recall that the longest delay of a D^2FA is the length of the longest path of default transitions in D^2.

Bounded Longest Delay by Constructing Small Trees We first review the algorithm by Kumar et al. [29]. The algorithm is based on a simple modification to the maximum spanning tree construction in Step 2 of the algorithm from Sect. 4.

Let L be a parameter. We modify Step 2 by constructing a maximum spanning forest with the constraint that each tree in the forest has a diameter of at most $\Delta = 2L$. To do so, we run Kruskal's algorithm but simply ignore any edges that would cause a tree diameter to exceed Δ. Also, among the edges with maximum similarity, we select one that causes a minimum increase to any tree diameter. After constructing the forest, we root each tree in a central node and direct edges toward each root.

Since each tree has diameter at most Δ and is rooted in a central node, the final D^2FA has a longest delay of at most $\lceil \Delta/2 \rceil = L$. To implement the modified Step 2, Kumar et al. [29] maintains the radius of the tree for each node during the maximum spanning forest construction. When we add an edge, we need to merge two trees and potentially update the radius for each node in the resulting tree. Hence, we may need to update $\Omega(n^2)$ radii in total during the maximum spanning forest construction.

We note that new Step 2 uses $\Omega(n^2)$ time, whether or not we run it on a sparse or a dense space reduction graph. Hence, we cannot directly apply our sparsification technique from Sect. 5 to obtain an efficient algorithm. We present a new algorithm in the next section that efficiently combines with sparsification.

Fast Compression with Bounded Longest Delay We now present a new algorithm that uses sparsification to efficiently construct D^2FAs with bounded longest delay. The idea is to construct a single large maximum spanning tree and then cut edges until each tree in the resulting forest has small diameter.

Given a DFA D and an integer parameter L, we proceed as follows.

Step 1: Construct SSRG Construct a sparse space reduction graph G for D as in Sect. 5.

Step 2: Construct MST Construct a maximum spanning tree T_0 over G using Kruskal's algorithm. Pick a central node v_0 in T_0 and then discard T_0. Construct a new maximum spanning tree T using Prim's algorithm [40] with v_0 as the initial node. When we queue a new edge (u, v), where u is the node already in T, assign weight $w'_{u,v} = sim(u, v) - 2^{d_v}$ where d_v is the distance from v_0 to v in T.

Step 3: Cut Edges Cut a minimum number of edges in T to obtain a forest with each tree diameter at most $\Delta = 2L$. We do so using the algorithm of Farley et al. [19] that cuts the necessary edges in a bottom-up traversal of T. Then, direct the edges of each tree towards the root.

Step 4: Construct Transitions As in Step 3 in Sect. 4, create default transitions along edges in the trees and then copy in every labeled transition from D that is not redundant in D^2.

After cutting, each tree has a diameter of at most $\Delta = 2L$. Since we direct edges toward each root, the longest delay is at most $\lceil \Delta/2 \rceil = L$. Step 1 and 4 uses $O(n|\Sigma|)$ time as before. Step 2 uses $O(n \log n)$ time, and Step 3 uses $O(n)$ time. In total, we use $O(n \log n + n|\Sigma|)$ time.

Each edge (u, v) we cut results in $sim(u, v) - 1$ more labeled transitions in D^2, as that default transition is then not constructed. Intuitively, the lower the diameter of T, the fewer edges we cut to get each tree below the bound. Therefore, we use an edge weight that trades similarity for lower diameter, as we observed that the fewer cuts outweighed the lost similarity.

We found the simple heuristic of the modified weight w' performed well in practice. A similar idea was used by Kumar et al. [29] in their solution.

Because T is not a maximum spanning tree w.r.t. similarity, the choice of initial node v_0 affects the total similarity of the final tree. We found that picking v_0 to be a central node in a MST w.r.t. similarity (T_0) yielded the best compression in practice.

Note that we cut the minimum *number* of edges to uphold the diameter constraint. Alternatively, we could cut edges of minimum *total similarity*, which might result in better compression. However, our approach is simple and fast in practice, and because each edge in the SRG has near-maximum weight, the difference in compression is negligible.

7 Compression with Bounded Matching Delay

We now consider the bounded matching delay variant of D^{2}FA, that is, given a DFA D, construct a D^{2}FA D^2 equivalent to D, such that the matching delay of D^2 is at most $|S|$ on any input string S.

Bounded Matching Delay by the A-DFA Algorithm We first review A-DFA algorithm by Becchi and Crowley [7,9]. Let D be an input DFA. For a state $v \in Q$, define the *depth* of v, denoted $d(v)$, to be the length of the shortest path from the initial state q_0 to v. The key idea is only to add default transitions from state v to state u if $d(u) < d(v)$. This implies that the matching delay is at most $|S|$ on any input string S (see, e.g., Aho and Corasick [2]).

Initialize a D^{2}FA D^2 with no default transitions. We proceed as follows.

Step 1: Calculate Depth Calculate the depth $d(v)$ of each state $v \in Q$ by a breadth-first traversal of D.

Step 2: Construct Default Transitions For each state $u \in Q$ add default transition (u, v) to D^2, where v is the state such that $sim(u, v)$ is maximum and $d(v) < d(u)$.

Step 3: Construct Labeled Transitions Copy every labeled transition from D that is not redundant in D^2.

Step 1 and 3 uses $O(n|\Sigma|)$ time to traverse D. Step 2 uses $O(n^2|\Sigma|)$ to compute the similarity of each pair of states. In total, we use $O(n^2|\Sigma|)$ time.

Fast Compression with Bounded Matching Delay. We now speed up the A-DFA algorithm using sparsification. Let D be the input DFA and let r and k be two constant, positive integer parameters. We initialize a D^{2}FA D^2 with no default transitions, i.e., we set $F(u) = u$ for each $u \in Q$. The algorithm runs in r rounds, where each round proceeds as follows.

First, pick k unique random characters $c_1, \ldots, c_k \in \Sigma$. Then, for each state $v \in Q$ we construct the sequence $V = \delta(v, c_1), \ldots, \delta(v, c_k)$, and hash V into a single hash value $h(v)$. We insert v into a table with key $h(v)$. For each unique hash value h_i, we consider the set of states C_i that hash to that value. For each state $u \in C_i$ we pick another state $v \in C_i$ uniformly at random. If v has lower depth and the default transition (u, v) compresses better than the current default transition of u, i.e., $d(v) < d(u)$ and $sim(u, v) > sim(u, F(u))$, we update the default transition of u to point to v in D^2, otherwise, we continue. After r rounds, the algorithm terminates and returns the resulting D^{2}FA D^2.

Hashing a state and computing the similarity of the potential new default transition uses $O(k|\Sigma|)$ time. Hence, the full algorithm uses $O(rkn + rn|\Sigma|) = O(n|\Sigma|)$ time.

8 Experimental Evaluation

We implemented our methods described in the previous section and measured their performance on regular expressions extracted from widely used intrusion detection systems. The implementation is available at https://github.com/MaxRishoj/fcomp-dfa.

Datasets. We extracted our datasets from regular expressions used in the popular Snort [43], Zeek(formerly Bro) [38], and Suricata [1] intrusion detection systems (see current homepages for these systems at www.snort.org, zeek.org, and suricata.io). The Snort and Zeek datasets are extracted from current versions of the datasets used in most of the previous work.

For each dataset, we extracted prefixes of the rules to generate DFAs of different sizes to explore the scalability of our algorithms on DFAs that have between roughly 1k to 1M states. The details are in the full version of the paper [10]. As in previous work, we have filtered out some rules that used advanced features. All regular expressions use an ASCII alphabet size of 256.

Algorithms Tested. We evaluate the following algorithms.

D^2FA. The algorithm of [29], described in Sect. 4.
D^2FA-LD. The algorithm of [29] for bounded longest delay, described in Sect. 6 with parameter $L = 2$.
D^2FA-LD-CUT. The algorithm in Sect. 6 for bounded longest delay using the SRG instead of the SSRG.
D^2FA-MD. The algorithm of [7,9] for bounded matching delay, described in Sect. 7.
SPARSE-D^2FA. The algorithm in Sect. 5.
SPARSE-D^2FA-LD. As D^2FA-LD using the SSRG instead of the SRG.
SPARSE-D^2FA-LD-CUT. The algorithm in Sect. 6 for bounded longest delay.
SPARSE-D^2FA-MD. The algorithm in Sect. 7 for bounded matching delay.

We use the locality-sensitive hashing scheme from Sect. 5 without replacement and parameters $k = 8$ and $r = 512$. We evaluated several locality-sensitive hashing schemes, including the one in Sect. 5 with replacement and minhash [13] over the set of outgoing transitions with one or k random permutations of the universe. We also evaluated several combinations of r and k and found that $k = 8$ gave the best compression size. Increasing r results in better compression size but increases the compression time linearly. Our chosen variant achieved the best combination of compression size and compression time.

In our longest delay variant, we report results for parameter $L = 2$. We have experimented with other values of L but did not observe significant differences in the relative performances of the algorithm for this variant. We note that the bounded matching delay variant also appears in a more general version, where we can tune the overhead of the default transitions (see the journal version of the result [9]). The version tested here is the simplest version and leads to the best compression size.

Setup. Experiments were run on a machine with an Intel Xeon Gold 6226R 2.9 GHz processor and 128 GB of memory. The operating system was Scientific Linux 7.9 kernel version 3.10.0-1160.80.1.el7.x86_64. Source code was compiled with `g++` version 9.4 with options `-Wall -O4`. The input to each algorithm is a DFA constructed from a set of regular expressions. We measured the time for constructing an equivalent D^2FA for the input DFA, using the `clock` function of the C standard library.

Results. We compare the algorithms across the datasets and measure compression time and compression size (number of states in D^2FA as a percent of the number of states in the input DFA) for each of the variants (general compression, longest delay, and bounded matching delay). The relative performance of our algorithms is similar across the datasets, so we focus on the results for the Snort dataset shown in Fig. 2. The corresponding results for the Suricata and Zeek datasets are in the full version of the paper [10]. We point out whenever there are significant differences between observed results across datasets. We note that the absolute compression size varies significantly across the datasets and variants. Most instances are highly compressible to around 10% of the original DFA (many even in the low single-digit percentages). At the same time, a single one (bounded matching delay on the Zeek dataset) compresses to around 50 percent on the largest DFAs.

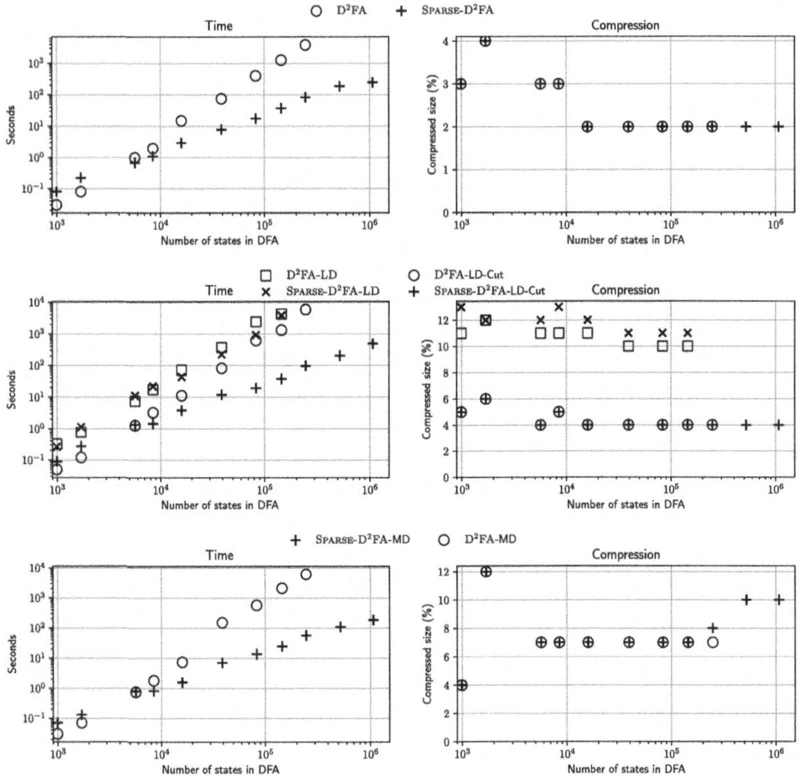

Fig. 2. Results for the Snort dataset on the algorithms for general compression (top), bounded longest delay (middle), and bounded longest matching delay (bottom). On the left, we show compression time in seconds vs. the number of states in the input DFA. On the right, we show the number of transitions in the D^2FA as a percent of the number of transitions in the input DFA.

General Compression. We compare the D^{2}FA and SPARSE-D^{2}FA general compression algorithms. We observe that SPARSE-D^{2}FA compresses up to an order of magnitude faster than D^{2}FA with either no or little loss of compression size. For DFAs with around 1k states, the compression time is comparable and increases gradually to roughly an order of magnitude for the largest DFAs.

Compression with Bounded Longest Delay. We compare the D^{2}FA-LD, SPARSE-D^{2}FA-LD, D^{2}FA-LD-CUT and SPARSE-D^{2}FA-LD-CUT compression algorithms with bounded longest delay. We observe that D^{2}FA-LD and SPARSE-D^{2}FA-LD achieve similar compression times as expected. The compression size achieved by SPARSE-D^{2}FA-LD is around 10–15% worse in the Snort dataset, roughly comparable in the Zeek dataset, and around 100% worse in the Suricata dataset. We believe that the bounded diameter approach in Kumar et al. [29] is highly sensitive to the greedy choice of edges at each step, leading to the observed differences in the compression size across the different datasets.

More importantly, we observe that SPARSE-D^{2}FA-LD-CUT compresses up to an order of magnitude faster than all other algorithms. SPARSE-D^{2}FA-LD-CUT also achieves a substantially better compression size than D^{2}FA-LD and SPARSE-D^{2}FA-LD (except for the Zeek dataset, where the compression size is comparable). Compared to the dense version D^{2}FA-LD-CUT, SPARSE-D^{2}FA-LD-CUT has either no or little loss of compression size.

Compression with Bounded Matching Delay. We compare the compression algorithms D^{2}FA-MD and SPARSE-D^{2}FA-MD with bounded matching delay. We observe that SPARSE-D^{2}FA-MD compresses up to an order of magnitude faster than D^{2}FA-MD. The compression size varies depending on the dataset. For the Snort dataset, SPARSE-D^{2}FA-MD achieves 10–15% worse compression size for DFAs with more than 10k states, for the Zeek dataset, SPARSE-D^{2}FA-MD is comparable, and for the Suricata dataset, SPARSE-D^{2}FA-MD achieves 100% worse compression size for DFAs with more than 10k states. We note that the absolute compression size varies significantly, which may explain this difference in relative compression size.

9 Conclusion and Acknowledgements

We presented a simple, practical framework for constructing D^{2}FA based on locality-sensitive hashing. On DFAs from widely used modern intrusion detection systems, we achieved compression times of up to an order of magnitude faster than existing solutions with either no or little loss of compression size. An interesting open problem is to explore if our new framework can be combined with other DFA compression techniques, such as the ones mentioned in Sect. 1.

This paper is inspired by earlier work in an MSc. thesis [24] supervised by two of the authors. We thank the anonymous reviewers of earlier drafts of this article for many valuable comments that improved the quality of the work.

References

1. https://www.suricata.io/
2. Aho, A.V., Corasick, M.J.: Efficient string matching: an aid to bibliographic search. Commun. ACM **18**(6), 333–340 (1975). https://doi.org/10.1145/360825.360855
3. Antonello, R., Fernandes, S.F.L., Sadok, D., Kelner, J., Szabó, G.: Deterministic finite automaton for scalable traffic identification: the power of compressing by range. In: NOMS 2012, pp. 155–162 (2012). https://doi.org/10.1109/NOMS.2012.6211894
4. Antonello, R., Fernandes, S.F.L., Sadok, D.F.H., Kelner, J., Szabó, G.: Design and optimizations for efficient regular expression matching in DPI systems. Comput. Commun. **61**, 103–120 (2015). https://doi.org/10.1016/j.comcom.2014.12.011
5. Becchi, M., Cadambi, S.: Memory-efficient regular expression search using state merging. In: Proceedings of 26th INFOCOM, pp. 1064–1072 (2007). https://doi.org/10.1109/INFCOM.2007.128
6. Becchi, M., Crowley, P.: A hybrid finite automaton for practical deep packet inspection. In: Proceedings of 3rd CoNEXT Conference, pp. 1–12 (2007)
7. Becchi, M., Crowley, P.: An improved algorithm to accelerate regular expression evaluation. In: Proceedings of ANCS 2007, pp. 145–154 (2007). https://doi.org/10.1145/1323548.1323573
8. Becchi, M., Crowley, P.: Efficient regular expression evaluation: theory to practice. In: Proceedings of ANCS 2008, pp. 50–59 (2008). https://doi.org/10.1145/1477942.1477950
9. Becchi, M., Crowley, P.: A-DFA: A time- and space-efficient DFA compression algorithm for fast regular expression evaluation. ACM Trans. Archit. Code Optim. **10**(1), 4:1–4:26 (2013). https://doi.org/10.1145/2445572.2445576
10. Bille, P., Gørtz, I.L., Pedersen, M.R.: Fast practical compression of deterministic finite automata. arXiv:2306.12771 (2024)
11. Bille, P., Gørtz, I.L., Puglisi, S.J., Tarnow, S.R.: Hierarchical relative lempel-ziv compression. In: Proceedings of 21st SEA (2023)
12. Black, J., Halevi, S., Krawczyk, H., Krovetz, T., Rogaway, P.: UMAC: fast and secure message authentication. In: Wiener, M. (ed.) CRYPTO 1999. LNCS, vol. 1666, pp. 216–233. Springer, Heidelberg (1999). https://doi.org/10.1007/3-540-48405-1_14
13. Broder, A.Z.: On the resemblance and containment of documents. In: Proceedings of SEQUENCES, pp. 21–29 (1997)
14. Broder, A.Z., Charikar, M., Frieze, A.M., Mitzenmacher, M.: Min-wise independent permutations. J. Comput. Syst. Sci. **60**(3), 630–659 (2000). https://doi.org/10.1006/jcss.1999.1690
15. Brodie, B.C., Taylor, D.E., Cytron, R.K.: A scalable architecture for high-throughput regular-expression pattern matching. In: Proceedings of 33rd ISCA, pp. 191–202 (2006). https://doi.org/10.1109/ISCA.2006.7
16. Charikar, M.: Similarity estimation techniques from rounding algorithms. In: Proceedings of 34th STOC, pp. 380–388 (2002). https://doi.org/10.1145/509907.509965
17. Ding, S., Attenberg, J., Suel, T.: Scalable techniques for document identifier assignment in inverted indexes. In: Proceedings of 19th WWW, pp. 311–320 (2010)
18. Douglis, F., Iyengar, A.: Application-specific delta-encoding via resemblance detection. In: Proceedings of USENIX ATC, General Track 2003, pp. 113–126 (2003). http://www.usenix.org/events/usenix03/tech/douglis.html

19. Farley, A.M., Hedetniemi, S.T., Proskurowski, A.: Partitioning trees: matching, domination, and maximum diameter. Int. J. Parallel Program. **10**(1), 55–61 (1981). https://doi.org/10.1007/BF00978378
20. Ficara, D., Giordano, S., Procissi, G., Vitucci, F., Antichi, G., Pietro, A.D.: An improved DFA for fast regular expression matching. Comput. Commun. Rev. **38**(5), 29–40 (2008). https://doi.org/10.1145/1452335.1452339
21. Ficara, D., Pietro, A.D., Giordano, S., Procissi, G., Vitucci, F., Antichi, G.: Differential encoding of dfas for fast regular expression matching. IEEE/ACM Trans. Netw. **19**(3), 683–694 (2011). https://doi.org/10.1109/TNET.2010.2089639
22. Gong, L., Wang, C., Xia, H., Chen, X., Li, X., Zhou, X.: Enabling fast and memory-efficient acceleration for pattern matching workloads: the lightweight automata processing engine. IEEE Trans. Comput. **72**(4), 1011–1025 (2023). https://doi.org/10.1109/TC.2022.3187338
23. Har-Peled, S., Indyk, P., Motwani, R.: Approximate nearest neighbor: towards removing the curse of dimensionality. Theory Comput. **8**(1), 321–350 (2012). https://doi.org/10.4086/toc.2012.v008a014
24. Hemmingsen, M., Lam, B.W.: Fast Compression of DFAs for Intrusion Detection Systems. Master's thesis, Tech. Uni. Denmark. (2021)
25. Kong, S., Smith, R., Estan, C.: Efficient signature matching with multiple alphabet compression tables. In: Proceedings of 4th SECURECOMM, p. 1 (2008). https://doi.org/10.1145/1460877.1460879
26. Krcál, L., Holub, J.: Incremental locality and clustering-based compression. In: DCC 2015, pp. 203–212 (2015). https://doi.org/10.1109/DCC.2015.23
27. Kruskal, J.B.: On the shortest spanning subtree of a graph and the traveling salesman problem. Proc. Am. Math. Soc. **7**(1), 48–50 (1956)
28. Kulkarni, P., Douglis, F., LaVoie, J.D., Tracey, J.M.: Redundancy elimination within large collections of files. In: Proceedings of USENIX ATC, General Track 2004, pp. 59–72 (2004)
29. Kumar, S., Dharmapurikar, S., Yu, F., Crowley, P., Turner, J.S.: Algorithms to accelerate multiple regular expressions matching for deep packet inspection. In: Proceedings of SIGCOMM 2006, pp. 339–350 (2006). https://doi.org/10.1145/1159913.1159952
30. Kumar, S., Turner, J.S., Williams, J.: Advanced algorithms for fast and scalable deep packet inspection. In: Proceedings of ANCS 2006, pp. 81–92 (2006). https://doi.org/10.1145/1185347.1185359
31. Liu, A.X., Torng, E.: An overlay automata approach to regular expression matching. In: Proceedings of 33rd INFOCOM, pp. 952–960 (2014). https://doi.org/10.1109/INFOCOM.2014.6848024
32. Liu, S., Su, S., Liu, D., Huang, Z., Xiao, M.: Efficient compression algorithm for ternary content addressable memory-based regular expression matching. Electron. Lett. **53**(3), 152–154 (2017)
33. Matousek, D., Kubis, J., Matousek, J., Korenek, J.: Regular expression matching with pipelined delayed input dfas for high-speed networks. In: Proceedings of ANCS 2018, pp. 104–110 (2018). https://doi.org/10.1145/3230718.3230730
34. Matousek, D., Matousek, J., Korenek, J.: High-speed regular expression matching with pipelined memory-based automata. In: Proceedings of 26th FCCM, p. 214 (2018). https://doi.org/10.1109/FCCM.2018.00048
35. Meiners, C.R., Patel, J., Norige, E., Torng, E., Liu, A.X.: Fast regular expression matching using small tcams for network intrusion detection and prevention systems. In: 19th USENIX Security, pp. 111–126 (2010)

36. Ouyang, Z., Memon, N.D., Suel, T., Trendafilov, D.: Cluster-based delta compression of a collection of files. In: Proceedings of 3rd WISE, pp. 257–268 (2002). https://doi.org/10.1109/WISE.2002.1181662
37. Patel, J., Liu, A.X., Torng, E.: Bypassing space explosion in high-speed regular expression matching. IEEE/ACM Trans. Netw. **22**(6), 1701–1714 (2014). https://doi.org/10.1109/TNET.2014.2309014
38. Paxson, V.: Bro: a system for detecting network intruders in real-time. Comput. Netw. **31**(23–24), 2435–2463 (1999)
39. Peel, A., Wirth, A., Zobel, J.: Collection-based compression using discovered long matching strings. In: Proceedings of 20th CIKM, pp. 2361–2364 (2011). https://doi.org/10.1145/2063576.2063967
40. Prim, R.C.: Shortest connection networks and some generalizations. Bell Syst. Tech. J. **36**(6), 1389–1401 (1957)
41. Prithi, S., Sumathi, S.: A survey on recent dfa compression techniques for deep packet inspection in network intrusion detection system. J. Electr. Eng. **17**(3), 14–14 (2017)
42. Qi, Y., et al.: FEACAN: front-end acceleration for content-aware network processing. In: Proceedings of 30th INFOCOM, pp. 2114–2122 (2011). https://doi.org/10.1109/INFCOM.2011.5935021
43. Roesch, M.: Snort: lightweight intrusion detection for networks. In: Proceedings of 13th LISA, pp. 229–238 (1999)
44. Roussev, V.: Data fingerprinting with similarity digests. In: IFIP International Conference Digital Forensics 2010, vol. 337, pp. 207–226 (2010). https://doi.org/10.1007/978-3-642-15506-2_15
45. Shankar, S.S., Lin, P., Herkersdorf, A., Wild, T.: A divide and conquer state grouping method for bitmap based transition compression. In: Proceedings of 18th PDCAT, pp. 400–406 (2017). https://doi.org/10.1109/PDCAT.2017.00071
46. Shilane, P., Huang, M., Wallace, G., Hsu, W.: Wan-optimized replication of backup datasets using stream-informed delta compression. ACM Trans. Storage **8**(4), 13:1–13:26 (2012). https://doi.org/10.1145/2385603.2385606
47. Tang, Q., Jiang, L., Dai, Q., Su, M., Xie, H., Fang, B.: RICS-DFA: a space and time-efficient signature matching algorithm with reduced input character set. Concurr. Comput. Pract. Exp. **29**(20) (2017). https://doi.org/10.1002/cpe.3940
48. Tuck, N., Sherwood, T., Calder, B., Varghese, G.: Deterministic memory-efficient string matching algorithms for intrusion detection. In: Proceedings of 23rd INFOCOM, pp. 2628–2639 (2004). https://doi.org/10.1109/INFCOM.2004.1354682
49. Xia, W., Jiang, H., Feng, D., Hua, Y.: Silo: a similarity-locality based near-exact deduplication scheme with low RAM overhead and high throughput. In: USENIX ATC 2011 (2011)
50. Xu, C., Chen, S., Su, J., Yiu, S., Hui, L.C.K.: A survey on regular expression matching for deep packet inspection: applications, algorithms, and hardware platforms. IEEE Commun. Surv. Tutorials **18**(4), 2991–3029 (2016). https://doi.org/10.1109/COMST.2016.2566669
51. Yu, F., Chen, Z., Diao, Y., Lakshman, T.V., Katz, R.H.: Fast and memory-efficient regular expression matching for deep packet inspection. In: Proceedings of ANCS 2006, pp. 93–102 (2006). https://doi.org/10.1145/1185347.1185360

Orienteering (with Time Windows) on Restricted Graph Classes

Kevin Buchin, Mart Hagedoorn(✉), Guangping Li, and Carolin Rehs

Technische Universität Dortmund, Dortmund, Germany
{kevin.buchin,mart.hagedoorn,guangping.li,carlin.rehs}@tu-dortmund.de

Abstract. Given a graph with edge costs and vertex profits and given a budget B, the Orienteering Problem asks for a walk of cost at most B of maximum profit. Additionally, each profit may be given with a time window within which it can be collected by the walk. While the Orienteering Problem and thus the version with time windows are NP-hard in general, it remains open on numerous special graph classes. Since in several applications, especially for planning a route from A to B with waypoints, the input graph can be restricted to tree-like or path-like structures, in this paper we consider orienteering on these graph classes. While the Orienteering Problem with time windows is NP-hard even on undirected paths and cycles, and remains so even if all profits must be collected, we show that for directed paths it can be solved in $\mathcal{O}(m \log m)$ time (where m is the total number of time windows), even if each profit can be collected in one of several time windows. The same case is shown to be NP-hard for directed cycles.

Particularly interesting is the Orienteering Problem on a directed cycle with one time window per profit. We give an efficient algorithm for the case where all time windows are shorter than the length of the cycle, resulting in a 2-approximation for the general setting. Based on the algorithm for directed paths, we further develop a polynomial-time approximation scheme for this problem. For the case where all profits must be collected, we present an $\mathcal{O}(n^4)$-time algorithm. For the Orienteering Problem with time windows for the edges, we give a quadratic time algorithm for undirected paths and observe that the problem is NP-hard for trees.

In the variant without time windows, we show that on trees and thus on graphs with bounded tree-width the Orienteering Problem remains NP-hard. We present, however, an FPT algorithm to solve orienteering with unit profits that we then use to obtain a $(1+\varepsilon)$-approximation algorithm on graphs with arbitrary profits and bounded tree-width, which improves current results on general graphs.

1 Introduction

The *Orienteering Problem* has a wide range of applications. The problem originates from the sport of orienteering: Competitors navigate between control points marked on a map, aiming to visit as many points as possible within a

R. Královič and V. Kůrková (Eds.): SOFSEM 2025, LNCS 15538, pp. 151–165, 2025.
https://doi.org/10.1007/978-3-031-82670-2_12

given time limit. With the addition of different profits for each control point, this is a frequently applied routing problem, for instance in logistics [16,27], tourism [25,30], and journey planning [8,22]. In such applications, the class of graphs to be considered can often be restricted. For instance, when planning a tour from A to B, one can assume that the tour stays close or even largely makes use of the main route between A and B, while taking small detours to visit interesting sites close to this main route. Thus, we can limit the problem to a path-like or tree-like subgraph of the travel network. This scenario is modelled well by graphs with bounded tree-width.

Formally, in the Orienteering Problem (OP) we are given a graph G with $n = |V(G)|$, a starting point $s \in V(G)$, a budget $B \in \mathbb{R}_{>0}$, a cost function $c : E(G) \to \mathbb{R}_{\geq 0}$, and a profit function $\pi : V(G) \to \mathbb{R}_{\geq 0}$. The goal is to find a walk $P = (e_1, \cdots, e_\ell)$, starting at s such that $\sum_{e \in P} c(e) \leq B$, which maximises the total profit, where the profit of each vertex is collected only once, i.e. $\sum_{v \in \bigcup \{v,w | (v,w) \in P\}} \pi(v)$.

The Orienteering Problem has been a popular topic of research in the last decades (see [29] for a survey). Considerable effort has been devoted to designing exact (exponential-time) algorithms [12,13,20,23] and practical heuristics [2,26]. Since –as a generalization of the travelling salesperson problem– the Orienteering Problem is NP-hard [16], theoretical work on the Orienteering Problem has mostly focused on approximation algorithms. The current best approximation algorithm is a $(2 + \varepsilon)$-approximation given by Checkuri et al. [10]. However, the best lower bound on the approximation ratio is at most 1481/1480 as given by Blum et al. [6]. Furthermore, when the sites lie in a fixed-dimensional Euclidean space and the underlying graph is the complete Euclidean graph, a $(1 + \varepsilon)$-approximation algorithm has been proposed by Chen et al. [11].

A natural extension of the Orienteering Problem are versions with *time windows*, i.e., intervals in time. In applications, sites often need to be reached in specific time windows due to the availability of certain services, hours of operation, or deadlines.

In the Orienteering Problem with time windows, additionally each profit is associated with one or more time windows, and it can only be collected if the walk visits the corresponding vertex during one of these windows. We distinguish between the version, in which every vertex has one single time window (OP-1TW) and the version with multiple time windows per vertex (OP-MTW).

A restricted version of the Orienteering Problem with time windows, which is closely related to the Travelling Salesperson Problem with time windows, asks if there is a walk P which collects the profit of every vertex. We denote these problems as the Covering Orienteering Problem with single/multiple time windows (COP-1TW/COP-MTW) and show several hardness results even for these restricted problems.

Problems closely related to the Orienteering Problem with time windows have been studied before. Most frequently considered is the Travelling Salesperson Problem with time windows [1,3,5]. Since NP-hardness for the Travelling Salesperson Problem implies NP-hardness for these versions with time windows, research has been primarily focused on heuristics (see [17] for a survey), approximation approaches [3], or algorithms on restricted graph classes [28]. Most notably, the work done by Tsitsiklis [28] is strongly related to the COP-1TW, since, in some variations discussed by Tsitsiklis, a walk is allowed to visit the same vertex multiple times. Moreover, Garg et al. [14] introduced the multiple time window model for Orienteering, proving it is APX-hard on trees.

While approximation algorithms for the Orienteering Problem with time windows have been considered before, there are no algorithmic results on restricted graph classes. This is particularly surprising since the problem with time windows is interesting from both an application and an algorithmic point of view even on very restricted graph classes: Here, an optimal route might pass an important point at a 'wrong' time and have to come back later.

In this paper, we investigate the Orienteering Problem with time windows. For an overview of the presented results, see Table 1. Even for undirected paths, COP-1TW and OP-1TW are NP-hard, what we can follow from a similar problem considered in [28]. Furthermore, we show that COP-MTW and thus OP-MTW are NP-hard even on directed cycles. Most interestingly this is not true for one single time window per vertex: We give a polynomial-time algorithm for the COP-1TW. Moreover, we present a dynamic program for OP-MTW on directed paths. While the question of NP-hardness remains open for OP-1TW on directed cycles, we give an $\mathcal{O}(n^2 \log n)$-time algorithm for short time windows (which admits an FPT algorithm with respect to the number of long intervals), a 2-approximation, and a polynomial-time approximation scheme.

Table 1. Results for the Orienteering Problem with one single or multiple time windows per vertex, where m is the total number of time windows.

	Single time window	Multiple time windows
Directed path	$\mathcal{O}(n \log n)$, Prop. 1	$\mathcal{O}(m \log m)$, Prop. 1
Directed cycle	COP-1TW: $\mathcal{O}(n^4)$, Thm. 1 OP-1TW: PTAS, Thm. 3 FPT, Cor. 1	NP-hard, Thm. 4
Undirected path	NP-hard [28]	NP-hard [28]

Another natural setting in applications, which is closely related to having time windows for the profits, is that edges cannot be used at all times, for example due to construction sights, road illumination or seasonal scenery. Formally, we obtain the Orienteering Problem in a setting of dynamic graphs (OP-D for short) by adding one or more intervals to each edge in which that edge is traversable. Surprisingly, while the setting seems very similar to the Orienteering Problem with time windows, we show that it is solvable in quadratic time on an undirected path. However, the Orienteering Problem on dynamic graphs remains NP-hard on trees with unit profit and unit cost.

We further consider the Orienteering Problem without time windows. We show that though this problem is NP-hard even on trees (and thus also on graphs with bounded tree-width), we can give a $(1+\varepsilon)$-approximation algorithm on graphs with bounded tree-width. This stands in contrast to the Orienteering Problem on general graphs that does not admit an efficient $(1+\varepsilon)$-approximation unless P = NP.

2 The Orienteering Problem with Time Windows

In this section, we study the Orienteering Problem with time windows. We distinguish between the version of the problem using a single time window per profit (OP-1TW) and using multiple time windows (OP-MTW). An instance of OP-1TW consists of an instance of the Orienteering Problem together with an interval $[r_i, d_i]$ as time window for each of the vertices $v_i \in V(G)$. A walk W through G is now defined by a sequence of tuples (v, t), where v is a vertex and $t \in \mathbb{N}$ is a *time step* such that if (v, t) is followed by (v', t') in W, then $t' \geq t + c(\{v, v'\})$. The profit of a vertex v_i can only be collected once by W and only if W passes v_i at time step t with $r_i \leq t \leq d_i$.

The OP-MTW generalized OP-1TW by replacing the time interval per vertex by a set $\mathcal{I}_{v_i} = \{[r_{i,1}, d_{i,1}], \cdots [r_{i,m_i}, d_{i,m_i}]\}$ of m_i time intervals. The profit of vertex v_i can then be collected at any time step t where there is an interval $[r_{i,j}, d_{i,j}] \in \mathcal{I}_{v_i}$, such that $r_{i,j} \leq t \leq d_{i,j}$. Furthermore, let d_{max} be the latest deadline of all time windows and w.l.o.g. assume that budget $B = d_{max}$.

As we will see in Sect. 4, the Orienteering Problem in general is NP-hard even on trees, and thus remains NP-hard in the version with time windows. However, in the setting of Orienteering Problem with time windows, the problem is NP-hard even for undirected paths, which follows directly from previous results:

The line-TSP problem as described and proven NP-complete by Tsitsiklis [28] is closely related to the OP-1TW on undirected paths even with unit edge cost and unit vertex profit. In line-TSP, each job $j \in J$ has an integer position x_j and time window $[r_j, d_j]$ and travel time between two jobs $j, j' \in J$ is defined as $|x_j - x_{j'}|$. The goal in line-TSP is to find a walk which collects the profit of all the jobs in J. The difference to the OP is that in line-TSP two jobs can have the same position, which is not possible when each vertex has exactly one associated profit and all edges have unit costs. However, it is possible to reduce line-TSP to the OP-1TW on an undirected path with unit edge cost and unit vertex profit by a straightforward reduction.[1]

Remark* 1. *The OP-1TW and thus the OP-MTW on undirected paths is weakly NP-hard even assuming unit edge cost and vertex profit.*

[1] * The proofs of the results marked with $*$ are deferred to the full version of this paper [9].

2.1 Directed Path

While on an undirected path the problem is NP-hard, the Orienteering Problem with time windows is tractable in polynomial time if the input graph is a directed path. This holds even for multiple time windows:

Proposition* 1. *Given an instance of the OP-MTW with a directed path G, arbitrary profits, and arbitrary edge weights, the OP-MTW can be solved in $\mathcal{O}(m \log m)$ time, where m is the total number of time windows.*

Proof (Sketch). Assume w.l.o.g. that each vertex has at least one time window, i.e. $m \geq |V(G)|$. Furthermore, let $V(G) = \{v_1, \cdots, v_n\}$ with $E(G) = \{\{v_i, v_{i+1}\} \mid 1 \leq i < n\}$. For $1 \leq i \leq n$ and $t \in [0, d_{max}]$, let $W_{i,t}$ be a walk with maximum profit starting at v_1 and ending in v_i at time t and let $\Pi_{i,t}$ be the profit of this walk. Moreover, for ease of notation let $c_i = c(\{v_{i-1}, v_i\})$.

For any $1 \leq i \leq n$ and $t, t' \in [0, d_{max}]$ such that $t' > t$, profit $\Pi_{i,t'} \geq \Pi_{i,t}$. Therefore, for any $1 \leq i \leq n$ and $t \in [0, d_{max}]$, if $\exists I \in \mathcal{I}_{v_i}$ s.t. $t \in I$, then $\Pi_{i,t} = \Pi_{i-1,t-c_i} + \pi(v_i)$. If $t \notin I$ for all $I \in \mathcal{I}_{v_i}$ and v_i is not collected by $W_{i,t}$, then $\Pi_{i,t} = \Pi_{i-1,t-c_i}$. If however v_i is collected, then $\Pi_{i,t} = \Pi_{i,d_{i,j}} + \pi(v_i)$, where $d_{i,j}$ is the latest deadline of v_i before time t.

We now give an algorithm that computes $W_{i,d_{max}}$ by maintaining a segment tree T. In the ith iteration of the algorithm, every leaf $[\ell_l, \ell_r)$ of T is augmented with walk $W_{i,l}$ and every node $w \in T$ is augmented with some profit value π_w. Define query $T(t)$ to return the sum of profit values stored in the nodes in the path from the root of T, to the leaf ℓ such that $t \in [\ell_l, \ell_r)$. While iterating, the algorithm maintains values Π_{max} and c. Value Π_{max} is used to keep track of and return the profit of an optimal walk. Cost c will be increased such that in iteration i, it stores the cost of the shortest walk that goes from v_1 to v_i.

For $i = 1$, the only critical time points are $t = 0$ and $t = r_{1,1}$. Hence, for some small constant $\varepsilon > 0$, initialise T with two leaves, i.e. intervals $[0, r_{1,1})$ and $[r_{1,1}, d_{max} + \varepsilon)$, storing profits 0 and $\pi(v_i)$. Then, for any time $t \in [0, d_{max}]$, $T(t)$ returns the optimal profit $\Pi_{1,t}$ for a walk ending in v_1 at time t. Thus, initialise $\Pi_{max} := T(d_{max})$ and $c := 0$.

For the following iterations $1 < i \leq n$, set $c := c + c_i$ and for each $1 \leq j \leq m_i$, update $r'_{i,j} = r_{i,j} - c$ and $d'_{i,j} = d_{i,j} - c$. Then, locate the leaves l_a and l_b in T, such that $r'_{i,j} \in [a_l, a_r)$ and $d'_{i,j} \in [b_l, b_r)$, respectively. Remove $[a_l, a_r)$ from T and insert $[a_l, r'_{i,j})$ and $[r'_{i,j}, a_r)$ into T and store the same profit as in l_a in both leaves. Thereafter, while $T(b_l) + \pi(v_i) \geq T(b_r)$, locate the leaf such that $b_r \in [b_r, b'_r)$ and remove this leaf from T. Update $b_r := b'_r$ and repeat until either $T(b_l) + \pi(v_i) < T(b_r)$ or $b_r = B + \varepsilon$. Then, identify the $\mathcal{O}(\log m)$ subtrees that cover the range $[r_{i,j}, b_r)$ and add $\pi(v)$ to the profit values of each root. After all time windows of v_i have been processed, update $\Pi_{max} := \max(\Pi_{max}, T(d_{max} - c))$.

Hence, after m iterations, Π_{max} can be returned. □

2.2 Directed Cycle

So far we have seen that on an undirected path, the Orienteering Problem with time windows is NP-hard, while on a directed path, it is solvable in polynomial time (regardless of the number of time windows). What seemingly makes the problem for directed path easier is that we cannot revisit vertices. A natural question now is, what happens for directed cycles. In this graph class, a walk can revisit vertices. But in contrast to undirected path, decisions are more limited because we can only walk in one direction (or wait at a vertex). We will see that this already makes the problem more complicated to solve than for directed path. Therefore, we start with the Covering Orienteering Problem (COP) with time windows, which asks if there is a walk such that the profit of every vertex can be collected.

Single Time Windows. We start by considering the COP-1TW model on directed cycles, i.e. every vertex v_i has exactly one time window $[r_i, d_i]$. Recall that a COP-1TW instance is a yes-instance if and only if there exists a walk where all profits are collected.

Lemma* 1. *Given an instance I of the COP-1TW, there is a reduced instance I' where the latest deadline is upper bounded by $4nC$ (where C is the length of cycle), such that I is a yes-instance if and only if I' is a yes-instance.*

Using Lemma 1 the following theorem can be proven.

Theorem 1. *Given an instance of the COP-1TW on a directed cycle G with arbitrary profits and edge weights, the COP-1TW can be solved in $\mathcal{O}(n^4)$.*

Proof. Let C be the total length of the directed cycle. W.l.o.g. by Lemma 1, assume that the latest deadline $d_{max} \leq 4nC$. We initialise a *schedule* T containing for every $1 \leq i \leq n$ and $1 \leq j \leq 4n$ a value $T_{i,j}$. Time $T_{i,j}$ is the time until which the corresponding walk W_T visits the ith vertex for the jth time. There is a natural order on the tuples: we define $i', j' \lhd i, j$ as the lexicographic order first on j' and j then on i' and i. Furthermore, for each $1 \leq i, j \leq n$ let $d(v_i, v_j)$ be the distance of the shortest walk travelling from v_i to v_j. Initially, T is a schedule, where W_T never waits, i.e. $T_{i,j} := d(v_1, v_i) + C \cdot (j-1)$ (see Fig. 1a). We call T a *valid* schedule, if the walk W_T *hits* all intervals, i.e. W_T visits all the vertices within their time intervals. Moreover, W_T is called *proper* if the walk is monotone with respect to time and always either waits at vertices or follows the direction of the cycle. We will modify T iteratively until it is either valid or a conflict occurs, meaning there is no valid schedule. Apply the following modification iteratively: First, schedule T is tested to see if it is valid. If T is valid, W_T can be returned, otherwise the algorithm continues with the following steps. If there exists a vertex v_i with time window $[r_i, d_i]$ that walk W_T does not hit, then we first check if $T_{i,1} > d_i$. If $T_{i,1} > d_i$, then we return that no valid solution exists. Otherwise, set

$$j := \max\{j | T_{i,j} < r_i\} \quad \text{and} \tag{1}$$
$$T_{i,j} := r_i. \tag{2}$$

Further, for all $T_{i',j'}$ with $i, j \lhd i', j'$ set

$$T_{i',j'} := \max(T_{i',j'}, r_i + d(v_i, v_{i'}) + C \cdot (j' - j)). \quad (3)$$

Then, we start again at the beginning of this loop and test if there exists a vertex with a time window that is not currently hit.

Two iterations of the algorithm are illustrated in Fig. 1. To show that the algorithm is correct we show that the algorithm maintains the following loop invariant: Walk W_T is a proper walk on the cycle and, for each $1 \leq i, j \leq n$, $T_{i,j} \leq S_{i,j}$ for any valid schedule S. In the first round of the algorithm, the invariant trivially holds; walk W_T is proper by definition and does not wait, thus if S exists, then for every $1 \leq i, j \leq n$, $T_{i,j} \leq S_{i,j}$.

Assume that the loop invariant holds at the beginning of an iteration and let there be some v_i for which $T_{i,1} > d_i$. Since, the loop invariant is true we know that $d_i < T_{i,1} \leq S_{i,j}$ for any valid schedule. This inequality is only possible if no valid schedule exists, as it directly contradicts the definition of a valid schedule. Thus, the algorithm is correct if it returns that there is no valid schedule.

Next, we show that after updating, W_T is still a proper walk on the cycle. The schedule remains unchanged for all $i', j' \lhd i, j$, thus up to time $T_{i,j}$ the path remains proper. Then, updating $T_{i,j}$ is a safe operation as $r_i > T_{i,j}$, which means that W_T will simply wait at vertex v_i in the jth round. Furthermore, for all $i, j \lhd i', j'$ the last visit times are updated based on the minimum travel time from v_i to $v_{i'}$ with $j' - j$ rounds in between. Thus, W_T remains proper since either $T_{i',j'}$ is the minimum travel time from leaving v_i at r_i or a larger value. As soon as, $T_{i',j'} \geq r_i + d(v_i, v_{i'}) + C \cdot (j' - j)$, for all $i', j' \lhd i'', j''$, value $T_{i'',j''}$ remains unchanged because this part was a proper path and therefore must be greater than or equal to the minimum travel time from v_i. Thus, after one update cycle, walk W_T is still proper.

Second, let T' be the schedule after T has been updated, we need to show that after an iteration the loop invariant holds for schedule T' assuming the invariant holds for T. Trivially, the loop invariant holds for the schedule up to $T'_{i,j}$, since this part has not been modified. Since, the loop invariant is true at the start of an iteration and v_i is not collected, we know $d_i < T_{i,j+1} \leq S_{i,j+1}$. So for any valid solution, vertex v_i must be collected strictly before the $j + 1$th round, thus $S_{i,j} \geq r_i = T'_{i,j}$. Then, for all $i, j \lhd i', j'$, where $r_i + d(v_i, v_{i'}) + C \cdot (j' - j) > T_{i',j'}$, since S corresponds to a proper walk W_S, $S_{i',j'} \geq r_i + d(v_i, v_{i'}) + C \cdot (j' - j) = T'_{i',j'}$. Finally, for all $i, j \lhd i'', j''$, where $r_i + d(v_i, v_{i''}) + C \cdot (j'' - j) \leq T_{i'',j''}$, it holds that $T_{i'',j''} = T'_{i'',j''}$ and hence $S_{i'',j''} \geq T'_{i'',j''}$.

Thus, by our loop invariant, termination occurs only if no valid schedule exists or a valid schedule is found. Due to the choice of j in Eq. 1, for every $1 \leq j \leq 4n$, the value of $T_{i,j}$ can be updated at most once in Eq. 2. Thus, after $\mathcal{O}(n^2)$ iterations, the algorithm must either report a valid schedule or that no such schedule exists. Furthermore, each iteration takes at most $\mathcal{O}(n^2)$ steps. Thus, a valid schedule will be computed in $\mathcal{O}(n^4)$ time if it exists. □

Then, we start again at the beginning of this loop and test if there exists a vertex with a time window that is not currently hit.

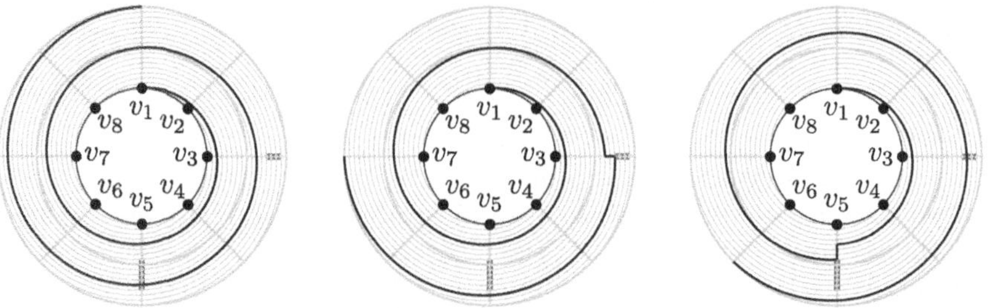

(a) The walk W_T directly after initialisation of schedule T which does not wait.

(b) Vertex v_3 is not collected, hence time $T_{3,2}$ was increased to $r_3 = 12$ and the rest of the schedule is updated sequentially.

(c) Now vertex v_5 is no longer collected, hence time $T_{5,1}$ was increased to $r_5 = 7$.

Fig. 1. An example of two iterations of the algorithm described in Theorem 1. The temporal graphs illustrate possible walks on a clockwise directed cycle with $n = 8$ and unit edge weights. The grey edges cannot be traversed in counterclockwise order and time windows are shown as red intervals on the radial edges.

Thus, COP-1TW can be solved in polynomial time. In Sect. 2.2, we will see that the version with multiple time windows, COP-MTW and thus OP-MTW is NP-hard. The hardness of OP-1TW remains open for future work. However, in this section a 2-approximation algorithm is given using the following theorem:

Theorem 2. *Given an instance of the OP-1TW on a directed cycle G with arbitrary profits and edge weights, and where every time window is shorter than the total length of the cycle, the OP-1TW can be solved in* $\mathcal{O}(n^2 \log n)$ *time.*

Proof. Let C be the cost of traversing cycle G exactly once. W.l.o.g. assume that $d_{max} \leq 4nC$ (Lemma 1). To solve the Orienteering Problem on a directed cycle with 'short' time windows the cycle can be unwrapped and concatenated to itself $4n$ times, resulting in a directed path G'. On this directed path the result of Lemma 1 can be applied.

More formally, for $1 \leq j \leq 4n$ and for each $1 \leq i \leq n$ create a vertex $v_{i,j}$ and add this to $V(G')$ with profit $\pi(v_i)$ and time window $[r_i, d_i]$, where $v_i \in V(G)$. Furthermore, for $1 \leq j \leq 4n$ and $1 \leq i < n$, add an edge $(v_{i,j}, v_{i+1,j})$ to $E(G')$ with the same cost as $(v_i, v_{i+1}) \in E(G)$. Finally, for $1 \leq j < 4n$, add an edge $(v_{n,j}, v_{1,j+1})$ to $E(G')$ with the cost of $(v_n, v_1) \in E(G)$. Let OPT be the profit of an optimal walk W in G and OPT$'$ be the profit of an optimal walk W' in G'. Then it, always holds that OPT $=$ OPT$'$.

Since all time windows are shorter than C, walk W' could not have collected two distinct vertices $v_{i,j}$ and $v_{i,j'}$ for some $1 \leq i \leq n$ and $1 \leq j, j' \leq n$. Therefore, any walk in G can be modified to collect the same profit in G' and vice versa. Hence, Lemma 1 can be used to solve G', which can directly be used as a solution for G. Since, G' has $4n^2$ intervals, the solution to the OP-1TW on cycles with no time windows longer than or equal to C can be calculated in $\mathcal{O}(n^2 \log n)$. □

The problem why a dynamic programming algorithm like the one in Theorem 2 does not extend to 'long' intervals is that when the algorithm encounters such an interval, it does not know whether it already collected the profit in a previous iteration. If all but k intervals are 'short', the algorithm could of course, at the cost of a factor of 2^k in the running time, keep track on the profit depending on which subset of the k 'long' intervals it already collected. Alternatively, we can consider 'short' and 'long' intervals separately, resulting in a 2-approximation.

Corollary 1. *Given an instance of the OP-1TW on a directed cycle with arbitrary profits and edge weights, the OP-1TW is fixed-parameter tractable in the number of time windows that are at least as long as the length of the cycle.*

Corollary* 2. *Given an instance of the OP-1TW on a directed cycle G with arbitrary profits and edge weights, there is a 2-approximation algorithm that runs in $\mathcal{O}(n^2 \log n)$ time.*

In the following we develop a polynomial-time approximation scheme for the OP-1TW. For this, we first consider (sub-)problems that can be optimally solved by a walk taking at most k rounds. A *round* is a walk starting and ending at v_0 visiting every other vertex once. The following lemma makes use generalization of the dynamic program as used in Proposition 1 for directed paths.

Lemma* 2. *Given an instance of the OP-1TW on a directed cycle with arbitrary profits and edge weights where a walk can take at most k rounds, the OP-1TW can be solved in $n^{\mathcal{O}(k)}$ time.*

Now suppose that we would compute for every pair of times $t < t'$ the maximum-profit walk that takes at most k rounds. Unfortunately, concatenating such walks is not optimal, since they might collect the profit of the same vertices. To circumvent this problem we restrict to walks that occasionally perform a *sprint*, that is, a round in which the walk does not wait at vertices. A sprint takes exactly C time.

We define a *k-sprint* as a sequence of at most k rounds of which the last round is a sprint. Given two times $t < t'$ we can compute the maximum-profit *k-sprint* between these times similarly to Lemma 2, and these can be optimally concatenated using dynamic programming. The overall result is not a maximum profit walk but by choosing k suitably depending on ε we obtain a $(1+\varepsilon)$-approximation

Theorem* 3. *Given an instance of the OP-1TW on a directed cycle G with arbitrary profits and edge weights, there is a $(1+\varepsilon)$-approximation algorithm that runs in $n^{\mathcal{O}(1/\varepsilon)}$ time.*

Multiple Time Windows. We show that the COP-MTW on directed cycles is NP-complete, implying NP-completeness for the OP-MTW.

Theorem* 4. *The COP-MTW on directed cycles is NP-complete.*

Proof (Sketch). Clearly this problem lies in NP as we can non-deterministically guess a permutation of vertices and check if we can collect all vertices according to this order.

In the following, we show the NP-hardness by a reduction from the well-known NP-complete problem 3SAT. Consider a 3SAT instance with a set X of n variables and a set Γ of m clauses. In 3SAT, each clause $\gamma \in \Gamma$ has exactly three literals, and the goal is to find an assignment of X such that the formula is satisfied. We reduce this problem to the COP-MTW on a directed cycle with $C = n+m$ vertices. Since in this proof for the time windows only discrete points instead of multiple intervals are needed, time windows will be described by sets of time points. For every $1 \leq i \leq n$, each variable $x_i \in X$ corresponds to v_i in the cycle with time points $\{2iC, 2iC+C-1\}$. Furthermore, for every $1 \leq j \leq m$, clause $\gamma_j \in \Gamma$ corresponds to vertex v_{n+j} with an initial empty time window $\{\}$ (see Fig. 2). Then, for every clause γ_j in which x_i appears as a positive literal, the time point $2iC + (n+j-i)$ is added to vertex v_{n+j}. Otherwise, if x_i appears as a negated literal in γ_j, the time point $2iC + C - (m - j + i) - 1$ is added to v_{n+j} (see Fig. 2 for an example). Note that in this construction, no profit can be collected at time $t \in [iC, 2iC)$ for any $0 \leq i \leq n$.

Then, if there exists a feasible solution in the 3SAT instance, the assignment could be used to choose when to visit the vertices corresponding to the variables. Since every clause is satisfied, this means that in our instance every clause vertex can be reached from at least one vertex variable, giving a valid walk that collects all vertices. Analogously, if a valid walk exists in the COP-MTW instance, we can use the time of collecting the vertices corresponding to the variables for an assignment of the 3SAT instance. □

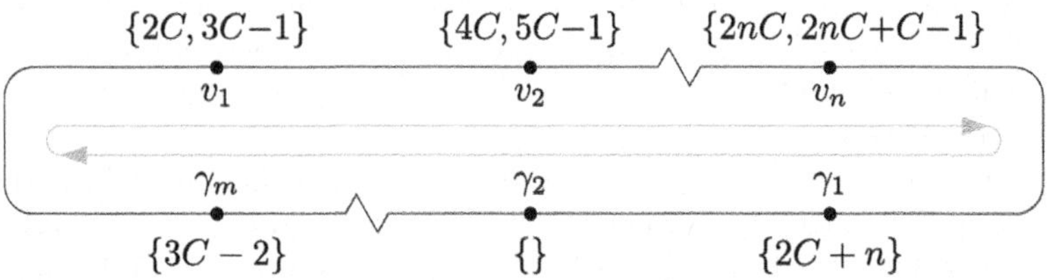

Fig. 2. Example reduction from 3SAT, where v_1 appears in γ_1 as a positive and in γ_m as a negated literal.

3 The Orienteering Problem on Dynamic Graphs

In the Orienteering Problem with time windows, profits are only collectable in certain time windows assigned to the corresponding vertex. A similar extension of the Orienteering Problem is to add time windows to the edges and thus consider the Orienteering Problem in a setting with dynamic graphs. For an instance of the Orienteering Problem on dynamic graphs with input graph $G = (V, E)$, the time windows for edges are sets of intervals $\mathcal{I}_e$ for all $e \in E$. The graph at time step i is then $G_i = (V, E_i)$ with $E_i = \{e \in E \mid \exists I \in \mathcal{I}_e \text{ s.t. } i \in I\}$. We say an edge e is *active* at time step i, if $e \in E_i$.

In contrast to the Orienteering Problem with time windows, directed paths or cycles can be solved trivially by simply advancing the walk whenever possible. Surprisingly, even dynamic undirected paths can be solved with a simple algorithm, as shown in Proposition 2. However, the problem immediately becomes NP-hard when dynamic trees are considered, as shown in Proposition 3.

Proposition* 2. *Given an instance of the Orienteering Problem with a dynamic undirected path* (V, E)*, this instance can be solved in* $\mathcal{O}(n^2)$ *time.*

Proposition* 3. *The Orienteering Problem is NP-hard even on dynamic trees.*

4 The Orienteering Problem Without Time Windows

In general, the Orienteering Problem is NP-hard even on trees and thus also on graphs with bounded tree-width. If vertex profits are unit, the Orienteering Problem is still NP-hard in general [6], but we show that in this case it is possible to give a dynamic program on a tree decomposition. We will extend this result to a $(1+\varepsilon)$-approximation for the Orienteering Problem with arbitrary vertex profits on graphs with bounded tree-width.

Note that, it is quite easy to show that the Orienteering Problem with unit vertex profits is polynomial for a tree, by giving a dynamic program: For every vertex we only need to distinct if it has been passed an even or an uneven number of times. Contrary, for graphs with bounded tree-width, we have to add an additional partition of special structures to the dynamic program.

Lemma* 3. *Given an instance of the Orienteering Problem with unit vertex profit and input tree* G*, the Orienteering Problem is solvable in* $\mathcal{O}(n^3)$ *time.*

Lemma* 4. *Given an instance of the Orienteering Problem with unit vertex profits and input graph* G *of bounded tree-widthk, the Orienteering Problem is solvable in* $\mathcal{O}(n^2) \cdot k^{\mathcal{O}(k)}$ *time.*

While we mainly need result of Lemma 4 to give a $(1+\varepsilon)$-approximation algorithm for the general Orienteering Problem on graphs with bounded tree-width, it also means that the Orienteering Problem with unit vertex profits is fixed-parameter tractable for the parameter tree-width, since a tree decomposition of a fixed width can be computed in linear time [7]. Note that this result has been obtained simultaneously to this work in [24], where the authors consider the P2P-Orienteering problem. This problem is very similar to the Orienteering Problem as we define it, but includes start and end point of the required walk, instead of only a start point, and only considers unit profits. They show that the P2P-Orienteering problem is fixed-parameter-tractable on graphs with bounded tree-width by giving a different dynamic program than ours.

Using similar techniques as in [10,15,18], Lemma 4 can be extended to an approximation for graphs with bounded tree-width and arbitrary profits.

Theorem 5. *There exists a $(1+\varepsilon)$-approximation algorithm for the Orienteering Problem on graphs with bounded tree-widthk that takes $\mathcal{O}(n^6) \cdot k^{\mathcal{O}(k)}$ time.*

Proof. We modify a given instance $((V,E),\pi,c,s)$ s.t. all vertices have unit profit and use Theorem 4 to solve this new instance. First guess the highest vertex profit π_{max} of an optimal walk and remove all vertices with a higher profit. Create a new instance (V',E') by copying (V,E) and adding, for each $v \in V$, $\lfloor n^2\pi(v)/\pi_{max} \rfloor$ additional vertices connected to v with an edge of cost 0. Let $\pi'(v) = \lfloor n^2\pi(v)/\pi_{max} \rfloor + 1$, i.e. when a walk includes v we assume that it also includes all vertices that are connected with 0 cost edges. Now consider a feasible walk W in (V,E) of distinct vertices $v_1, \cdots, v_\ell$ which has profit $\pi(W) = \sum_{i=1}^{\ell} \pi(v_i)$. The same walk W' in (V',E') (including our newly added vertices) collects $\pi'(W') = \sum_{i=1}^{\ell} \pi'(v_i) > n^2\pi(W)/\pi_{max}$. Let OPT$'$ and OPT be the scores of the optimal walks in the modified and original instance, respectively. Then, OPT$'$ $>$ (n^2/π_{max})OPT, however, $\pi'(W) \leq \sum_{i=1}^{\ell} n^2\pi(v_i)/\pi_{max} + 1$. Hence, $(n^2/\pi_{max})\pi(W) \geq \pi'(W') - n \geq \pi'(W') - n \cdot \mathrm{OPT}/\pi_{max}$, and thus we obtain

$$\pi(W) \geq \frac{\pi_{max}}{n^2}\pi'(W') - \frac{1}{n}\mathrm{OPT}.$$

As the aforementioned transformations do not increase the tree-width and the result of Theorem 4 can be applied to obtain solution W' with running time $\mathcal{O}(n^6) \cdot k^{\mathcal{O}(k)}$, solution W' has profit $\pi(W') \geq (1 - \frac{1}{n})$OPT. Hence, when $\frac{1}{n} < \varepsilon$, we can brute force a solution, otherwise the above reduction can be followed. □

Note that we cannot expect to find an exact polynomial-time algorithm for the Orienteering Problem on graphs with bounded tree-width (unless P=NP), since it is already NP-hard for $k = 1$, i.e. on trees.

Proposition* 4. *The Orienteering Problem is NP-hard even on trees.*

5 Conclusion and Open Problems

In this paper, we explore various versions of the Orienteering Problem on tree-like and path-like graph structures. The best known approximation factor for Orienteering on general undirected graphs is $(2+\varepsilon)$ [10]. For graphs with bounded tree-width we achieve a $(1+\varepsilon)$-approximation. Can this approximation factor be extended to a wider class of graphs, in particular planar graphs? For closely related problems like the Prize-Collecting Stroll Problem the $(1+\varepsilon)$-approximation algorithm on planar graphs builds upon an algorithm for bounded tree-width [4].

For the Orienteering Problem with time windows we focus on the case of walks on directed cycles and one time window per profit. This is motivated by the fact that while the problem is easy to solve on directed paths, it is already NP-hard for undirected paths. For directed cycles we successfully give efficient algorithms for several fundamental cases, including the case that all profits need to be collected, but leave the complexity of this problem open in general. We do give a $(1+\varepsilon)$-approximation, and it would be interesting to extend this result to a wider class of graphs, even at the cost of a larger approximation factor, since for general graphs not even an $\mathcal{O}(\log n)$ approximation is known [19].

Finally, in applications of the Orienteering Problem, frequently the Edge Orienteering Problem, where profits are given for travelling over edges instead of vertices, is considered [15,22,31], also with time windows [21]. All our results could be extended to Edge Orienteering and its versions with time windows and dynamic graphs.

References

1. Abeledo, H.G., Fukasawa, R., Pessoa, A.A., Uchoa, E.: The time dependent traveling salesman problem: polyhedra and algorithm. Math. Program. Comput. **5**(1), 27–55 (2013). https://doi.org/10.1007/S12532-012-0047-Y
2. Archetti, C., Speranza, M.G., Vigo, D.: Vehicle routing problems with profits. In: Vehicle Routing: Problems, Methods, and Applications, 2nd edn., pp. 273–297. SIAM (2014). https://doi.org/10.1137/1.9781611973594.ch10
3. Bansal, N., Blum, A., Chawla, S., Meyerson, A.: Approximation algorithms for deadline-TSP and vehicle routing with time-windows. In: Babai, L. (ed.) Proceedings of the 36th Annual ACM Symposium on Theory of Computing, Chicago, IL, USA, 13–16 June 2004, pp. 166–174. ACM (2004). https://doi.org/10.1145/1007352.1007385
4. Bateni, M., Chekuri, C., Ene, A., Hajiaghayi, M.T., Korula, N., Marx, D.: Prize-collecting Steiner problems on planar graphs. In: Proceedings of 22nd Annual ACM-SIAM Symposium on Discrete Algorithms, pp. 1028–1049. SIAM (2011). https://doi.org/10.1137/1.9781611973082.79
5. Bigras, L., Gamache, M., Savard, G.: The time-dependent traveling salesman problem and single machine scheduling problems with sequence dependent setup times. Disc. Optim. **5**(4), 685–699 (2008). https://doi.org/10.1016/J.DISOPT.2008.04.001
6. Blum, A., Chawla, S., Karger, D.R., Lane, T., Meyerson, A., Minkoff, M.: Approximation algorithms for orienteering and discounted-reward TSP. SIAM J. Comput. **37**(2), 653–670 (2007). https://doi.org/10.1137/050645464
7. Bodlaender, H.L.: A linear-time algorithm for finding tree-decompositions of small treewidth. SIAM J. Comput. **25**(6), 1305–1317 (1996). https://doi.org/10.1137/S0097539793251219
8. Buchin, K., Hagedoorn, M., Li, G.: Tour4me: a framework for customized tour planning algorithms. In: Proceedings of the 30th International Conference on Advances in Geographic Information Systems. SIGSPATIAL 2022. Association for Computing Machinery, New York (2022). https://doi.org/10.1145/3557915.3560992
9. Buchin, K., Hagedoorn, M., Li, G., Rehs, C.: Orienteering (with time windows) on restricted graph classes (2024). https://arxiv.org/abs/2410.12401
10. Chekuri, C., Korula, N., Pál, M.: Improved algorithms for orienteering and related problems. ACM Trans. Algor. (TALG) **8**(3), 1–27 (2012). https://doi.org/10.1145/2229163.2229167

11. Chen, K., Har-Peled, S.: The euclidean orienteering problem revisited. SIAM J. Comput. **38**(1), 385–397 (2008). https://doi.org/10.1137/060667839
12. Feillet, D., Dejax, P., Gendreau, M.: Traveling salesman problems with profits. Transp. Sci. **39**(2), 188–205 (2005). https://doi.org/10.1287/trsc.1030.0079
13. Fischetti, M., Gonzalez, J.J., Toth, P.: Solving the orienteering problem through branch-and-cut. INFORMS J. Comput. **10**(2), 133–148 (1998). https://doi.org/10.1287/ijoc.10.2.133
14. Garg, N., Khanna, S., Kumar, A.: Hardness of approximation for orienteering with multiple time windows. In: Marx, D. (ed.) Proceedings of the 2021 ACM-SIAM Symposium on Discrete Algorithms, SODA 2021, Virtual Conference, 10–13 January 2021, pp. 2977–2990. SIAM (2021). https://doi.org/10.1137/1.9781611976465.177
15. Gavalas, D., Konstantopoulos, C., Mastakas, K., Pantziou, G., Vathis, N.: Approximation algorithms for the arc orienteering problem. Inf. Process. Lett. **115**(2), 313–315 (2015). https://doi.org/10.1016/j.ipl.2014.10.003
16. Golden, B.L., Levy, L., Vohra, R.: The orienteering problem. Naval Res. Logist. (NRL) **34**(3), 307–318 (1987). https://doi.org/10.1002/1520-6750(198706)34:3⟨307::AID-NAV3220340302⟩3.0.CO;2-D
17. Gunawan, A., Lau, H.C., Vansteenwegen, P.: Orienteering problem: a survey of recent variants, solution approaches and applications. Eur. J. Oper. Res. **255**(2), 315–332 (2016). https://doi.org/10.1016/J.EJOR.2016.04.059
18. Korula, N.: Approximation algorithms for network design and orienteering. University of Illinois at Urbana-Champaign (2010). https://www.ideals.illinois.edu/items/16799
19. Korula, N.: Orienteering problems. In: Encyclopedia of Algorithms, pp. 1481–1484. Springer, Heidelberg (2016). https://doi.org/10.1007/978-1-4939-2864-4_540
20. Laporte, G., Martello, S.: The selective travelling salesman problem. Disc. Appl. Math. **26**(2), 193–207 (1990). https://doi.org/10.1016/0166-218X(90)90100-Q
21. Lu, Y., et al.: Scenic routes now: efficiently solving the time-dependent arc orienteering problem. In: Proceedings of the 2017 ACM on Conference on Information and Knowledge Management, pp. 487–496 (2017). https://doi.org/10.1145/3132847.3132874
22. Lu, Y., Shahabi, C.: An arc orienteering algorithm to find the most scenic path on a large-scale road network. In: Proceedings of 23rd SIGSPATIAL International Conference on Advances in Geographic Information Systems, pp. 1–10 (2015). https://doi.org/10.1145/2820783.2820835
23. Ramesh, R., Yoon, Y.S., Karwan, M.H.: An optimal algorithm for the orienteering tour problem. ORSA J. Comput. **4**(2), 155–165 (1992). https://doi.org/10.1287/ijoc.4.2.155
24. Ren, K., Salavatipour, M.R.: Approximation schemes for orienteering and deadline tsp in doubling metrics. arXiv preprint arXiv:2405.00818 (2024)
25. Souffriau, W., Vansteenwegen, P., Vertommen, J., Vanden Berghe, G.: A personalized tourist trip design algorithm for mobile tourist guides. Appl. Artif. Intell. **22**, 964–985 (2008). https://doi.org/10.1080/08839510802379626
26. Stavropoulou, F., Repoussis, P.P., Tarantilis, C.D.: The vehicle routing problem with profits and consistency constraints. Eur. J. Oper. Res. **274**(1), 340–356 (2019). https://doi.org/10.1016/j.ejor.2018.09.046
27. Tsiligirides, T.: Heuristic methods applied to orienteering. J. Oper. Res. Soc. **35**(9), 797–809 (1984). https://doi.org/10.2307/2582629

28. Tsitsiklis, J.N.: Special cases of traveling salesman and repairman problems with time windows. Networks **22**(3), 263–282 (1992). https://doi.org/10.1002/NET.3230220305
29. Vansteenwegen, P., Souffriau, W., Oudheusden, D.V.: The orienteering problem: a survey. Eur. J. Oper. Res. **209**(1), 1–10 (2011). https://doi.org/10.1016/j.ejor.2010.03.045
30. Vansteenwegen, P., Van Oudheusden, D.: The mobile tourist guide: an or opportunity. OR insight **20**, 21–27 (2007). https://doi.org/10.1057/ori.2007.17
31. Verbeeck, C., Vansteenwegen, P., Aghezzaf, E.H.: An extension of the arc orienteering problem and its application to cycle trip planning. Transp. Res. Part E **68**, 64–78 (2014). https://doi.org/10.1016/j.tre.2014.05.006

Massively Parallel Maximum Coverage Revisited

Thai Bui and Hoa T. Vu(✉)

San Diego State University, San Diego, CA 92182, USA
{tbui8182,hvu2}@sdsu.edu

Abstract. We study the maximum set coverage problem in the massively parallel model. In this setting, m sets that are subsets of a universe of n elements are distributed among m machines. In each round, these machines can communicate with each other, subject to the memory constraint that no machine may use more than $\tilde{O}(n)$ memory. The objective is to find the k sets whose coverage is maximized. We consider the regime where $k = \Omega(m)$ (i.e., $k = m/100$), $m = O(n)$, and each machine has $\tilde{O}(n)$ memory[1].

Maximum coverage is a special case of the submodular maximization problem subject to a cardinality constraint. This problem can be approximated to within a $1 - 1/e$ factor using the greedy algorithm, but this approach is not directly applicable to parallel and distributed models. When $k = \Omega(m)$, to obtain a $1 - 1/e - \epsilon$ approximation, previous work either requires $\tilde{O}(mn)$ memory per machine which is not interesting compared to the trivial algorithm that sends the entire input to a single machine, or requires $2^{O(1/\epsilon)}n$ memory per machine which is prohibitively expensive even for a moderately small value ϵ.

Our result is a randomized $(1-1/e-\epsilon)$-approximation algorithm that uses

$$O(1/\epsilon^3 \cdot \log m \cdot (\log(1/\epsilon) + \log m))$$

rounds. Our algorithm involves solving a slightly transformed linear program of the maximum coverage problem using the multiplicative weights update method, classic techniques in parallel computing such as parallel prefix, and various combinatorial arguments.

1 Introduction

Maximum coverage is a classic NP-Hard problem. In this problem, we have m sets $S_1, S_2, \ldots, S_m$ that are subsets of a universe of n elements $[n] = \{1, 2, \ldots, n\}$. The goal is to find k sets that cover the maximum number of elements. In the offline model, the greedy algorithm achieves a $1 - 1/e$ approximation and assuming P $\neq$ NP, this approximation is the best possible in polynomial time [8].

This work is supported by the National Science Foundation under Grant No. 2342527.
[1] The input size is $O(mn)$ and each machine has the memory enough to store a constant number of sets.

R. Královič and V. Kůrková (Eds.): SOFSEM 2025, LNCS 15538, pp. 166–178, 2025.
https://doi.org/10.1007/978-3-031-82670-2_13

However, the greedy algorithm for maximum coverage and the related set cover problem is not friendly to streaming, distributed, and massively parallel computing. A large body of work has been devoted to designing algorithms for these problems in these big data computation models. An incomplete list of work includes [3–7,9,11–13,16,21,22,24,25].

Some example applications of maximum coverage includes facility and sensor placement [17], circuit layout and job scheduling [10], information retrieval [1], market design [15], data summarization [24], and social network analysis [13].

The MPC Model. We consider the massively parallel computation model (MPC) in which m sets $S_1, S_2, \ldots, S_m \subseteq [n]$ are distributed among m machines. Each machine has memory $\tilde{O}(n)$ and holds a set. In each round, each machine can communicate with others with the constraint that no machine receives a total message of size more than $\tilde{O}(n)$. Similar to previous work in the literature, we assume that $m \leq n$.

The MPC model, introduced by Karloff, Suri, and Vassilvitskii [14] is an abstraction of various modern computing paradigms such as MapReduce, Hadoop, and Spark.

Previous Work. This problem is a special case of submodular maximization subject to a cardinality constraint. The results of Liu and Vondrak [20], Barbosa et al. [23], Kumar et al. [18] typically require that each machine has enough memory to store $O(\sqrt{km})$ items which are sets in our case (and storing a set requires $\tilde{O}(n)$ memory) with $\sqrt{m/k}$ machines. When $k = \Omega(m)$ (e.g., $k = m/100$), this means that a single machine may need $\tilde{O}(mn)$ memory. This is not better than the trivial algorithm that sends the entire input to a single machine and solves the problem in 1 round.

Assadi and Khanna gave a randomized $1 - 1/e - \epsilon$ approximation algorithm in which each machine has $\tilde{O}\left(m^{\delta/\epsilon} n\right)$ memory and the number of machines is $m^{1-\delta/\epsilon}$ for any $\epsilon, \delta \in (0, 1)$ (see Corollary 10 in the full paper of [4]). Setting $\delta = \Theta(1/\log m)$ gives us a $1 - 1/e - \epsilon$ approximation in $O(1/\epsilon \cdot \log m)$ rounds with $O(m)$ machines each of which uses $\tilde{O}\left(2^{1/\epsilon} n\right)$ memory. While Assadi and Khanna's result is nontrivial in this regime, the dependence on ϵ is exponential and if n is large, then even a moderately small value of $\epsilon = 0.01$ can lead to a prohibitively large memory requirement $\approx 2^{100} n$. Their work however can handle the case where $k = o(m)$.

Our Result. We present a relatively simple randomized algorithm that achieves a $1 - 1/e - \epsilon$ approximation in $O(1/\epsilon^3 \cdot \log m \cdot (\log(1/\epsilon) + \log m))$ rounds with $\tilde{O}(n)$ memory per machine assuming $k = \Omega(m)$. Our space requirement does not depend on ϵ compared to the exponential dependence in Assadi and Khanna's result.

We note that assuming $k = \Omega(m)$ does not make the problem any easier since there are still exponentially many solutions to consider. In practice, one can think of many applications where one can utilize a constant fraction of the available sets (e.g., 10% or 20%). We state our main result as a theorem below.

Theorem 1. *Assume* $k = \Omega(m)$ *and there are* m *machines each of which has* $\tilde{O}(n)$ *memory. There exists an algorithm that with high probability finds* k *sets that cover at least* $(1-1/e-\epsilon)\text{OPT}$ *elements in* $O(1/\epsilon^3 \cdot \log m \cdot (\log(1/\epsilon)+\log m))$ *rounds.*

If the maximum frequency f (the maximum number of sets that any element belongs to) is bounded, we can drop the assumption that $k = \Omega(m)$, and parameterize the number of rounds based on f. In particular, we can obtain a $1 - 1/e - \epsilon$ approximation in $O(f^3/\epsilon^6 \cdot \log^2(\frac{kf}{\epsilon}))$ rounds.

Remark. We could easily modify our algorithm so that each machine uses $\tilde{O}(1/\epsilon \cdot n)$ memory and the number of rounds is $O(1/\epsilon^2 \cdot \log m \cdot (\log(1/\epsilon)+\log m))$. At least one $\log m$ factor is necessary based on the lower bound given by Corollary 9 of [4].

Randomization appears in two parts of our algorithms: the rounding step and the subsampling step to reduce the number of rounds from $\log m \cdot \log n$ to $\log m \cdot (\log(1/\epsilon) + \log m)$. If we only need to compute an approximation to the optimal coverage value such that the output is in the interval $[(1-\epsilon)\text{OPT}, \text{OPT}/(1-1/e-\epsilon)]$, then we have a deterministic algorithm that runs in $O(1/\epsilon^3 \cdot \log n \cdot \log m)$ rounds. The algorithm by Assadi and Khanna [4] combines the sample-and-prune framework with threshold greedy. This strategy requires sampling sets. It is unclear how to derandomize their algorithm even just to compute an approximation to the optimal coverage value.

Our Techniques and Paper Organization. In Sect. 2.1, we transform the standard linear program for the maximum coverage problem into an equivalent packing linear program that can be solved "approximately" by the multiplicative weights update method. At a high level, the multiplicative weights update method gives us a fractional solution that is a $1 - 1/e - O(\epsilon)$ bi-criteria approximation where $(1 + O(\epsilon))k$ "fractional" sets cover $(1 - 1/e - O(\epsilon))\text{OPT}$ "fractional" elements. We then show how to find k sets covering $(1 - 1/e - O(\epsilon))\text{OPT}$ elements from this fractional solution through a combinatorial argument and parallel prefix.

Section 2.2 outlines the details to solve the transformed linear program in the MPC model. While this part is an adaptation of the standard multiplicative weights, an implementation in the MPC model requires some additional details such as the number of bits to represent the weights. All missing proofs and detailed calculation can be found in the full version.

Preliminaries. In this work, we will always consider the case where each machine has $\tilde{O}(n)$ memory and $m \le n$. Without loss of generality, we may assume the non-central machine j stores the set S_j. For each element $i \in [n]$, we use f_i to denote the number of sets that i is in. This is also referred to as the frequency of i. Assume each machine has $\tilde{O}(n)$ space. The vector $\mathbf{f}$ can be computed in $O(\log m)$ rounds and broadcasted to all machines. Each machine j starts with the characteristic vector $v_j \in \{0,1\}^n$ of the set S_j that it holds. The vector $\mathbf{f}$ is just the sum of the characteristic vectors of the sets. We can aggregate the vectors $\{v_j\}$ in $O(\log m)$ rounds using the standard binary tree aggregation algorithm.

Since in this work, the dependence on $1/\epsilon$ is polynomial, an $\alpha - O(\epsilon)$ approximation can easily be translated to an $\alpha - \epsilon$ approximation by scaling ϵ by a constant factor. We can also assume that $1/\epsilon < n/10$; otherwise, we can simulate the greedy algorithm in $O(1/\epsilon)$ rounds. For the sake of exposition, we will not attempt to optimize the constants in our algorithm and analysis. Finally, in this work we consider $1 - 1/\operatorname{poly}(m)$ as a "high probability". We use $[E]$ to denote the indicator variable of the event E that is 1 if E happens and 0 otherwise.

2 Algorithm

2.1 The Main Algorithm

Linear Programming (re)formulation. We first recall the relaxed linear program (LP) for the maximum coverage problem Π_0:

$$\begin{aligned}
\text{maximize} \quad & \sum_{i\in[n]} x_i \\
\text{(s.t.)} \quad & x_i \le \sum_{S_j \ni i} y_j && \forall i \in [n] \\
& \sum_{j\in[m]} y_j = k \\
& x_i, y_j \in [0,1] && \forall i \in [n], j \in [m].
\end{aligned}$$

We first reformulate this LP and then approximately solve the new LP using the multiplicative weights update method [2]. For each $j \in [m]$, let $z_j := 1 - y_j$. We have the following fact.

Fact 1. *For each* $i \in [n]$, $x_i \le \sum_{S_j \ni i} y_j \iff x_i + \sum_{S_j \ni i} z_j \le \sum_{S_j \ni i} (y_j + z_j) = f_i$.

Note that if $\mathbf{y} \in [0,1]^m$ and $\sum_j y_j = k$, then $\mathbf{z} \in [0,1]^m$ and $\sum_j z_j = m - k$. Thus, it is not hard to see that the original LP is equivalent to the following LP which we will refer to as Π_1.

$$\begin{aligned}
\text{maximize} \quad & \sum_{i\in[n]} x_i \\
\text{(s.t.)} \quad & \frac{x_i}{f_i} + \frac{1}{f_i} \cdot \sum_{S_j \ni i} z_j \le 1 && \forall i \in [n] \\
& \sum_{j\in[m]} z_j = m - k \\
& x_i, z_j \in [0,1] && \forall i \in [n], j \in [m].
\end{aligned}$$

In this section, we will assume the existence an MPC algorithm that approximately solves the linear program Π_1 in $O(1/\epsilon^3 \cdot \log n \cdot \log m)$ rounds. The proof will be deferred to Sect. 2.2.

Theorem 2. *There is an algorithm that finds* $\boldsymbol{x} \in [0,1]^n, \boldsymbol{z} \in [0,1]^m$ *such that*

1. $\sum_{i\in[n]} x_i \geq (1-\epsilon)\text{OPT}$,
2. $\sum_{j\in[m]} z_j = m - k$, *and*
3. $\frac{x_i}{f_i} + \frac{1}{f_i} \cdot \sum_{S_j \ni i} z_j \leq 1 + \epsilon \quad \forall i \in [n]$.

in $O(\epsilon^{-3} \log n \cdot \log m)$ *rounds.*

Let $\mathbf{x}$ and $\mathbf{z}$ be the be the output given by Theorem 2. Then, let $\mathbf{x}' = \mathbf{x}/(1+\epsilon)$, $\mathbf{z}' = \mathbf{z}/(1+\epsilon)$, and $\mathbf{y}' = \mathbf{1} - \mathbf{z}'$. We have

$$\sum_{i=1}^{n} x'_i = \frac{1}{1+\epsilon}\sum_{i=1}^{n} x_i \geq \frac{1-\epsilon}{1+\epsilon}\text{OPT} > (1-4\epsilon)\text{OPT}, \tag{1}$$

$$\sum_{j=1}^{m} y'_i = \sum_{j=1}^{m}\left(1 - \frac{z_j}{1+\epsilon}\right) = m - \frac{m-k}{1+\epsilon} \leq m - (1-2\epsilon)(m-k) < k + 2\epsilon m, \tag{2}$$

$$x'_i + \sum_{S_j \ni i} z'_j \leq f_i \iff x'_i \leq \sum_{S_j \ni i} y'_j, \quad \forall i \in [n], \text{ by Fact 1.} \tag{3}$$

Thus, by setting $\mathbf{x} \leftarrow \mathbf{x}/(1+\epsilon)$, and $\mathbf{y} \leftarrow \mathbf{y}'$, we have an approximate solution $\mathbf{x} \in [0,1]^n, \mathbf{y} \in [0,1]^n$ to the LP Π_0 such that

$$\sum_{i=1}^{n} x_i \geq (1-4\epsilon)\text{OPT}, \qquad \sum_{j=1}^{m} y_j \leq k + 2\epsilon m, \text{ and}$$
$$x_i \leq \sum_{S_j \ni i} y_j, \forall i \in [n].$$

We can then apply the standard randomized rounding to find a sub-collection of at most $k + 2\epsilon m$ sets that covers at least $(1-4\epsilon)\text{OPT}$ elements. For the sake of completeness, we will provide the rounding algorithm in the MPC model in the following lemma.

Lemma 1. *Suppose* $\boldsymbol{x} \in [0,1]^n$ *and* $\boldsymbol{y} \in [0,1]^m$ *satisfy:*

1. $\sum_{i\in[n]} x_i \geq L$,
2. $x_i \leq \sum_{S_j \ni i} y_j$ *for all* $i \in [n]$,
3. $\sum_{j\in[m]} y_j = k$,
4. $x_i, y_j \in [0,1]$ *for all* $i \in [n]$ *and* $j \in [m]$.

Then there exists a rounding algorithm that finds a sub-collection of k sets that in expectation cover at least $(1-1/e)L$ elements in $O(1)$ round. To obtain a high probability guarantee, the algorithm requires $O(1/\epsilon \cdot \log m)$ rounds to find k sets that cover least $(1-1/e-O(\epsilon))L$ elements.

Applying Lemma 1 to $\mathbf{x}$ and $\mathbf{y}$ with $k + 2\epsilon m$ in place of k, we obtain a sub-collection of at most $k + 2\epsilon m$ sets that covers at least $(1 - 1/e - O(\epsilon))$OPT elements. Since we assume that $k = \Omega(m)$, that means we have found $k + O(\epsilon)k$ sets that cover at least $(1 - 1/e - O(\epsilon))$OPT elements. The next lemma shows that we can find k sets among these that cover at least $(1 - 1/e - O(\epsilon))$OPT elements. The proof is a combination of a counting argument and the well-known parallel prefix algorithm [19].

Algorithm 1: Parallel prefix coverage

Compute $|S_1|, |S_2 \setminus S_1|, |S_3 \setminus (S_1 \cup S_2)|, \ldots, |S_k \setminus (S_1 \cup \ldots \cup S_{k-1})|$ in $O(\log k)$ rounds.

Function `PrefixCoverage`($S_1, S_2, \ldots, S_k$):

  `//` `Compute` $|S_1|, |S_2 \cup S_1|, |S_3 \cup S_2 \cup S_1|, \ldots, |S_k \cup S_{k-1} \cup \ldots \cup S_1|$

  if $k = 1$ **then**

    return $|S_1|$.

  else

    `// Assume` k `is even.`

    In one round, machine $2j - 1$ sends S_{2j-1} to machine $2j$, then machine $2j$ computes $Q_j = S_{2j-1} \cup S_{2j}$.

    Run `PrefixCoverage`($Q_1, Q_2, \ldots, Q_{k/2}$) on machines $2, 4, 6, \ldots, k$.

    Machine j now has $S_1 \cup S_2 \cup S_3 \cup \ldots \cup S_j$ for $j = 2, 4, 6, \ldots, k$.

    In one round, machine $j = 1, 3, 5, \ldots, k - 1$ communicates with machine $j - 1$ which has $S_1 \cup S_2 \cup \ldots S_{j-1}$ and computes $S_1 \cup S_2 \cup \ldots \cup S_j$.

    `// If` k `is odd, run the above algorithm on` $S_1, S_2, \ldots, S_{k-1}$ `and then compute` $S_1 \cup S_2 \cup \ldots \cup S_k$ `in one round.`

    Each machine j communicates with machine $j - 1$ to compute $|S_1 \cup S_2 \cup \ldots \cup S_j| - |S_1 \cup S_2 \cup \ldots \cup S_{j-1}|$ in one round.

We rely on the following result which is a simulation of the parallel prefix.

Lemma 2. *Suppose there are k sets and machine j holds the set S_j. Then Algorithm 1 computes $|S_1|, |S_2 \setminus S_1|, |S_3 \setminus (S_1 \cup S_2)|, |S_4 \setminus (S_1 \cup S_2 \cup S_3)|, \ldots, |S_k \cup S_{k-1} \cup \ldots \cup S_1|$ in $O(\log k)$ rounds.*

Proof. We first show how to compute $(S_1), (S_1 \cup S_2), (S_1 \cup S_2 \cup S_3), \ldots, (S_1 \cup S_2 \cup \ldots \cup S_k)$ in $O(\log k)$ rounds where machine j holds $S_1 \cup S_2 \cup \ldots \cup S_j$ at the end. Once this is done, machine j can send $S_1 \cup S_2 \cup \ldots \cup S_j$ to machine $j + 1$ and machine $j + 1$ can compute $|S_1 \cup S_2 \cup \ldots \cup S_{j+1}| - |S_1 \cup S_2 \cup \ldots \cup S_j| = |S_{j+1} \setminus (S_1 \cup S_2 \cup \ldots \cup S_j)|$.

The algorithm operates recursively. In one round, machine $2j - 1$ sends S_{2j-1} to machine $2j$, then machine $2j$ computes $Q_j = S_{2j-1} \cup S_{2j}$. Assuming k is even, the algorithm recursively computes $(Q_1), (Q_1 \cup Q_2), (Q_1 \cup Q_2 \cup Q_3), \ldots, (Q_1 \cup Q_2 \cup \ldots \cup Q_k)$ on machines $2, 4, \ldots, k$. After recursion, machines with even indices $2j$ has the set $S_1 \cup S_2 \cup \ldots \cup S_{2j}$. Then, in one round, machines with odd indices $2j + 1$ communicate with machine $2j$ to learn about $S_1 \cup S_2 \cup \ldots \cup S_{2j+1}$. If k is odd, we just do the same on $S_1, S_2, \ldots, S_{k-1}$ and then compute $S_1 \cup S_2 \cup \ldots \cup S_k$ in one round.

There are $O(\log k)$ recursion levels and therefore, the total number of rounds is $O(\log k)$.

Lemma 3. *Let $\mathcal{S} = \{S_1, \ldots, S_r\}$ be a collection of $r = (1+\gamma)k$ sets whose union contains L elements where $\gamma \in [0, 1)$, then there exist k sets in $\mathcal{S}$ whose union contains at least $(1-\gamma)L$ elements. Furthermore, we can find these k sets in $O(\log r)$ rounds.*

Proof. Consider the following quantities $\phi_1 = |S_1|, \phi_2 = |S_1 \cup S_2| - |S_1|, \phi_3 = |S_1 \cup S_2 \cup S_3| - |S_1 \cup S_2|, \ldots$

Clearly, $\sum_{j=1}^{r} \phi_j = L$. We say S_j is responsible for element i if $i \in S_j \setminus (\bigcup_{l<j} S_l)$. This establishes a one-to-one correspondence between the sets $S_1, \ldots, S_r$ and the elements they cover. S_j is responsible for exactly ϕ_j elements. Furthermore, if we remove some sets from $\mathcal{S}$, and an element becomes uncovered, the set responsible for that element must have been removed. Thus, if we remove the γk sets corresponding to the γk smallest ϕ_j, then at most γL elements will not have a responsible set. Thus, the number of elements that become uncovered is at most γL.

To find these sets, we apply Lemma 2 with r in place of k and $O(\epsilon)$ in place of γ to learn about $\phi_1, \phi_2, \ldots, \phi_r$ in $O(\log r) = O(\log k)$ rounds. We then remove the $\gamma k = O(\epsilon)k$ sets corresponding to the γk smallest ϕ_j and output the remaining k sets.

Putting it all Together. We spend $O(1/\epsilon^3 \cdot \log n \cdot \log m)$ rounds to approximately solve the linear program Π_1. From there, we can round the solution to find a sub-collection of $k + O(\epsilon)k$ sets that cover at least $(1 - 1/e - O(\epsilon))\text{OPT}$ elements with high probability in $O(1/\epsilon \cdot \log m)$ rounds. We then apply Lemma 3 to find k sets among these that cover at least $(1 - 1/e - O(\epsilon))\text{OPT}$ elements in $O(\log k)$ rounds. The total number of rounds is therefore $O(1/\epsilon^3 \cdot \log n \cdot \log m)$.

Reducing the Number of Rounds to $O(1/\epsilon^3 \cdot \log m \cdot (\log m + \log(1/\epsilon)))$. The described algorithm runs in $O(1/\epsilon^3 \cdot \log n \cdot \log m)$ rounds. Our main result in Theorem 1 states a stronger bound $O(1/\epsilon^3 \cdot \log m \cdot (\log m + \log(1/\epsilon)))$ rounds. We achieve this by adopting the sub-sampling framework of McGregor and Vu [22].

Without loss of generality, we may assume that each element is covered by some set. If not, we can remove all of the elements that are not covered by any set using $O(\log m)$ rounds. Specifically, let $\mathbf{v}_j$ be the characteristic vector of S_j. We can compute $\mathbf{v} = \sum_{i=1}^{j} \mathbf{v}_j$ in $O(\log m)$ rounds using the standard converge-cast binary tree algorithm. We can then remove the elements that are not covered by any set (elements corresponding to 0 entries in $\mathbf{v}$).

We now have m sets covering n elements. Since $k = \Omega(m)$, we must have that $\text{OPT} = \Omega(n)$. McGregor and Vu showed that if one samples each element in the universe $[n]$ independently with probability $p = \Theta\left(\frac{\log \binom{m}{k}}{\epsilon^2 \text{OPT}}\right)$ then with high probability, if we run a β approximation algorithm on the subsampled universe, the solution will correpond to a $\beta - \epsilon$ approximation on the original universe. We have just argued that $\text{OPT} = \Omega(n)$ and therefore with high probability, we sample $O\left(\frac{\log \binom{m}{k}}{\epsilon^2}\right) = O(1/\epsilon^2 \cdot m)$ elements by appealing to Chernoff bound and the fact that $\binom{m}{k} \leq 2^m$.

As a result, we may assume that $n = O(1/\epsilon^2 \cdot m)$. This results in an $O(1/\epsilon^2 \cdot \log m \cdot (\log m + \log(1/\epsilon)))$ round algorithm.

Bounded Frequency. Assuming $f = \max_i f_i$ is known, we can lift the assumption that $k = \Omega(m)$ and parameterize our algorithm based on f instead. McGregor et al. [21] showed that the largest $\lceil kf/\eta \rceil$ sets contain a solution that covers at least $(1-\eta)\text{OPT}$

elements. We therefore can assume that $m = O(kf/\eta)$ by keeping only the largest $\lceil kf/\eta \rceil$ sets which can be identified in $O(1)$ rounds.

We set $\epsilon = \eta^2/f$ and proceed to obtain a solution that covers at least $(1-\eta^2/f)(1-\eta)\mathrm{OPT} = (1-O(\eta))\mathrm{OPT}$ elements using at most $k + O(\epsilon m) = k + O(\eta^2/f \cdot kf/\eta) = k + O(\eta k)$ sets as in the discussion above. Appealing to Lemma 3, we can find k sets that cover at least $(1-O(\eta))\mathrm{OPT}$ elements. The total number of rounds is $O(f^3/\eta^6 \cdot \log \frac{kf}{\eta} \cdot (\log \frac{1}{\eta} + \log \frac{kf}{\eta})) = O(f^3/\eta^6 \cdot \log^2 \frac{kf}{\eta})$.

2.2 Approximate the LP's Solution via Multiplicative Weights

Fix an objective value L. Let P be a convex region defined by

$$P = \{(\mathbf{x}, \mathbf{z}) \in [0,1]^n \times [0,1]^m : \sum_{i \in [n]} x_i = L \text{ and } \sum_{j \in [m]} z_j = m - k\}.$$

Note that if $(\mathbf{x}_1, \mathbf{z}_1), (\mathbf{x}_2, \mathbf{z}_2), \ldots, (\mathbf{x}_T, \mathbf{z}_T) \in P$ then $\left(\frac{1}{T}\sum_{t=1}^{T} \mathbf{x}_t, \frac{1}{T}\sum_{t=1}^{T} \mathbf{z}_t\right) \in P$. Consider the following problem Ψ_1 that asks to either correctly declare that

$$\nexists(\mathbf{x}, \mathbf{z}) \in P : \frac{x_i}{f_i} + \frac{1}{f_i} \cdot \sum_{S_j \ni i} z_j \leq 1, \quad \forall i \in [n]$$

or to output a solution $(\mathbf{x}, \mathbf{z}) \in P$ such that

$$\frac{x_i}{f_i} + \frac{1}{f_i} \cdot \sum_{S_j \ni i} z_j \leq 1 + \epsilon \quad \forall i \in [n].$$

Once we have such an algorithm, we can try different values of $L = \lfloor (1+\epsilon)^0 \rfloor, \lfloor (1+\epsilon)^1 \rfloor, \lfloor (1+\epsilon)^2 \rfloor, \ldots, n$ and return the solution corresponding to the largest L that has a feasible solution. There are $O(1/\epsilon \cdot \log n)$ such guesses. We know that the guess L where $\mathrm{OPT}/(1+\epsilon) \leq L \leq \mathrm{OPT}$ must result in a feasible solution.

To avoid introducing a $\log n$ factor in the number of rounds, we partition these $O(1/\epsilon \cdot \log n)$ guesses into batches of size $O(1/\epsilon)$ where each batch corresponds to $O(\log n)$ guesses. Algorithm copies that correspond to guesses in the same batch will run in parallel. This will only introduce a $\log n$ factor in terms of memory used by each machine. By returning the solution corresponding to the largest feasible guess L, one attains Theorem 2.

Oracle Implementation. Given a weight vector $\mathbf{w} \in \mathbb{R}^n$ in which $w_i \geq 0$ for all $i \in [n]$. We first consider an easier feasibility problem Ψ_2. It asks to either correctly declares that

$$\nexists(\mathbf{x}, \mathbf{z}) \in P : \sum_{i=1}^{n} w_i \cdot \left(\frac{x_i}{f_i} + \sum_{S_j \ni i} \frac{z_i}{f_i} \right) \leq \sum_{i=1}^{n} w_i, \quad \forall i \in [n]$$

or to outputs a solution $(\mathbf{x}, \mathbf{z}) \in P$ such that

$$\sum_{i=1}^{n} w_i \cdot \left(\frac{x_i}{f_i} + \sum_{S_j \ni i} \frac{z_i}{f_i} \right) \leq \sum_{i=1}^{n} w_i + \frac{1}{n^5}, \quad \forall i \in [n]. \tag{4}$$

That is, if the input is feasible, then output the corresponding $(\mathbf{x},\mathbf{z}) \in P$ that approximately satisfy the constraint. Otherwise, correctly conclude that the input is infeasible. In the multiplicative weights update framework, this is known as the approximate oracle. Note that if there is a feasible solution to Ψ_1, then there is a feasible solution to Ψ_2 since

$$\frac{x_i}{f_i} + \frac{1}{f_i} \cdot \sum_{S_j \ni i} z_j \leq 1 \quad \forall i \in [n] \implies \sum_{i=1}^{n} w_i \cdot \left(\frac{x_i}{f_i} + \sum_{S_j \ni i} \frac{z_i}{f_i} \right) \leq \sum_{i=1}^{n} w_i.$$

We can implement an oracle that solves the above feasibility problem Ψ_2 as follows. First, observe that

$$\sum_{i=1}^{n} \frac{w_i}{f_i} \cdot \left(x_i + \sum_{S_j \ni i} z_j \right) \leq \sum_{i=1}^{n} w_i$$

$$\iff \sum_{i=1}^{n} \frac{w_i}{f_i} \cdot x_i + \sum_{j=1}^{m} z_j \cdot \sum_{i \in S_j} \frac{w_i}{f_i} \leq \sum_{i=1}^{n} w_i.$$

To ease the notation, define

$$p_i := \frac{w_i}{f_i}, \quad \forall i \in [n], \text{ and } q_j := \sum_{i \in S_j} \frac{w_i}{f_i} = \sum_{i \in S_j} p_i, \quad \forall j \in [m].$$

We therefore want to check if there exists $(\mathbf{x},\mathbf{z}) \in P$ such that

$$LHS(\mathbf{x},\mathbf{z}) := \sum_{i=1}^{n} x_i p_i + \sum_{j=1}^{m} z_j q_j \leq \sum_{i=1}^{n} w_i.$$

We will minimize the left hand side by minimizing each sum separately. We can indeed do this exactly. However, there is a subtle issue where we need to bound the number of bits required to represent $p_i = \frac{w_i}{f_i}$ and $q_j = \sum_{i \in S_j} \frac{w_i}{f_i} = \sum_{i \in S_j} p_i$ given the memory constraint. To do this, we truncate the value of each p_i after the $(10 \log_2 n)$-th bit following the decimal point. Note that this will result in an underestimate of p_i by at most $1/n^{10}$. In particular, let $\hat{p}_i$ be p_i after truncating the value of p_i at the $(10 \log_2 n)$-th bit after the decimal point and $\hat{q}_j = \sum_{i \in S_j} \hat{p}_i$. For any $(\mathbf{x},\mathbf{z}) \in [0,1]^n \times [0,1]^m$, we can show that

$$\widehat{LHS}(\mathbf{x},\mathbf{z}) := \sum_{i=1}^{n} \hat{p}_i x_i + \sum_{j=1}^{m} z_j \sum_{i \in S_j} \hat{p}_i > LHS(\mathbf{x},\mathbf{z}) - \frac{1}{n^5}$$

Therefore, $LHS(\mathbf{x},\mathbf{z}) - 1/n^5 \leq \widehat{LHS}(\mathbf{x},\mathbf{z}) \leq LHS(\mathbf{x},\mathbf{z})$. Note that since $\sum_{i=1}^{n} x_i = L$ and $\sum_{j=1}^{m} z_j = m - k$, to minimize $\widehat{LHS}(\mathbf{x},\mathbf{z})$ over $(\mathbf{x},\mathbf{z}) \in P$, we simply set

$$x_i = [\hat{p}_i \text{is among the} L \text{smallest values of} \{\hat{p}_t\}_{t=1}^{n}\}],$$
$$z_j = [\hat{q}_j \text{is among the} m - k \text{smallest values of} \{\hat{q}_t\}_{t=1}^{m}\}].$$

After setting $\mathbf{x},\mathbf{z}$ as above, if $\widehat{LHS}(\mathbf{x},\mathbf{z}) > \sum_{i=1}^{n} w_i \implies LHS(\mathbf{x},\mathbf{z}) > \sum_{i=1}^{n} w_i$, then it is safe to declare that the system is infeasible. Otherwise, we have found $(\mathbf{x},\mathbf{z}) \in P$ such that $\widehat{LHS}(\mathbf{x},\mathbf{z}) \leq \sum_{i=1}^{n} w_i \implies LHS(\mathbf{x},\mathbf{z}) \leq \sum_{i=1}^{n} w_i + 1/n^5$ as required by Equation (4).

Lemma 4. *Assume that all machines have the vector* $\boldsymbol{w}$*. We can solve the feasibility problem* Ψ_2 *in* $O(1)$ *rounds, where all machines either learn that the system is infeasible or obtain an approximate solution* $(\boldsymbol{x}, \boldsymbol{z}) \in P$ *that satisfies Equation (4).*

Solving the LP via Multiplicative Weights. Once the existence of such an oracle is guaranteed, we can follow the multiplicative weights framework to approximately solve the LP. We will first explain how to implement the MWU algorithm in the MPC model. See Algorithm 2.

Algorithm 2: Multiplicative weights for solving the LP

Input: Objective value L, $\epsilon \leq 1/4$

Initialize $w_i^{(0)} = 1$ for all $i \in [n]$.

for *iteration* $t = 1, 2, \ldots, T = O(1/\epsilon^2 \cdot \log n)$ **do**

- Run the oracle in Lemma 4 with $\mathbf{w}^{(t-1)}$ to check if there exists a feasible solution. If the answer is INFEASIBLE, stop the algorithm. If the answer is FEASIBLE, let $\mathbf{x}^{(t)}$ and $\mathbf{z}^{(t)}$ be the output of the oracle that are now stored in all machines.
- Each machine j constructs $Y_j = \{Y_{j1}, Y_{j2}, \ldots, Y_{jn}\}$ where $Y_{ji} = z_j^{(t)} \cdot [\{i \in S_j\}]$.
- Compute $W = \sum_{j \in [m]} Y_j$ in $O(\log m)$ rounds using the a converge-cast binary tree and send W to the central machine. Note that $W_i = \sum_{S_j \ni i} z_j^{(t)}$.
- For each $i \in [n]$, the central machine computes $E_i^{(t)} = f_i \cdot \text{error}_i^{(t)}$ where $\text{error}_i^{(t)} := 1 - \frac{x_i^{(t)}}{f_i} - \frac{W_i}{f_i} = 1 - \frac{x_i^{(t)}}{f_i} - \sum_{S_j \ni i} \frac{z_j^{(t)}}{f_i}$ $\forall i \in [n]$ and sends $\sum_{d=1}^{t} E_i^{(d)}$ to all other machines.
- For each $i \in [n]$, each machine computes $w_i^{(t)} = 2^{-\epsilon \cdot \sum_{d=1}^{t} E_i^{(d)}/f_i} = 2^{-\epsilon \cdot \sum_{d=1}^{t} \text{error}_i^{(d)}} = 2^{-\epsilon \cdot \text{error}_i^{(t)}} w_i^{(t-1)}$.

After T iterations, output $\mathbf{x} = \frac{1}{T} \sum_{t=1}^{T} \mathbf{x}^{(t)}$ and $\mathbf{z} = \frac{1}{T} \sum_{t=1}^{T} \mathbf{z}^{(t)}$.

Lemma 5. *Algorithm 2 can be implemented in* $O(1/\epsilon^2 \cdot \log n \cdot \log m)$ *rounds.*

The next lemma is an adaptation of the standard multiplicative weights algorithm.

Lemma 6. *The output of Algorithm 2 satisfies the following property. If there exists a feasible solution, then the output satisfies:*

$$\frac{x_i}{f_i} + \sum_{S_j \ni i} \frac{z_j}{f_i} \leq 1 + \epsilon \quad \forall i \in [n], \text{ and } \sum_{i=1}^{n} x_i = L.$$

Otherwise, the algorithm correctly concludes that the system is infeasible.

Proof. (sketch) If the algorithm does not output INFEASIBLE, this implies that in each iteration t, $\sum_{i=1}^{n} x_i^{(t)} = L$ and $\sum_{j=1}^{m} z_j^{(t)} = m - k$. Hence, the output $(\mathbf{x}, \mathbf{z}) \in P$. Define the potential function $\Phi^{(t)} := \sum_{i=1}^{n} w_i^{(t)}$.

We will make use of the fact $\exp(-\eta x) \leq 1 - \eta x + \eta^2 x^2$ for $|\eta x| \leq 1$. Note that for all i and t, we always have $\text{error}_i^{(t)} = 1 - \frac{x_i^{(t)}}{f_i} - \sum_{S_j \ni i} \frac{z_j^{(t)}}{f_i} \in [-1, 1]$. Let $\alpha = \epsilon \cdot \ln(2)$, as long as $\epsilon \leq 1/4$, we have $|\alpha \cdot \text{error}_i^{(t)}| < 1$ and therefore

$$w_i^{(t)} = \exp\left(-\alpha \cdot \text{error}_i^{(t)}\right) \cdot w_i^{(t-1)} \leq \left(1 - \alpha \cdot \text{error}_i^{(t)} + \alpha^2 \cdot \left(\text{error}_i^{(t)}\right)^2\right) \cdot w_i^{(t-1)}.$$

Summing over i gives:

$$\begin{aligned}\Phi^{(t)} &\leq \sum_{i=1}^{n} \left(1 - \alpha \cdot \text{error}_i^{(t)} + \alpha^2\right) \cdot w_i^{(t-1)} \\ &= (1+\alpha^2) \sum_{i=1}^{n} w_i^{(t-1)} - \alpha \sum_{i=1}^{n} \text{error}_i^{(t)} \cdot w_i^{(t-1)}.\end{aligned}$$

The first inequality follows from the fact that $\left(\text{error}_i^{(t)}\right)^2 \in [0, 1]$. Note that

$$\begin{aligned}\sum_{i=1}^{n} \text{error}_i^{(t)} w_i^{(t-1)} &= \sum_{i=1}^{n} w_i^{(t-1)} \left(1 - \frac{x_i^{(t-1)}}{f_i} - \sum_{S_j \ni i} \frac{z_j^{(t-1)}}{f_i}\right) \\ &= \sum_{i=1}^{n} w_i^{(t-1)} - \sum_{i=1}^{n} w_i^{(t-1)} \left(\frac{x_i^{(t-1)}}{f_i} + \sum_{S_j \ni i} \frac{z_j^{(t-1)}}{f_i}\right) \geq -\frac{1}{n^5}.\end{aligned}$$

The last inequality follows from the oracle's guarantee. Thus, $\Phi^{(t)} \leq (1 + \alpha^2) \sum_{i=1}^{n} w_i^{(t-1)} + \frac{\alpha}{n^5} = (1+\alpha^2)\Phi^{t-1} + \frac{\alpha}{n^5}$. We can show by induction that

$$\Phi^{(T)} \leq (1+\alpha^2)^T \Phi^{(0)} + \frac{1}{\alpha n^5}(1+\alpha^2)^T.$$

Recall that $\alpha = \epsilon \ln 2$ and we assume $1/\epsilon < n/10$. Thus, $1/(\alpha n^5) < \epsilon^4 / \ln 2 < 1$. Furthermore, recall that $\Phi^{(0)} = n$. We have,

$$\begin{aligned} w_i^{(T)} &\leq (1+\alpha^2)^T (n+1) < (1+\alpha^2)^T 2n \\ \exp\left(-\alpha \sum_{t=1}^{T} \text{error}_i^{(t)}\right) &\leq (1+\alpha^2)^T (2n) \\ -\alpha \sum_{t=1}^{T} \text{error}_i^{(t)} &\leq \ln(2n) + T \ln(1+\alpha^2).\end{aligned}$$

We use the fact that $\ln(1+x) \leq x$ for $x \in \mathbb{R}$ to get

$$\begin{aligned}\sum_{t=1}^{T} \text{error}_i^{(t)} &\geq -\frac{\ln(2n)}{\alpha} - T\frac{\ln(1+\alpha^2)}{\alpha} \\ \sum_{t=1}^{T} \left(1 - \frac{x_i^{(t)}}{f_i} - \sum_{S_j \ni i} \frac{z_j^{(t)}}{f_i}\right) &\geq -\frac{\ln(2n)}{\alpha} - T\frac{\alpha^2}{\alpha} \\ \frac{1}{T}\sum_{t=1}^{T} \left(1 - \frac{x_i^{(t)}}{f_i} - \sum_{S_j \ni i} \frac{z_j^{(t)}}{f_i}\right) &\geq -\frac{\ln(2n)}{T\alpha} - \alpha\end{aligned}$$

$$1 - \frac{x_i}{f_i} - \sum_{S_j \ni i} \frac{z_j}{f_i} \geq -\frac{\ln(2n)}{T\alpha} - \alpha$$

$$\frac{x_i}{f_i} + \sum_{S_j \ni i} \frac{z_j}{f_i} \leq 1 + \frac{\ln(2n)}{T\alpha} + \alpha \implies \frac{x_i}{f_i} + \sum_{S_j \ni i} \frac{z_j}{f_i} \leq 1 + O(\epsilon).$$

The last inequality follows from choosing $T = \Theta(1/\epsilon^2 \cdot \log n)$ and the fact that $\alpha = \epsilon \ln(2)$; furthermore, recall that the final solution $x_i = \frac{1}{T}\sum_{t=1}^{T} x_i^{(t)}$ and $z_j = \frac{1}{T}\sum_{t=1}^{T} z_j^{(t)}$. Thus, the output of the algorithm satisfies the desired properties.

References

1. Anagnostopoulos, A., Becchetti, L., Bordino, I., Leonardi, S., Mele, I., Sankowski, P.: Stochastic query covering for fast approximate document retrieval. ACM Trans. Inf. Syst. **33**(3), 11:1–11:35 (2015)
2. Arora, S., Hazan, E., Kale, S.: The multiplicative weights update method: a meta-algorithm and applications. Theory Comput. **8**(1), 121–164 (2012)
3. Assadi, S.: Tight space-approximation tradeoff for the multi-pass streaming set cover problem. In: PODS, pp. 321–335. ACM (2017)
4. Assadi, S., Khanna, S.: Tight bounds on the round complexity of the distributed maximum coverage problem. In: SODA, pp. 2412–2431. SIAM (2018)
5. Assadi, S., Khanna, S., Li, Y.: Tight bounds for single-pass streaming complexity of the set cover problem. SIAM J. Comput. **50**(3) (2021)
6. Cervenjak, P., Gan, J., Umboh, S.W., Wirth, A.: Maximum unique coverage on streams: improved FPT approximation scheme and tighter space lower bound. In: APPROX/RANDOM. LIPIcs, vol. 317, pp. 25:1–25:23. Schloss Dagstuhl - Leibniz-Zentrum für Informatik (2024)
7. Chakrabarti, A., McGregor, A., Wirth, A.: Improved algorithms for maximum coverage in dynamic and random order streams. CoRR abs/2403.14087 (2024)
8. Feige, U.: A threshold of ln n for approximating set cover. J. ACM **45**(4), 634–652 (1998)
9. Har-Peled, S., Indyk, P., Mahabadi, S., Vakilian, A.: Towards tight bounds for the streaming set cover problem. In: PODS, pp. 371–383. ACM (2016)
10. Hochbaum, D.S., Pathria, A.: Analysis of the greedy approach in problems of maximum k-coverage. Naval Res. Logistics (NRL) **45**(6), 615–627 (1998)
11. Indyk, P., Mahabadi, S., Rubinfeld, R., Ullman, J.R., Vakilian, A., Yodpinyanee, A.: Fractional set cover in the streaming model. In: APPROX-RANDOM. LIPIcs, vol. 81, pp. 12:1–12:20. Schloss Dagstuhl - Leibniz-Zentrum für Informatik (2017)
12. Indyk, P., Vakilian, A.: Tight trade-offs for the maximum k-coverage problem in the general streaming model. In: PODS, pp. 200–217. ACM (2019)
13. Jaud, S., Wirth, A., Choudhury, F.M.: Maximum coverage in sublinear space, faster. In: SEA. LIPIcs, vol. 265, pp. 21:1–21:20. Schloss Dagstuhl - Leibniz-Zentrum für Informatik (2023)
14. Karloff, H.J., Suri, S., Vassilvitskii, S.: A model of computation for MapReduce. In: SODA, pp. 938–948. SIAM (2010)
15. Kempe, D., Kleinberg, J.M., Tardos, É.: Maximizing the spread of influence through a social network. Theory Comput. **11**, 105–147 (2015)
16. Khanna, S., Konrad, C., Alexandru, C.: Set cover in the one-pass edge-arrival streaming model. In: PODS, pp. 127–139. ACM (2023)

17. Krause, A., Guestrin, C.: Near-optimal observation selection using submodular functions. In: AAAI, pp. 1650–1654. AAAI Press (2007)
18. Kumar, R., Moseley, B., Vassilvitskii, S., Vattani, A.: Fast greedy algorithms in MapReduce and streaming. ACM Trans. Parallel Comput. **2**(3), 14:1–14:22 (2015)
19. Ladner, R.E., Fischer, M.J.: Parallel prefix computation. J. ACM **27**(4), 831–838 (1980)
20. Liu, P., Vondrák, J.: Submodular optimization in the MapReduce model. In: SOSA. OASIcs, vol. 69, pp. 18:1–18:10. Schloss Dagstuhl - Leibniz-Zentrum für Informatik (2019)
21. McGregor, A., Tench, D., Vu, H.T.: Maximum coverage in the data stream model: Parameterized and generalized. In: ICDT. LIPIcs, vol. 186, pp. 12:1–12:20. Schloss Dagstuhl - Leibniz-Zentrum für Informatik (2021)
22. McGregor, A., Vu, H.T.: Better streaming algorithms for the maximum coverage problem. Theory Comput. Syst. **63**(7), 1595–1619 (2019)
23. da Ponte Barbosa, R., Ene, A., Nguyen, H.L., Ward, J.: A new framework for distributed submodular maximization. In: FOCS, pp. 645–654. IEEE Computer Society (2016)
24. Saha, B., Getoor, L.: On maximum coverage in the streaming model & application to multi-topic blog-watch. In: SDM, pp. 697–708. SIAM (2009)
25. Warneke, R., Choudhury, F.M., Wirth, A.: Maximum coverage in random-arrival streams. In: ESA. LIPIcs, vol. 274, pp. 102:1–102:15. Schloss Dagstuhl - Leibniz-Zentrum für Informatik (2023)

Distance Vector Domination

Gennaro Cordasco[1(✉)], Luisa Gargano[2], and Adele A. Rescigno[2]

[1] Department of Psychology, University of Campania "L.Vanvitelli", Caserta, Italy
gennaro.cordasco@unicampania.it

[2] Department of Computer Science, University of Salerno, Fisciano, Italy
{lgargano,arescigno}@unisa.it

Abstract. Identifying and mitigating the spread of fake information is a challenging issue that has become dominant with the rise of social media. We consider a generalization of the Domination problem that can be used to detect a set of individuals who, once immunized, can prevent the spreading of fake narratives. The considered problem, named *Distance Vector Domination* generalizes both distance and multiple domination, at individual (i.e., vertex) level. We study the parameterized complexity of the problem according to several standard and structural parameters. We prove the W[1]-hardness of the problem with respect to neighborhood diversity, even when all the distances are 1. We also give fixed-parameter algorithms for some variants of the problem and parameter combinations.

Keywords: Distance Domination · Vector Domination · Parameterized complexity · Neighborhood diversity · Modular-width · Treewidth

1 Introduction

Domination is a fundamental concept in graph theory, which deals with the idea of dominating sets within a graph. In this problem, you seek to find the smallest set of vertices in a graph in such a way that every vertex in the graph is either in the dominating set or adjacent to a vertex in the dominating set. Dominating sets are critical in various real-world applications across fields that involve networks, connections, and coverage [3,5,9,18,24].

The Domination Problem in graph theory has several important variants that focus on different aspects of the problem. We focus on generalizations to distance domination and multiple domination. Distance and multiple domination provide ways to address practical concerns related to the physical or operational limitations of networks. We refer to [33,34] for a survey of domination in graphs.

This paper is motivated by an application aimed at combating the spread of misinformation using epidemiological principles. Graph-based information diffusion algorithms offer a means to analyze the dissemination of both genuine and

Work partially supported by project SERICS (PE00000014) under the MUR National Recovery and Resilience Plan funded by the European Union - NextGenerationEU.

R. Královič and V. Kurková (Eds.): SOFSEM 2025, LNCS 15538, pp. 179–194, 2025.
https://doi.org/10.1007/978-3-031-82670-2_14

fake information within a network. Sociologists widely employ threshold models to characterize collective behaviors [31], and their application to studying the propagation of innovations through networks was initially proposed in [40]. The linear threshold model has then been extensively employed in the literature to study influence maximization, a critical problem in network analysis, which aims at identifying a small subset of vertices capable of maximizing the diffusion of content throughout the network [4,11–17,19]. While these algorithms are not designed for intercepting fakes, they can be used as a component in a broader strategy for identifying and mitigating the spread of fake information.

Controlling the spread of misinformation/disinformation is an ongoing challenge. Strategies for reducing the spread size by either blocking some links, so that they cannot contribute to the diffusion process [41], or by immunizing/removing vertices were considered in several papers [1,46]. In this paper, we focus on the second strategy: limit the spread by immunizing a bounded number of vertices in the network. We consider a population of interconnected individuals that can potentially be misinformed by a word-of-mouth diffusion strategy. We assume that when an individual is reached by a sufficient amount of debunking information, he/she becomes immunized. With more details, an individual gets immunized if he/she receives the debunking information from a number of neighbors at least equal to its threshold. Moreover, each individual has a certain level of trust in the others (circle of trust) described by a radius around it. Only debunking information coming from within the circle of trust is considered reliable. In particular, we consider the use of generalized dominating sets to detect a group of individuals who, by setting a debunking (or prebunking) campaign, can prevent the spreading of negative narratives. In the presence of a debunking (or prebunking) campaign, the *immunization* operation on a vertex inhibits the contamination of the vertex itself. Thus, to avoid the diffusion of malicious items, we are looking for a small subset T of vertices (immunizing set) that, by spreading the debunking information, enables us to stop the misinformation diffusion. The immunizing set should be able to cover each vertex multiple times (based on the vertex threshold) within a maximum distance (depending on the radius that specifies the circle of trust of the vertex). We propose the Distance Vector Domination problem, which includes both multiplicity and distance.

The Problem. Let $G = (V, E)$ be a graph. We denote by $n = |V|$ the number of vertices in G. For a set of vertices $X \subseteq V$, we denote by $G[X]$ the induced subgraph of G generated by X. Given two vertices $u, v \in V$, we denote by $\delta_G(u, v)$ the distance between u and v in G. Moreover, for a vertex $v \in V$, we denote by $N_G(v) = \{u \in V \mid (u, v) \in E\}$ the neighborhood of v and by $N_{G,d}(v) = \{u \in V \mid u \neq v \wedge \delta_G(u, v) \leq d\}$ the *neighborhood of radius d around v*. Clearly, $N_{G,1}(v) = N_G(v)$. We also define the distance between a vertex $v \in V$ and a set $U \subseteq V$ as $\delta_G(v, U) = \min_{u \in U} \delta_G(v, u)$. We omit the subscript G whenever the graph is clear from the context.

Definition 1. *Given a graph* $G = (V, E)$ *and vectors* $\mathbf{t} = (t_v \mid t_v \in \mathbb{N}, v \in V)$ *and* $\mathbf{d} = (d_v \mid d_v \in \mathbb{N}, v \in V)$*, where* $t_v \leq |N_{d_v}(v)|$*, a Distance Vector Dominating set S is a set* $S \subseteq V$ *such that* $|N_{d_v}(v) \cap S| \geq t_v$*, for all* $v \in V \setminus S$*.*

We will consider the following problem.

DISTANCE VECTOR DOMINATION (DVD):
Input: A graph $G = (V, E)$, vectors $\mathbf{t} = (t_v \mid t_v \in \mathbb{N}, v \in V)$ and $\mathbf{d} = (d_v \mid d_v \in \mathbb{N}, v \in V)$.
Output: A Distance Vector Dominating set of minimum size.

For each vertex $v \in V$, we will refer to t_v and d_v as the *demand* of v and the *radius around* v, respectively. Furthermore, given a set $S \subseteq V$, we say that a vertex $v \in V \setminus S$ is *dominated by* S if $|N_{d_v}(v) \cap S| \geq t_v$. Since the problem can be solved independently in each connected component of the input graph, from now on, we assume that the input graphs are connected.

DVD generalizes several well-known and widely studied problems. Consider an input graph $G = (V, E)$ together with vectors $\mathbf{t} = (t_v \mid t_v \in \mathbb{N}, v \in V)$ and $\mathbf{d} = (d_v \mid d_v \in \mathbb{N}, v \in V)$:

- When $\mathbf{t} = \mathbf{d} = \mathbf{1} = (1, \ldots, 1)$, DVD becomes the classical DOMINATING SET (DS) problem [33].
- If $\mathbf{d} = \mathbf{1}$ and $\mathbf{t} = (t_v \mid t_v \in \mathbb{N}, v \in V)$, then DVD becomes the VECTOR DOMINATION (VD) problem, which asks for a minimum size set $S \subseteq V$ such that $|N(v) \cap S| \geq t_v$, for each $v \in V \setminus S$. Vector Domination, introduced in [32], has been extensively studied [10,35,44,50] and was recently studied from the parameterized complexity point of view in [38,47]. The special case $\mathbf{t} = (r, \ldots, r)$ for some positive integer r has been studied under the name of r-DOMINATION [27].
- The problem corresponding to the case $\mathbf{t} = \mathbf{1}$ and $\mathbf{d} = (d_v \mid d_v \in \mathbb{N}, v \in V)$ was introduced by Slater in [49] under the name of R-DOMINATION (RD). The special case $\mathbf{d} = (d, \ldots, d)$, for some positive integer d, has been studied under the name of DISTANCE DOMINATION (DD) [36].

Knowing that DVD generalizes the VD problem [10], we immediately have

Theorem 1. DVD *cannot be approximated in polynomial time to within a factor of* $0.2267 \log n$ *, unless P=NP.*

Moreover, following the lines of the proof of Theorem 1 in [10], one can easily get a logarithmic approximation algorithm.

Theorem 2. DVD *can be approximated in polynomial time by a factor* $\log n + 2$.

1.1 Parameterized Algorithms

Parameterized complexity is a refinement to classical complexity in which one takes into account, not only the input size but also other aspects of the problem given by a parameter p. We recall that a problem with input size m and parameter p is called *fixed parameter tractable (FPT)* if it can be solved in time $f(p) \cdot m^c$, where f is a computable function only depending on p and c is a constant. A problem is in *XP* parameterized by p if it can be solved in time $m^{f(p)}$, where f is a computable function only depending on p.

Known Results. The DVD problem, as well as each of the special cases described in Sect. 1, is W[2]-hard with respect to the size k of the solution since they all generalise the W[2]-complete DOMINATING SET problem [25].

It is shown in [7] that VD is W[1]-hard with respect to treewidth, thus implying that DVD is W[1]-hard with respect to treewidth. Recently, Lafond and Luo [43] presented an FPT algorithm for r-DOMINATION parameterized by neighborhood diversity and proved that this problem is W[1]-hard with respect to modular-width. In [8] the authors consider the DD problem parameterized by the radius (d) and the treewidth ($\mathtt{tw}$). They show an FPT ($O((2d+1)^{\mathtt{tw}}n)$) algorithm. Moreover, they also show that the running time dependence on d and $\mathtt{tw}$ is the best possible under Strong Exponential Time Hypothesis (SETH) [37]. This lower bound applies also to the RD problem which generalizes DD. In [39] the authors study the (k, r)-center problem, which, given a graph G, asks if there exists a set K of at most k vertices of G, so that $\min_{v \in K} \delta(v, u) \leq r$ for each $u \notin K$. The (k, r)-Center problem represents the decision version of the DD problem and is W[1]-hard when parameterized by $\mathtt{fvs} + k$, where $\mathtt{fvs}$ represents the feedback vertex set parameter [39]. Since the treewidth of a graph G is upper bounded by the feedback vertex set number of G plus one, this result implies that RD is W[1]-hard when parameterized by the treewidth.

A *XP* algorithm, with running time $O(n+1)^{O(\mathtt{cw})}$, for a generalization of VD on graphs of bounded clique-width ($\mathtt{cw}$) has been provided in [11]. Since $\mathtt{cw} \leq 2^{\mathtt{tw}+1}+1$ [22], this result implies the *XP* solvability of the VD problem for graphs of bounded treewidth. Assuming to have a branch decomposition of the input graph of width $\mathtt{bw}$, a FPT algorithm with running time $O((\tau+2)^{\mathtt{bw}}[(\tau+1)^2+1]^{\mathtt{bw}/2}n^2)$ for VD is given in [38], where τ is the largest demand of the vertices, i.e., $\tau = \max_{v \in V} t_v$. Since $\max\{\mathtt{bw}, 2\} \leq \mathtt{tw}+1 \leq \max\{3\mathtt{bw}/2, 2\}$, this result implies an $O((\tau+2)^{\mathtt{tw}+1}[(\tau+1)^2+1]^{(\mathtt{tw}+1)/2}n^2)$ time algorithm for VD. An algorithm for DOMINATING SET problem parameterized by modular-width ($\mathtt{mw}$) was given by Romanek [48]; it requires $O(2^{\mathtt{mw}}n^2)$ time. However, little work has been done to design FPT algorithms for DD, VD and RD problems with respect to the neighborhood diversity and/or modular-width parameters.

Our Results. We give some positive and negative results with respect to some structural parameters of the input graph: modular-width, neighborhood diversity, and treewidth. The definitions of the parameters are given in Sections 2, 3, and 4, respectively. It is worth mentioning that modular decomposition parameters, which comprise modular-width, neighborhood diversity, and tree-like parameters, such as treewidth and pathwidth, are two incomparable classes that can be viewed as representing dense and sparse graphs, respectively.

- We prove the W[1]-hardness of DVD with respect to neighborhood diversity even when all the radii are equal to 1; namely, we show that VD is W[1]-hard with respect to neighborhood diversity. This negative result also applies to any generalization of neighborhood diversity and, in particular to modular-width and clique-width.
- On the positive side, we present FPT algorithms parameterized by:
  (i) Modular-width for RD, with running time $O(\mathtt{mw}\, 2^{\mathtt{mw}}\, n)$;

(ii) Modular-width plus the size k of the solution for VD, with running time $O(\mathtt{mw}\ k(k+1)^{\mathtt{mw}}\ n^2)$;
(iii) Modular-width plus the size k of the solution for DVD, with running $O(\mathtt{mw}^2\ k(k+1)^{2\mathtt{mw}}\ n^2)$;
(iv) Treewidth plus the maximum radius $\delta = \max_{v \in V} d_v$ for RD, with running time $O(\mathtt{tw}(2\delta+1)^{\mathtt{tw}}(\mathtt{tw}^2+n)\ n^2 \log n)$;
(v) Treewidth plus $\tau = \max_{v \in V} t_v$ for VD, with running time $O(\mathtt{tw}^2 2^{\mathtt{tw}} (\tau+1)^{\mathtt{tw}}\ n)$.
This last result improves the above-described result achieved in [38].

Table 1. Parameterized complexity results with respect to neighborhood diversity (nd), modular-width (mw) and treewidth (tw).

Parameters\Problem	DVD	VD	RD
nd	**W[1]-hard** [Cor 1]	**W[1]-hard** [Th 3]	**FPT** [Th 5]
mw	**W[1]-hard** [Cor 1]	**W[1]-hard** [Th 3]	**FPT** [Th 5]
mw and k	**FPT** [Th 7]	**FPT** [Th 6]	**FPT** [Th 5]
tw	W[1]-hard [7]	W[1]-hard [7]	W[1]-hard [39]
tw and δ	open	Not applicable ($\delta = 1$)	**FPT** [Th 9]
tw and τ	open	**FPT** [38] [Th 8]	Not applicable ($\tau = 1$)

2 Hardness

In this section, we prove that VD, and as a consequence its generalization DVD, is W[1]-hard on graphs of bounded neighborhood diversity.

Neighborhood Diversity. Given a graph $G = (V, E)$, two vertices $u, v \in V$ are said to have the same *same type* if $N_G(v) \setminus \{u\} = N_G(u) \setminus \{v\}$.
The *neighborhood diversity* of a graph G, introduced by Lampis [45] and denoted by $\mathtt{nd}(G)$, is the smallest integer nd such that there exists a partition $V_1, \ldots, V_{\mathtt{nd}}$, of the vertex set V, where all the vertices in V_i have the same type, for $i \in [\mathtt{nd}]$[1]. The unique family $\{V_1, \ldots, V_{\mathtt{nd}}\}$ is called the *type partition* of G.

Theorem 3. VD *is W[1]-hard with respect to neighborhood diversity.*

Proof. We use a reduction from MULTI-COLORED CLIQUE (MQ): *Given a graph $G = (V, E)$ and a proper vertex-coloring* $\mathbf{c} : V \to [q]$ *for G, does G contain a clique of size q?*

It is worth noticing that a multi-colored clique of size q has one vertex of each color. Hence, a vertex v can belong to a multi-colored clique only if $N_G(v) \cup \{v\}$ contains at least one vertex from each color class. Hence, in the following, we

[1] For a positive integer a, we use $[a]$ to denote the set of integers $[a] = \{1, 2, \ldots, a\}$.

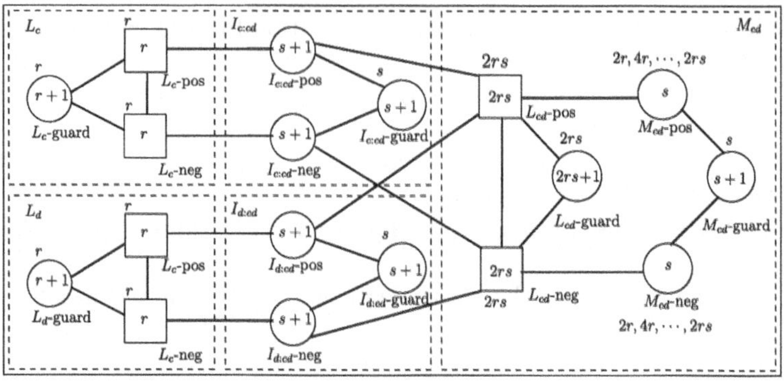

Fig. 1. An overview of the reduction. Each circle represents a bag. Each square represents a clique. The number inside a bag (resp. clique) is the number of vertices of the bag (resp. clique). The value t_v for a vertex v is displayed in red. (Color figure online)

will assume that all the vertices that do not satisfy such a property are removed from G since they are irrelevant to the problem.

Given an instance $\langle G, \mathbf{c}, q\rangle$ of MQ, we construct $\langle G' = (V', E'), \mathbf{t}, \mathbf{1}, k\rangle$, an instance of the decision version of VD. Our goal is to guarantee that any solution of VD in G' of size k encodes a multi-colored clique in G of size q and vice-versa.

For a color $c \in [q]$, we denote by V_c the class of vertices in G of color c and for a pair of distinct colors $c, d \in [q]$, we let E_{cd} represent the edges in G connecting a vertex in V_c and one in V_d. We use the fact that MQ remains W[1]-hard even if each color class has the same size, and between every pair of color classes we have the same number of edges [23]. We then denote by $r+1$ the size of each color class V_c and by $s+1$ the size of each set E_{cd} (notice that $r, s \geq 1$ since otherwise G is a clique). We use the following notation

$$V_c = \{v_0^c, v_1^c, \ldots, v_r^c\}, \qquad E_{cd} = \{e_0^{cd}, \ldots, e_s^{cd}\} \qquad c, d \in [q], c \neq d \tag{1}$$

and refer to v_i^c and e_j^{cd} as the i-th vertex in V_c and j-th edge in E_{cd}, respectively.

Let $\langle G, \mathbf{c}, q\rangle$ be an instance of MQ. We describe a reduction from $\langle G, \mathbf{c}, q\rangle$ to an instance $\langle G', \mathbf{t}, \mathbf{1}, k\rangle$ of VD such that $\mathtt{nd}(G')$ is $O(q^2)$. To this aim, we introduce some gadgets for the construction of G', inspired by those used in [26]. The rationale behind the construction is the following. First, we create two sets of gadgets (Selection and Multiple), which encode in G' the selection of vertices and edges as part of a potential multi-colored clique in G. Then we create another set of gadgets (Incidence gadgets) that is used to check whether the selected sets of vertices and edges form a multi-colored clique. Our goal is to guarantee that any solution of VD in G' encodes a multi-colored clique in G and vice-versa.

In the following we call *bag* an independent set of vertices sharing all neighbors. Hence, a connection between two bags implies that the vertices in the bags induce a complete bipartite graph. We will also use *cliques*; the connection between two cliques is a complete bipartite graph among the vertices in the cliques. Figure 1 shows the gadgets we are going to introduce and how they are connected.

Selection Gadget. For each $c \in [q]$, the selection gadget L_c consists of two cliques, L_c-neg and L_c-pos of r vertices each, and one bag L_c-guard of $r+1$ vertices. The cliques L_c-neg and L_c-pos are connected, and the bag L_c-guard is connected to both L_c-neg and L_c-pos. The selection gadget L_c is connected to the rest of the graph G' using only vertices from $L_c\text{-neg} \cup L_c\text{-pos}$. We set now the value $t_v = r$ for each vertex $v \in L_c\text{-pos} \cup L_c\text{-neg} \cup L_c\text{-guard}$.

Multiple Gadget. For each $c, d \in [q]$ with $c \neq d$, we create a multiple gadget M_{cd} consisting of four bags, L_{cd}-guard of $2rs+1$ vertices, M_{cd}-pos and M_{cd}-neg of s vertices each, and M_{cd}-guard of $s+1$ vertices, and two cliques L_{cd}-pos and L_{cd}-neg of $2rs$ vertices each. M_{cd}-guard is connected to M_{cd}-pos and M_{cd}-neg. M_{cd}-pos is connected to L_{cd}-pos, and M_{cd}-neg is connected to L_{cd}-neg. The two cliques L_{cd}-pos and L_{cd}-neg are connected. Finally, the bag L_{cd}-guard is connected to both L_{cd}-pos and L_{cd}-neg. The rest of graph G' is connected only to the cliques L_{cd}-pos and L_{cd}-neg. We set now the value t_v of each v in M_{cd} as:

$$t_v = \begin{cases} 2rs & \text{if} v \in L_{cd}\text{-neg} \cup L_{cd}\text{-pos} \cup L_{cd}\text{-guard} \\ 2rj & \text{if} v = x_j \text{where} M_{cd}\text{-pos} = \{x_1, \dots, x_s\} \\ 2rj & \text{if} v = y_j \text{where} M_{cd}\text{-neg} = \{y_1, \dots, y_s\} \\ s & \text{if} v \in M_{cd}\text{-guard} \end{cases} \tag{2}$$

Incidence Gadget. For each pair of distinct $c, d \in [q]$, we construct two incidence gadgets: $I_{c:cd}$ (connected with the gadgets L_c and M_{cd}) and $I_{d:cd}$ (connected with the gadgets L_d and M_{cd}). In the following, we present the gadget $I_{c:cd}$, which has the same structure as the gadget $I_{d:cd}$. The incidence gadget $I_{c:cd}$ has three bags $I_{c:cd}$-pos, $I_{c:cd}$-neg and $I_{c:cd}$-guard of $s+1$ vertices each. We connect $I_{c:cd}$-guard to $I_{c:cd}$-pos and $I_{c:cd}$-neg. Furthermore, we connect $I_{c:cd}$-pos to L_c-pos and L_{cd}-pos. Similarly, we connect $I_{c:cd}$-neg to L_c-neg and L_{cd}-neg.

Recalling that there are $s+1$ edges in the set E_{cd}, and that there are $s+1$ vertices in $I_{c:cd}$-pos and $I_{c:cd}$-neg, we create one-to-one correspondences between E_{cd} and $I_{c:cd}$-pos and between E_{cd} and $I_{c:cd}$-neg. Namely, for each $j = 0, \dots s$, we associate the j-th edge e_j^{cd} in E_{cd} (cfr. (1)) to a vertex $u_j \in I_{c:cd}$-pos and to a vertex $w_j \in I_{c:cd}$-neg (with $u_j \neq u_{j'}$ and $w_j \neq w_{j'}$, for $j \neq j'$). Moreover, if the endpoint of e_j^{cd} of color c is the ith vertex v_i^c of V_c (cfr. (1)) then we set the value t_v of each vertex v in $I_{c:cd}$ as:

$$t_v = \begin{cases} 2rj + i & \text{if} v = u_j \text{where} I_{c:cd}\text{-pos} = \{u_0, \dots, u_s\} \\ 2r(s-j) + r - i & \text{if} v = w_j \text{where} I_{c:cd}\text{-neg} = \{w_0, \dots, w_s\} \\ s & \text{if} v \in I_{c:cd}\text{-guard} \end{cases} \tag{3}$$

It is worth observing that the vertices in $I_{c:cd}$-pos (respectively, $I_{c:cd}$-neg) have different demands. Indeed, the numbers $2rj + i$ (respectively, $2r(s-j) + r - i$) are all different, for $0 \le i \le r$ and $0 \le j \le s$.

The budget k is set to $k = qr + \binom{q}{2}(2r+3)s$.

Lemma 1. *$\langle G, \mathbf{c}, q\rangle$ is a* YES *instance of* MQ *iff $\langle G', \mathbf{t}, 1, k\rangle$ is a* YES *instance.*

We complete the proof by showing that G' has neighborhood diversity $O(q^2)$. Since each bag in G' is a type set in the type partition of G' and, since for each $c \in [q]$, there are two cliques and one bag in L_c and, for each $c, d \in [q]$ with $c \neq d$ there are four bags and two cliques in M_{cd}, and three bags in both $I_{c:cd}$ and $I_{d:cd}$, we have that the neighborhood diversity of G' is $3q + 12\binom{q}{2}$. □

Corollary 1. DVD *is W[1]-hard with respect to neighborhood diversity.*

3 FPT Algorithms for Graphs of Bounded Modular-Width

The notion of modular decomposition of graphs was introduced by Gallai in [29], as a tool to define hierarchical decompositions of graphs. A *module* of a graph $G = (V, E)$ is a subgraph $G[M]$ induced by a set $M \subseteq V$ such that all the vertices of M share the same neighbors in $V \setminus M$. The *modular-width* parameter has been proposed in [28].

Definition 2. *Consider graphs obtainable by using (in any order or number) the following operations.*

- *(O1) The creation of an isolated vertex.*
- *(O2) $G_1 \oplus G_2$, called* disjoint union *of two graphs: $G_1 \oplus G_2$ is the graph with vertex set $V(G_1) \cup V(G_2)$ and edge set $E(G_1) \cup E(G_2)$.*
- *(O3) $G_1 \otimes G_2$, called* complete join*: $G_1 \otimes G_2$ is the graph with vertex set $V(G_1) \cup V(G_2)$ and edges $E(G_1) \cup E(G_2) \cup \{(u, w) \mid u \in V(G_1),\ w \in V(G_2)\}$.*
- *(O4) $H(G_1, \ldots, G_p)$, the substitution of the vertices $v_1, \ldots, v_p$ of a graph H by the graphs (modules) $G_1, \ldots, G_p$: $H(G_1, \ldots, G_p)$ is the graph with vertices $\bigcup_{1\leq \ell \leq p} V(G_\ell)$ and edges $\bigcup_{1\leq \ell \leq p} E(G_\ell) \cup \{(u, w) \mid u \in V(G_i),\ w \in V(G_j),\ (v_i, v_j) \in E(H)\}$.*

The *modular-width* of a graph G, denoted $\mathtt{mw}(G)$, is the least integer p such that G can be obtained by using only the operations (O1)–(O4) (in any number and order) and where each operation (O4) has at most p modules. A hierarchical decomposition of G that is an expression using only the operations (O1)–(O4) of width $\mathtt{mw}(G)$ can be constructed in linear time [20].
Notice that any module G_i of $G = H(G_1, \ldots, G_p)$ is such that all the vertices in $V(G_i)$ have the same neighborhood in $V(G) \setminus V(G_i)$; that is, for each vertex $u \in V(G) \setminus V(G_i)$ either $V(G_i) \subseteq N_G(u)$ or $V(G_i) \cap N_G(u) = \emptyset$. Moreover, operations (O2) and (O3) are special cases of (O4) for H being K_2 or its complement. Hence, any graph G can be written as $G = H(G_1, \ldots, G_p)$ with $p \leq \max\{2, \mathtt{mw}(G)\}$. Consider then the parse-tree of an expression describing G, according to the operations (O1)–(O4). The leaves of the parse-tree are the isolated vertex modules, created by (O1) and representing the vertices in G. Any internal vertex in the parse-tree is obtained through (O2)-(O4): Each such an operation corresponds to a vertex $H(G_1, \ldots, G_p)$ with $p \geq 2$ children $G_1, \ldots, G_p$.

Observation 4. *If* $G = H(G_1, \ldots, G_p)$ *is a connected undirected graph, then* $\delta_G(u, v) \leq 2$, *for each* $u, v \in V(G_i)$ *for any* $i \in [p]$.

3.1 RD with Parameter mw

Algorithm 1: $\mathrm{RD}(G = H(G_1, \ldots, G_p), \mathbf{d}, S_H)$

$S = \emptyset$
for *each* $i \in [p]$ **do** $\ell_i = \min_{j \in S_H \setminus \{i\}} \delta_H(i, j)$
for *each* $i \in [p] \setminus S_H$ **do**
  if $\ell_i > \min_{v \in V(G_i)} d_v$ **then return** (R=`false`, S=$\emptyset$)
for *each* $i \in S_H$ **do**
  $A_i = \{v \in V(G_i) \mid 1 = d_v < \ell_i\}$
  if $A_i = \emptyset$ **then** $S = S \cup \{u_i\}$, where u_i is any vertex in $V(G_i)$
  else
    if $\bigcap_{v \in A_i} N_{G_i}(v) = \emptyset$ **then return** (R=`false`, S=$\emptyset$)
    else $S = S \cup \{u_i\}$, where u_i is any vertex in $\bigcap_{v \in A_i} N_{G_i}(v)$
return (R = `true`, S)

This section is devoted to proving the following result.

Theorem 5. RD *can be solved in time* $O(\mathtt{mw}\ 2^{\mathtt{mw}}\ n)$.

Lemma 2. *Let* $G = H(G_1, \ldots, G_p)$. *There exists a solution* S *for the instance* $\langle G, \mathbf{1}, \mathbf{d}\rangle$ *of the* RD *problem such that* $|S \cap V(G_i)| \leq 1$, *for each* $i \in [p]$.

By exploiting Lemma 2, our algorithm proceeds by considering all the subsets $S_H \subseteq [p]$, ordered by size, and checking whether it is possible to find a vertex $u_i \in V(G_i)$ for each $i \in S_H$ such that $S = \{u_i \mid i \in S_H\}$ is a solution for the instance $\langle G, \mathbf{1}, \mathbf{d}\rangle$ of the RD problem. Algorithm 1 implements this check.

Lemma 3. *Given any set* $S_H \subseteq [p]$, *let* (R, S) *be the pair returned by Algorithm* $\mathrm{RD}(G = H(G_1, \ldots, G_p), \mathbf{d}, S_H)$. *If* $R = \mathtt{true}$ *then* S *is a solution for the instance* $\langle G, \mathbf{1}, \mathbf{d}\rangle$ *of the* RD *problem, otherwise the problem has no solution with exactly one vertex selected from each* $V(G_i)$ *with* $i \in S_H$.

Now we evaluate the running time of our algorithm. First of all, we can obtain $\min_{v \in V(G_i)} d_v$ for $i \in [p]$ in time $O(n)$. Then, for at most each $S_H \subseteq [p]$, we use algorithm $\mathrm{RD}(G, \mathbf{d}, S_H)$ to verify if a solution S with exactly one vertex selected from each $V(G_i)$ with $i \in S_H$ exists. Considering that Algorithm $\mathrm{RD}(G, \mathbf{d}, S_H)$ requires time $O(p\, n)$ and that the number of modules of G is $p \leq \mathtt{mw}$, overall we have time complexity $O(\mathtt{mw}\ 2^{\mathtt{mw}}\ n)$. □

Algorithm 2: VD-MW($\hat{G} = \hat{H}(\hat{G}_1, \ldots, \hat{G}_{\hat{p}}), \hat{\mathbf{t}}, \hat{b}$)

if $\hat{G} = \hat{H}(\hat{G}_1) = (\{v\}, \emptyset)$ **then** // $\hat{H}$ is a single vertex graph
  if $\hat{b} = 0 \wedge t_v \geq 1$ **then return** ($R =$ `false`, $\hat{S} = \emptyset$)
  if $\hat{b} = 0 \wedge t_v = 0$ **then return** ($R =$ `true`, $\hat{S} = \emptyset$)
  if $\hat{b} = 1$ **then return** ($R =$ `true`, $\hat{S} = \{v\}$)
else
  for *each* $(s_1, \ldots, s_{\hat{p}}) \mid \sum_{i=1}^{\hat{p}} s_i = \hat{b}$ *and* $0 \leq s_i \leq \min\{\hat{b}, |V(\hat{G}_i)|\}$ **do**
    for $i = 1, \ldots, \hat{p}$ **do**
      for *each* $v \in V(\hat{G}_i)$ **do** $\hat{t}'_v = \max\{0,\ \hat{t}_v - \sum_{j|(i,j)\in E(\hat{H})} s_j\}$
      $(R_i, S_i) =$VD-MW$(\hat{G}_i, \hat{\mathbf{t}}', s_i)$
    if $\bigwedge_{i=1}^{\hat{p}} R_i =$ **true then return** ($R =$ **true**, $\hat{S} = \bigcup_{i=1}^{\hat{p}} S_i$)
  return ($R =$ `false`, $\hat{S} = \emptyset$)

3.2 VD with Parameters mw and the solution size k

This section is devoted to proving the following result.
Theorem 6. VD *can be solved in time* $O(\texttt{mw}\ k(k+1)^{\texttt{mw}}\ n^2)$.

Consider the parse-tree of an expression describing the input graph G, according to the operations (O1)-(O4) in Definition 2. We design a recursive algorithm that computes a *Vector Dominating* set for the instance $\langle G, \mathbf{t}, \mathbf{1}\rangle$ based on the parse-tree of G.

Except for the leaves of the parse-tree (representing (O1)) and thus graphs consisting of exactly one vertex, i.e., $\hat{H}(\hat{G}_1) = (\{v\}, \emptyset)$), for all the other vertices of the parse-tree we just need to focus on the operation (O4), that is $\hat{G} = \hat{H}(\hat{G}_1, \ldots, \hat{G}_{\hat{p}})$ such that $\hat{p} \leq \max\{2, \texttt{mw}(G)\}$.

For the instance $\langle G, \mathbf{t}, \mathbf{1}\rangle$ of the VD problem, our algorithm checks if there exists a solution for the decision version of the problem, with instance $\langle G, \mathbf{t}, \mathbf{1}, b\rangle$, that asks for a *Vector Dominating* set of size b of G with respect to the demand vector $\mathbf{t}$. The minimum positive integer b for which the instance $\langle G, \mathbf{t}, \mathbf{1}, b\rangle$ has a solution is the size k of the solution of the instance $\langle G, \mathbf{t}, \mathbf{1}\rangle$ of the VD problem. The algorithm uses a recursive approach along the parse-tree of G and for each vertex $\hat{G} = \hat{H}(\hat{G}_1, \ldots, \hat{G}_{\hat{p}})$ of the parse-tree and the relative instance $\langle \hat{G}, \hat{\mathbf{t}}, \mathbf{1}, \hat{b}\rangle$ with $\hat{b} \leq |V(\hat{G})|$, constructs an equivalent instance of the problem on each $\hat{G}_i$ obtained by partitioning the budget $\hat{b}$ among the $\hat{p}$ modules $\hat{G}_1, \ldots, \hat{G}_{\hat{p}}$ and appropriately reducing the values in the demand vector. The solution set $\hat{S}$ for $\langle \hat{G}, \hat{\mathbf{t}}, \mathbf{1}, \hat{b}\rangle$ is reconstructed by using the solutions recursively obtained for each $\hat{G}_i$ (cf. Algorithm 2).

Algorithm 3: DVD-MW$(G = H(G_1, \ldots, G_p), \mathbf{t}, \mathbf{d}, b)$

for *each* $(s_1, \ldots, s_p) \mid \sum_{i=1}^{\hat{p}} s_i = b$ *and* $0 \le s_i \le \min\{b, |V(G_i)|\}$ **do**
  for $i = 1, \ldots, p$ **do**
    if $\exists\, v \in V(G_i)$: $(d_v \ge 2)$ *and* $(t_v - s_i - \sum_{j \mid\, j \neq i\, \wedge\, \delta_H(i,j) \le d_v} s_j > 0)$ **then return** $(R = \mathtt{false}, S = \emptyset)$
    for *each* $v \in V(G_i)$ **do**
      $t'_v = \begin{cases} 0 & \text{if } d_v \ge 2 \\ \max\{0,\ t_v - \sum_{j \mid (i,j) \in E_H} s_j\} & \text{if } d_v = 1 \end{cases}$
    $(R_i, S_i) =$VD-MW$(G_i, \mathbf{t}', s_i)$
  if $\bigwedge_{i=1}^{p} R_i = \mathtt{true}$ **then return** $(R = \mathtt{true},\ S = \bigcup_{i=1}^{p} S_i)$
return $(R = \mathtt{false}, S = \emptyset)$

3.3 DVD with Parameters mw and the solution size k

In this section we present an algorithm to solve the DVD problem by using the algorithm VD-MW given in the previous section. We prove the following result.

Theorem 7. DVD *can be solved in time* $O(\mathtt{mw}^2\, k(k+1)^{2\mathtt{mw}}\, n^2)$.

Let $G = H(G_1, \ldots, G_p)$ be the input graph and let $\langle G, \mathbf{t}, \mathbf{d}, k \rangle$ be an instance of the decision version of the DVD problem, asking for a *Distance Vector Dominating* set of size k of G with respect to the demand vector $\mathbf{t}$ and the radius vector $\mathbf{d}$. Our algorithm DVD-MW, which checks for a solution for the decision version of the problem with instance $\langle G, \mathbf{t}, \mathbf{d}, b \rangle$, is based on the following easy considerations, for any vertex $v \in V(G_i)$ and $i \in [p]$:

- If $d_v \ge 2$ then any vertex in $V(G_i)$ that is selected in the solution dominates v (recall Observation 4) together with any vertex in $V(G_j)$ that is selected in the solution, for j such that $j \neq i$ and $\delta_H(i,j) \le d_v$.
- If $d_v = 1$ then any vertex in $V(G_j)$ with $(i,j) \in E(H)$, that is selected in the solution, dominates v.

Consider a partition $(s_1, \ldots, s_p)$ of b and select s_i vertices in G_i, for each $i \in [p]$. If there exists $v \in V(G_i)$ with $d_v \ge 2$ and demand $t_v > s_i + \sum_{j \mid\, j \neq i\, \wedge\, \delta_H(i,j) \le d_v} s_j$ then the partition has to be discarded (since there are not enough selected vertices to dominate v). Otherwise, all the vertices with $d_v \ge 2$ are dominated by any choice of vertices satisfying the partition and we only have to worry about each vertex v with $d_v = 1$. In particular, the selection in each $V(G_i)$ has to be accurate in order to have a vector dominating set for $\langle G_i, \mathbf{t}', 1, s_i \rangle$, where t'_v is defined in Algorithm 3.

4 FPT Algorithms for Graphs of Bounded Treewidth

Definition 3. *A tree decomposition of a graph* $G = (V, E)$ *is a pair* $(T, \{W_i\}_{i \in V(T)})$, *where* T *is a tree and each* i *in* T *is assigned a* $W_i \subseteq V$ *such that:*

1. $\bigcup_{i \in V(T)} W_i = V$.
2. *For each* $e = (v, u) \in E$, *there exists* i *in* T *s.t.* W_i *contains both* v *and* u.
3. *For each* $v \in V$, *the tree induced by* $T_v = \{i \in V(T) \mid v \in W_i\}$ *is connected.*

The width of a tree decomposition $(T, \{W_i\}_{i \in V(T)})$ of a graph G, is defined as $\max_{i \in V(T)} |W_i| - 1$. The treewidth of G, denoted by $\mathtt{tw}(G)$, is the minimum width over all tree decompositions of G. Deciding whether a graph has tree decomposition of treewidth at most k is NP-complete [2] and proved fpt in [6].

Definition 4. *[42] A tree decomposition* $(T, \{W_i\}_{i \in V(T)})$ *is called nice if it satisfies conditions 1. and 2.:*

1. $W_r = \emptyset$, *for* r *the root of* T *and* $W_i = \emptyset$, *for every leaf* i *of* T.
2. *Every non-leaf vertex of* T *is of one of the following three types:*
   Introduce: *a vertex* i *with one child* j *s.t.* $W_i = W_j \cup \{v\}$ *for a vertex* $v \notin W_j$.
   Forget: *a vertex* i *with one child* j *s.t.* $W_j = W_i \cup \{v\}$ *for a vertex* $v \notin W_i$.
   Join: *a vertex* i *with two children* i_1, i_2 *s.t.* $W_i = W_{i_1} = W_{i_2}$.

Consider a graph $G = (V, E)$. Given a tree decomposition of G of width $\mathtt{tw}$, one can compute in polynomial time a nice tree decomposition $(T, \{W_i\}_{i \in V(T)})$ of G of treewidth at most $\mathtt{tw}$ having $O(\mathtt{tw}|V(G)|)$ vertices [42]. Let T be rooted in r. For any i in T, denote by $T(i)$ the subtree of T rooted at i, by $W(i) = \bigcup_{j \in T(i)} W_j$ the union of the bags in $T(i)$, and by $s_i = |W_i|$ the size of W_i.

A FPT algorithm parameterized by $\mathtt{tw}$ plus τ for VD. We give a dynamic programming algorithm which, exploiting a nice tree decomposition, recursively solves the Vector Domination (VD) problem. Fix $i \in V(T)$, to recursively reconstruct the solution, we calculate optimal solutions under different hypotheses based on the following considerations: For each vertex $v \in W_i$ we have two cases: $v \in S$, $v \notin S$. We are going to consider all the 2^{s_i} combinations of such states with respect to some solution S of the problem. We denote each combination with a binary vector L_i of size s_i indexed by the elements of W_i, where for each $v \in W_i$, $L_i(v) = 1$, if $v \in S$ and $L_i(v) = 0$, otherwise. The configuration $L_i = \emptyset$ denotes the vector of length 0 corresponding to an empty bag. We denote by $\mathcal{L}_i$ the family of all the 2^{s_i} possible state vectors of the s_i vertices in W_i.
We consider all the possible contributions to the VD problem, of vertices in $V \setminus W(i)$; that is, for each $v \in W_i$, we consider all the possible demands among $t_v, t_v{-}1, \ldots, 0$. As a consequence, we will have up to $(\tau{+}1)^{s_i}$ demand combinations, where $\tau{=} \max_{v \in V} t_v$. We denote each possible demand combination with a vector K_i, indexed by the s_i elements in W_i. The configuration $K_i = \emptyset$ denotes the demand vector of length 0 corresponding to an empty bag. Moreover, $\mathcal{K}_i$ represents the family of all the possible demand combinations of vertices in W_i.

The following definition introduces the values that will be computed by the algorithm in order to keep track of all the above cases.

Definition 5. *For each vertex* $i \in V(T)$, *each* $L_i \in \mathcal{L}_i$ *and each* $K_i \in \mathcal{K}_i$, *we define* $B_i(L_i, K_i)$ *as the minimum number of vertices to be selected in* $G[W(i)]$

in order to dominate all the remaining vertices in $G[W(i)]$, where the states and the demands of vertices in W_i are given by L_i and K_i.

By noticing that the root r of a nice tree decomposition has $W_r = \emptyset$, we have that the solution of the VD problem $\langle G, \mathbf{t}, \mathbf{1}\rangle$ can be obtained by computing $B_r(\emptyset, \emptyset)$.

Lemma 4. *For each $i \in T$, the computation of $B_i(L_i, K_i)$, for each $L_i \in \mathcal{L}_i$ and $K_i \in \mathcal{K}_i$, comprises $O(2^{s_i}(\tau+1)^{s_i})$ values, each of which can be computed recursively in time $O(s_i)$.*

Theorem 8. *If a tree decomposition of G with width $\mathtt{tw}$ is given then* VD *is solvable in time $O(\mathtt{tw}^2 2^{\mathtt{tw}}(\tau+1)^{\mathtt{tw}}\, n)$.*

Proof. The decomposition tree has at most $O(\mathtt{tw}\ n)$ vertices [42]. Hence, the desired value $B_r(\emptyset, \emptyset)$, which corresponds to the solution of the VD instance $\langle G, \mathbf{t}, \mathbf{1}\rangle$, can be computed in time $O(\mathtt{tw}^2 2^{\mathtt{tw}}(\tau+1)^{\mathtt{tw}}\, n)$. The optimal set S can be computed in the same time by standard backtracking technique. □

A FPT algorithm parameterized by $\mathtt{tw}$ plus δ for RD. Exploiting a nice tree decomposition of the input graph G and a strategy similar to the one adopted in [8] we obtain the following result.

Theorem 9. *If a tree decomposition of G with width $\mathtt{tw}$ is given then* RD *is solvable in time $O(\mathtt{tw}(2\delta+1)^{\mathtt{tw}}(n+\mathtt{tw}^2)\, n^2 \log n)$.*

5 Concluding Remarks

We introduced the Distance Vector Domination problem which generalizes both distance and multiple domination, at individual (i.e., vertex) level. The problem is motivated by the development of strategies to mitigate the spread of fake information. Indeed the set identified by the problem can be used to detect a set of individuals who, disseminating debunking information, can prevent the spreading, of misinformation. We analyzed the parameterized complexities of the problem according to several standard and structural parameters. It eluded us the design of an FPT algorithm for the DVD problem parameterized by the combination of treewidth and one of the other problem parameters, such as the size k of the solution, the largest demand τ or the largest radius δ, which we leave as an open problem. Additionally, it would be interesting to investigate the complexity of the RD problem with respect to the clique-width parameter.

References

1. Albert, R., Jeong, H., Barabási, A.-L.: Error and attack tolerance of complex networks. Nature **404**, 378–382 (2000)
2. Arnborg, S., Corneil, D.G., Proskurowski, A.: Complexity of finding embeddings in a k-tree. SIAM J. Alg. Disc. Meth. **8**, 277–284 (1987)

3. Banerjee, S., Jenamani, M., Pratihar, D.K.: A survey on influence maximization in a social network. Knowl. Inf. Syst. **62**, 3417–3455 (2020). https://doi.org/10.1007/s10115-020-01461-4
4. Ben-Zwi, O., Hermelin, D., Lokshtanov, D., Newman, I.: Treewidth governs the complexity of target set selection. Discret. Optim. **8**(1), 87–96 (2011)
5. Bermond, J.-C., Gargano, L., Rescigno, A.A.: Gathering with minimum delay in tree sensor networks. In: Proceedings of SIROCCO'08, LNCS 5058, pp. 262–276 (2008)
6. Bodlaender, H.L., Kloks, T.: Better algorithms for the pathwidth and treewidth of graphs. In: Proceedings of ICALP'91, LNCS 510, pp. 544–555 (1991)
7. Betzler, N., Bredereck, R., Niedermeier, R., Uhlmann, J.: On Bounded-Degree Vertex Deletion parameterized by treewidth. Discret. Appl. Math. **160**(1–2), 53–60 (2012)
8. Borradaile, G., Le, H.: Optimal dynamic program for R-domination problems over tree decompositions. In: Proceedings of 11th International Symposium on Parameterized and Exact Computation, IPEC, LIPIcs, vol. 63, pp. 8:1–8:23 (2016)
9. Chong, C.-Y., Kumar, S.P.: Sensor networks: Evolution, opportunities, and challenges. Proc. IEEE **91**(8), 1247–1256 (2003)
10. Cicalese, F., Milanic, M., Vaccaro, U.: On the approximability and exact algorithms for vector domination and related problems in graphs. Discrete Appl. Math. **161**, 750–767 (2013)
11. Cicalese, F., Cordasco, G., Gargano, L., Milanic, M., Vaccaro, U.: Latency-bounded target set selection in social networks. Theoret. Comput. Sci. **535**, 1–15 (2014)
12. Cordasco, G., Gargano, L., Mecchia, M., Rescigno, A.A., Vaccaro, U.: A fast and effective heuristic for discovering small target sets in social networks. In: Proceedings of the 9th International Conference on Combinatorial Optimization and Applications (COCOA) (2015)
13. Cordasco, G., Gargano, L., Rescigno, A.A.: Influence propagation over large scale social networks. In: Proceedings of IEEE/ACM International Conference on Advances in Social Networks Analysis and Mining (ASONAM'15), pp. 1531–1538 (2015)
14. Cordasco, G., Gargano, L., Rescigno, A.A.: On finding small sets that influence large networks. Soc. Netw. Anal. Min. **6**(11) (2016)
15. Cordasco, G., Gargano, L., Mecchia, M., Rescigno, A.A., Vaccaro, U.: Discovering small target sets in social networks: a fast and effective algorithm. Algorithmica **80**(6), 1804–1833 (2018)
16. Cordasco, G., Gargano, L., Rescigno, A.A., Vaccaro, U.: Evangelism in social networks: algorithms and complexity. Networks **71**(4), 346–357 (2018)
17. Cordasco, G., Gargano, L., Rescigno, A.A.: Active influence spreading in social networks. Theoret. Comput. Sci. **764**, 15–29 (2019)
18. Cordasco, G., et al.: Whom to befriend to influence people. Theoret. Comput. Sci. **810**, 26–42 (2020)
19. Cordasco, G., Gargano, L., Rescigno, A.A.: Parameterized complexity for iterated type partitions and modular-width. Discrete Appl. Math., 350 (2024)
20. Corneil, D., Habib, M., Paul, C., Tedder, M.: A recursive linear time modular decomposition algorithm via LexBFS. arXiv:0710.3901 [cs.DM] (2024)
21. Courcelle, B.: The monadic second-order logic of graphs recognizable sets of finite graphs. Inform. Comput. **85**(1), 12–75 (1990)
22. Courcelle, B., Olariu, S.: Upper bounds to the clique width of graphs. Discret. Appl. Math. **101**(1), 77–114 (2000)

23. Cygan, M., et al.: Parameterized Algorithms. Springer International Publishing, Cham (2015). https://doi.org/10.1007/978-3-319-21275-3
24. Dasgupta, K., Kukreja, M., Kalpakis, K.: Topology-aware placement and role assignment for energy-efficient information gathering in sensor networks. In Proceedings of IEEE Symposium on Computers and Communications, pp. 341–348 (2003)
25. Downey, R.G., Fellows, M.R.: Parameterized Complexity. Springer, Heidelberg (1999). https://doi.org/10.1007/978-1-4612-0515-9
26. Dvorák, P., Knop, D., Toufar, T.: Target set selection in dense graph classes. In: Proceedings of 29th International Symposium on Algorithms and Computation (ISAAC'18) (2018). https://doi.org/10.4230/LIPIcs.ISAAC.2018.18
27. Fink, J.F., Jacobson, M.S.: n-Domination in Graphs. Graph Theory with Applications to Algorithms and Computer Science. John Wiley & Sons, pp. 283–300 (1985)
28. Gajarský, J., Lampis, M., Ordyniak, S.: Parameterized algorithms for modular-width. In: Proceedings of 8th International Symposium on Parameterized and Exact Computation (IPEC 2013), LNCS 8246, pp. 163–176 (2013)
29. Gallai, T.: Transitiv orientierbare Graphen. Acta Mathematica Academiae Scientiarum Hungarica **18**, 26–66 (1967)
30. Ganian, R.: Using Neighborhood Diversity to Solve Hard Problems. arXiv 2012 (2012). arXiv:1201.3091
31. Granovetter, M.: Threshold models of collective behaviors. Am. J. Sociol. **83**(6), 1420–1443 (1978)
32. Goddard, W., Henning, M.A.: Restricted domination parameters in graphs. J. Comb. Optim. **13**, 353–363 (2007)
33. Haynes, T.W., Hedetniemi, S., Slater, P.: Fundamentals of Domination in Graphs, Marcel Dekker (1998)
34. Haynes, T.W., Hedetniemi, S., Slater, P. (Eds.). Domination in Graphs: Advanced Topics, Marcel Dekker (1998)
35. Harant, J., Prochnewski, A., Voigt, M.: On dominating sets and independent sets of graphs. Comb. Probab. Comput. **8**, 547–553 (1999)
36. Henning, M.A.: Distance domination in graphs. In: Topics in Domination in Graphs. Developments in Mathematics, vol. 64 (2020). https://doi.org/10.1007/978-3-030-51117-3_7
37. Impagliazzo, R., Paturi, R.: On the complexity of k-SAT. J. Comput. Syst. Sci. **62**(2), 367–375 (2001). https://doi.org/10.1006/jcss.2000.1727
38. Ishii, T., Ono, H., Uno, Y.: (Total) Vector domination for graphs with bounded branchwidth. Discret. Appl. Math. **207**, 80–89 (2016)
39. Katsikarelis, I., Lampis, M., Paschos, V.T.: Structural parameters, tight bounds, and approximation for (k, r)-center. Discr. App. Math. **264**, 90–117 (2019)
40. Kempe, D., Kleinberg, J., Tardos, E.: Maximizing the spread of influence through a social network. In: Proceedings of KDD'03, pp. 137–146 (2003)
41. Kimura, M., Saito, K., Motoda, H.: Blocking links to minimize contamination spread in a social network. ACM Trans. Knowl. Discov. Data **3**(2) (2009)
42. Kloks, T.: Treewidth Computations and Approximations. LNCS 842, Springer-Verlag Berlin, Heidelberg (1994). ISSN 0302-9743, https://doi.org/10.1007/BFb0045375
43. Lafond, M., Luo, W.: Parameterized complexity of domination problems using restricted modular partitions. In: MFCS 2023, pp. 61:1–61:14 (2023)
44. Lamblet Mafort, R., Protti, F.: Vector Domination in split-indifference graphs. Inf. Process. Lett., 155 (2020). ISSN 0020-0190

45. Lampis, M.: Algorithmic meta-theorems for restrictions of treewidth. Algorithmica **64**, 19–37 (2012)
46. Newman, M.E.J., Forrest, S., Balthrop, J.: Email networks and the spread of computer viruses. Phys. Rev. E **66** (2002)
47. Raman, V., Saurabh, S., Srihari, S.: Parameterized algorithms for generalized domination. In: Yang, B., Du, D.-Z., Wang, C.A. (eds.) COCOA 2008. LNCS, vol. 5165, pp. 116–126. Springer, Heidelberg (2008). https://doi.org/10.1007/978-3-540-85097-7_11
48. Romanek, M.: Parameterized algorithms for modular-width. Bachelor's Thesis (2016)
49. Slater, P.J.: R-domination in graphs. J. Assoc. Comp. Mach. **23**(3), 446–450 (1976)
50. Li, P., Wang, A., Shang, J.: A simple optimal algorithm for k-tuple dominating problem in interval graphs. J. Comb. Optim. **45**(14) (2023)
51. Wolsey, L.A.: An analysis of the greedy algorithm for the submodular set covering problem. Combinatorica **2**, 385–393 (1982)

Sufficient Conditions for Polynomial-Time Detection of Induced Minors

Clément Dallard[1], Maël Dumas[2,4], Claire Hilaire[3], and Anthony Perez[4](✉)

[1] Department of Informatics, University of Fribourg, Fribourg, Switzerland
clement.dallard@unifr.ch
[2] Institute of Informatics, University of Warsaw, Warsaw, Poland
mdumas@mimuw.edu.pl
[3] FAMNIT and IAM, University of Primorska, Koper, Slovenia
claire.hilaire@upr.si
[4] Université d'Orléans, INSA CVL, LIFO, UR 4022, Orléans, France
anthony.perez@univ-orleans.fr

Abstract. The H-Induced Minor Containment problem (H-IMC) consists in deciding if a fixed graph H is an induced minor of a graph G given as input, that is, whether H can be obtained from G by deleting vertices and contracting edges. Several graphs H are known for which H-IMC is NP-complete, even when H is a tree. In this paper, we investigate which conditions on H and G are sufficient so that the problem becomes polynomial-time solvable. Our results identify three infinite classes of graphs such that, if H belongs to one of these classes, then H-IMC can be solved in polynomial time. Moreover, we show that if the input graph G excludes long induced paths, then H-IMC is polynomial-time solvable for any fixed graph H. As a byproduct of our results, this implies that H-IMC is polynomial-time solvable for all graphs H with at most 5 vertices, except for three open cases.

1 Introduction

The notion of *graph containment* has been intensively studied in the literature from both the algorithmic and structural viewpoints. There are many ways to define whether a graph G *contains* a graph H, usually in terms of operations allowed on G to obtain H. The most common operations are vertex deletion, edge deletion, and edge contraction. Any combination of these operations defines a *graph containment relation.* The *subgraph* relation only allows vertex and edge deletions, while the *minor* relation also allows edge contractions. Their induced counterparts, namely the *induced subgraph* and *induced minor* relations, are defined analogously but without edge deletions. These relations can then be used to define classes of graphs that *exclude* a fixed collection $\mathcal{H}$ of graphs with respect to some fixed relation. Well-known graph classes can be characterized

This works is partly financed by the French *Fédération de Recherche ICVL* (Informatique Centre-Val de Loire) and by the *Slovenian Research and Innovation Agency* (research project J1-4008).

R. Královič and V. Kůrková (Eds.): SOFSEM 2025, LNCS 15538, pp. 195–208, 2025.
https://doi.org/10.1007/978-3-031-82670-2_15

in terms of forbidden graphs. For example: weakly sparse graphs (sometimes simply referred to as sparse graphs) are those that exclude some fixed complete bipartite graph as a subgraph (see, *e.g.*, [4,5]); cographs are defined as graphs that exclude P_4 as an induced subgraph; planar graphs are characterized by excluding K_5 and $K_{3,3}$ as minors; graphs with bounded treewidth are those that exclude a planar graph as a minor; and chordal graphs correspond to graphs that exclude C_4 as an induced minor (see, *e.g.*, [6]).

A natural question that arises in this context is the complexity of determining whether a given graph $G = (V, E)$ *contains* another graph H. If H is part of the input, then the problem of determining if H is a subgraph, induced subgraph, minor, or induced minor of a given graph G is NP-complete.[1] However, if we consider H as fixed, then some of these problems can be solved in polynomial time. For the subgraph and induced subgraph relations, the corresponding problems can be solved in polynomial time by a simple brute-force approach, enumerating all (induced) subgraphs with $|V(H)|$ vertices. For the minor relation, there is the famous $\mathcal{O}(|V(G)|^3)$ algorithm by Robertson and Seymour [19], later improved to $\mathcal{O}(|V(G)|^2)$ by Kawarabayashi, Kobayashi, and Reed [13], and recently improved to almost linear time $\mathcal{O}((|V(G)| + |E(G)|)^{1+o(1)})$ by Korhonen, Pilipczuk, and Stamoulis [16]. In sharp contrast, Fellows, Kratochvíl, Middendorf, and Pfeiffer [10] proved that when considering the induced minor relation, the problem is NP-hard for some fixed graph H on 68 vertices. It is thus natural to wonder for which graphs H this problem is tractable.

In this paper, we focus on the *induced minor* relation and consider, for several choices of H, the following problem:

H-INDUCED MINOR CONTAINMENT (H-IMC)

Input: A graph G.
Question: Does G admit H as an induced minor?

Related Work. Fellows, Kratochvíl, Middendorf, and Pfeiffer [10] asked whether H-IMC can be solved in polynomial if H is a tree or a planar graph. Motivated by this question, Fiala, Kamiński, and Paulusma [11] showed that H-IMC can be solved in polynomial time for all but one forest H on at most 7 vertices. The complexity of the remaining forest (obtained from two claws after identifying one of their leaves) is still open (and has recently been the subject of an open question at a workshop [7]). They also showed that, when H is a subdivided star or obtained by adding at least two leaves to both of the endpoints of an edge, the problem remains polynomial-time solvable. Recently, Korhonen and Lokshtanov [15] eventually settled the complexity of the problem when H is a tree and showed that there exists a tree (with over 2^{300} vertices)

[1] The HAMILTONIAN CYCLE problem, with input graph J, can be reduced in polynomial-time to the considered problem, fixing G to be the graph obtained from J after subdividing every edge once and setting H to be the $2|V(J)|$-vertex cycle.

for which H-IMC is NP-hard. In the same paper, the authors also gave a randomized polynomial-time algorithm that, given two graphs G and H, outputs either an induced minor model of H in G or a balanced separator of G of size $\mathcal{O}(\min(\log|V(G)|, |V(H)|^2) \cdot \sqrt{|V(H)| + |E(H)|} \cdot \sqrt{|E(G)|})$. In particular, if H is fixed, this implies subexponential-time algorithms for several NP-hard problems on H-induced-minor-free graphs. Previous to that, Korhonen [14] showed that graphs with large treewidth and bounded degree contain a large grid as an induced minor, which implies that, for every planar H, there is a subexponential time algorithm for MAX WEIGHT INDEPENDENT SET on H-induced-minor-free graphs. Another recent result, by Nguyen, Scott, and Seymour [18], states that finding $s \geqslant 1$ pairwise anticomplete, disjoint cycles can be done in polynomial time. This in particular implies that if H is a disjoint union of s triangles, then H-IMC can be solved in polynomial time, thus answering the open question about the complexity of $2C_3$-IMC asked in [11].

Several results have also been obtained when restricting the input graph. For instance, Fellows, Kratochvíl, Middendorf, and Pfeiffer [10] showed that H-IMC is polynomial-time solvable, for any graph H, in the class of planar graphs (note that if H is not planar, then the problem becomes trivial). Then, van 't Hof *et al.* [20] extended this result by showing that H-IMC can be solved efficiently on proper minor-closed graph classes,[2] for any planar graph H. In similar fashion, Golovach, Kratsch, and Paulusma [12] proved that the problem is polynomial-time solvable in AT-free graphs while Belmonte *et al.* [1] showed the same result for chordal graphs.

Our Results. In a companion work [9], a superset of the authors proved that $K_{2,3}$-IMC is polynomial-time solvable, where $K_{2,3}$ is the complete bipartite graph with 2 and 3 vertices on each part. We carry on this line of research by settling the complexity status of H-IMC for all but three graphs with up to five vertices. For each such graph H, we show that H-IMC can be solved in polynomial-time. Many cases actually follow from more general results, which we prove in this paper, that settle the complexity of H-IMC for some infinite classes of graphs H. In particular, we show that H-IMC is polynomial-time solvable if H is a flower, a generalized bull or house, or a complete split graph. Formal definitions of these families can be found in Sect. 3; see also Fig. 1 for a representation of such graphs with five vertices. Informally, flowers are the intersection of paths, cycles and diamonds in one vertex.

Theorem 1.1. *If H is a flower, then H-IMC is polynomial-time solvable.*

The generalized houses and bulls are obtained from houses and bulls by subdividing the edges not in the triangle.

Theorem 1.2. *If H is a generalized house or a generalized bull, then H-IMC is polynomial-time solvable.*

[2] A graph class $\mathcal{G}$ is *minor-closed* whenever any minor of graph $G \in \mathcal{G}$ belongs to $\mathcal{G}$. It is *proper* if it is not the class of all graphs.

The graph $S_{k,p}$ is the graph obtained by adding all edges between a clique of size k and an independent set of size p.

Theorem 1.3. *Let $k \leqslant 3$ and p be positive integers. Then $S_{k,p}$-IMC is polynomial-time solvable.*

We emphasize that the class of flowers contains all subdivided stars. Therefore, Theorem 1.1 generalizes the result by Fiala, Kamiński, and Paulusma [11]. Let us also mention that Milanič and Pivač [17] independently showed that House-IMC can be solved in polynomial time (see Fig. 1 for a representation of the house). Their approach, different from ours, relies on an algorithm for detecting the house as an induced topological minor, and then reducing the induced minor case to the former. In the same paper, the authors make use of our structural results for the flowers to detect butterflies (two triangles sharing one vertex) as an induced minor in polynomial time.

When considering restricted input graphs, we broaden the complexity landscape by showing the following result on P_t-free graphs, that is, graphs without induced paths on t vertices. We refer the reader to Fig. 1 for a representation of all graphs mentioned hereafter.

Theorem 1.4. *For any graph H and any positive integer t, H-IMC is polynomial-time solvable in P_t-free graphs.*

Note that *Gem*-induced-minor-free graphs and $\widehat{K_4}$-induced-minor-free graphs may contain arbitrarily long induced paths. Nonetheless, leveraging their structure, we show that *Gem*-IMC and $\widehat{K_4}$-IMC on general graphs can be reduced to graphs without long induced paths. Thus, the two problems are polynomial-time solvable as a consequence of Theorem 1.4 (see Theorems 4.1 and 4.3).[3]

2 Preliminaries

We consider simple, undirected graphs $G = (V, E)$, where V denotes the *vertex set* and E the *edge set*. We may also use $V(G)$ to denote the vertex set of G and $E(G)$ its edge set to clarify the context. Given a vertex $u \in V$, the *open neighborhood* of u is the set $N_G(u) = \{v \in V : uv \in E\}$. The *closed neighborhood* of u is defined as $N_G[u] = N_G(u) \cup \{u\}$. Given a subset of vertices $S \subseteq V$, $N_G[S]$ is the set $\cup_{v \in S} N_G[v]$ and $N_G(S)$ is the set $N_G[S] \setminus S$. We do not necessarily mention G if the context is clear. Given a set of vertices $S \subseteq V$, the subgraph of G *induced by* S, denoted $G[S]$, is the graph (S, E_S) where $E_S = \{uv \colon \{u, v\} \in S \times S\}$. In a slight abuse of notation, we use $G \setminus S$ to denote the graph induced $G[V \setminus S]$. Given two sets of vertices $A, B \subseteq V$, we say that A and B are *adjacent* if there exist $u \in A$ and $v \in B$ such that uv is an edge of G.

[3] The polynomial-time solvability of *Gem*-IMC and $\widehat{K_4}$-IMC also follows from the fact that *Gem*-induced-minor-free graphs and $\widehat{K_4}$-induced-minor-free graphs have bounded clique-width (see [2]).

Given an edge uv of G, we define the *contraction* of uv as the graph obtained from G by removing u and v and by adding a new vertex w with neighborhood $N(\{u,v\})$. Similarly, the *subdivision* of uv is obtained by removing the edge uv from E and inserting a new vertex w and edges wu, wv.

We may denote a path P with ℓ vertices by a sequence $p_1 \dots p_\ell$ of vertices such that two consecutive vertices in the sequence are adjacent. The path on ℓ vertices is denoted by P_ℓ. The vertices $\{p_2, \dots, p_{\ell-1}\}$ are called the *internal vertices* of P_ℓ. Similarly, a sequence $p_1 \dots p_\ell p_1$ describes a cycle C with ℓ vertices such that two consecutive vertices in the sequence are adjacent. The cycle on ℓ vertices is denoted by C_ℓ. The edges of a path, or of a cycle, are the edges between consecutive vertices of the sequence and the *length* of a path, or cycle, is the number of edges it has. Given a path P and some vertices u, v of P, we let uPv be the subpath of P with extremities u and v. If w is adjacent to u, then $wuPv$ is the path obtained by adding the edge wu to uPv, and similarly, if w is adjacent to v, then $uPvw$ is the path obtained by adding the edge vw to uPv.

Induced Minor Models. A graph H is an *induced minor* of G, sometimes denoted $H \subseteq_{im} G$, whenever H can be obtained from G by removing vertices and contracting edges. An *induced minor model* of H in G, or simply a *model* of H, is a collection $\mathcal{X}_H = \{X_u : u \in V(H)\}$ of pairwise disjoint non-empty subsets of $V(G)$ such that:

- for $u \in V(H)$, $G[X_u]$ is connected, and
- for $u \neq v \in V(H)$, X_u and X_v are adjacent if and only if $uv \in E(H)$.

Each set $X_u \in \mathcal{X}_H$ is called a *bag* of $\mathcal{X}_H$. The subgraph of G *induced by* $\mathcal{X}_H$ is the subgraph induced by the union of the bags of $\mathcal{X}_H$. We say that a bag X_u is trivial if $|X_u| = 1$.

A model $\mathcal{X}'_H$ of H is said to be *included* in another model $\mathcal{X}_H$ of H if the union of the bags of $\mathcal{X}'_H$ is included in the union of the bags of $\mathcal{X}_H$. Note that it is not required that each bag of $\mathcal{X}'_H$ is a subset of a bag of $\mathcal{X}_H$. Given $S \subset V(H)$, we say that $\mathcal{X}_H$ *minimizes the size of the bags of* S (or just *minimizes the bags of* S) if there is no model $\mathcal{X}'_H = \{X'_u : u \in V(H)\}$ of H included in $\mathcal{X}_H$ such that $\sum_{v \in S} |X'_v| < \sum_{v \in S} |X_v|$. In particular, we say that $\mathcal{X}_H$ is a *minimal model of* H if $\mathcal{X}_H$ minimizes the bags of $V(H)$. Finally, we say that a bag X_u of $\mathcal{X}_H$ is *minimal* if there is no strict subset X'_u of X_u such that replacing X_u by X'_u results in a model of H. Note in particular that if $\mathcal{X}_H$ is a minimal model of H, then each bag of $\mathcal{X}_H$ is minimal.

A *premodel* is a collection of disjoint subset of vertices of G, $\mathcal{X} = \{X_u : u \in V(H)\}$, that is not necessarily a model of H. In particular, X_u can be the empty set. We say that a model $\mathcal{X}^* = \{X^*_u : u \in V(H)\}$ of H in G *extends* a premodel $\mathcal{X} = \{X_u : u \in V(H)\}$ if, for each $u \in V(H)$, we have $X_u \subseteq X^*_u$.

2.1 Graphs with at Most 5 vertices

Before diving into our more general proofs, we provide Fig. 1 an exhaustive list of graphs with 5 vertices and recall known and new results regarding the complexity of H-IMC, for various graphs H.

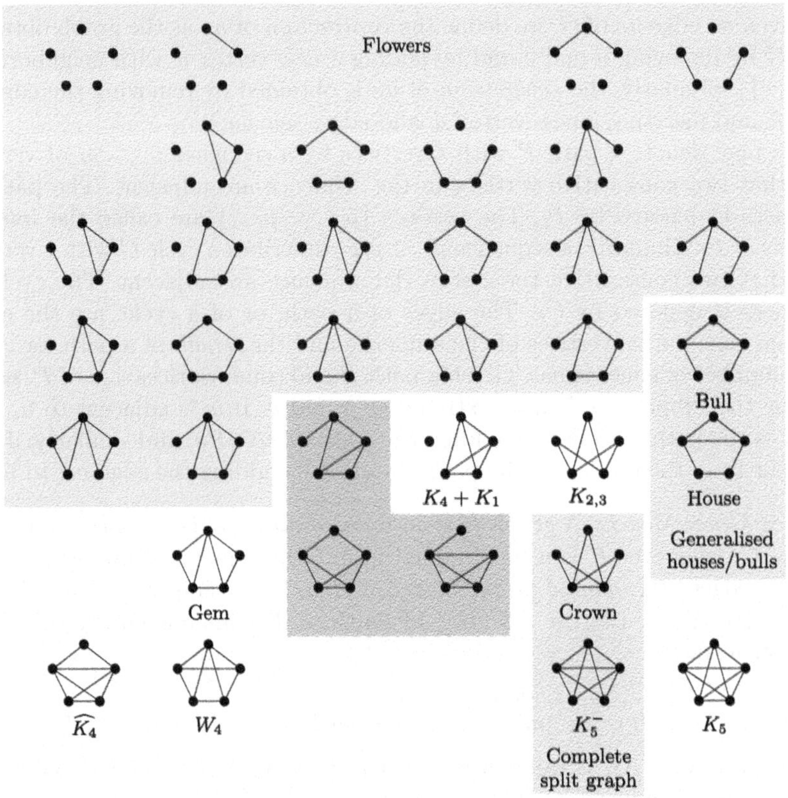

Fig. 1. Exhaustive list of graphs with 5 vertices. The group of graphs with green background belongs to infinite families studied in this paper. The ones with blue background are the ones for which the complexity of H-IMC remains open.

We first note that whenever H is a clique, having H as an induced minor is equivalent to having H as a minor. Hence, H-IMC is polynomial-time solvable whenever H is a clique [13,16,19]. A similar observation can be made for K_t+K_1-IMC: for each choice of vertex x of G as a bag for K_1, all that remains is to try to find a model of K_t in $G \setminus N[x]$. Moreover, it is easily noticed that a graph contains a path P as an induced minor if and only if it contains P as an induced subgraph. More generally, if H is a disjoint union of paths, then a graph contains H as an induced minor if and only if it contains H as an induced subgraph; see Lemma 3.3. Therefore, in such a case, H-IMC can be solved in polynomial time. For graphs H with at most four vertices, a short discussion in the introduction of [9] explains that determining if H is an induced minor can be done efficiently.

For $K_{2,3}$, a recent result by a superset of the authors of the current paper, based on a characterization in terms of forbidden induced subgraphs and the so-called shortest-path detectors technique, lead to a polynomial-time algorithm for $K_{2,3}$-IMC [9]. The same (super)set of authors, in an ongoing project, was able to extend these results to show that, among others, W_4-IMC is polynomial-time solvable.

All other results can be deduced from our work. In particular, we show Sect. 3 that H-IMC is polynomial-time solvable whenever H is a flower (Sect. 3.1), the bull or the house (Sect. 3.2), and the two complete split graphs (Sect. 3.3). Let us mention once again that our results apply to some generalizations of aforementioned graphs, and thus imply polynomial-time algorithms for graphs with more vertices. Moreover, the result for flowers encompasses known results of Fiala, Kamiński, and Paulusma [11] who proved polynomial-time solvability of H-IMC whenever H is a subdivided star.

The cases where H is the *Full House* (denoted also $\widehat{K_4}$, which is a K_4 plus a vertex adjacent to two vertices of that K_4) or the *Gem* (a P_4 plus a vertex complete to it) are discussed in Sect. 4.

3 Almost Trivial Models

Note that when considering induced minor models of H in G, we can focus on models with as few vertices as possible, and thus on models that minimizes the bags of $V(H)$ (recall that we minimize over all the model, not each bag individually). In particular, if this size amounts to $|V(H)|$, then $\mathcal{X}_H$ induces a subgraph in G isomorphic to H.

Based on this observation, we can restate the induced subgraph relation by saying that for every graph G admitting H as an induced subgraph there exists a model of H in G such that every bag is trivial (*i.e.* contains exactly one vertex). In this section, we consider a natural generalization of this observation, where we allow only a subset of vertices of H to have non-trivial bags.

Definition 3.1 (S-non-trivia). *Let H be a graph and S be a (potentially empty) set of vertices of H. We say that H is* S-non-trivial, *or* S-NT *for short, if for every graph G such that $H \subseteq_{im} G$ there exists a model $\mathcal{X}_H = \{X_u : u \in V(H)\}$ of H in G such that for each $v \in V(H) \setminus S$, $|X_v| = 1$.*

Observe in particular that if H is $\emptyset$-non-trivial, then H is an induced minor of some graph G if and only if H is an induced subgraph of G. Recall that, in this case, the problem can be trivially solved in polynomial time.

In the remaining of this section, we prove that H-IMC is polynomial-time solvable for S-NT graphs H with $|S| \leqslant 1$. We first give some structural properties on the bags of a model for vertices of small degree in H. The properties in Lemma 3.1 are already known and used in other papers (see, for instance, [9,11]). We restate them with our notations for the sake of readability. A consequence of the following Lemma is that paths are $\emptyset$-NT.

Lemma 3.1. *Let G and H be two graphs such that $H \subseteq_{im} G$. Let $\mathcal{X}_H$ be a model of H in G, such that X_u is minimal for a vertex $u \in V(H)$. Then:*

- *if $\deg_H(u) \leqslant 1$, then $|X_u| = 1$;*

- *if* $\deg_H(u) = 2$, *with neighbors* v, w, *then there is a unique vertex* x_v *in* $N(X_v) \cap X_u$ *and a unique vertex* x_w *in* $N(X_w) \cap X_u$, *and* X_u *induces in* G *a path whose extremities are* x_v, x_w.

Using this lemma, we get the following result that guarantees that, if a graph admits some graph H as an induced minor, then there exists a model $\mathcal{X}_H$ of H such that for every connected component of H that is not a cycle, every vertex of H of degree at most 2 has a trivial bag in $\mathcal{X}_H$. Moreover, if a connected component of H is a cycle, then at most one bag of it is non-trivial in $\mathcal{X}_H$.

Lemma 3.2. *Let* H *be a graph and* P *be a path in* H *such that the internal vertices of* P *have degree 2 (potentially the extremities of* P *can be adjacent). Let* G *be a graph such that* $H \subseteq_{im} G$, *and* $\mathcal{X}_H$ *a model of* H *in* G. *Then there is a model of* H *in* G *included in* $\mathcal{X}_H$ *such that the internal vertices of* P *have trivial bags. Moreover, only the bag of one of the extremities of* P *is bigger than it was in* $\mathcal{X}_H$, *and the bags of* $H \setminus V(P)$ *are the same as in* $\mathcal{X}_H$.

From the above lemmas, we can deduce the following result.

Lemma 3.3. H *is* $\emptyset$*-NT if and only if* H *is a disjoint union of paths.*

Fiala, Kamiński, and Paulusma [11] observed that subdivided stars of center u are $\{u\}$-NT, and gave a polynomial time algorithm for detecting them [11, Proposition 2]. We generalize their result for every $\{u\}$-NT graph H.

Theorem 3.1. *If* H *is* S*-NT,* $|S| \leqslant 1$, *then* H*-IMC is polynomial-time solvable.*

Proof. If $S = \emptyset$ then by Lemma 3.3 it is equivalent to testing if H is a disjoint union of paths, which can clearly be done in polynomial time. Suppose that H is $\{u\}$-NT for one of its vertex u. Let G be the input graph and assume that $H \subseteq_{im} G$. Since H is $\{u\}$-NT, then there is a model $\mathcal{X}_H$ where only the bag of u is non-trivial. Observe that $G[X_u]$ induces a connected graph. Moreover, for each vertex v not adjacent to u in H, X_u is not adjacent to X_v, so $X_u \cap N(X_v) = \emptyset$, and similarly, X_u must contain a vertex of $N(X_w)$ for each w adjacent to u.

This gives us the following polynomial strategy to detect if H is an induced minor of G, and output a model in the positive case: we enumerate all the premodels of H where the bags contain exactly one vertex of G, except for the bag of u which is empty. There are $\mathcal{O}(n^{|V(H)|-1})$ possibilities. Given such a premodel $\mathcal{X} = \{X_v \colon w \in V(H) \setminus \{u\}\}$, we first check if for each v, w in $H \setminus \{u\}$, the vertices in the trivial bags X_v and X_w are adjacent if and only if v, w are adjacent. If this condition is not satisfied, we can reject the premodel.

Otherwise, let $Y = \bigcup_{v \in N_H(u)} X_v$ and $Z = \bigcup_{v \notin N_H(u)} X_v$. We enumerate the connected components $C_1, \ldots, C_r$ of $G \setminus (Y \cup N[Z])$, which can be done by Breadth-First Search in time $\mathcal{O}(|V(G)| + |E(G)|)$. If there is one connected component C_i containing a vertex of $N_G(v)$ for each $v \in Y$, then we have found a model of H in G with $X_u = C_i$. If for every possible premodel we did not find a suitable connected component, then we can conclude that H is not an induced minor of G. The algorithm described here takes polynomial time, since H is fixed. □

3.1 Flowers

We say that a graph H is a *flower* if there is a vertex $u \in V(H)$ such that $H \setminus \{u\}$ is a disjoint union of paths and for each path P, either $|V(P)| = 3$ and P is complete to u (*sepal*), or P is connected only by 0, 1 (*stamens*) or 2 (*petal*) of its extremities to u. The vertex u is called the *center* of H. We refer the reader to Fig. 1 for an exhaustive list of flowers with 5 vertices.

Lemma 3.4. *If H is a flower of center u, then H is $\{u\}$-NT.*

Proof. Suppose H is a flower of center u, let G be a graph such that $H \subseteq_{im} G$ and let $\mathcal{X}_H$ be a model of H in G that minimizes the size of the bags of $H \setminus \{u\}$ (*i.e.* such that there is no model $\mathcal{X}'_H$ included in $\mathcal{X}_H$ such that the sum of the sizes of the bags of $H \setminus \{u\}$ is strictly smaller in $\mathcal{X}'_H$ than in $\mathcal{X}_H$). In particular, every bag of $H \setminus \{u\}$ is minimal. Let us show that X_u is the only bag that is non-trivial.

Suppose first that H contains a sepal, *i.e.* a path $P = abc$ that is complete to u, and suppose that the bag of a vertex of P is not trivial. Let y_a, y_u, y_c be three vertices adjacent to X_b, respectively belonging to X_a, X_u, X_c. Let P_{ac} be a shortest path in $G[X_b \cup \{y_a, y_c\}]$ from y_a to y_c. Note that y_a and y_c are not adjacent, and thus P_{ac} admits at least one internal vertex in X_b. Let P_u be a shortest path in $G[X_b \cup \{y_u\}]$ from y_u to some internal vertex of P_{ac}. Then the vertex x_b at the intersection of P_{ac} and P_u has degree at least 3 in $G[V(P_{ac}) \cup V(P_u)]$ and belongs to X_b. Let x_u be the neighbor of x_b on $P_u y_u$, and similarly let x_a be the neighbor of x_b on $x_b P_{ac} y_a$ and x_c be the neighbor of x_b on $x_b P_{ac} y_c$. Note that it is possible to have $x_a = y_a$, $x_b = y_b$ or $x_u = y_u$, but by construction, $\{x_a, x_b, x_c, x_u\}$ are all distinct. Similarly, our construction allows x_u being adjacent to x_a or x_c, but the fact that P_{ac} is a shortest path in $G[X_b \cup \{y_a, y_c\}]$ prevents x_a and x_c from being adjacent. Therefore, if we replace in $\mathcal{X}_H$ the bags X_a, X_b, X_c and X_u by respectively $\{x_a\}, \{x_b\}, \{x_c\}$ and $X'_u = X_u \cup X_a \cup X_b \cup X_c \setminus \{x_a, x_b, x_c\}$ (in particular X'_u contains x_u that is adjacent to x_b), we obtain a new model of H included in $\mathcal{X}_H$ in which the bags of a, b and c are trivial, and this contradicts the choice of $\mathcal{X}_H$.

We showed that all the vertices of H that are in a sepal (except u) have trivial bags. Observe that every vertex $v \neq u$ that is in a stamen or petal either has degree 1 or is an internal vertex of a path with degree 2. By Lemmas 3.1 and 3.2, every such vertex has a trivial bag in $\mathcal{X}_H$. Hence, every vertex that is not u has a trivial bag in $\mathcal{X}_H$, and thus H is $\{u\}$-NT. □

Combining Lemma 3.4 and Theorem 3.1, we obtain Theorem 1.1, which we restate here for convenience.

Theorem 1.1. *If H is a flower, then H-IMC is polynomial-time solvable.*

3.2 Generalized Houses and Bulls

We say that a graph is a *generalized house* if it consists of a triangle a, u, v, vertices b and c adjacent to u and v respectively, and a path $R = b_1 b_2 \dots b_r$ from

$b = b_1$ to $c = b_r$, with $r > 1$ (see Fig. 2). A generalized bull is defined similarly where R has a missing edge. More formally, in a generalized bull u, v, a, b, c and their adjacencies are defined similarly but R is replaced by two paths $R_b = b_1 \dots b_s$ and $R_c = b_{s+1} \dots b_r$, with $1 \leqslant s < r$, with still $b = b_1$ and $c = b_r$.

We show that if H is a generalized house or a generalized bull, then H-IMC can be solved in polynomial time. The main idea here is that these graphs are $\{u, v\}$-NT, with one of the bags having a specific structure (Lemma 3.5). It also allows us to prove that generalized houses are $\{u\}$-NT (Lemma 3.6).

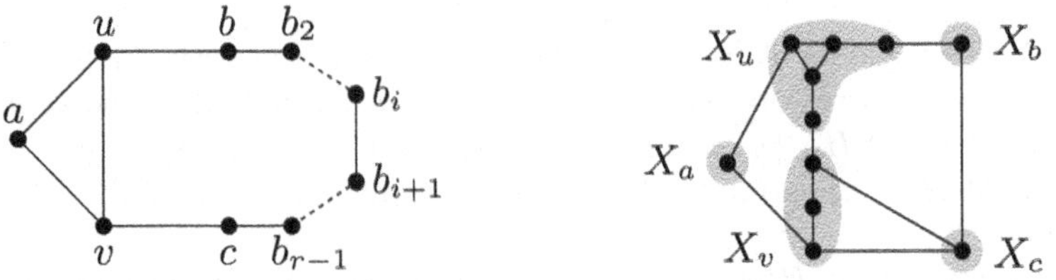

(a) A generalized house (b) A model of a house

Fig. 2. Note that removing one edge $b_i b_{i+1}$ for some $1 \leqslant i \leqslant r-1$ results in a generalized bull.

The idea of the proof of Lemma 3.5 is similar to that of Lemma 3.4: we start from a model that minimizes the bag of v and deduce the structure of the bags.

Lemma 3.5. *If H is a generalized house or a generalized bull, then H is $\{u, v\}$-NT. Moreover, if a graph G admits H as induced minor, then there exists a model $\mathcal{X}_H$ such that $G[X_v]$ is a path from a vertex adjacent to both X_a and X_c to a vertex adjacent to both X_c and X_u. Furthermore, these vertices are the unique vertices in X_v adjacent to X_a and X_u respectively.*

From a model of the generalized house with X_u and X_v non-trivial, we can construct a model with only u having a non-trivial bag (illustrated in Fig. 3).

Lemma 3.6. *If H is a generalized house, then H is $\{u\}$-NT.*

Unfortunately, this construction does not extend to generalized bulls, as some of them are not $\{u\}$-NT, see Fig. 3. However, Theorem 1.2 presents a polynomial time algorithm for detecting generalized bulls, constructing models with possibly two non-trivial bags. The idea behind the algorithm is, given H, to compute first a premodel of H where each bag contains one vertex, and such that the bags are adjacent if and only if the vertices are adjacent in H, except between the bags X_u, X_v and X_u, X_b that might not be adjacent yet. Then, if this premodel can be extended into a model of H, we show that we can connect X_u, X_b by choosing an arbitrary path between their respective vertices, then connecting the vertex of X_v to this path.

Theorem 1.2. *If H is a generalized house or a generalized bull then H-IMC is polynomial-time solvable.*

Fig. 3. On the left, the construction of a model for the house with only one non-trivial bag. On the right, an example of a graph that admits the bull with subdivided horns as induced minor, but with at least two big bags in *every* model.

3.3 Complete Split Graphs

Let $k, p \in \mathbb{N}$. The graph $S_{k,p}$, obtained by adding all possible edges between a clique of size k and an independent set of size p, is called a *complete split graph.* For $k = 2$ and $p = 3$, $S_{k,p}$ is also known as the *Crown*, and for $k = 3$ and $p = 2$, it corresponds to K_5^- (K_5 minus an edge); see Fig. 1 for a graphical representation. In this section, we show that if $k \leqslant 3$, then $S_{k,p}$-IMC can be solved in polynomial time. The idea of the algorithm is to first guess the p vertices in the independent set. Then we try to guess k pairwise disjoint sets, each containing at least one neighbor of each of the p vertices, and check if we can construct a model of a clique using these sets. The last part can be done in quasi-linear time with the algorithm of [16] for the Rooted Minor Containment problem. In this problem, given graphs G and H and a premodel (a *root*) of H in G, the goal is to find a model of H in G that extends the given premodel. In the theorem below, the $O_{H,|X|}(\cdot)$-notation hides factors that depend on H and X and are computable. The set X is the set of vertices in the root.

Theorem 3.2. *([16, Theorem 1.1]) The* Rooted Minor Containment *problem can be solved in time $O_{H,|X|}((|V(G)| + |E(G)|)^{1+o(1)})$. In case of a positive answer, the algorithm also provides a model with the same running time.*

Using this result, we can prove the following theorem. We say that a clique K of a graph G is *universal* if for every $x \in K$, $N[x] = V(G)$.

Theorem 3.3. *Let H be a graph with some universal clique K. If H is K-NT, then H-IMC is polynomial-time solvable.*

Lemma 3.7. *The graph $S_{k,p}$ with $k \leqslant 3$ is K-NT, where K is the clique of size k of $S_{k,p}$.*

Lemma 3.7 can be proved in a similar way as Lemma 3.4. Observe that the above result is not true if $k > 3$. Indeed, there might be no vertex of G of degree at least k in the bags of the vertices of the independent part of $S_{k,p}$. Combining Lemma 3.7 and Theorem 3.3, we thus obtain Theorem 1.3, restated below.

Theorem 1.3. *Let $k \leqslant 3$ and p be positive integers. Then $S_{k,p}$-IMC is polynomial-time solvable.*

4 H-IMC on graphs with no long induced paths

In this section, we show that if the input graph does not contain long induced paths, then H-IMC can be solved in polynomial time, for any fixed graph H. This allows us to develop polynomial-time algorithms for *Gem*-IMC and $\widehat{K_4}$-IMC. In what follows, we write that a graph is P_t-free if it excludes the path on t vertices as an induced subgraph. The idea of the following result is that a minimal induced minor model in a P_t-free graph contains a bounded number of vertices. Hence, an exhaustive search of the possible models of H can be done in polynomial time.

Theorem 1.4. *For any graph H and any positive integer t, H-IMC is polynomial-time solvable in P_t-free graphs.*

Błasiok, Kamiński, Raymond, and Trunck [3] showed that the class of H-induced minor-free graphs are well-quasi-ordered by induced minors if and only if H is an induced minor of the *Gem* or $\widehat{K_4}$. Moreover, they showed decomposition theorems for these two classes of graphs. We make use of these theorems to test H-IMC in polynomial time for the *Gem* and $\widehat{K_4}$.

Theorem 4.1. *([3, Theorem 3]) Let G be a 2-connected graph such that Gem $\not\subseteq_{im}$ G. Then G has a subset $X \subseteq V(G)$ of at most six vertices such that every connected component of $G \setminus X$ is either a cograph or a path whose internal vertices are of degree two in G.*

The idea for the algorithm is the following. If a graph G does not have the structure of *Gem*-induced-minor-free graphs, then we conclude that *Gem* $\subseteq_{im} G$; otherwise, we show that we can check if *Gem* $\subseteq_{im} G$ in polynomial time in the restricted structure of *Gem*-induced minor-free graphs.

Theorem 4.2. *Gem-IMC is polynomial-time solvable.*

Proof. First, since the *Gem* is 2-connected, we may assume that G is 2-connected; otherwise, we can consider the 2-connected components of G independently.

The algorithm is as follows: We test all subsets $X \subseteq V$ of size at most six and check whether the connected components of $G \setminus X$ meet the requirements of Theorem 4.1, that is, are cographs or paths whose internal vertices have degree 2 in G. Note that cographs, which are exactly P_4-free graphs, can be recognized in linear time [8]. If such a set X does not exist, then *Gem* $\subseteq_{im} G$. Hence, we may assume that the algorithm finds such a set X. We contract the internal vertices of components of $G \setminus X$ that are paths of length at least 3 to P_3. Let G' be the obtained graph and observe that *Gem* $\subseteq_{im} G'$ if and only if *Gem* $\subseteq_{im} G$. Since $|X| \leqslant 6$, the longest induced path in G' is of length at most 26, obtained by alternating between paths on at most 3 vertices in $G \setminus X$ and vertices of X. Therefore, G' is P_{28}-free and we can use Theorem 1.4 to conclude. □

We use a similar approach for the $\widehat{K_4}$, but the structure of $\widehat{K_4}$-induced minor-free graphs is more subtle, leading to more cases to consider.

Theorem 4.3. *$\widehat{K_4}$-IMC is polynomial-time solvable.*

References

1. Belmonte, R., Golovach, P.A., Heggernes, P., van 't Hof, P., Kamiński, M., Paulusma, D.: Finding contractions and induced minors in chordal graphs via disjoint paths. In: Asano, T., Nakano, S., Okamoto, Y., Watanabe, O. (eds.) ISAAC 2011. LNCS, vol. 7074, pp. 110–119. Springer, Heidelberg (2011). https://doi.org/10.1007/978-3-642-25591-5_13
2. Belmonte, R., Otachi, Y., Schweitzer, P.: Induced minor free graphs: isomorphism and clique-width. Algorithmica **80**, 29–47 (2018). https://doi.org/10.1007/S00453-016-0234-8
3. Błasiok, J., Kamiński, M., Raymond, J.F., Trunck, T.: Induced minors and well-quasi-ordering. J. Comb. Theory Ser. B **134**, 110–142 (2019). https://doi.org/10.1016/j.jctb.2018.05.005
4. Bonamy, M., et al.: Sparse graphs with bounded induced cycle packing number have logarithmic treewidth. J. Comb. Theory Ser. B **167**, 215–249 (2024). https://doi.org/10.1016/j.jctb.2024.03.003
5. Bonnet, É.: Sparse induced subgraphs of large treewidth (2024). https://doi.org/10.48550/arXiv.2405.13797
6. Brandstädt, A., Le, V.B., Spinrad, J.P.: Graph Classes: A Survey. SIAM (1999)
7. Chudnovsky, M., Misra, N., Paulusma, D., Schaudt, O., Agrawal, A.: Vertex partitioning in graphs: from structure to algorithms (Dagstuhl seminar 22481). Dagstuhl Rep. **12**(11), 109–123 (2022). https://doi.org/10.4230/DAGREP.12.11.109
8. Corneil, D.G., Perl, Y., Stewart, L.K.: A linear recognition algorithm for cographs. SIAM J. Comput. **14**(4), 926–934 (1985)
9. Dallard, C., Dumas, M., Hilaire, C., Milanic, M., Perez, A., Trotignon, N.: Detecting $K_{2,3}$ as an induced minor. In: Proceedings of the 35th International Workshop on Combinatorial Algorithms, IWOCA. Lecture Notes in Computer Science, vol. 14764, pp. 151–164. Springer, Heidelberg (2024). https://doi.org/10.1007/978-3-031-63021-7_12
10. Fellows, M.R., Kratochvíl, J., Middendorf, M., Pfeiffer, F.: The complexity of induced minors and related problems. Algorithmica **13**(3), 266–282 (1995). https://doi.org/10.1007/BF01190507
11. Fiala, J., Kamiński, M., Paulusma, D.: Detecting induced star-like minors in polynomial time. J. Disc. Algor. **17**, 74–85 (2012). https://doi.org/10.1016/J.JDA.2012.11.002
12. Golovach, P.A., Kratsch, D., Paulusma, D.: Detecting induced minors in AT-free graphs. Theor. Comput. Sci. **482**, 20–32 (2013). https://doi.org/10.1016/J.TCS.2013.02.029
13. Kawarabayashi, K., Kobayashi, Y., Reed, B.A.: The disjoint paths problem in quadratic time. J. Comb. Theory Ser. B **102**(2), 424–435 (2012). https://doi.org/10.1016/J.JCTB.2011.07.004
14. Korhonen, T.: Grid induced minor theorem for graphs of small degree. J. Comb. Theory Ser. B **160**, 206–214 (2023). https://doi.org/10.1016/j.jctb.2023.01.002

15. Korhonen, T., Lokshtanov, D.: Induced-minor-free graphs: separator theorem, subexponential algorithms, and improved hardness of recognition. In: Proceedings of the 2024 ACM-SIAM Symposium on Discrete Algorithms, SODA, pp. 5249–5275. SIAM (2024). https://doi.org/10.1137/1.9781611977912.188
16. Korhonen, T., Pilipczuk, M., Stamoulis, G.: Minor containment and disjoint paths in almost-linear time. CoRR arxiv:2404.03958 (2024). https://doi.org/10.48550/ARXIV.2404.03958
17. Milanič, M., Pivač, N.: A tame vs. feral dichotomy for graph classes excluding an induced minor or induced topological minor. CoRR arxiv:2405.15543 (2024). https://doi.org/10.48550/ARXIV.2405.15543
18. Nguyen, T., Scott, A.D., Seymour, P.D.: Induced paths in graphs without anticomplete cycles. J. Comb. Theory B **164**, 321–339 (2024). https://doi.org/10.1016/J.JCTB.2023.10.003
19. Robertson, N., Seymour, P.: Graph minors. XIII. The disjoint paths problem. J. Comb. Theory Series B **63**(1), 65–110 (1995). https://doi.org/10.1006/jctb.1995.1006
20. van 't Hof, P., Kaminski, M., Paulusma, D., Szeider, S., Thilikos, D.M.: On graph contractions and induced minors. Disc. Appl. Math. **160**(6), 799–809 (2012). https://doi.org/10.1016/J.DAM.2010.05.005

Pathways to Tractability for Geometric Thickness

Thomas Depian, Simon Dominik Fink, Alexander Firbas(✉), Robert Ganian, and Martin Nöllenburg

TU Wien, Vienna, Austria
{tdepian,sfink,afirbas,rganian,noellenburg}@ac.tuwien.ac.at

Abstract. We study the classical problem of computing geometric thickness, i.e., finding a straight-line drawing of an input graph and a partition of its edges into as few parts as possible so that each part is crossing-free. Since the problem is NP-hard, we investigate its tractability through the lens of parameterized complexity. As our first set of contributions, we provide two fixed-parameter algorithms which utilize well-studied parameters of the input graph, notably the vertex cover and feedback edge numbers. Since parameterizing by the thickness itself does not yield tractability and the use of other structural parameters remains open due to general challenges identified in previous works, as our second set of contributions, we propose a different pathway to tractability for the problem: extension of partial solutions. In particular, we establish a full characterization of the problem's parameterized complexity in the extension setting depending on whether we parameterize by the number of missing vertices, edges, or both.

1 Introduction

The *thickness* of a graph G is the minimum integer ℓ such that there exists a drawing Γ of G and a partitioning of its edges into ℓ layers such that no two edges assigned to the same layer cross in Γ. Thickness is a classical generalization of graph planarity; its early study dates back to the seventies [6,43] and yet it remains a prominent topic of research to this day [13,22,23,34]. An overview of classical results on thickness and its initial applications can be found, e.g., in the dedicated survey of Mutzel, Odenthal, and Scharbrodt [36].

While the distinction of whether Γ represents edges as curves or line segments is immaterial when determining whether G is planar (i.e., whether it has thickness 1), this is far from true for higher thickness values. In fact, a seminal paper of Eppstein established that the gap between *geometric thickness* (with edges represented as straight-line segments) and *graph-theoretic thickness* (where edges are curves) can be arbitrarily large [26]. Both of these notions are

All authors acknowledge support from the Vienna Science and Technology Fund (WWTF) [10.47379/ICT22029]. Robert Ganian and Alexander Firbas also acknowledge support from the Austrian Science Fund (FWF) [10.55776/Y1329].

R. Královič and V. Kůrková (Eds.): SOFSEM 2025, LNCS 15538, pp. 209–224, 2025.
https://doi.org/10.1007/978-3-031-82670-2_16

now often studied independently, with a number of articles focusing on various combinatorial properties of geometric or graph-theoretic thickness [21–23,34].

In terms of computational complexity, determining whether a graph has graph-theoretic thickness 2 was shown to be NP-hard already 40 years ago [35]. The analogous question for geometric thickness was resolved only in 2016 [22]—16 years after it was posed as an open question by Dillencourt, Eppstein, and Hirschberg [19]; see also the selected open questions in the proceedings of GD 2003 [12]. While this sets both problems on the same level in terms of lower bounds, there still remains a huge gap between the two notions in the complementary setting of identifying the boundaries of tractability for computing the thickness of a graph.

Indeed, in 2007 Dujmović and Wood [21] showed that both notions of thickness are upper-bounded by $\lceil \frac{k}{2} \rceil$ in every graph of *treewidth* k. For graph-theoretic thickness, this result in combination with the well-established machinery of Courcelle's Theorem [14] immediately implies that graph-theoretic thickness can be computed in linear time on all graphs of bounded treewidth; stated in terms of the more modern *parameterized complexity* paradigm, the problem is *fixed-parameter tractable* when parameterized by treewidth. While this provides a robust and classical graph parameter that can be exploited to compute graph-theoretic thickness, the same approach cannot be replicated when aiming for geometric thickness. In fact, no parameterized algorithms were previously known for computing the geometric thickness at all. The overarching aim of this article is to change this via a detailed investigation of the parameterized complexity of computing geometric thickness.

1.1 Contributions

There are two classical perspectives through which one typically applies parameterized analysis to identify tractable fragments for a problem of interest: parameterizing by the solution quality (e.g., the size of a sought-after set), or by structural properties of the input (such as various graph parameters). The NP-hardness of verifying whether a graph has geometric thickness 2 [22] effectively rules out the former approach, and so we begin our investigation by focusing on the second perspective.

Treewidth [40] typically represents a natural first choice for a structural graph parameter one could use to overcome the general intractability of a problem of interest. Indeed, there are numerous examples of combinatorial problems that are known to be fixed-parameter tractable parameterized by treewidth, and the same holds for the aforementioned problem of computing graph-theoretic thickness. Yet, when dealing with graph problems that involve geometric aspects—such as the computation of the *obstacle number* [3] or *Right-Angle Crossing (RAC) drawings* [11]—one often finds that treewidth and other "decomposition-based" graph parameters (including, e.g., *pathwidth* [39] and *treedepth* [38]) are incredibly challenging to use. We refer readers to the survey of Zehavi [44] for further examples of the many open questions surrounding the algorithmic application of such parameters in the context of graph drawing.

Given the above, for our first result we instead consider the computation of the geometric thickness when parameterized by the *vertex cover number*—i.e., the size of a minimum vertex cover of the input graph. In particular, we show:

Theorem 1. GEOMETRIC THICKNESS *is fixed-parameter tractable when parameterized by the vertex cover number of the input graph.*

We prove Theorem 1 through the kernelization technique, which iteratively reduces the size of the input instance while preserving its geometric thickness. Kernelization may be considered the "staple" approach for vertex cover number, but its application here is non-trivial and specific to the problem at hand: it relies on a sequence of arguments establishing that every bounded-thickness drawing of a sufficiently large instance must contain a certain configuration of "similar" vertices drawn in the plane. We use this to prove that the geometric thickness will not decrease if we remove a specific vertex from the graph.

For our second result, we turn towards the *feedback edge number*—i.e., the edge deletion distance to acyclicity—as a second candidate parameter. The feedback edge and vertex cover numbers can be seen as complementary to each other: the two parameters are pairwise incomparable and form basic restrictions on the structure of the input graph. We show:

Theorem 2. GEOMETRIC THICKNESS *is fixed-parameter tractable when parameterized by the feedback edge number of the input graph.*

The proof of Theorem 2 also relies on kernelization, but the approach is entirely different from the one used for Theorem 1. In particular, after some simple preprocessing steps, we show that either the instance already has bounded size or one can identify a path P of degree-2 vertices such that deleting P preserves the geometric thickness of the input graph.

While the vertex cover and feedback edge numbers have found successful applications in graph drawing [3,5,8,10,11] as well as numerous other settings [2,27,28], they are still highly "restrictive" in the sense that they only attain low values on graphs which exhibit rather simple structural properties. In light of the aforementioned obstacles standing in the way of developing efficient algorithms that rely on less restrictive structural parameters such as treewidth, the second half of our article takes a different approach. There, we investigate a third perspective to identify meaningful tractable fragments for computing geometric thickness: instead of targeting well-structured graphs, we consider instances where a large part of the solution is already pre-determined (i.e., provided by a user or as the output of some preceding process). This is formally captured through the setting of *solution extension* problems, which was pioneered in the seminal paper on extending planar drawings [1] and has since then led to a new perspective for studying the complexity of classical graph drawing problems; in this setting, fixed-parameter algorithms were obtained for extending, e.g., *1-planar* [24,25], *crossing-optimal* [30], *planar orthogonal* [7] and *linear* [17] drawings.

For GEOMETRIC THICKNESS EXTENSION (GTE in short), we consider the input to also contain an edge-partitioned drawing of a subgraph as the partial solution. As our first (and also technically simplest) result in the extension setting, we provide a linear-time algorithm that can add a constant number of missing edges while preserving a bound on the geometric thickness of the drawing:

Theorem 3. GTE *when only k edges are missing from the provided partial drawing is fixed-parameter tractable when parameterized by k.*

A major restriction in the above setting is that the partial drawing must already contain all of the vertices of the graph—i.e., one is not allowed to add new vertices. Indeed, typically one allows the addition of vertices and edges in a drawing extension problem, and a common parameter is simply the number of elements missing from the drawing. Surprisingly, we show that a result analogous to the previously mentioned FPT-algorithms cannot be obtained for geometric thickness: while the problem is polynomial-time solvable when the task is to add a constant number of missing edges and vertices (which we show by providing and analyzing a formulation of the problem in the *Existential Theory of Reals* [41]), we rule out fixed-parameter tractability via a non-trivial reduction.

Theorem 4. GTE *when only k edges and vertices are missing from the provided drawing is* XP*-tractable and* W[1]*-hard when parameterized by k.*

As our final result, we complete our analysis of the extension setting by showing that the problem remains NP-hard even if the partial drawing is only missing two vertices and their incident edges, i.e., if the vertex deletion distance to the original graph is 2:

Theorem 5. GTE *is* NP*-hard even if the drawn subgraph can be obtained from the input graph by deleting only two vertices.*

We remark that Theorem 5 contrasts parameterized algorithms that exploit the vertex deletion distance to solve the drawing extension problem, e.g., for *IC-planar* [25] and *1-planar* [24] drawings.

Due to space constraints, we provide full proofs and details of results marked with ★ in the full version of this paper [16].

1.2 Preliminaries

We assume familiarity with standard graph terminology [18] and the basic concepts of the parameterized complexity paradigm, specifically *fixed-parameter tractability*, XP-*tractability*, W[1]-*hardness* and *kernelization* [15,20]. For an integer p, we set $[p] = \{1, \ldots, p\}$. For a function $f\colon A \to B$ and $X \subseteq A$, let $f(X)$ denote the set $\{f(x) \mid x \in X\}$.

Let G be a simple graph with vertex set $V(G)$ and edge set $E(G)$. We use $\overline{E(G)}$ to denote the complement edge set, i.e., $\overline{E(G)} = \binom{V(G)}{2} \setminus E(G)$, and $N_G(v)$ to denote the neighborhood of a vertex v in G excluding v. A *straight-line drawing* Γ of G is a mapping from vertices to distinct points in $\mathbb{R}^2$, where we consider edges to be drawn as straight-line segments connecting their end vertices.

A *geometric ℓ-layer drawing* of a graph G is a straight-line drawing Γ of G accompanied with a function χ mapping each edge of G to a *planar layer* (also called *color*) from $[\ell]$ with the property that no pair of edges assigned to the same planar layer cross each other in Γ. We equivalently say that (Γ, χ) is free of *monochromatic crossings*. We can now define GEOMETRIC THICKNESS formally:

GEOMETRIC THICKNESS (GT)
Input: A graph G and an integer ℓ.
Question: Does there exist a geometric ℓ-layer drawing (Γ, χ) of G?

GT on multi-graphs is known to be $\exists\mathbb{R}$-complete [29], and thereby it is in particular decidable on simple graphs. We will frequently make use of the following folklore observation:

Observation 1. *Let G be a graph. If G admits a geometric ℓ-layer drawing, it also admits a geometric ℓ-layer drawing in general position, i.e., where no three vertices are collinear and no three edges cross at the same point.*

A geometric ℓ-layer drawing (Γ_G, χ_G) of G is an *extension* of a geometric ℓ-layer drawing (Γ_H, χ_H) of a subgraph H of G if Γ_G and χ_G are extensions of Γ_H and χ_H, respectively. That is, (Γ_G, χ_G) and (Γ_H, χ_H) assign the same values to the vertices and edges, respectively, of H. With this, we can formalize GT in the extension setting.

GEOMETRIC THICKNESS EXTENSION (GTE)
Input: A graph G, an integer ℓ, a geometric ℓ-layer drawing (Γ_H, χ_H) of some subgraph H of G.
Question: Does there exist a geometric ℓ-layer drawing (Γ_G, χ_G) of G which extends (Γ_H, χ_H)?

Observe that GT can be seen as a special case of GTE where H is the empty graph. We remark that while the problems are formalized as decision problems for complexity-theoretic purposes, every algorithmic result presented within this article is constructive and can be extended via standard techniques to also output a geometric ℓ-layer drawing as a witness.

2 Parameterizing by the Vertex Cover Number

In this section, we show that GEOMETRIC THICKNESS parameterized by the vertex cover number is fixed-parameter tractable (Theorem 1). On a high level, our proof strategy is as follows. We say two vertices are *clones* in a geometric ℓ-layer drawing of a graph G if they have the same neighborhood in G and each edge incident to one vertex is colored the same as the respective edge incident to the other vertex. Given a vertex cover of the input graph and a drawing

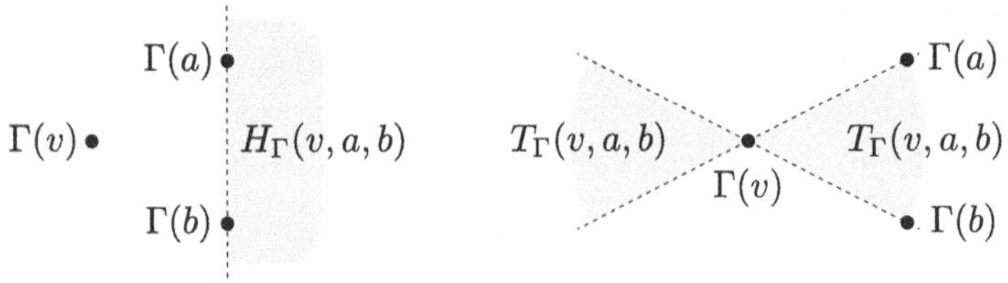

Fig. 1. Illustration of Definition 1.

of it, we partition the plane into a small number of *cells*. We first formulate a characterization that gives necessary and sufficient conditions for when a vertex in the drawing can be "cloned" within the cell it is positioned in without changing the outcome of the instance.

Towards subsequently obtaining a kernel, we partition the vertices outside the vertex cover into equivalence classes based on their neighborhood w.r.t. the vertex cover. If an equivalence class is sufficiently large, we delete an arbitrary vertex x of the class. Clearly, deleting a vertex preserves positive instances. Conversely, in a drawing of the instance without x, using the pigeon-hole principle, we will always be able to find at least two mutual clones c_1, c_2 of x's class that share a cell. Using our characterization, this will imply that c_1 can be cloned within that cell, i.e., x can always be re-inserted into the drawing, positioned inside the cell, and its incident edges can be colored to match those incident to c_1. We thereby obtain a kernel with size dependent on the vertex cover number and the number of colors, and later show that the latter can be dropped from the parameterization. Note that, similar to the work of Bannister et al. [4], as the graphs of bounded vertex cover number are well-quasi-ordered under induced subgraphs [37], the weaker notion of nonuniform fixed-parameter tractability [15] for this problem is already implied, although this neither yields an actual FPT algorithm nor provides concrete runtime bounds.

A Characterization of Cloneability within a Cell. To derive the characterization, we start by defining two simple geometric constructions (see Fig. 1).

Definition 1. *Given a graph G, a drawing Γ of G in general position, and three distinct vertices $v, a, b \in V(G)$, we use $H_\Gamma(v,a,b)$ to denote the interior of the half plane defined by the line through $\Gamma(a), \Gamma(b)$ that does not contain $\Gamma(v)$. Additionally, we use $T_\Gamma(v,a,b)$ to denote the interior of the "tie"-shape defined by the center point $\Gamma(v)$ and directions $\overrightarrow{\Gamma(v)\Gamma(a)}$ and $\overrightarrow{\Gamma(v)\Gamma(b)}$.*

The characterization is based upon two observations regarding the possible positions where a vertex (or a clone thereof) can be placed while avoiding monochromatic crossings, each time considering different sets of edges. The two observations follow directly from the definitions; see also Fig. 2.

Observation 2. *Let G be a graph, (Γ, χ) be a geometric ℓ-layer drawing of G in general position, and $a, b \in N_G(v)$ be distinct such that $\chi(va) = \chi(vb)$.*

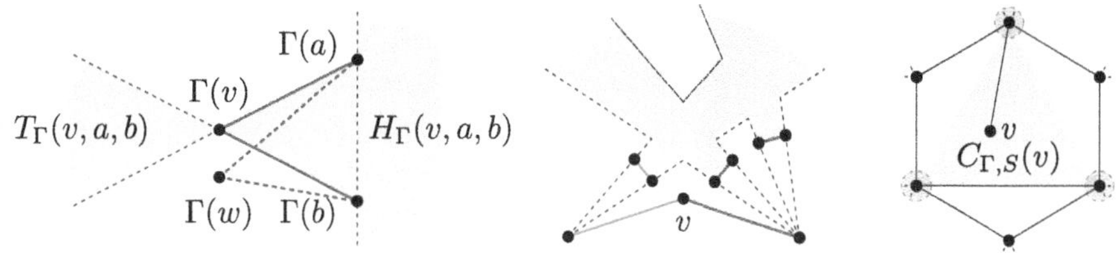

Fig. 2. Illustrations for Observation 2 (left), Observation 3 (middle, the gray area is the complement of B), and Definition 2 (right, the set S is shown in blue, the cells of Γ induced by S are shown in gray).

Consider the tuple (Γ', χ') obtained from (Γ, χ) by cloning v into a new vertex w placed in general position. Then, there is no monochromatic crossing between two edges e_1, e_2 with $e_1 \in \{va, vb\}$ and $e_2 \in \{wa, wb\}$ in (Γ', χ') if and only if $\Gamma'(w) \in T_\Gamma(v,a,b) \cup H_\Gamma(v,a,b)$.

The next observation implies there is a small disk of "safe placements" around each vertex v in a drawing, where a clone of v can be safely inserted, ignoring crossings with v's edges:

Observation 3. *Let G be a graph, (Γ, χ) be a geometric ℓ-layer drawing of G in general position, and $v \in V(G)$. Then, the set $B \subseteq \mathbb{R}^2$ where v can be moved to in (Γ, χ) without introducing monochromatic crossings, is open.*

With the above setup in hand, we can now turn to the notion of cells and admissible regions. The intuition behind the admissible region of a vertex is that we consider the restrictions on the position of a clone imposed by all pairs of same-colored edges incident to the vertex (for a single pair, this is precisely Observation 2) after disregarding the "half plane portion", since we are only interested in potential positions "close" to the vertex (i.e., in the same cell). An illustration for the notion of a cell is provided in Fig. 2.

Definition 2. *Let G be a graph, S a vertex cover of G, and Γ a straight-line drawing of G in general position. For a vertex $v \in V(G) \setminus S$, we define the* cell *of v* in *Γ* with respect to *S as the complement of the union of half-planes $H_\Gamma(v,\cdot,\cdot)$ induced by $\Gamma(S)$, i.e.:*

$$C_{\Gamma,S}(v) := \mathbb{R}^2 \setminus \bigcup_{a,b \in S, a \neq b} H_\Gamma(v,a,b).$$

Furthermore, we call the set obtained by intersecting tie-shapes defined by same-colored edges

$$A_{\Gamma,\chi}(v) := \bigcap_{\substack{a,b \in N_G(v), a \neq b, \\ \chi(va)=\chi(vb)}} T_\Gamma(v,a,b)$$

the admissible region *of v in (Γ, χ).*

We can now provide a characterization of when vertices may be cloned inside their cell:

Lemma 1. *Let G be a graph, (Γ, χ) be a geometric ℓ-layer drawing of G in general position, S a vertex cover of G, and $v \in V(G) \setminus S$. Then, cloning v and placing its clone into the cell $C_{\Gamma,S}(v)$ can yield a geometric ℓ-layer drawing if and only if $A_{\Gamma,\chi}(v) \neq \emptyset$.*

Proof. (Sketch) ($\Rightarrow$): Let (Γ', χ') be a geometric ℓ-layer drawing in general position obtained from (Γ, χ) by cloning v into the cell $C_{\Gamma,S}(v)$; call the clone w. Applying Observation 2 for all distinct $a, b \in N_G(v)$ where $\chi(va) = \chi(vb)$ yields

$$\Gamma'(w) \in \bigcap_{\substack{a,b \in N_G(v), a \neq b, \\ \chi(va)=\chi(vb)}} T_\Gamma(v, a, b) \cup H_\Gamma(v, a, b).$$

Observe that $H_\Gamma(v, a, b) \cap C_{\Gamma,S}(v) = \emptyset$ for all distinct $a, b \in N_G(v)$. Hence, using $\Gamma'(w) \in C_{\Gamma,S}(v)$, we obtain

$$\Gamma'(w) \in \bigcap_{\substack{a,b \in N_G(v), a \neq b, \\ \chi(va)=\chi(vb)}} T_\Gamma(v, a, b) = A_{\Gamma,\chi}(v).$$

Therefore, $A_{\Gamma,\chi}(v) \neq \emptyset$.

($\Leftarrow$): We find a placement for the clone inside the cell of v that is "very close" to v and in $A_{\Gamma,\chi}(v)$. Correctness then follows from Observation 2 and Observation 3. For the full proof, see the full version of this paper [16]. □

Deriving the Kernel. Using Theorem 28.1.1 of the Handbook of Discrete and Computational Geometry [32] to bound the number of distinct cells in a drawing, we have all prerequisites to derive a kernel parameterized by the vertex cover number plus the number of colors, ℓ:

Lemma 2. Geometric Thickness *parameterized by $\ell + k$, where k denotes the vertex cover number of the input graph and ℓ the number of allowed colors in the drawing, admits a problem kernelization mapping each instance $I = (G, \ell)$ to an equivalent instance $I' = (G^*, \ell)$ where $|V(G^*)| = \ell^{\mathcal{O}(k)}$.*

Theorem 1 now follows directly from Lemma 2 in combination with the previously established fact that the geometric thickness ℓ is upper-bounded by $\lceil \frac{k}{2} \rceil$ in the class of graphs of treewidth at most k (which also contains all graphs of vertex cover number k) [21].

Theorem 1. Geometric Thickness *is fixed-parameter tractable when parameterized by the vertex cover number of the input graph.*

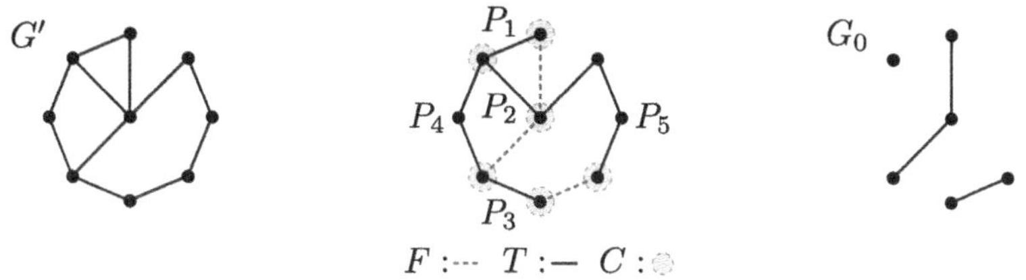

Fig. 3. A graph G' without degree one and zero vertices (left), a corresponding feedback edge set F and forest T as well as the vertex set C (middle), and the graph G_0 (right).

3 Parameterizing by the Feedback Edge Set Number

In this section, we show that GT is also fixed-parameter tractable when parameterized by the feedback edge number (Theorem 2). Our proof strategy is as follows: After some trivial preprocessing steps and handling of special cases, we obtain a feedback edge set F of a preprocessed input graph G', where G' can be decomposed into a "small" part G_0 essentially consisting of F, and a "small" set of potentially "long" paths connecting pairs of vertices from G_0. We sort the paths by length in ascending order and add them to G_0 sequentially, obtaining G_1, G_2 and so forth. If the i'th path is not significantly longer than the number of edges in G_{i-1}, we add the path to G_{i-1} and the resulting graph G_i is still "small". Otherwise, we stop and set $G_j{:=}G_{i-1}$ as our kernel (while in the case where we never encounter a "long" path, G' itself is a kernel). The centerpiece of the correctness argument is a proof that paths that are significantly longer than the previous ones can be removed from the graph without impacting its geometric thickness.

To formalize this notion, let (G, ℓ) be an instance of Geometric Thickness and let F be a minimum feedback edge set of G of size k, which can be computed in polynomial time by computing a spanning tree of G. First, we describe how to compute the kernel. Later, we will argue about its size, correctness, and running time. For the following, we assume $\ell \geq 2$, since for $\ell \leq 1$ the problem is polynomial-time solvable.

Let G' be obtained by exhaustively removing degree-0 and degree-1 vertices from G. Consider the forest T with $V(T){:=}V(G')$ and $E(T){:=}E(G') \setminus F$. We define C to be the union of the set $\{v \in V(T) \mid \deg_T(v) \geq 3\}$ and the set of the endpoints of edges in F. Next, we consider the decomposition of T into x edge-disjoint paths whose endpoints are in C; see Fig. 3. Observe that the decomposition is unique as $V(T) \setminus C$ only contains degree-2 vertices. We refer to these paths, sorted by length in ascending order, by $P_1, \ldots, P_x$. Let G_0 be the graph with $V(G_0){:=}C$ and $E(G_0){:=}F$ and define G_i as $G_0 \cup P_1 \cup \ldots P_i$ for $i \in [x]$. Observe that $G_x = G'$. Finally, the kernel is given by (G_j, ℓ), where j is the smallest element of $\{0, \ldots, x-1\}$ such that $|E(P_{j+1})| > 2\,(|E(G_j)| + x)$ if such an element exists. Otherwise, we set $j{:=}x$.

Clearly, the kernel G_j can be computed in polynomial time and is decidable since GT is in $\exists\mathbb{R}$ [29]. Furthermore, using the recursive bounds on the graphs $G_0, \ldots, G_j$, one can show that the kernel has at most $10k \cdot 81^k$ vertices, and that the instance (G, ℓ) is positive if and only if (G_j, ℓ) is. For the latter, the key to derive the non-trivial direction of the correctness proof, i.e., how to reintroduce the removed paths without increasing the thickness, is to draw the removed paths as subdivided straight lines. Since these paths are "sufficiently long", all potential monochromatic crossings can be resolved by subdividing and coloring the segments appropriately. In total, we have derived Theorem 2.

Theorem 2. Geometric Thickness *is fixed-parameter tractable when parameterized by the feedback edge number of the input graph.*

4 Geometric Thickness Extension

We analyze the complexity of GTE under three natural parameterizations. When only edges are missing, using their count as parameter, we obtain an FPT algorithm via a straightforward branching argument. In the general setting where also vertices are missing, on one hand we establish NP-hardness even if all that is missing from the graph are two vertices and their incident edges. This is shown by developing a geometric analogue of the technique of Depian et al. [17], and excludes fixed-parameter as well as XP-tractability when parameterizing by the vertex deletion distance. On the other hand, in Theorem 4 we show that the problem is in XP and W[1]-hard parameterized by the total number of missing vertices and edges.

First, we show XP-membership. Our strategy is to express GT and GTE as an Existential Theory of the Reals (ETR) formula and to decide said formula using the algorithm by Grigoryev et al. [33]. We recall that while membership of GT in $\exists\mathbb{R}$ was recently established [29], the proof does not provide an explicit formula (which is required for our approach). Intuitively, an ETR formula is a first order formula where only existential quantification over the reals is allowed, the terms are real polynomials and $\{<, \leq, =, >, \geq\}$ are the allowed predicates. We refer to [42] for a formal definition and introduction to the existential theory of the reals.

Lemma 3. *An instance (G, ℓ) of* GT *with $n = |V(G)|$ and $m = |E(G)|$ can be expressed as an* ETR *formula in $2n + m$ variables with polynomials of total degree at most 6, using $\mathcal{O}(n^4)$ polynomial (in)equalities.*

To complete the proof of Theorem 4, we reduce from the Multicolored Clique problem [15]. Note that we obtain W[1]-hardness even when additionally parameterizing by the total number of layers. An instance of said problem consists of a k-partite graph X with a vertex partition $V(X) = V^1 \mathbin{\dot\cup} V^2 \mathbin{\dot\cup} \ldots \mathbin{\dot\cup} V^k$ and an integer k as parameter. The instance is positive if and only if X contains a k-clique.

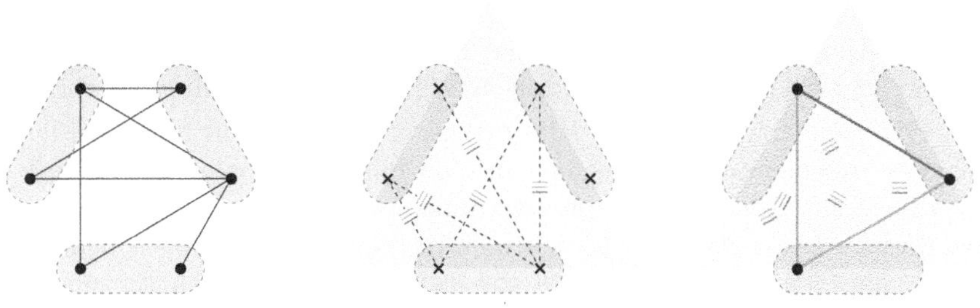

Fig. 4. An instance $(X, k = 3)$ of MULTICOLORED CLIQUE (left), the resulting instance of GTE where crosses denote possible vertex positions and non-edges of X are drawn as dashed lines (middle), and a valid extension showing that X contains a K_3 (right).

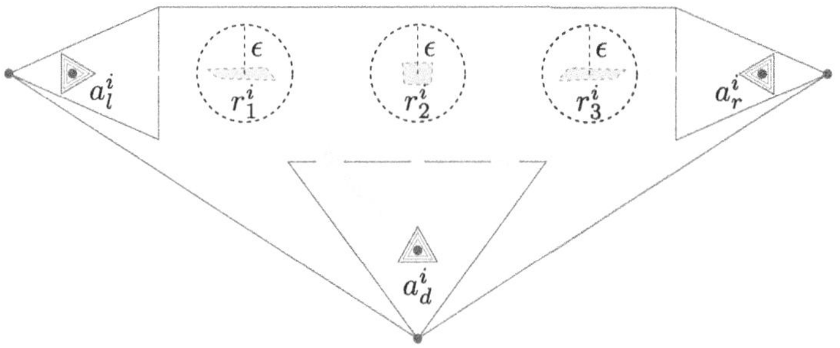

Fig. 5. A choice gadget for V^i with $|V^i| = 3$ and the clique vertex c^i missing.

Our reduction is based on the following idea, illustrated in Fig. 4: Suppose GTE allowed us to specify for each missing vertex a set of possible positions. We construct an instance of GTE where a drawing is to be extended by a k-clique. For the i'th vertex of the k-clique, we allow precisely $|V^i|$ possible positions, all placed on the i'th side of a regular k-gon. To avoid monochromatic crossings between clique-edges in the extension, we allow for one color for each edge of the k-clique. For each non-edge of X, we "block the visibility" between the two potential positions corresponding to endpoints of the non-edge via predrawn edges. Now, if it is possible to draw the clique without monochromatic crossings, the set of chosen positions directly gives the desired k-clique in X. We refer to the full version of this paper [16] for the full details of our construction.

As GTE does not allow us to specify possible vertex positions, our approach is to instead attach k "choice gadgets" around the k-gon, which allow us to constrain the position of a vertex to a set of disjoint regions. Intuitively, this construction (illustrated in Fig. 5), works as follows: The i'th vertex of the clique, call it c^i, is connected to three so-called *anchor vertices* a_l^i, a_d^i, a_r^i in the target graph G, but not in the predrawn graph H, as $c^i \notin V(H)$. Around each anchor vertex, we closely position triangles in all colors except one, say red. This forces the edges from $\{a_l^i, a_d^i, a_r^i\}$ to c^i to be drawn in red in any extension of the given drawing. Around each anchor vertex, we draw a red triangle with specific "holes" on the perimeter. For each hole, this induces a cone where c^i can be drawn without creating a red crossing with that triangle. Altogether, the intersection of all cones yields a union of disjoint regions $r^i_1, \ldots, r^i_{|V^i|}$, modelling the choice of $v \in V^i$. One can show that every valid extension requires c^i to be placed inside one of these regions and each region must be fully contained within a disk of sufficiently small radius ϵ with a predetermined center.

We arrange the k choice gadgets around a regular k-gon. Subsequently, we insert the *blocking edges* that model the structure of X. With the positions constrained to disks instead of points, this becomes non-trivial, as the "lines of sight" to block obtain a non-zero area. See ?? for an illustration of the lines of sight and the complete reduction. In particular, we need to be careful not to block "too much" (i.e., ensuring edges of X are preserved), while avoiding monochromatic crossings when inserting the blocking edges (i.e., ensuring non-edges of X are preserved).

Leveraging trigonometry, we calculate a set of permissible positions where we may insert the blocking edges. To ensure that this set is non-empty, we calculate a scaling factor that, when applied to the entire construction, ensures this set to be non-empty. One can show that this construction is always possible. The correctness of the reduction follows from the aforementioned properties. Combined with the XP-tractability result shown in the previous subsection, we have derived Theorem 4.

Theorem 4. GTE *when only k edges and vertices are missing from the provided drawing is* XP*-tractable and* W[1]*-hard when parameterized by k.*

5 Concluding Remarks

Our investigation provides the first systematic investigation of the parameterized complexity of computing the geometric thickness of graphs, a fundamental concept which has been studied from a variety of perspectives since its introduction over sixty years ago. The main open questions that arise from our work are resolving the complexity of GT when parameterized by treewidth `tw` and treedepth `td`—in particular, does there exist a computable function f such that GT can be solved at least in time $|V(G)|^{f(\mathtt{tw})}$ or $|V(G)|^{f(\mathtt{td})}$?

The main obstacle in the way of obtaining such algorithms seems to be a lack of understanding of solutions for these "well-structured" graphs. In particular, if one could show that every k-treewidth graph with geometric thickness ℓ admits a geometric ℓ-layer drawing with some "suitable combinatorial properties", this would open the door towards solving GT via the classical dynamic programming approach typically used on graphs of bounded treewidth. On the other hand, if one could show that even bounded-treewidth graphs may require the use of a wide range of "combinatorially ill-behaved" geometric ℓ-layer drawings, this would likely open the door to establishing hardness. The issue described above is essentially the reason why several other prominent graph drawing problems remain open when parameterized by treewidth and treedepth, with examples including the computation of *stack*, *queue* and *obstacle numbers* of graphs [3, 8, 9, 31, 44]; the first of these is in fact equivalent to the variant of GT where vertices are forced to lie in convex position. To make this issue more concrete, we remark that the staple approach for finding a treedepth-kernel (see e.g. [4]), that is, contracting repeated subtrees with a similar structure into single vertices, cannot be applied in a straightforward manner, as e.g., there may be unrelated interfering edges passing through the to-be contracted subgraph in a given drawing.

As an alternative approach to identify meaningful tractable fragments, we investigated GT in the extension setting. Surprisingly, the problem turned out to be NP-complete even in the case where only two vertices are missing from a given solution. For GTE, future work could target the question of whether there are natural circumstances under which one can achieve fixed-parameter tractability even if vertices are missing from the partial drawing—for instance, is the problem FPT when parameterized by the *vertex+edge deletion distance* [17, 24, 25] plus the number of layers?

References

1. Angelini, P., et al.: Testing planarity of partially embedded graphs. ACM Trans. Algorithms **11**(4), 32:1–32:42 (2015). https://doi.org/10.1145/2629341
2. Balabán, J., Ganian, R., Rocton, M.: Computing twin-width parameterized by the feedback edge number. In: Beyersdorff, O., Kanté, M.M., Kupferman, O., Lokshtanov, D. (eds.) 41st International Symposium on Theoretical Aspects of Computer Science, STACS 2024, March 12-14, 2024, Clermont-Ferrand, France. LIPIcs, vol. 289, pp. 7:1–7:19. Schloss Dagstuhl - Leibniz-Zentrum für Informatik (2024). https://doi.org/10.4230/LIPICS.STACS.2024.7
3. Balko, M., Chaplick, S., Ganian, R., Gupta, S., Hoffmann, M., Valtr, P., Wolff, A.: Bounding and computing obstacle numbers of graphs. In: Chechik, S., Navarro, G., Rotenberg, E., Herman, G. (eds.) 30th Annual European Symposium on Algorithms, ESA 2022, September 5-9, 2022, Berlin/Potsdam, Germany. LIPIcs, vol. 244, pp. 11:1–11:13. Schloss Dagstuhl - Leibniz-Zentrum für Informatik (2022). https://doi.org/10.4230/LIPICS.ESA.2022.11
4. Bannister, M.J., Cabello, S., Eppstein, D.: Parameterized complexity of 1-planarity. In: Dehne, F., Solis-Oba, R., Sack, J.R. (eds.) Algorithms and Data Structures, pp. 97–108. Springer Berlin Heidelberg, Berlin, Heidelberg (2013). https://doi.org/10.1007/978-3-642-40104-6_9
5. Bannister, M.J., Cabello, S., Eppstein, D.: Parameterized complexity of 1-planarity. J. Graph Algorithms Appl. **22**(1), 23–49 (2018). https://doi.org/10.7155/JGAA.00457
6. Beineke, L.W., Harary, F.: The thickness of the complete graph. Can. J. Math. **17**, 850–859 (1965)

7. Bhore, S., Ganian, R., Khazaliya, L., Montecchiani, F., Nöllenburg, M.: Extending orthogonal planar graph drawings is fixed-parameter tractable. In: Chambers, E.W., Gudmundsson, J. (eds.) 39th International Symposium on Computational Geometry, SoCG 2023, June 12-15, 2023, Dallas, Texas, USA. LIPIcs, vol. 258, pp. 18:1–18:16. Schloss Dagstuhl - Leibniz-Zentrum für Informatik (2023). https://doi.org/10.4230/LIPICS.SOCG.2023.18
8. Bhore, S., Ganian, R., Montecchiani, F., Nöllenburg, M.: Parameterized algorithms for book embedding problems. J. Graph Algorithms Appl. **24**(4), 603–620 (2020). https://doi.org/10.7155/JGAA.00526
9. Bhore, S., Ganian, R., Montecchiani, F., Nöllenburg, M.: Parameterized algorithms for queue layouts. J. Graph Algorithms Appl. **26**(3), 335–352 (2022). https://doi.org/10.7155/JGAA.00597
10. Binucci, C., et al.: On the complexity of the storyplan problem. J. Comput. Syst. Sci. **139**, 103466 (2024). https://doi.org/10.1016/J.JCSS.2023.103466
11. Brand, C., Ganian, R., Röder, S., Schager, F.: Fixed-parameter algorithms for computing RAC drawings of graphs. In: Bekos, M.A., Chimani, M. (eds.) Graph Drawing and Network Visualization - 31st International Symposium, GD 2023, Isola delle Femmine, Palermo, Italy, September 20-22, 2023, Revised Selected Papers, Part II. Lecture Notes in Computer Science, vol. 14466, pp. 66–81. Springer (2023). https://doi.org/10.1007/978-3-031-49275-4_5
12. Brandenburg, F., Eppstein, D., Goodrich, M.T., Kobourov, S., Liotta, G., Mutzel, P.: Selected open problems in graph drawing. In: Liotta, G. (ed.) GD 2003. LNCS, vol. 2912, pp. 515–539. Springer, Heidelberg (2004). https://doi.org/10.1007/978-3-540-24595-7_55
13. Cheong, O., Pfister, M., Schlipf, L.: The thickness of fan-planar graphs is at most three. In: Angelini, P., von Hanxleden, R. (eds.) Graph Drawing and Network Visualization - 30th International Symposium, GD 2022, Tokyo, Japan, September 13-16, 2022, Revised Selected Papers. Lecture Notes in Computer Science, vol. 13764, pp. 247–260. Springer (2022). https://doi.org/10.1007/978-3-031-22203-0_18
14. Courcelle, B.: The monadic second-order logic of graphs. i. recognizable sets of finite graphs. Inf. Comput. **85**(1), 12–75 (1990). https://doi.org/10.1016/0890-5401(90)90043-H
15. Cygan, M., et al.: Parameterized Algorithms. Springer (2015). https://doi.org/10.1007/978-3-319-21275-3
16. Depian, T., Fink, S.D., Firbas, A., Ganian, R., Nöllenburg, M.: Pathways to tractability for geometric thickness (2024). https://arxiv.org/abs/2411.15864
17. Depian, T., Fink, S.D., Ganian, R., Nöllenburg, M.: The parameterized complexity of extending stack layouts. In: Proceedings of the 32nd International Symposium on Graph Drawing and Network Visualization (GD 2024). Vienna, Austria (2024). https://doi.org/10.4230/LIPIcs.GD.2024.12
18. Diestel, R.: Graph Theory, 4th Edition, Graduate texts in mathematics, vol. 173. Springer (2012)
19. Dillencourt, M.B., Eppstein, D., Hirschberg, D.S.: Geometric thickness of complete graphs. J. Graph Algorithms Appl. **4**(3), 5–17 (2000). https://doi.org/10.7155/JGAA.00023
20. Downey, R.G., Fellows, M.R.: Fundamentals of Parameterized Complexity. Texts in Computer Science, Springer (2013). https://doi.org/10.1007/978-1-4471-5559-1
21. Dujmović, V., Wood, D.R.: Graph treewidth and geometric thickness parameters. Discret. Comput. Geom. **37**(4), 641–670 (2007). https://doi.org/10.1007/S00454-007-1318-7

22. Durocher, S., Gethner, E., Mondal, D.: Thickness and colorability of geometric graphs. Comput. Geom. **56**, 1–18 (2016). https://doi.org/10.1016/J.COMGEO.2016.03.003
23. Durocher, S., Mondal, D.: Relating graph thickness to planar layers and bend complexity. SIAM J. Discret. Math. **32**(4), 2703–2719 (2018). https://doi.org/10.1137/16M1110042
24. Eiben, E., Ganian, R., Hamm, T., Klute, F., Nöllenburg, M.: Extending nearly complete 1-planar drawings in polynomial time. In: Esparza, J., Král', D. (eds.) 45th International Symposium on Mathematical Foundations of Computer Science, MFCS 2020, August 24-28, 2020, Prague, Czech Republic. LIPIcs, vol. 170, pp. 31:1–31:16. Schloss Dagstuhl - Leibniz-Zentrum für Informatik (2020). https://doi.org/10.4230/LIPICS.MFCS.2020.31
25. Eiben, E., Ganian, R., Hamm, T., Klute, F., Nöllenburg, M.: Extending partial 1-planar drawings. In: Czumaj, A., Dawar, A., Merelli, E. (eds.) 47th International Colloquium on Automata, Languages, and Programming, ICALP 2020, July 8-11, 2020, Saarbrücken, Germany (Virtual Conference). LIPIcs, vol. 168, pp. 43:1–43:19. Schloss Dagstuhl - Leibniz-Zentrum für Informatik (2020). https://doi.org/10.4230/LIPICS.ICALP.2020.43
26. Eppstein, D.: Separating thickness from geometric thickness. In: Kobourov, S.G., Goodrich, M.T. (eds.) Graph Drawing, 10th International Symposium, GD 2002, Irvine, CA, USA, August 26-28, 2002, Revised Papers. Lecture Notes in Computer Science, vol. 2528, pp. 150–161. Springer (2002). https://doi.org/10.1007/3-540-36151-0_15
27. Fellows, M.R., Lokshtanov, D., Misra, N., Rosamond, F.A., Saurabh, S.: Graph layout problems parameterized by vertex cover. In: Hong, S., Nagamochi, H., Fukunaga, T. (eds.) Algorithms and Computation, 19th International Symposium, ISAAC 2008, Gold Coast, Australia, December 15–17, 2008. Proceedings. Lecture Notes in Computer Science, vol. 5369, pp. 294–305. Springer (2008). https://doi.org/10.1007/978-3-540-92182-0_28
28. Fomin, F.V., Liedloff, M., Montealegre, P., Todinca, I.: Algorithms parameterized by vertex cover and modular width, through potential maximal cliques. Algorithmica **80**(4), 1146–1169 (2018). https://doi.org/10.1007/S00453-017-0297-1
29. Förster, H., Kindermann, P., Miltzow, T., Parada, I., Terziadis, S., Vogtenhuber, B.: Geometric thickness of multigraphs is $\exists\mathbb{R}$-complete. In: Soto, J.A., Wiese, A. (eds.) LATIN 2024: Theoretical Informatics - 16th Latin American Symposium, Puerto Varas, Chile, March 18-22, 2024, Proceedings, Part I. Lecture Notes in Computer Science, vol. 14578, pp. 336–349. Springer (2024). https://doi.org/10.1007/978-3-031-55598-5_22
30. Ganian, R., Hamm, T., Klute, F., Parada, I., Vogtenhuber, B.: Crossing-optimal extension of simple drawings. In: Bansal, N., Merelli, E., Worrell, J. (eds.) 48th International Colloquium on Automata, Languages, and Programming, ICALP 2021, July 12-16, 2021, Glasgow, Scotland (Virtual Conference). LIPIcs, vol. 198, pp. 72:1–72:17. Schloss Dagstuhl - Leibniz-Zentrum für Informatik (2021). https://doi.org/10.4230/LIPICS.ICALP.2021.72
31. Ganian, R., Montecchiani, F., Nöllenburg, M., Zehavi, M.: Parameterized complexity in graph drawing (dagstuhl seminar 21293). Dagstuhl Rep. **11**(6), 82–123 (2021). https://doi.org/10.4230/DAGREP.11.6.82
32. Goodman, J.E., O'Rourke, J., Tóth, C.D. (eds.): Handbook of discrete and computational geometry. Discrete Mathematics and its Applications (Boca Raton), CRC Press, Boca Raton, FL, third edn. (2018)

33. Grigoryev, D.Y., Vorobjov, N.N., Jr.: Counting connected components of a semialgebraic set in subexponential time. Comput. Complex. **2**, 133–186 (1992). https://doi.org/10.1007/BF01202001
34. Jain, R., Ricci, M., Rollin, J., Schulz, A.: On the geometric thickness of 2-degenerate graphs. In: Chambers, E.W., Gudmundsson, J. (eds.) 39th International Symposium on Computational Geometry, SoCG 2023, June 12-15, 2023, Dallas, Texas, USA. LIPIcs, vol. 258, pp. 44:1–44:15. Schloss Dagstuhl - Leibniz-Zentrum für Informatik (2023). https://doi.org/10.4230/LIPICS.SOCG.2023.44
35. Mansfield, A.: Determining the thickness of graphs is np-hard. In: Mathematical Proceedings of the Cambridge Philosophical Society, vol. 93, pp. 9–23. Cambridge University Press (1983)
36. Mutzel, P., Odenthal, T., Scharbrodt, M.: The thickness of graphs: a survey. Graphs Comb. **14**(1), 59–73 (1998). https://doi.org/10.1007/PL00007219
37. Nešetřil, J., de Mendez, P.O.: Sparsity. Algorithms and Combinatorics **28**, xxiv+–457 (2012). https://doi.org/10.1007/978-3-642-27875-4
38. Nesetril, J., de Mendez, P.O.: Sparsity - Graphs, Structures, and Algorithms, Algorithms and Combinatorics, vol. 28. Springer (2012). https://doi.org/10.1007/978-3-642-27875-4
39. Robertson, N., Seymour, P.D.: Graph minors. i. excluding a forest. J. Comb. Theory, Ser. B **35**(1), 39–61 (1983). https://doi.org/10.1016/0095-8956(83)90079-5
40. Robertson, N., Seymour, P.D.: Graph minors. II. algorithmic aspects of tree-width. J. Algorithms **7**(3), 309–322 (1986). https://doi.org/10.1016/0196-6774(86)90023-4
41. Schaefer, M.: Complexity of some geometric and topological problems. In: Eppstein, D., Gansner, E.R. (eds.) Graph Drawing (GD'09). LNCS, vol. 5849, pp. 334–344. Springer (2009). https://doi.org/10.1007/978-3-642-11805-0_32
42. Toth, C.D., O'Rourke, J., Goodman, J.E.: Handbook of discrete and computational geometry. CRC Press (2017)
43. Tutte, W.T.: The thickness of a graph. Indag. Math. **25**, 561–577 (1963)
44. Zehavi, M.: Parameterized analysis and crossing minimization problems. Comput. Sci. Rev. **45**, 100490 (2022). https://doi.org/10.1016/J.COSREV.2022.100490

Minimum Monotone Spanning Trees

Emilio Di Giacomo[1], Walter Didimo[1], Eleni Katsanou[3(✉)], Lena Schlipf[2], Antonios Symvonis[3], and Alexander Wolff[4]

[1] Università degli Studi di Perugia, Perugia, Italy
[2] Universität Tübingen, Tübingen, Germany
[3] National Technical University of Athens, Athens, Greece
ekatsanou@mail.ntua.gr
[4] Universität Würzburg, Würzburg, Germany

Abstract. Computing a Euclidean minimum spanning tree of a set of points is a seminal problem in computational geometry and geometric graph theory. We combine it with another classical problem in graph drawing, namely computing a monotone geometric representation of a given graph. More formally, given a finite set S of points in the plane and a finite set $\mathcal{D}$ of directions, a geometric spanning tree T with vertex set S is *$\mathcal{D}$-monotone* if, for every pair $\{u, v\}$ of vertices of T, there exists a direction $d \in \mathcal{D}$ for which the unique path from u to v in T is monotone with respect to d. We provide a characterization of $\mathcal{D}$-monotone spanning trees. Based on it, we show that a $\mathcal{D}$-monotone spanning tree of minimum length can be computed in polynomial time if the number $k = |\mathcal{D}|$ of directions is fixed, both when (i) the set $\mathcal{D}$ of directions is prescribed and when (ii) the objective is to find a minimum-length $\mathcal{D}$-monotone spanning tree over all sets $\mathcal{D}$ of k directions. For $k = 2$, we describe algorithms that are much faster than those for the general case. Furthermore, in contrast to the classical Euclidean minimum spanning tree, whose vertex degree is at most six, we show that for every even integer k, there exists a point set S_k and a set $\mathcal{D}$ of k directions such that any minimum-length $\mathcal{D}$-monotone spanning tree of S_k has maximum vertex degree $2k$.

1 Introduction

We study a problem that combines the notion of minimum spanning tree of a set of points in the plane with the notion of monotone drawings of graphs.

The problem of computing a (Euclidean) *minimum spanning tree (MST)* of a set of points in the plane is a well-established topic with a long history in computational geometry [26]. An MST of a finite set S of points is a geometric tree T such that: (*i*) T *spans* S, i.e., the vertices of T are the points of S, and (*ii*) T has minimum length subject to property (*i*), where the length of T is the sum of the lengths of its edges and the length of an edge is the Euclidean distance of its endpoints. Equivalently, the MST is the minimum spanning tree of

This work was initiated at the GNV Workshop 2022 at Heiligkreuztal.

R. Královič and V. Kůrková (Eds.): SOFSEM 2025, LNCS 15538, pp. 225–240, 2025.
https://doi.org/10.1007/978-3-031-82670-2_17

the complete graph on S where the weight of each edge is the Euclidean distance of its incident vertices. It is known that an MST is a subgraph of a Delaunay triangulation [45] (see Figs. 1a and 1b). Given a set S of n points, its Delaunay triangulation has at most $3n-6$ edges, hence an MST of S can be computed in $O(n \log n)$ time (in the real RAM model of computation) via standard MST algorithms. Eppstein [18] has a survey on MSTs.

Monotone drawings of graphs have been introduced by the authors of [4] and have received considerable attention in recent years. They are related to other types of drawings of graphs, such as angle-monotone [10–12,14,33], upward [16, 23], greedy [3,6,14,15,41,43], self-approaching [1,9,38], and increasing-chord drawings [8,14,36,38]. Computing monotone drawings is also related to the geometric problem of finding monotone trajectories between two given points in the plane avoiding convex obstacles [7]. A plane path is *monotone with respect to a direction* d if the order of its vertices along the path coincides with the order of their projections on a line parallel to d. Any monotone path is necessarily crossing-free [4]. A straight-line drawing of a graph G in the plane is *monotone* if there exists a monotone path (with respect to some direction) between any two vertices of G; the direction of monotonicity may be different for each path. If the directions of monotonicity for the paths are restricted to a set $\mathcal{D}$ of directions, then the drawing is $\mathcal{D}$*-monotone*. Results about monotone drawings include algorithms for different graph classes [2,4,5,19] and the study of the area requirement of such drawings (see [29,32,39] for monotone drawings of trees and [30,31,40] for different classes of planar graphs).

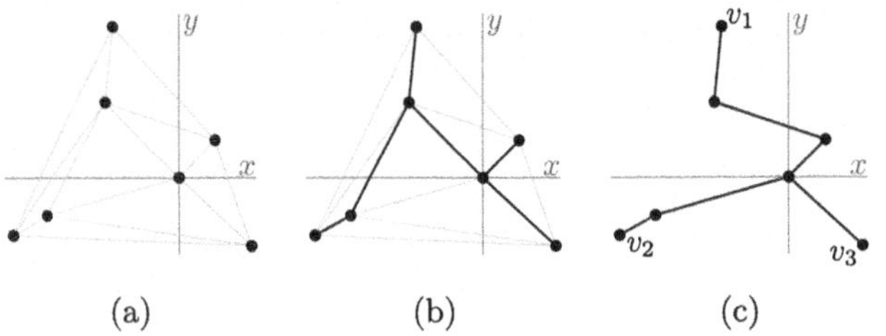

Fig. 1. (a) A point set S with its Delaunay triangulation, (b) MST of S, (c) MMST of S w.r.t. $\{\binom{1}{0}, \binom{0}{1}\}$. The v_2–v_3 path is x-monotone; the v_1–v_2 and v_1–v_3 paths are y-monotone.

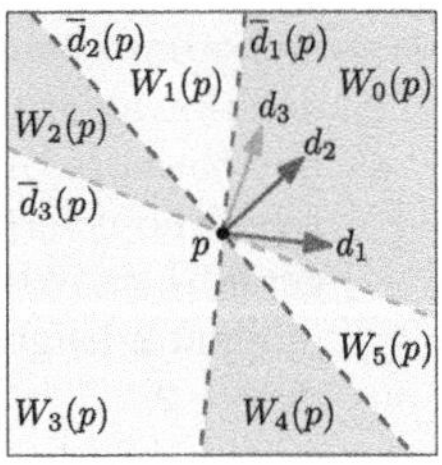

Fig. 2. The set $\mathcal{W}_{\mathcal{D}}(p)$ for the point p and the set $\mathcal{D} = \{d_1, d_2, d_3\}$.

Our Setting. In this paper, we study a natural setting that combines the benefits of spanning trees of minimum length with the benefits of monotone drawings. Namely, given a set S of n points in the plane and a prescribed set $\mathcal{D}$ of directions, we study the problem MMST(S, $\mathcal{D}$) of computing a $\mathcal{D}$-monotone spanning tree of S of minimum length (see Fig. 1c). We call such a tree a *minimum* $\mathcal{D}$*-monotone spanning tree*. For a point set S and an integer $k \geq 1$, we also address the problem MMST(S, k) of computing a minimum *k-directional monotone spanning tree* of S, i.e., a $\mathcal{D}$-monotone spanning tree of minimum length among all possible sets $\mathcal{D}$ of k directions. In this variant, the choice of the directions of monotonicity adjusts to the given point set, which can lead to shorter monotone spanning trees.

We remark that there are other prominent attempts in the literature to couple the MST problem with an additional property. For example, the *Euclidean degree-Δ MST* asks for an MST whose maximum degree is bounded by a given integer Δ [20,42]. Seo, Lee, and Lin [44] studied MSTs of smallest diameter or smallest radius. Finding the k smallest spanning trees [17,21,22] or dynamic MSTs [13,46] are further problems related to spanning trees.

Particularly relevant to our study is the *Rooted Monotone MST problem* introduced by Mastakas and Symvonis [37] and further studied by Mastakas [35]. In that problem, given a set S of n points with a designated root $r \in S$, the task is to compute an MST such that the path from r to any other point of S is monotone. Mastakas [34] extended this setting to multiple roots.

Contribution. The main results in this paper are as follows:

- We provide a characterization of $\mathcal{D}$-monotone spanning trees; see Sect. 4. Based on it, we show how to solve MMST$(S, \mathcal{D})$ in $O(f(|\mathcal{D}|)n^{2|\mathcal{D}|-1} \log n)$ time for some function f of $|\mathcal{D}|$; see Sect. 5. In other words, MMST$(S, \mathcal{D})$ is in XP (that is, slicewise polynomial) when parameterized by $|\mathcal{D}|$. For $|\mathcal{D}| = 2$, we show how to solve MMST$(S, \mathcal{D})$ in $O(n^2)$ time.
- Regarding MMST(S, k), we describe $O(n^2 \log n)$- and $O(n^6)$-time algorithms for $k = 1$ and $k = 2$, respectively. For $k \geq 3$, we present an XP-algorithm that runs in $O(f(k)n^{2k(2k-1)} \log n)$ time; see Sect. 5.
- We show that, in contrast to the MST, whose vertex degree is at most six [42], for every even integer $k \geq 2$, there exists a point set S_k and a set $\mathcal{D}$ of k directions such that any minimum-length $\mathcal{D}$-monotone spanning tree of S_k has maximum vertex degree $2k$; see Sect. 6.

Missing proofs (marked "$\star$") can be found in the full version of this paper [25].

2 Basic Definitions

Let C denote the unit circle centered at the origin o of $\mathbb{R}^2$. Any segment oriented from the center of C to a point of C defines a *direction vector* or simply a *direction*. Two directions are *opposite* if the two segments that define them belong to the same line and lie on opposite sides of the origin. Given a direction d and a set S of points in the plane, we say that S is in *d-general position* if no two points in S lie on a line orthogonal to d. If S is in d-general position, let ord(S, d) be the linear ordering of the orthogonal projections of the points of S on any line parallel to d and directed as d; note that ord(S, d) is uniquely defined. Given a direction d and a point set $S = \{p_1, \ldots, p_n\}$ in d-general position, we say that the geometric path $\langle p_1, \ldots, p_n\rangle$ is *d-monotone* if ord$(S, d) = \langle p_1, \ldots, p_n\rangle$ or

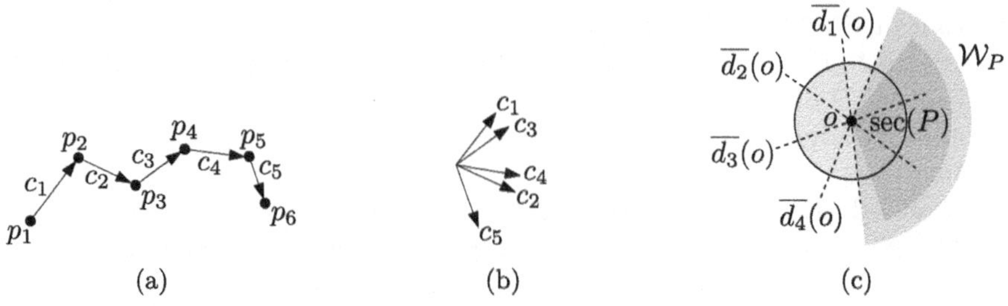

Fig. 3. (a) A directed geometric path P, (b) its sector of directions sec(P) (in dark gray) and (c) the wedge set $\mathcal{W}_P$ of path P (in blue). (Color figure online)

$\text{ord}(S,d) = \langle p_n,\ldots,p_1\rangle$; in this case, all projections of the oriented segments $\overrightarrow{p_ip_{i+1}}$ (for $i \in \{1,\ldots,n-1\}$) on a line parallel to d point towards the same direction. A path is *monotone* if it is d-monotone with respect to some direction d.

Let S be a finite set of points, and let $\mathcal{D}$ be a finite set of directions such that no two of them are opposite. A spanning tree T of S is *$\mathcal{D}$-monotone* if, for every pair of vertices $\{u,v\}$ of T, there is a $d \in \mathcal{D}$ such that the unique geometric path from u to v in T is d-monotone (which requires that the subset of points on the path from u to v is in d-general position).

A *minimum $\mathcal{D}$-monotone spanning tree* of S is a $\mathcal{D}$-monotone spanning tree of S of minimum length among all $\mathcal{D}$-monotone spanning trees of S; we call MMST($S,\mathcal{D}$) the problem of computing such a tree. For a positive integer k, we say that a spanning tree T of S is *k-directional monotone* if there exists a set $\mathcal{D}$ of k directions such that T is $\mathcal{D}$-monotone. A *minimum k-directional monotone spanning tree* of S is a k-directional monotone spanning tree of S of minimum length among all k-directional monotone spanning trees of S; we call MMST(S,k) the problem of computing such a tree. To solve this problem, it turns out that it is sufficient to consider only sets $\mathcal{D}$ of directions such that S is in *$\mathcal{D}$-general position*, i.e., S is in d-general position for every $d \in \mathcal{D}$.

Given two points u and v, let $l_{u,v}$ be the line passing through u and v. Given a direction d and a point x, let $d(x)$ be the line parallel to d passing through x and let $\overline{d}$ be the direction orthogonal to d obtained by rotating d counterclockwise (ccw.) by an angle of 90°. Accordingly, $\overline{d}(x)$ is the line orthogonal to $d(x)$ and $\overline{l_{u,v}}(x)$ is the line orthogonal to $l_{u,v}$ passing through x. Given two vertices u and v of a geometric tree T, let $P_{u,v}$ denote the path of T from u to v.

Given a sorted set $\mathcal{D} = \{d_1,d_2,\ldots,d_k\}$ of $k \geq 1$ pairwise non-opposite directions (assumed to be sorted with respect to the directions' slopes) and a point p in the plane, let $\mathcal{W}_\mathcal{D}(p) = \{W_0(p),W_1(p),\ldots,W_{2k-1}(p)\}$ be the set of $2k$ wedges determined by the lines $\overline{d_1}(p),\overline{d_2}(p),\ldots,\overline{d_k}(p)$. See Fig. 2 for an illustration where $k=3$. We fix the numbering of the wedges by starting with an arbitrary wedge $W_0(p)$ and then continue with $W_1(p),W_2(p),\ldots,W_{2k-1}(p)$ in ccw. order around p. Whenever we refer to a wedge $W_i(p)$ for some integer i, we assume that i is taken modulo $2k$. If p coincides with the origin o, we just write $\mathcal{W}_\mathcal{D} = \{W_0,W_1,\ldots,W_{2k-1}\}$ instead of $\mathcal{W}_\mathcal{D}(o) = \{W_0(o),W_1(o),\ldots,W_{2k-1}(o)\}$.

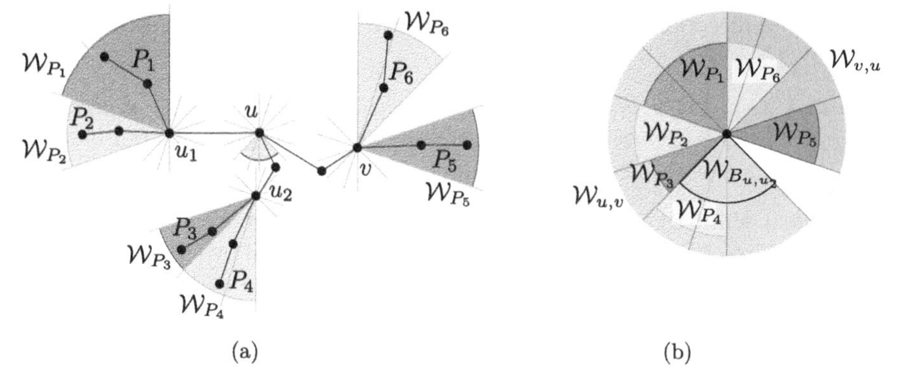

Fig. 4. (a) A monotone tree and its sets of utilized wedges for each leaf path. (b) All sets of utilized wedges drawn on the same unit circle. Set $\mathcal{W}_{u\setminus v}$ (resp. $\mathcal{W}_{v\setminus u}$) consists of all wedges in the blue (resp. gray) region. (Color figure online)

For a directed geometric path $P = \langle p_1, \ldots, p_r \rangle$ and $i \in [r-1]$, let $c_i(P)$ be the oriented segment starting from the origin o that is parallel to and has the same orientation as $\overrightarrow{p_i p_{i+1}}$. Define $\sec(P)$, the *sector of directions* of path P, to be the smallest sector of the unit circle that includes the oriented segment $c_i(P)$ for every $i \in [r-1]$; see Figs. 3a and 3b.

Moreover, let $\mathcal{W}_P \subseteq \mathcal{W}_\mathcal{D}$ be the *wedge set of the directed path* P, i.e., the smallest set of consecutive wedges in counterclockwise (ccw.) order whose union contains $\sec(P)$; see Fig. 3c. For a point p, let $\mathcal{W}_P(p)$ be the region of the plane determined by $\mathcal{W}_P$ translated such that p is its apex. If $\overleftarrow{P}$ is the reverse path of P, then $\mathcal{W}_{\overleftarrow{P}}$ consists of the wedges opposite to those in $\mathcal{W}_P$. We say that path P *utilizes* wedge set $\mathcal{W}_P$.

In a $\mathcal{D}$-monotone spanning tree T, a *branching vertex* is a vertex of degree at least 3 and a *leaf path* is a path of degree-2 vertices from a branching vertex to a leaf. Given two adjacent branching vertices u and v in T, the *branch* $B_{u,v}$ is the unique path that connects u and v via a sequence of degree-2 vertices. Both a leaf path and a branch may consist of a single edge. Further, for any pair of (not necessarily adjacent) vertices u and v, let $T_{u\setminus v}$ be the subtree of T consisting of u and all subtrees hanging from u except for the one containing v. Let $\mathcal{W}_{u\setminus v} \subseteq \mathcal{W}_\mathcal{D}$, the wedge set of $T_{u\setminus v}$, be the smallest set of consecutive wedges that contains all wedges utilized by either leaf paths or branches oriented away from u in $T_{u\setminus v}$ and that does not contain the wedge utilized by the edge out of u that leads to vertex v; see Fig. 4. Note that if u and/or v is a leaf, then $\mathcal{W}_{u\setminus v} = \emptyset$ and/or $\mathcal{W}_{v\setminus u} = \emptyset$. Let $\mathcal{W}_{u\setminus v}(u)$ be the region defined by the wedges in $\mathcal{W}_{u\setminus v}$ translated such that u is their apex.

3 Properties of Monotone Paths and Trees

We describe basic properties of monotone paths and $\mathcal{D}$-monotone trees, which we use in Sect. 4. Unless otherwise stated, we assume that $\mathcal{D}$ consists of pairwise non-opposite directions and that the point set S is always in $\mathcal{D}$-general position.

Lemma 1 (⋆). *Let S be a set of points, and let $P = \langle u, x, v\rangle$ be a geometric path on S. Let d be a direction such that S is in d-general position. If u and v lie in the same half-plane determined by $\overline{d}(x)$, then the path P is not d-monotone.*

The next lemma generalizes Lemma 1. It concerns the wedges formed by a set of $k > 1$ directions (in contrast to the half-plane formed by the perpendicular to a single direction) and two arbitrary points in the same wedge.

Lemma 2 (⋆). *Let S be a set of points, let T be a spanning tree of S, and let $\mathcal{D}$ be a set of k directions. Let x, u, and v be points in S such that $x \in P_{u,v}$. If u and v lie in the same wedge in $\mathcal{W}_{\mathcal{D}}(x)$, then the path $P_{u,v}$ is not $\mathcal{D}$-monotone.*

For any vertex x of T, the set of lines $\{\overline{d}(x) : d \in \mathcal{D}\}$ partitions the plane into $2k$ wedges with apex x, each wedge containing at most one neighbor of x.

Lemma 3 (⋆). *Let S be a set of points, let $\mathcal{D}$ be a set of k directions, and let T be a $\mathcal{D}$-monotone spanning tree of S. Let $\Delta(T)$ denote the maximum degree of tree T. Then, $\Delta(T) \leq 2k$.*

The authors of [4] gave the following characterization.

Lemma 4 ([4]). *Let P be a directed geometric path. Then, P is monotone if and only if the angle of its sector of directions* $\sec(P)$ *is smaller than π.*

While Lemma 4 can be used to recognize monotone paths, it does not specify a direction of monotonicity. This is rectified by Lemma 5.

Lemma 5 (⋆). *Given a direction d, a monotone directed geometric path P is d-monotone if and only if $\overline{d}(o)$ does not intersect* $\sec(P)$, *where o is the origin.*

The following corollary is an immediate consequence of Lemma 5.

Corollary 1. *Let S be a set of points, $\mathcal{D}$ be a set of k directions, and T be a $\mathcal{D}$-monotone spanning tree of S. Let P be a directed path in T. Given a direction $d \in \mathcal{D}$, P is d-monotone if and only if $\overline{d}(o)$ does not intersect the interior of $\mathcal{W}_P$, where o is the origin.*

Additional properties concerning paths of $\mathcal{D}$-monotone spanning trees and their corresponding wedge sets are presented in the following lemma.

Lemma 6 (⋆). *Let S be a set of points, let $\mathcal{D}$ be a set of k directions, and let T be a $\mathcal{D}$-monotone spanning tree of S. Then, T has the following properties: (i) Let P be a directed path originating at vertex u of T. Then, P lies in $\mathcal{W}_P(u)$. (ii) Let P_1 and P_2 be two edge-disjoint directed paths originating at internal vertices u and v of T and terminating at leaves of T. Then, sets $\mathcal{W}_{P_1}$ and $\mathcal{W}_{P_2}$ are disjoint and regions $\mathcal{W}_{P_1}(u)$ and $\mathcal{W}_{P_2}(v)$ are disjoint.*

Lemma 6 immediately implies the next bound on the number of leaves of $\mathcal{D}$-monotone trees.

Lemma 7. *Let S be a set of points, and let $\mathcal{D}$ be a set of k directions. If T is a $\mathcal{D}$-monotone spanning tree of S, then T has at most $2k$ leaves.*

The following lemma generalizes Lemma 6 (which concerns paths) for subtrees of a $\mathcal{D}$-monotone spanning tree T.

Lemma 8 ($\star$). *Let S be a set of points, let $\mathcal{D}$ be a set of k directions, let T be a $\mathcal{D}$-monotone spanning tree of S, and let u and v be two vertices of T. Then, it holds that: (i) Subtree $T_{u\setminus v}$ of T lies in $\mathcal{W}_{u\setminus v}(u)$. (ii) Sets $\mathcal{W}_{u\setminus v}$ and $\mathcal{W}_{v\setminus u}$ are disjoint, and regions $\mathcal{W}_{u\setminus v}(u)$ and $\mathcal{W}_{v\setminus u}(v)$ are disjoint.*

Let $B_{u,v}$ be a branch of a $\mathcal{D}$-monotone tree T connecting branching vertices u and v. Recall that $|\mathcal{W}_{B_{u,v}}| \leq k$, due to monotonicity of $B_{u,v}$. Let $R_{u,v} = \mathcal{W}_{B_{u,v}}(u) \cap \mathcal{W}_{B_{v,u}}(v)$. If $|\mathcal{W}_{B_{u,v}}| < k$, then $R_{u,v}$ is a parallelogram; see Fig. 5a. Otherwise (i.e., if $|\mathcal{W}_{B_{u,v}}| = k$), $R_{u,v}$ is a strip bounded by the parallel lines $\overline{d}(u)$ and $\overline{d}(v)$, where d is the direction of monotonicity of $B_{u,v}$; see Fig. 5b. We call $R_{u,v}$ the *region of branch* $B_{u,v}$. Similarly, if $P_{u,\lambda}$ is a leaf path from u to λ, then we define the *region of the leaf path* $R_{u,\lambda} = \mathcal{W}_{P_{u,\lambda}}(u) \cap \mathcal{W}_{P_{\lambda,u}}(\lambda)$.

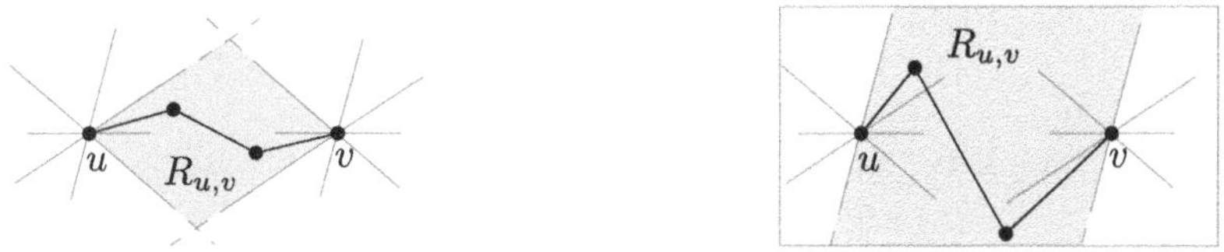

(a) $R_{u,v}$ is a parallelogram if $|\mathcal{W}_{B_{u,v}}| < k$.

(b) $R_{u,v}$ is a strip if $|\mathcal{W}_{B_{u,v}}| < k$.

Fig. 5. The different shapes of $R_{u,v}$ depending on $|\mathcal{W}_{B_{u,v}}|$.

Lemma 9 ($\star$). *Let S be a set of points, let $\mathcal{D}$ be a set of k (pairwise non-opposite) directions such that S is in $\mathcal{D}$-general position, and let T be a $\mathcal{D}$-monotone spanning tree of S. If $P_{u,v}$ is either a branch or a leaf path of T, then $\mathcal{W}_{P_{u,v}} \cap \mathcal{W}_{u\setminus v} = \emptyset$ and $R_{u,v} \cap \mathcal{W}_{u\setminus v}(u) = \emptyset$.*

4 A Characterization of $\mathcal{D}$-Monotone Spanning Trees

In this section we provide a characterization of $\mathcal{D}$-monotone spanning trees. It is the basis for our algorithm that solves MMST$(S, \mathcal{D})$; see Sect. 5.

Theorem 1. *Let S be a set of points, let $\mathcal{D}$ be a set of k (pairwise non-opposite) directions such that S is in $\mathcal{D}$-general position, and let T be a spanning tree of S. Then, T is $\mathcal{D}$-monotone if and only if:*

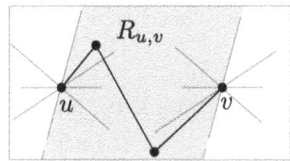

(a) Every leaf path and every branch P in T is $\mathcal{D}$-monotone.
(b) For every two leaf paths P_1 and P_2 incident to branching vertices u and v, respectively, $\mathcal{W}_{P_1}$ and $\mathcal{W}_{P_2}$ are disjoint.
(c) For every branch or leaf path $P_{u,v}$ of T it holds that $R_{u,v} \cap \mathcal{W}_{u\backslash v}(u) = \emptyset$.

Proof. ($\Rightarrow$) Since T is a $\mathcal{D}$-monotone tree, any subtree of T is also $\mathcal{D}$-monotone and, hence statement (a) holds. Statement (b) follows from Lemma 6(ii) since any two leaf paths are edge-disjoint. Statement (c) follows from Lemma 9.
($\Leftarrow$) For the monotonicity of T, it suffices to show that, for any two leaves λ and μ, the path $P_{\lambda,\mu}$ is $\mathcal{D}$-monotone. Let $P_{u,\lambda}$ and $P_{v,\mu}$ be the leaf paths to λ and μ where u and v are the branching vertices they are incident to, respectively. Suppose first that $P_{u,\lambda}$ and $P_{v,\mu}$ are incident to the same vertex, i.e., $u = v$. Due to (a), both leaf paths are $\mathcal{D}$-monotone; hence, $|\mathcal{W}_{P_{u,\lambda}}| \leq k$ and $|\mathcal{W}_{P_{v,\mu}}| \leq k$. Also, due to (b), $\mathcal{W}_{P_{u,\lambda}}$ and $\mathcal{W}_{P_{v,\mu}}$ are disjoint. Hence, there exists a direction d in $\mathcal{D}$ such that $\overline{d}(u)$ separates $\mathcal{W}_{P_{u,\lambda}}(u)$ and $\mathcal{W}_{P_{v,\mu}}(v)$ and does not intersect the interior of either of them. By Corollary 1, $P_{u,\lambda}$ and $P_{v,\mu}$ are both d-monotone and, additionally, they lie in different halfplanes with respect to $\overline{d}(u)$. Hence, the path from λ to μ is d-monotone, and thus $\mathcal{D}$-monotone.

Suppose now that $u \neq v$. Let $\mathcal{B} = \{u = b_1, \ldots, b_r = v, b_{r+1} = \mu\}$ be the sequence of the branching vertices on $P_{\lambda,\mu}$ in order of appearance where, for convenience, μ is treated as a branching vertex. By Corollary 1, it suffices to show that there is a direction d such that $\overline{d}(\mu)$ does not intersect the interior of $\mathcal{W}_{P_{\mu,\lambda}}(\mu)$. Let $\mathcal{P}_i = P_{b_i,\lambda}$ denote the subpath of $P_{\mu,\lambda}$ from vertex b_i to leaf λ. We show by induction on the size of $\mathcal{B}$ that *for every* $i \in \{1, \ldots, r+1\}$, $|\mathcal{W}_{\mathcal{P}_i}| \leq k$. Since $\mathcal{P}_{r+1}$ is by definition the oriented path from μ to λ, the fact that $|\mathcal{W}_{\mathcal{P}_{r+1}}| \leq k$ together with Corollary 1 guarantee that there exists a direction $d \in \mathcal{D}$ such that the path from λ to μ is d-monotone. For the base of the induction, observe that $\mathcal{P}_1$ is the leaf path $P_{u,\lambda}$, which is $\mathcal{D}$-monotone by (a). For the induction hypothesis, assume that $|\mathcal{W}_{\mathcal{P}_i}| \leq k$ for $i \leq m$. We show that $|\mathcal{W}_{\mathcal{P}_{m+1}}| \leq k$. Assume, for a contradiction, that $|\mathcal{W}_{\mathcal{P}_{m+1}}| > k$. Since $\mathcal{P}_{m+1}$ consists of $\mathcal{P}_m$ and of the branch $B_{b_m,b_{m+1}}$, the wedges of $\mathcal{W}_{\mathcal{P}_{m+1}} \setminus \mathcal{W}_{\mathcal{P}_m}$ are due to branch $B_{b_m,b_{m+1}}$. Let W_1^m and W_2^m be the leading and the trailing wedges (in ccw. order) of $\mathcal{W}_{\mathcal{P}_m}$ and let W_1^{m+1} and W_2^{m+1} be the leading and the trailing wedges (in ccw. order) of $\mathcal{W}_{\mathcal{P}_{m+1}}$. Observe first that either $W_1^m = W_1^{m+1}$ or $W_2^m = W_2^{m+1}$. If this was not the case, then $|\mathcal{W}_{B_{b_{m+1},b_m}}| > k$ which contradicts the fact that all branches are $\mathcal{D}$-monotone (refer to Fig. 6a).

Now assume, without loss of generality, that $W_2^m = W_2^{m+1}$ (see Fig. 6b). The leading wedge of $\mathcal{W}_{B_{b_{m+1},b_m}}$ is W_1^{m+1}, and the branch B_{b_{m+1},b_m} uses at most k wedges as it is $\mathcal{D}$-monotone. Also, $\mathcal{W}_{B_{b_{m+1},b_m}}(b_{m+1})$ contains vertex b_m as otherwise it would not be $\mathcal{D}$-monotone. Consider now the utilized wedge set $\mathcal{W}_{B_{b_m,b_{m+1}}}$ of $B_{b_m,b_{m+1}}$ consisting of the opposite of $\mathcal{W}_{B_{b_{m+1},b_m}}$. Its leading wedge is the opposite of W_1^{m+1}, it is located before W_2^m (in ccw. order), and its trailing wedge is located after W_2^m (in ccw. order). Thus, $\mathcal{W}_{\mathcal{P}_m}(b_m)$ intersects the region of the branch (the green parallelogram in Fig. 6b). This is a contradiction,

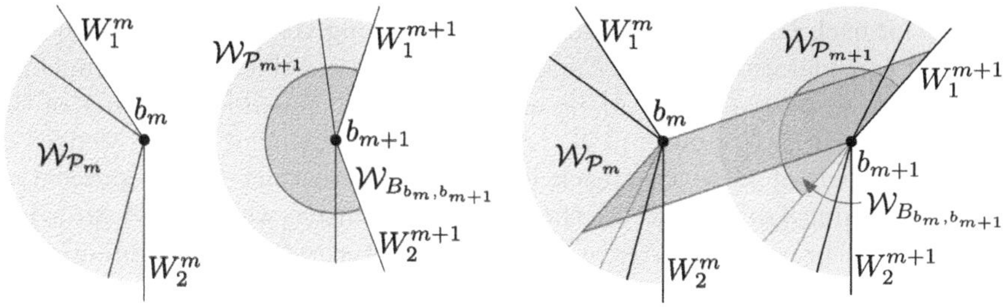

(a) $W_1^m = W_1^{m+1}$ or $W_2^m = W_2^{m+1}$. (b) $|\mathcal{W}_{\mathcal{P}_{m+1}}| > k$ contradicts condition (c).

Fig. 6. Different cases examined in the proof of Theorem 1

as $\mathcal{W}_{\mathcal{P}_m}(b_m) \subset \mathcal{W}_{b_m,b_{m+1}}(b_m)$ due to (c). Note that considering μ as a branching vertex does not affect the correctness of the proof. □

5 Algorithms for MMST$(S, \mathcal{D})$

In this section we prove that the problem MMST$(S, \mathcal{D})$ is in XP with respect to $|\mathcal{D}|$, that is, it can be solved in polynomial time for any fixed value of $|\mathcal{D}|$. An *embedding* of a tree is prescribed by the clockwise circular order of the edges incident to each vertex of the tree. A tree with a given embedding is an *embedded tree*. A *homeomorphically irreducible tree (HIT)*, is an embedded tree without vertices of degree two [27]. Let T_1 and T_2 be two trees; we say that T_1 and T_2 have the same *topology* if they are (possibly different) subdivisions of the same HIT H. Two trees with the same topology have the same *embedding* if the circular order of the edges around the vertices is the same in both trees. Given a HIT H and any embedded tree T that is a subdivision of H, we say that H *corresponds to* T. Since for a vertex of degree two the circular order of its incident edges is unique, the embedding of a tree T uniquely defines the embedding of the corresponding HIT. Note that, given an embedded tree T and the corresponding HIT H, an internal vertex of H corresponds to a branching vertex of T, a leaf of H to a leaf path of T, and an edge between two internal vertices of H to a branch of T.

Let n_ℓ be the numbers of HITs with at most ℓ leaves. We can use a result of Harary, Robinson, and Schwenk [28] concerning the number of (non-embedded) trees with $2\ell - 2$ vertices to derive a bound for n_ℓ. However, this does not yield an algorithm to generate all different HITs with at most ℓ leaves. For this reason we give an upper bound that is based on a generation scheme. Note that our scheme may generate the same HIT several times.

Lemma 10 ($\star$). *The number of different HITs with at most ℓ leaves is $O(7^\ell \cdot \ell!)$, and these HITs can be enumerated in $O(7^\ell \cdot \ell!)$ time.*

We now present an overview of the algorithm for solving the MMST$(S, \mathcal{D})$ problem. It examines every HIT with at most $2k$ leaves. Since there are many ($\mathcal{D}$-monotone) spanning trees that are subdivisions of the same HIT, the algorithm examines for each HIT all of its $\mathcal{D}$-monotone spanning trees on S. Let H be the HIT under consideration, and let ℓ and b be the numbers of leaves and branching vertices of H, respectively. Let M be one of the $O(n^b)$ possible mappings of the b branching vertices to points in S. Let A be an assignment of the wedges of $\mathcal{W}_\mathcal{D}$ to the leaves of H so that each leaf receives a distinct set of consecutive wedges. Assigning (as part of A) the set of consecutive wedges $\mathcal{W}^A$ to a leaf λ incident to a branching vertex v of H can be interpreted as our intention to cover all points in region $\mathcal{W}^A(v)$ by the monotone leaf path P that ends at λ. As shown in Fig. 7, the monotone leaf path may utilize a set of consecutive wedges $\mathcal{W}_P \subseteq \mathcal{W}^A$, i.e., some of the leading and/or trailing wedges of $\mathcal{W}^A$ may not be utilized by P.

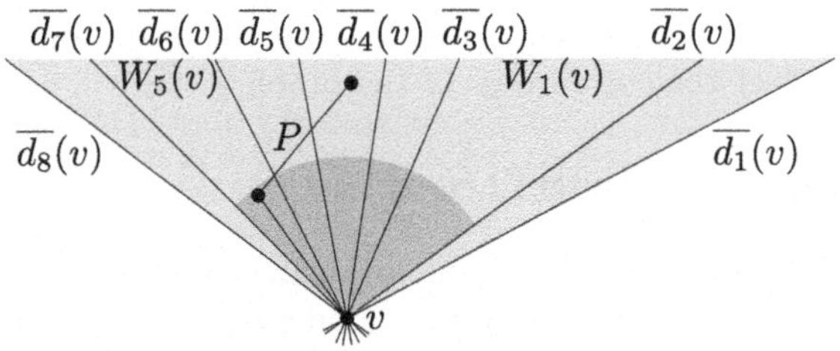

Fig. 7. A leaf path P that is assigned seven wedges but utilizes only five of them (shaded darkgray): It is not monotone with respect to $\{d_3, d_4, d_5, d_6\}$.

The point set S, the set $\mathcal{D}$ of k (pairwise non-opposite) directions, the HIT H, together with mapping M and assignment A, form an instance of a restricted problem that asks for a minimum $\mathcal{D}$-monotone spanning tree having H as its HIT and respecting M and A. Let $\text{MMST}(S, \mathcal{D}, H, M, A)$ denote this problem instance. Note that such a monotone spanning tree may not exist. If it exists, it turns out that it is unique (see Lemma 11). The algorithm for solving instances of type $\text{MMST}(S, \mathcal{D}, H, M, A)$ is repeatedly used by the algorithm that proves Theorem 2.

Lemma 11 ($\star$). *Let S be a set of n points, let $\mathcal{D}$ be a set of k (pairwise non-opposite) directions, let H be a HIT, let M be a mapping of the internal vertices of H to points of S, and let A be an assignment of $\mathcal{W}_\mathcal{D}$ to the leaves of H so that each leaf receives a distinct set of consecutive wedges. Then,* $\text{MMST}(S, \mathcal{D}, H, M, A)$ *can be solved in $O(n \log n + nk + k)$ time. Moreover, if a solution to the* $\text{MMST}(S, \mathcal{D}, H, M, A)$ *exists, then it is unique.*

Proof (sketch). Based on the characterization in Theorem 1, the algorithm checks whether point set S admits a $\mathcal{D}$-monotone spanning tree whose associated HIT is H, respecting mapping M and assignment A. Condition (b) of Theorem 1 is satisfied by definition since A is a valid assignment. For condition (c), we first compute the set $\mathcal{R}$ that consists of all path regions and branch regions and for every branch $B_{u,v}$, we compute $\mathcal{W}_{u \setminus v}$ and $\mathcal{W}_{v \setminus u}$. These computations take $O(k)$

time since HIT H has size $O(k)$. Then, the algorithm verifies, for every edge (u, v) of H, whether regions $\mathcal{W}_{u\setminus v}(u)$ and $R_{u,v}$ are disjoint, in $O(k)$ time. For condition (a), we compute, for every remaining point p in S, the region of $\mathcal{R}$ that contains p, in $O(nk)$ time. We then check, for every region in $\mathcal{R}$, whether there exists a path that (i) is monotone with respect to the two directions that are orthogonal to its boundaries and (ii) spans all points in the region. This can be done in $O(n \log n)$ time by sorting the points according to both directions. If the spanning tree exists, its uniqueness follows from the fact that each region in $\mathcal{R}$ contains a unique $\mathcal{D}$-monotone path. □

Theorem 2 (⋆). *Let S be a set of n points, and let $\mathcal{D}$ be a set of k (pairwise non-opposite) distinct directions. There exists a function $f\colon \mathbb{N} \to \mathbb{N}$ such that, if S is in $\mathcal{D}$-general position, then we can compute a minimum $\mathcal{D}$-monotone spanning tree of S in $O(f(k) \cdot n^{2k-1} \log n)$ time. In other words, the problem* MMST$(S, \mathcal{D})$ *is in XP when parameterized by k.*

Proof (sketch). The given set $\mathcal{D}$ of k directions yields a set of $2k$ wedges. Hence, a $\mathcal{D}$-monotone spanning tree has at most $2k$ leaves and at most $2k-2$ branching vertices. We enumerate the at most $7^{2k} \cdot (2k)!$ HITs according to Lemma 10. Let H be the current HIT, let ℓ be the number of leaves, and let $b \le \ell - 2$ be the number of branching vertices of H. We go through each of the $O(n^b) = O(n^{2k-2})$ subsets of cardinality $b \le 2k-2$ of S. Let M be the mapping of the branching vertices of H to points in S. Let A be the assignment of a set of consecutive wedges in $\mathcal{W}_{\mathcal{D}}$ to the leaves of H. There are at most $2k \cdot \binom{2k-1}{\ell-1} \le 2k \cdot 2^{2k}$ many such assignments since we have $2k$ choices for mapping the first leaf to some wedge, and then we select $\ell - 1$ out of the $2k-1$ remaining wedges that we attribute to a different leaf than the preceding wedge (in circular order). For each of the $n^{2k-2} \cdot f_0(k)$, (with $f_0(k) = 7^{2k} \cdot (2k)! \cdot 2k \cdot 4^k \in 2^{O(k \log k)}$) choices of a HIT H, mapping M and assignment A, we run the algorithm presented in the proof of Lemma 11 for the MMST$(S, \mathcal{D}, H, M, A)$, which terminates in $O(n \log n + nk + k)$ time. Finally, we return the shortest tree that we have found (if any). The total runtime is $O(f(k) \cdot n^{2k-1} \log n)$, where $f(k) = f_0(k) \cdot k \in 2^{O(k \log k)}$. We argue the correctness of the algorithm in the full version [25]. □

Speed-Up for $|\mathcal{D}| = 2$: For $|\mathcal{D}| = 2$, the algorithm from Sect. 5 computes a $\mathcal{D}$-monotone spanning tree of a set of n points in the plane in $O(n^3 \log n)$ time. In the full version [25], we show how to speed this up to $O(n^2)$ time.

Solving MMST(S, k): When the set of directions $\mathcal{D}$ is not prescribed and we are asked to search over all possible sets of k directions, a minimum k-directional monotone spanning tree of a point set S can be identified in $O(n^2 \log n)$ and in $O(n^6)$ time for $k = 1$ and $k = 2$, respectively. For $k \ge 3$, we describe an XP algorithm that runs in $2^{O(k \log k)} \cdot n^{2k(2k-1)} \log n$ time w.r.t. k [25].

6 Maximum Degree of the Minimum k-Directional MST

Since the (Euclidean) MST has maximum degree at most six [24], it is natural to ask whether this upper bound carries over to minimum k-directional monotone spanning trees. We prove that this is not the case by presenting a set $\mathcal{D}$ of k specific directions and a set S_k of $2k+1$ points such that the unique monotone k-directional spanning tree of S_k has degree $2k$.

Let k be an even positive integer, and let $\mathcal{D} = \{d_1, d_2, \ldots, d_k\}$ be the set of k distinct (pairwise non-opposite) directions (in ccw. order) such that d_1 is defined by the vector $(1,0)$ and, for $1 \leq i < k$, $\angle d_i d_{i+1} = \frac{\pi}{k}$. Since k is even, it holds that $\mathcal{W}_{\mathcal{D}} = \mathcal{W}_{\overline{\mathcal{D}}}$ where $\overline{\mathcal{D}} = \{\overline{d_1}, \overline{d_2}, \ldots, \overline{d_k}\}$. For simplicity, we consider W_0 to be the wedge defined by d_1 and d_2. We define $S_k = \{o\} \cup \{v_0, v_1, \ldots, v_{2k-1}\}$ to be the set of $2k+1$ points, where o is the origin and, for $i \in [k-1]$, v_i is placed on the unit circle in the (ccw.) second angle-trisection of wedge W_i of $\mathcal{W}_{\mathcal{D}}$; see Fig. 8a. By construction $S_k \setminus \{o\}$ is the vertex set of a regular $2k$-gon centered at o and the star with edges $ov_0, \ldots, ov_{2k-1}$ is a valid monotone spanning tree for S_k of length $2k$. Thus, any solution of MMST$(S_k, \mathcal{D})$ has length at most $2k$.

Let T be a tree that spans S_k. We call *polygon vertices* the vertices of T distinct from o. We refer to edges of T connecting adjacent polygon vertices as *external*, to edges incident to o as *rays* and to all other edges as *chords*. To show that the unique solution to the instance MMST$(S_k, \mathcal{D})$ is the $2k$-star centered at o, we first establish that polygon vertices have degree at most 2.

Lemma 12 ($\star$). *Let T be a solution to the* MMST$(S_k, \mathcal{D})$ *problem, and let x be a polygon vertex. Then,* $\deg_T(x) \leq 2$.

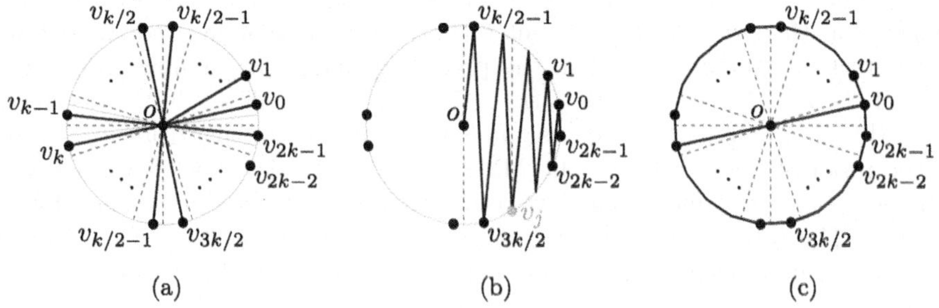

Fig. 8. (a) The point set S_k is defined based on the set $\mathcal{W}_{\mathcal{D}}$ of wedges (red dashed). (b) The path setting exploited in the proof of Theorem 3. (c) A monotone spanning graph of the point set in Fig. 8a whose length is much smaller than the $2k$-star in (a). (Color figure online)

Theorem 3 ($\star$). *The only solution to the* MMST$(S_k, \mathcal{D})$ *problem is the star $T^\star$ with center o and* $\deg_{T^\star}(o) = 2k \in \Omega(|S_k|)$.

Proof (sketch). Let tree T be a solution to the instance MMST$(S_k, \mathcal{D})$ and assume that T is not the $2k$-star with o at its center. It is easy to show that a leaf of T cannot be the endpoint of a chord. By using this property together with the fact that all polygon vertices of T have degree at most 2 (Lemma 12), we can show that T must contain the path $P = \langle v_{2k-1}, v_0, v_{2k-2}, \ldots, v_{2k-1-i}, v_i, \ldots, v_{\frac{k}{2}-1}, o\rangle$, (Fig. 8b). Consider the tree T' formed by replacing the edges of P by rays from o to the path vertices. Clearly, T' is also monotone. To show that T is not optimal, it suffices to show that the length $\|P\|$ of P is greater than the total length of the rays that replaced the edges of P in T' or, equivalently, that $\|P\| > k$. Indeed, using geometry, we show that $\|P\| = 1 + \sum_{i=1}^{k-1} 2\sin\left(\frac{\pi}{2k}i\right) = \cot\left(\frac{\pi}{4k}\right) > k$. □

7 Open Problems

We have presented an XP algorithm for solving MMST(S, k). It is natural to ask whether this problem is NP-hard if k is part of the input (rather than a fixed constant).

Another research direction is to study, for a given point set S and a set $\mathcal{D}$ of directions, the problem of computing a minimum $\mathcal{D}$-monotone spanning *graph* for S. Note that such a graph can have much smaller total length than a solution to MMST($S, \mathcal{D}$). Indeed, Theorem 3 shows that there is a point set S_k (Fig. 8a) and a set $\mathcal{D}$ of k directions such that the only solution to MMST($S_k, \mathcal{D}$) is the $2k$-star, which has a total length of $2k$. A monotone spanning *graph* of S_k (see Fig. 8c) has a total length of at most $2(\pi + 1)$.

References

1. Alamdari, S., Chan, T.M., Grant, E., Lubiw, A., Pathak, V.: Self-approaching graphs. In: Didimo, W., Patrignani, M. (eds.) GD 2012. LNCS, vol. 7704, pp. 260–271. Springer, Heidelberg (2013). https://doi.org/10.1007/978-3-642-36763-2_23
2. Angelini, P.: Monotone drawings of graphs with few directions. Inform. Process. Lett. **120**, 16–22 (2017). https://doi.org/10.1016/j.ipl.2016.12.004
3. Angelini, P., et al.: Greedy rectilinear drawings. Theor. Comput. Sci. **795**, 375–397 (2019). https://doi.org/10.1016/J.TCS.2019.07.019
4. Angelini, P., Colasante, E., Di Battista, G., Frati, F., Patrignani, M.: Monotone drawings of graphs. J. Graph Algorithms Appl. **16**(1), 5–35 (2012). https://doi.org/10.7155/jgaa.00249
5. Angelini, P., et al.: Monotone drawings of graphs with fixed embedding. Algorithmica **71**, 233–257 (2015). https://doi.org/10.1007/s00453-013-9790-3
6. Angelini, P., Frati, F., Grilli, L.: An algorithm to construct greedy drawings of triangulations. J. Graph Algorithms Appl. **14**(1), 19–51 (2010). https://doi.org/10.7155/jgaa.00197
7. Arkin, E.M., Connelly, R., Mitchell, J.S.B.: On monotone paths among obstacles with applications to planning assemblies. In: Proceedings of the 5th Ann. ACM Symposium on Computational Geometry (SoCG), pp. 334–343 (1989). https://doi.org/10.1145/73833.73870
8. Bahoo, Y., Durocher, S., Mehrpour, S., Mondal, D.: Exploring increasing-chord paths and trees. In: Gudmundsson, J., Smid, M. (eds.) Proceedings of 29th Canadian Conference on Computational Geometry (CCCG), pp. 19–24 (2017). https://2017.cccg.ca/proceedings/Session1B-paper1.pdf
9. Bakhshesh, D., Farshi, M.: (Weakly) Self-approaching geometric graphs and spanners. Comput. Geom. **78**, 20–36 (2019). https://doi.org/10.1016/j.comgeo.2018.10.002

10. Bakhshesh, D., Farshi, M.: Angle-monotonicity of Delaunay triangulation. Comput. Geom. **94**, 101711 (2021). https://doi.org/10.1016/j.comgeo.2020.101711
11. Bakhshesh, D., Farshi, M.: On the plane angle-monotone graphs. Comput. Geom. **100**, 101818 (2022). https://doi.org/10.1016/j.comgeo.2021.101818
12. Bonichon, N., Bose, P., Carmi, P., Kostitsyna, I., Lubiw, A., Verdonschot, S.: Gabriel triangulations and angle-monotone graphs: local routing and recognition. In: Hu, Y., Nöllenburg, M. (eds.) GD 2016. LNCS, vol. 9801, pp. 519–531. Springer, Cham (2016). https://doi.org/10.1007/978-3-319-50106-2_40
13. Chin, F., Houck, D.: Algorithms for updating minimal spanning trees. J. Comput. Syst. Sci. **16**(3), 333–344 (1978). https://doi.org/10.1016/0022-0000(78)90022-3
14. Dehkordi, H.R., Frati, F., Gudmundsson, J.: Increasing-chord graphs on point sets. In: Duncan, C., Symvonis, A. (eds.) GD 2014. LNCS, vol. 8871, pp. 464–475. Springer, Heidelberg (2014). https://doi.org/10.1007/978-3-662-45803-7_39
15. Dhandapani, R.: Greedy drawings of triangulations. Discrete. Comput. Geom. **43**, 375–392 (2010). https://doi.org/10.1007/s00454-009-9235-6
16. Didimo, W.: Upward graph drawing. In: Kao, M.-Y. (ed.) Encyclopedia of Algorithms, pp. 2308–2312. Springer (2016). https://doi.org/10.1007/978-1-4939-2864-4_653
17. Eppstein, D.: Finding the k smallest spanning trees. BIT **32**, 237–248 (1992). https://doi.org/10.1007/BF01994879
18. Eppstein, D.: Spanning trees and spanners. In: Sack, J.-R., Urrutia, J. (eds.) Handbook of Computational Geometry, pp. 425–461. North-Holland, Amsterdam (2000). https://doi.org/10.1016/B978-044482537-7/50010-3
19. Felsner, S., Igamberdiev, A., Kindermann, P., Klemz, B., Mchedlidze, T., Scheucher, M.: Strongly monotone drawings of planar graphs. In: Fekete, S., Lubiw, A. (eds.) Proceedings of 32nd International Symposium on Computational Geometry (SoCG). LIPIcs, vol. 51, pp. 37:1–37:15. Schloss Dagstuhl – Leibniz-Zentrum für Informatik (2016). https://doi.org/10.4230/LIPIcs.SoCG.2016.37
20. Francke, A., Hoffmann, M.: The Euclidean degree-4 minimum spanning tree problem is NP-hard. In: Proceedings of 25th Annual ACM Symposium on Computational Geometry (SoCG), pp. 179–188 (2009). https://doi.org/10.1145/1542362.1542399
21. Frederickson, G.N.: Ambivalent data structures for dynamic 2-edge-connectivity and k smallest spanning trees. SIAM J. Comput. **26**(2), 484–538 (1997). https://doi.org/10.1137/S0097539792226825
22. Gabow, H.N.: Two algorithms for generating weighted spanning trees in order. SIAM J. Comput. **6**, 139–150 (1977). https://doi.org/10.1137/0206011
23. Garg, A., Tamassia, R.: Upward planarity testing. Order **12**(2), 109–133 (1995). https://doi.org/10.1007/BF01108622
24. Georgakopoulos, G., Papadimitriou, C.H.: The 1-Steiner tree problem. J. Algorithms **8**(1), 122–130 (1987). https://doi.org/10.1016/0196-6774(87)90032-0
25. Giacomo, E.D., Didimo, W., Katsanou, E., Schlipf, L., Symvonis, A., Wolff, A.: Minimum monotone spanning trees. arXiv report (2024). http://arxiv.org/abs/2411.14038
26. Graham, R.L., Hell, P.: On the history of the minimum spanning tree problem. Ann. Hist. Comput. **7**(1), 43–57 (1985). https://doi.org/10.1109/MAHC.1985.10011
27. Harary, F., Prins, G.: The number of homeomorphically irreducible trees, and other species. Acta Math. **101**(1–2), 141–162 (1959). https://doi.org/10.1007/BF02559543

28. Harary, F., Robinson, R.W., Schwenk, A.J.: Twenty-step algorithm for determining the asymptotic number of trees of various species. J. Austral. Math. Soc. **20**(4), 483–503 (1975). https://doi.org/10.1017/S1446788700016190
29. He, D., He, X.: Optimal monotone drawings of trees. SIAM J. Discrete Math. **31**(3), 1867–1877 (2017). https://doi.org/10.1137/16M1080045
30. He, X., He, D.: Monotone drawings of 3-connected plane graphs. In: Bansal, N., Finocchi, I. (eds.) ESA 2015. LNCS, vol. 9294, pp. 729–741. Springer, Heidelberg (2015). https://doi.org/10.1007/978-3-662-48350-3_61
31. Hossain, M.I., Rahman, M.S.: Good spanning trees in graph drawing. Theoret. Comput. Sci. **607**, 149–165 (2015). https://doi.org/10.1016/j.tcs.2015.09.004
32. Kindermann, P., Schulz, A., Spoerhase, J., Wolff, A.: On monotone drawings of trees. In: Duncan, C., Symvonis, A. (eds.) GD 2014. LNCS, vol. 8871, pp. 488–500. Springer, Heidelberg (2014). https://doi.org/10.1007/978-3-662-45803-7_41
33. Lubiw, A., Mondal, D.: Construction and local routing for angle-monotone graphs. J. Graph Algorithms Appl. **23**(2), 345–369 (2019). https://doi.org/10.7155/jgaa.00494
34. Mastakas, K.: Uniform 2D-monotone minimum spanning graphs. In: Durocher, S., Kamali, S. (eds.) Proceedings of 30th Canadian Conference on Computational Geometry (CCCG), pp. 318–325 (2018). https://arxiv.org/abs/1806.08770
35. Mastakas, K.: Drawing a rooted tree as a rooted y-monotone minimum spanning tree. Inform. Process. Lett. **166**, 106035 (2021). https://doi.org/10.1016/j.ipl.2020.106035
36. Mastakas, K., Symvonis, A.: On the construction of increasing-chord graphs on convex point sets. In: Proceedings of 6th International Conference on Information, Intelligence, Systems and Applications (IISA), pp. 1–6 (2015). https://doi.org/10.1109/IISA.2015.7388028
37. Mastakas, K., Symvonis, A.: Rooted uniform monotone minimum spanning trees. In: Fotakis, D., Pagourtzis, A., Paschos, V.T. (eds.) CIAC 2017. LNCS, vol. 10236, pp. 405–417. Springer, Cham (2017). https://doi.org/10.1007/978-3-319-57586-5_34
38. Nöllenburg, M., Prutkin, R., Rutter, I.: On self-approaching and increasing-chord drawings of 3-connected planar graphs. J. Comput. Geom. **7**(1), 47–69 (2016). https://doi.org/10.20382/jocg.v7i1a3
39. Oikonomou, A., Symvonis, A.: Simple compact monotone tree drawings. In: Frati, F., Ma, K.-L. (eds.) GD 2017. LNCS, vol. 10692, pp. 326–333. Springer, Cham (2018). https://doi.org/10.1007/978-3-319-73915-1_26
40. Oikonomou, A., Symvonis, A.: Monotone drawings of k-inner planar graphs. In: Biedl, T., Kerren, A. (eds.) GD 2018. LNCS, vol. 11282, pp. 347–353. Springer, Cham (2018). https://doi.org/10.1007/978-3-030-04414-5_24
41. Papadimitriou, C.H., Ratajczak, D.: On a conjecture related to geometric routing. Theor. Comput. Sci. **344**(1), 3–14 (2005). https://doi.org/10.1016/j.tcs.2005.06.022
42. Papadimitriou, C.H., Vazirani, U.V.: On two geometric problems related to the travelling salesman problem. J. Algorithms **5**(2), 231–246 (1984). https://doi.org/10.1016/0196-6774(84)90029-4
43. Rao, A., Ratnasamy, S., Papadimitriou, C., Shenker, S., Stoica, I.: Geographic routing without location information. In: Proceedings of 9th Annual ACM Conference on Mobile Computing and Networking (MobiCom), pp. 96–108 (2003). https://doi.org/10.1145/938985.938996

44. Seo, D.Y., Lee, D.T., Lin, T.-C.: Geometric minimum diameter minimum cost spanning tree problem. In: Dong, Y., Du, D.-Z., Ibarra, O. (eds.) ISAAC 2009. LNCS, vol. 5878, pp. 283–292. Springer, Heidelberg (2009). https://doi.org/10.1007/978-3-642-10631-6_30
45. Shamos, M.I., Hoey, D.: Closest-point problems. In: Proceedings of 16th Annual IEEE Symposium on Foundations of Computer Science (FOCS), pp. 151–162 (1975). https://doi.org/10.1109/SFCS.1975.8
46. Spira, P.M., Pan, A.: On finding and updating spanning trees and shortest paths. SIAM J. Comput. **4**(3), 375–380 (1975). https://doi.org/10.1137/0204032

Representing Hypergraphs by Point-Line Incidences

Alexander Dobler[1(✉)], Stephen Kobourov[2], Debajyoti Mondal[3], and Martin Nöllenburg[1]

[1] TU Wien, Vienna, Austria
{adobler,noellenburg}@ac.tuwien.ac.at
[2] TUM School of Computation, München, Germany
stephen.kobourov@tum.de
[3] University of Saskatchewan, Saskatoon, Canada
d.mondal@usask.ca

Abstract. We consider hypergraph visualizations that represent vertices as points in the plane and hyperedges as curves passing through the points of their incident vertices. Specifically, we consider several different variants of this problem by (a) restricting the curves to be lines or line segments, (b) allowing two curves to cross if they do not share an element, or not; and (c) allowing two curves to overlap or not. We show $\exists\mathbb{R}$-hardness for six of the eight resulting decision problem variants and describe polynomial-time algorithms in some restricted settings.

Keywords: Hypergraph visualization · Point-line incidence · ETR-hardness

1 Introduction

Hypergraphs, or equivalently set systems, appear in many domains and are challenging to visualize. Classical methods like Venn and Euler diagrams [2,17,20] do not scale to large instances. Recent experimental work [25] has shown that representing hypergraphs with polylines for hyperedges and common intersection points for vertices improves the speed and accuracy of hypergraph-related tasks.In particular, the LineSets [1] and the MetroSets approach [16] use the metro map metaphor, where each hyperedge is a metro line and each vertex an interchange station. While LineSets connects pre-embedded vertices with arbitary curves, MetroSets optimizes vertex positions and aims to visualize the result in an octolinear style; see Fig. 1. Minimizing the visual complexity (which depends on the total number of bends along the metro lines), makes the representations simpler to understand and work with. A natural question is: which hypergraphs can be represented with just one bendless line segment per hyperedge?

A. Dobler and M. Nöllenburg—Supported by the Vienna Science and Technology Fund [10.47379/ICT19035]. D. Mondal was supported by the Natural Sciences and Engineering Research Council of Canada (NSERC).

R. Královič and V. Kůrková (Eds.): SOFSEM 2025, LNCS 15538, pp. 241–254, 2025.
https://doi.org/10.1007/978-3-031-82670-2_18

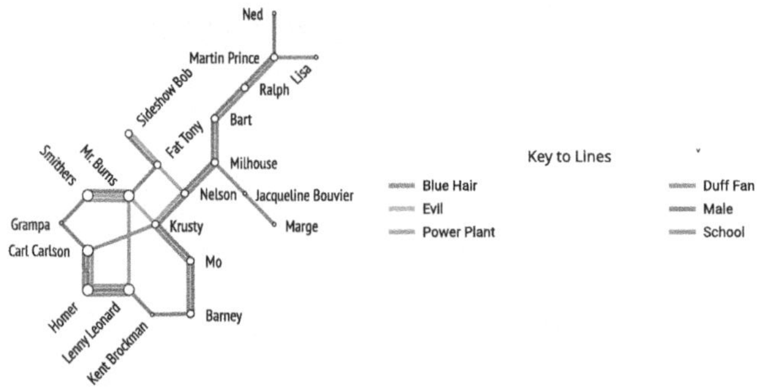

Fig. 1. A visualization of a Simpsons hypergraph with the MetroSets metaphor [16]. Hyperedges are represented by metro lines and elements are represented by stations.

With this in mind, we study the problem of representing a hypergraph with vertices as points in the plane and hyperedges as straight lines through incident vertices. Specifically, we consider several problems by varying one of the following requirements (see also Fig. 2 and formal definitions in Sect. 2):

(a) The curves must be line segments or (infinite) lines.
(b) Two curves may cross if they do not share an element, or not.
(c) Two curves may overlap (i.e. share a line segment), or not (called *strict*).

Table 1. Complexity results for deciding whether a representation that possibly has crossings exists. The value n.b. means unbounded max-degree or rank

	lines			line segments					
	strict (= non-strict)			strict			non-strict		
complexity	rank	max-deg	ref	rank	max-deg	ref	rank	max-deg	ref
$\exists\mathbb{R}$-hard	≥ 3	n.b.	Theorem 2	≥ 3	≥ 10	Theorem 5	≥ 3	≥ 6	Theorem 3
	n.b.	≥ 3	Cor 1				≥ 5	≥ 2	Theorem 4
Poly-time	≤ 2	n.b.	Observation 2	≤ 2	n.b.	Cor 2	≤ 2	n.b.	Cor 2
	n.b.	≤ 2	Observation 2	n.b.	≤ 2	Cor 2	≤ 3	≤ 2	Theorem 6

Contributions. In an extensive complexity study, we investigate for which hyperedge cardinalities and vertex degrees each of the eight problems are $\exists\mathbb{R}$-hard or solvable in polynomial time. We also study special graph classes that always admit such representations. Our contributions are detailed in Table 1 and 2. Our results are structured by distinguishing representations, where crossings are permitted (Sect. 4, Table 1) or not permitted (Sect. 5, Table 2). Section 5 further presents two hypergraph classes that always admit a crossing-free representation with segments.

Table 2. Complexity results for deciding whether a crossing-free representation exists. The value n.b. means unbounded max-degree or rank

	lines			line segments					
	strict (= non-strict)			strict			non-strict		
complexity	rank	max-deg	ref	rank	max-deg	ref	rank	max-deg	ref
∃ℝ-hard	?	?	-	≥ 5	≥ 12	Theorem 9	≥ 3	≥ 6	Cor 3
				n.b.	≥ 2	Theorem 10	≥ 5	≥ 2	Cor 4
Poly-time	≤ 2	n.b.	Theorem 7	≤ 2	n.b.	Observation 3	≤ 2	n.b.	Observation 3
	n.b.	≤ 2	Theorem 8						

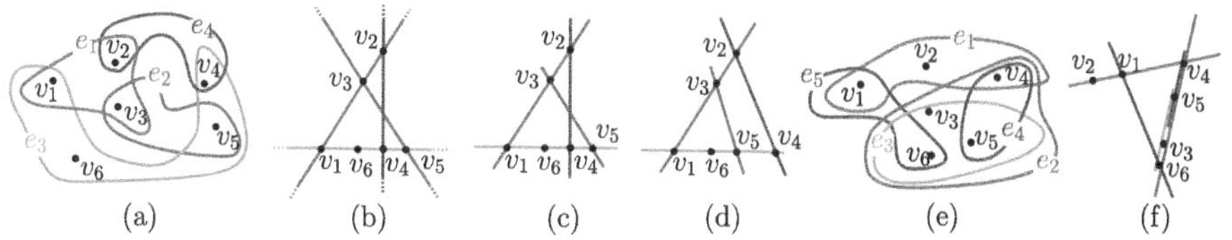

Fig. 2. (a) A linear hypergraph H. (b) A strict line representation of H. (c) A strict segment representation of H. (d) A crossing-free strict segment representation of H. (e)–(f) A hypergraph and its non-strict segment representation

Due to space constraints (⋆)-marked statements are proven in the full version [7].

Related Work. Representing hypergraphs with one line per hyperedge relates to classical geometric problems dating back to the 19th century, particularly in the study of configurations [15]. A *combinatorial* (or *geometric*) *configuration* is an abstract (or Euclidean) incidence structure with points and lines where each point is incident to the same number of lines, and each line to the same number of points. A *realization* of a combinatorial configuration is a geometric one, embedding points and lines in the Euclidean plane with the same incidences. This was central to Steinitz's PhD thesis [23] and studied in notable configurations like those of Desargues, Pappus, and Möbius-Kantor [15]. Steinitz [23] claimed that in a 3-uniform 3-regular linear hypergraph (where each point lies on three lines, each line passes through three points, and no two lines share more than one point), removing one point allows for a realization. This claim was later proven incorrect [8,15]. Gropp had already noted the connection between configurations and hypergraph drawings in his 1995 paper [13], arguing that using straight lines for hyperedges improves readability. He later collected classical results and posed hypergraph realizability as an open question [14].

Deciding whether the representation of a hypergraph can be a non-crossing straight-line drawing of a tree can be done in polynomial time [24]. A *support graph* G of a hypergraph H is a graph with the same vertex set where each hyperedge of H is connected in G. Techniques for drawing hypergraphs via support graphs [3–5] have a different focus and do not take into account whether vertices along hyperedges are collinear. Minimizing the number of crossings in a path-based support graph (hyperedges form a path in G) is NP-hard [9].

The intersection graph of line segments [19] can be seen as a strict segment representations of a linear hypergraph. Gonçalves [12] showed that some planar linear hypergraphs (see [12] for a definition) cannot be represented with straight line segments (in contrast to planar graphs). Segment contact representations of planar graphs can sometimes help in finding crossing-free strict line representations for linear hypergraphs. A necessary and sufficient condition for representing a graph as a contact system of segments is known [10,11], but no polynomial-time algorithm exists to test it. The hypergraph visualization problem is also similar to the stretchability problem [21,22]: given an arrangement of pseudolines, is there a combinatorially equivalent arrangement of lines? Unlike hypergraph representations, the order of vertices along each pseudoline is fixed in the stretchability problem. Matroid representability [18] is another related problem, asking whether elements of a matroid can be represented as vectors in $\mathbb{R}^3$ such that independent sets are retained, becoming similar to hypergraph representations through projective transformation onto the plane.

2 Preliminaries

A hypergraph $H = (V, E)$ is defined by a vertex set V and a hyperedge set E, where each $e \in E$ is a distinct non-empty subset of V. The *degree* of a vertex v in H is the number of hyperedges containing v. The *rank* of H is the maximum cardinality of a hyperedge $|e|$ over all e in E. A hypergraph H is k-*uniform* if every hyperedge has cardinality k and it is k-regular if every vertex has degree k. It is *linear* if $|e \cap e'| \leq 1$ for every pair of distinct hyperedges $e, e' \in E$. The *dual hypergraph* of H is obtained by interchanging the role of vertices and hyperedges, i.e., the dual of $H = (V, E)$ is $H_d = (E, E')$ where E' consists of hyperedges $\{e \mid v \in e\}$ for each $v \in V$. The *hyperedge intersection graph* G of H is a graph with E as its vertex set, where two vertices are adjacent if and only if the corresponding hyperedges share a common element.

A point x is *incident* to a line segment/linesegment ℓ if and only if ℓ contains x. A *line representation* of a hypergraph consists of an injective mapping α of vertices to points in $\mathbb{R}^2$ and an injective mapping β of hyperedges to lines in $\mathbb{R}^2$ such that $v \in e$ if and only if $\alpha(v) \in \beta(e)$ for $v \in V, e \in E$. A *segment representation* of a hypergraph is defined as a line representation with line segments instead of lines. A line representation/segment representation is *strict* if every pair of lines/line segments share at most one point. It is crossing-free if a pair of lines/line segments $\beta(e)$ and $\beta(e')$ in the representation share a point if and only if $e \cap e' \neq \emptyset$. We are concerned with the problem of deciding whether a hypergraph has a representation. We ask the following question.

Problem 1. *(H-Representation) Given a hypergraph H, $\mathcal{S} \in$ {strict, non-strict}, $\mathcal{C} \in$ {non-crossing-free, crossing-free}, and $\mathcal{X} \in$ {line, segment}, does H have an $\mathcal{S}$ $\mathcal{C}$ $\mathcal{X}$ representation?*

In fact, this leads to eight problem variants for the combinations of $\mathcal{S}, \mathcal{C}, \mathcal{X}$, which we all discuss. We often omit *non-crossing-free* and *non-strict*, i.e., by a segment representation we mean a non-strict non-crossing-free segment representation.

For simplicity of our presentation, we disallow two hyperedges to contain the same set of vertices. With this in mind, let us point out the following observation, which helps combine some of our complexity results.

Observation 1. *Let $H = (V, E)$ be a hypergraph. If H contains a pair of hyperedges e, e' with $|e \cap e'| \geq 2$ and $e \neq e'$, then H has no line representation, no strict line representation, no crossing-free line representation, and no strict crossing-free line representation. Otherwise, H is linear. Hence, H*

- *has a line representation if and only if it has a strict line representation, and*
- *has a crossing-free line representation if and only if it has a crossing-free strict line representation.*

We investigate complexity properties of all these problems, and show several of them are $\exists\mathbb{R}$-hard. Here the class $\exists\mathbb{R}$ is a complexity class between NP and PSPACE that contains all problems that can be reduced to solving an existentially quantified formula of polynomial equations and inequalities with integer coefficients; this means that $\exists\mathbb{R}$-hardness implies NP-hardness.

3 Pseudoline Stretchability and the Pappus Configuration

We often reduce from the $\exists\mathbb{R}$-hard problem PSEUDOLINE STRETCHABILITY [21] to prove $\exists\mathbb{R}$-hardness of our problems. A *pseudoline* is an x-monotone curve in $\mathbb{R}^2$ and a *simple pseudoline arrangement* is a set of pseudolines where every pair of pseudolines intersects exactly once and no three pseudolines meet at a common point. The PSEUDOLINE STRETCHABILITY problem takes a simple pseudoline arrangement as input and seeks a combinatorially equivalent drawing where each pseudoline is drawn as a straight line segment, i.e., a homeomorphic line segment arrangement (see Fig. 3(a)–(b)).

Many of our reductions make use of the *Pappus hypergraph* or *Pappus configuration*, and its ability to force three points to be collinear, as stated below.

Definition 1. The Pappus hypergraph or Pappus gadget $H_P = (V_P, E_P)$ is given as $V_P = \{p_1, \dots, p_9\}$ and E_P consists of the following hyperedges.

$$\{p_1, p_2, p_3\}, \{p_4, p_5, p_6\}, \{p_1, p_4, p_8\}, \{p_1, p_5, p_9\},$$
$$\{p_2, p_4, p_7\}, \{p_2, p_6, p_9\}, \{p_3, p_5, p_7\}, \{p_3, p_6, p_8\}$$

We call $\{p_1, \dots, p_6\}$ the *fixers* and $\{p_7, p_8, p_9\}$ the *anchors*.

For a line representation of the Pappus gadget see Fig. 3(c). Notice that all hyperedges have size three, all fixers have degree three, and all anchors have degree two. The most important property is given next, and follows directly from the well-known Pappus theorem.

Theorem 1. *[6, Chapter 3.5] In every line or line segment representation (strict and non-strict) of the Pappus gadget the anchors are collinear.*

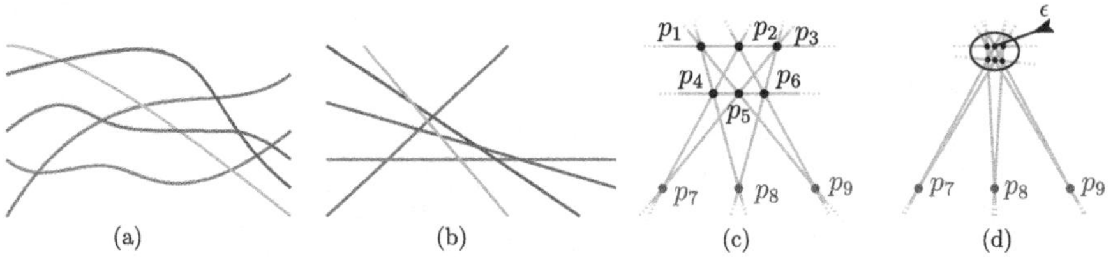

Fig. 3. (a) A simple pseudoline arrangement and (b) its stretched line segment representation. (c) A line representation of the Pappus gadget. The three anchors p_7, p_8, p_9 are always collinear. (d) "Pushing" the points p_1 and p_3 very close to p_2 forces all fixers p_1–p_6 to be inside a disk of arbitrarily small radius ϵ

Intuitively, our reductions use newly introduced vertices as fixers to force existing vertices which are the anchors to be represented in a collinear way. We also observe that, given a fixed embedding of the anchors, the remaining part of the Pappus gadget can be represented somewhat flexibly, such that it can avoid a finite set of predefined points and lines by representing the fixers in an arbitrary small radius disk (see Fig. 3(d)). The formal statement is in the full version [7].

4 Representations Without Bends

Table 1 shows our results for deciding whether a representation exists (irrespective of line or line segment crossings), which are proven in this section. The equivalence between results for line representations and strict line representations in Table 1 comes from Observation 1.

4.1 Complexity Results for Lines

Hardness Results. We show that it is $\exists\mathbb{R}$-hard to decide whether there exists a line representation for a given rank-3 hypergraph H. We reduce from MATROID REPRESENTABILITY [18]. For the purposes of a simple description, we give here a simplified description of a variant of that problem that is still $\exists\mathbb{R}$-hard [18]. We start with definitions. A *matroid* M is given as $M = (X, \mathcal{I})$ where X is the finite *ground set* and $\mathcal{I} \subseteq 2^X$ is the set of independent sets with

- $\emptyset \in \mathcal{I}$,
- $I' \subset I \in \mathcal{I}$ implies $I' \in \mathcal{I}$, and
- $I_1, I_2 \in \mathcal{I}$ with $|I_1| < |I_2|$ implies that there is an $x \in I_2 \setminus I_1$ with $I_1 \cup \{x\} \in \mathcal{I}$.

A *representation* of M is an injective mapping $f(X) : X \to \mathbb{R}^3$ such that for any $Y \subseteq X$ we have $Y \in \mathcal{I}$ if and only if $f(Y)$ forms a set of linearly independent vectors in $\mathbb{R}^3$. The $\exists\mathbb{R}$-hard problem MATROID REPRESENTABILITY is given as input a matroid and the question is whether there is a representation f of M. For the vectors $v \in \mathbb{R}^3$ we call the first, second, and third coordinate the x, y, and z-coordinates, respectively. We start by making some normalizations to M.

1. First, we can assume that every independent set $I \in \mathcal{I}$ has cardinality of at most 3, as there is otherwise no representation.
2. Second, we can assume that each pair $\{x, x'\} \in \binom{X}{2}$ forms an independent set, i.e. $\{x, x'\} \in \mathcal{I}$. Otherwise, $f(x) = cf(x')$ for some $c \in \mathbb{R}$ must hold for any representation. We can remove x' from X and replace any occurrence of x' in $\mathcal{I}$ by x, and obtain an equivalent instance w.r.t. representability.

Theorem 2 ($\star$). *It is $\exists\mathbb{R}$-hard to decide whether a rank-3 linear hypergraph has a line representation.*

Proof Sketch. The reduction is from MATROID REPRESENTABILITY. We are given a matroid $M = (X, \mathcal{I})$ (we assume all above normalizations were applied already) and transform it to a hypergraph $H = (V, E)$ as follows. To construct H, we first add to H the set X as vertices. Now consider a triple $t := \{x, x', x''\} \in \binom{X}{3}$.

- If t does not form an independent set in M, we introduce a new Pappus gadget and let t be its anchor (see Fig. 3 (c)). This forces t to be collinear.
- Otherwise, we add a new vertex d, force d being collinear with x and x' using a Pappus gadget with anchors d, x and x', and lastly add the hyperedge $\{d, x''\}$ to E (see Fig. 4(a)). With this t cannot be represented collinearly.

We argue that H has a strict line representation if and only if M has a representation. The forward direction is because we constructed H such that t has to be represented collinearly if and only if t does not form an independent set in M. The representation of X can be transformed into a representation of M by taking the representation of X as the x- and y-coordinate and setting the z-coordinate to one. The backward direction is due to a projective transformation of the matroid representation into a representation of the vertices $X \subseteq V(H)$ and because of the flexibility of the Pappus gadget (see [7]). □

The reduction in Theorem 2 produces a linear hypergraph without two degree-1 vertices being part of the same hyperedge. Hence, applying a standard point-line duality transformation to a line representation H results in a representation of its max-degree-3 dual hypergraph and vice versa. This is enough to show that $\exists\mathbb{R}$-hardness also holds for max-degree-3 hypergraphs.

Corollary 1. *It is $\exists\mathbb{R}$-hard to decide whether a max-degree-3 hypergraph has a line representation.*

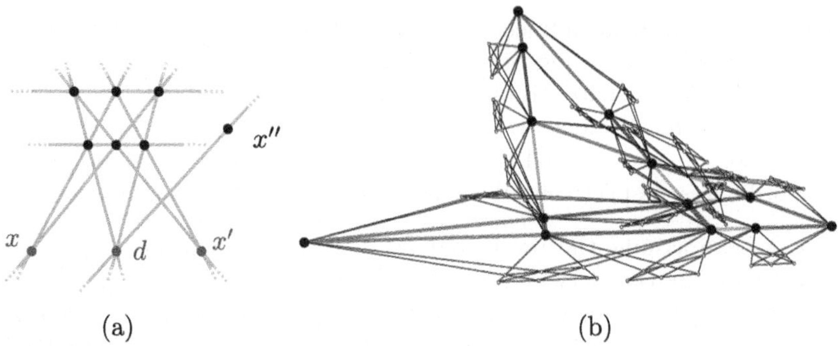

Fig. 4. (a) A gadget forcing x, x', and x'' not to be collinear. (b) A strict segment representation of the hypergraph resulting from the reduction in Theorem 5 applied to the pseudoline arrangement in Fig. 5(a)

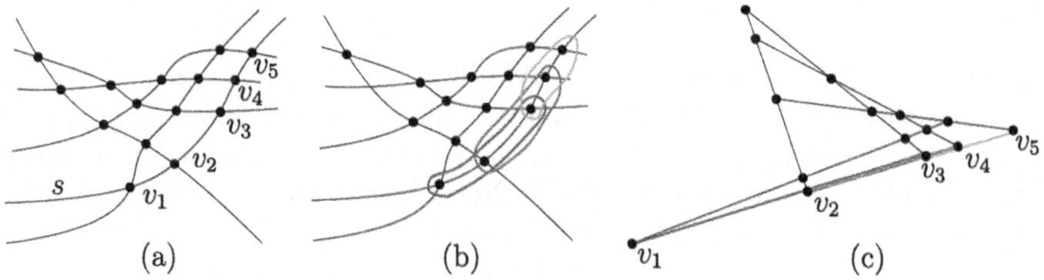

Fig. 5. (a) A pseudoline arrangement. (b) the hyperedges corresponding to pseudoline s as in the proof of Theorem 3. (c) A corresponding segment representation

Solvable Cases. By definition, non-linear hypergraphs do not have line representations. However, every rank-2 hypergraph (which is linear) and every linear max-degree-2 hypergraph has a strict line representation by starting with points or lines in general position, respectively.

Observation 2. *Every rank-2 hypergraph and every linear max-degree-2 hypergraph has a line representation.*

4.2 Complexity Results for Line Segments

Hardness Results. The first result is that deciding whether a hypergraph has a segment representation is $\exists\mathbb{R}$-hard, even for hypergraphs of constant rank and constant maximum degree.

Theorem 3. *It is $\exists\mathbb{R}$-hard to decide whether a rank-3 max-degree-6 hypergraph has a segment representation.*

Proof. We reduce from PSEUDOLINE STRETCHABILITY [21]. Given an instance I of PSEUDOLINE STRETCHABILITY, we construct a hypergraph H by taking intersection points as vertices. For each pseudoline with the intersection points $v_1, v_2, \ldots, v_t$ in this order, we construct hyperpaths consisting of the hyperedges $\{v_1, v_2, v_3\}, \{v_2, v_3, v_4\}, \ldots, \{v_{t-2}, v_{t-1}, v_t\}$ (Fig. 5(a)–(b)).

By the construction of H, a solution to the PSEUDOLINE STRETCHABILITY instance corresponds to a representation with straight line segments (Fig. 5(c)): We assume that H has a representation with straight line segments. Observe that then the union of hyperedges forming the hyperpaths as above appear together as one single line segment with the vertices in the correct order. Conversely, if I is stretchable, then H has a segment representation. □

We show later that the problem becomes tractable for rank 2 instead of rank 3. Surprisingly, this is not the case for maximum degree 2 even for constant rank, as shown with a similar but more involved reduction.

Theorem 4 ($\star$). *It is $\exists\mathbb{R}$-hard to decide whether a rank-5 max-degree-2 hypergraph has a segment representation.*

By careful use of the Pappus gadget we obtain the following result.

Theorem 5. *It is $\exists\mathbb{R}$-hard to decide whether a rank-3 max-degree-12 hypergraph has a strict segment representation.*

Proof. Consider an instance of PSEUDOLINE STRETCHABILITY. We build a hypergraph H with intersection points as vertices. For each pseudoline let its intersection points be $v_1, v_2, \ldots, v_t$. We need to force (1) all these points collinear, and (2) fix the correct order. For (1), we add for each triple v_i, v_{i+1}, v_{i+2} $(i = 1, \ldots, t-2)$ a new Pappus gadget with the triple as its anchors. For (2), we add the hyperedges $\{v_1, v_2\}, \{v_2, v_3\}, \ldots, \{v_{t-1}, v_t\}$. We claim that I is stretchable if and only if H has a strict segment representation, see Fig. 4(b).

"$\Rightarrow$": Consider the placement of intersection points as they appear in the stretched pseudoline arrangement. Size-two hyperedges can clearly be represented. We also need to represent the Pappus gadgets, which is possible because of the freedom of the Pappus gadget (see [7]).

"$\Leftarrow$" We built H such that in a strict segment representation the intersection points of a polyline are collinear and ordered correctly, making I stretchable. □

Solvable Cases. The first result is a direct consequence of Observation 2 and by the fact that rank-2 hypergraphs are linear by our assumptions.

Corollary 2. *Every rank-2 hypergraph and every linear max-degree-2 hypergraph has a (strict) segment representation.*

As we have seen, testing whether a max-degree-2 hypergraph has a segment representation is $\exists\mathbb{R}$-hard (Theorem 4). However, we can at least characterize all rank-3 max-degree-2 hypergraphs that have a segment representation by a set of forbidden subhypergraphs:

- The *rigid triangle* consists of vertices a, b, c, d and hyperedges $e_1 = \{a, b, c\}$ $e_2 = \{b, c, d\}$, and $e_3 = \{a, d\}$. The hyperedge e_3 can furthermore also contain another vertex f, i.e., $e_3 = \{a, d, f\}$. See Fig. 6 (a)–(b).

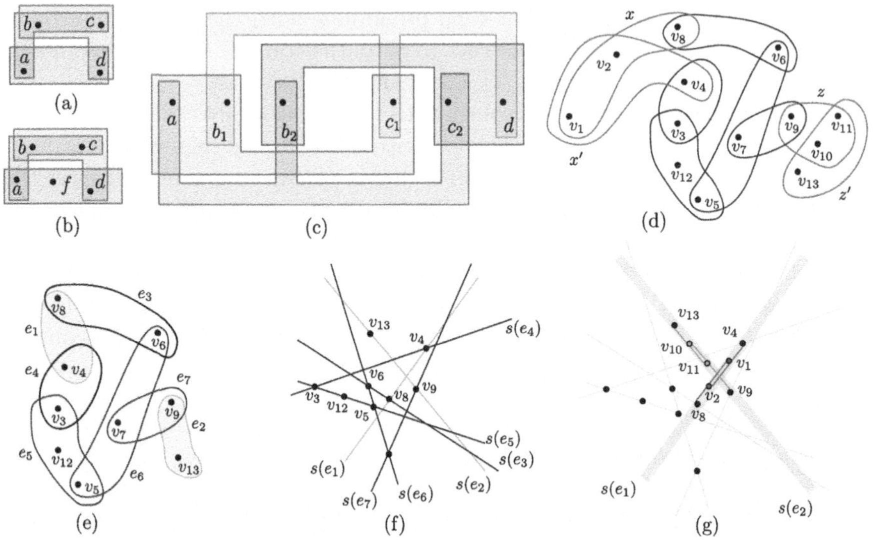

Fig. 6. (a)–(b) Illustration for rigid triangles. (c) A rigid parallel 2-path. (d) Illustration for hyperedges sharing two vertices in purple. (e) Illustration for E_c, where the hyperedges corresponding to sibling pairs are shown in gray. (f) A segment representation for E_c, where sibling pairs x, x' and z, z' are mapped to $s(e_1)$ and $s(e_2)$, respectively. (g) Splitting $s(e_1)$ and $s(e_2)$ to create $\beta(x), \beta(x')$ and $\beta(z), \beta(z')$, respectively

- The *rigid parallel 2-path* consists of the vertices a, b_1, b_2, c_1, c_2, d, and the hyperedges $e_1 = \{a, b_1, c_1\}$, $e'_1 = \{b_1, c_1, d\}$, $e_2 = \{a, b_2, c_2\}$, $e'_2 = \{b_2, c_2, d\}$. See Fig. 6 (c).

Theorem 6 ($\star$). *A rank-3 max-degree-2 hypergraph has a segment representation if and only if it does not contain a set of hyperedges that together form a rigid triangle or rigid parallel 2-path.*

Proof Sketch. For the forward direction, it is enough to verify that rigid triangles and rigid parallel 2-paths (Fig. 6(a)–(c)) themselves do not have a segment representation. Assume now that we are given a hypergraph $H = (V, E)$ without any of the two forbidden substructures. We first create an auxiliary hypergraph $H' = (V', E_c)$ that replaces every pair of hyperedges x, x' having two common vertices (purple in Fig. 6(d)) with a new hyperedge e that contains the remaining elements in $(x \cup x') \setminus (x \cap x')$, see Fig. 6(d)–(e). We refer to such a pair of hyperedges as a *sibling pair.* We then show that H' has a segment representation which can be constructed from a set of lines in general position (Fig. 6(f)). Finally, we split the segment $s(e)$ of each new hyperedge e to create the segments $\beta(x)$ and $\beta(x')$ that would represent the sibling pair corresponding to e (Fig. 6(g)). □

Notice that testing whether H contains a rigid triangle or a rigid parallel 2-path can be done in polynomial time, so we have identified a problem variant that is polynomial-time solvable. We do not know the complexity for hypergraphs with maximum degree 2 and rank 4.

5 Representations Without Bends and Without Crossings

In this section, we give some results on deciding the existence of crossing-free representations, which are summarized in Table 2. Again, the equivalence between strict and non-strict is because of Observation 1.

5.1 Complexity Results for Lines

We have two positive results for rank-2 and max-degree-2 hypergraphs, respectively. The first follows due to an easy case distinction, the second by an equivalence of crossing-free line representability of a linear hypergraph H (non-linear hypergraphs are not line representable) and its hyperedge intersection graph being a complete k-partite graph.

Theorem 7 ($\star$). *We can decide in polynomial time whether a rank-2 hypergraph has a crossing-free line representation.*

Theorem 8 ($\star$). *We can decide in polynomial-time whether a max-degree-2 hypergraph has a crossing-free line representation.*

5.2 Line Segments

For line segments, we first present hardness results and then cases that we solve, in particular this will include classes of hypergraphs that always have a crossing-free segment representation.

Hardness Results. The reduction in the proof of Theorem 3 constructs a hypergraph that admits a crossing-free segment representation if and only if the corresponding pseudoline arrangement is stretchable. Thus, the same reduction gives the following result.

Corollary 3. *It is* $\exists\mathbb{R}$*-hard to decide whether a rank-*3 *max-degree-*6 *hypergraph has a crossing-free segment representation.*

The same holds for the reduction in the proof of Theorem 4.

Corollary 4. *It is* $\exists\mathbb{R}$*-hard to decide whether a rank-*5 *max-degree-*2 *hypergraph has a crossing-free segment representation.*

With a slightly more involved reduction from stretchability using an *extended Pappus gadget* (intersection points are replaced by vertices), we can show that it is $\exists\mathbb{R}$-hard to decide whether a rank-5 max-degree-10 hypergraph has a strict crossing-free segment representation.

Theorem 9 (⋆). *It is $\exists\mathbb{R}$-hard to decide whether a rank-5 max-degree-10 hypergraph has a strict crossing-free segment representation.*

A correspondence between strict crossing-free segment representations of degree-2 hypergraphs and segment intersection graphs gives the following result.

Theorem 10 (⋆). *It is $\exists\mathbb{R}$-hard to decide whether a max-degree-2 hypergraph has a strict crossing-free segment representation.*

Solvable Cases. We now present some cases where the existence of a crossing-free segment representation can be decided in polynomial time.

The rank-2 case corresponds to deciding whether a graph is planar.

Observation 3. *Let H be a rank-2 hypergraph without hyperedges of size 1. Then H is a simple graph, and every crossing-free segment representation can be made strict. Furthermore, H has such a representation if and only if it is planar.*

As hyperedges of size 1 do not affect segment representability, this solves the problem for rank-2 hypergraphs.

We now present some hypergraph classes for which we can decide the existence of a strict crossing-free segment representation in polynomial time. The first class is based on *permutation graphs* which are graphs with vertex set $\{v_1, \ldots, v_n\}$ and there exists a permutation π such that v_i, v_j, $i < j$, are adjacent if and only if i comes after j in π.

Theorem 11 (⋆). *Let H be a degree-2 linear hypergraph and let G be its hyperedge intersection graph. If G is a permutation graph (which is verifiable in linear time), then H admits a crossing-free strict segment representation.*

The *vertex-edge incidence graph* of a hypergraph $H = (V, E)$ is a bipartite graph $G = (V \cup E, E')$, where an edge $(v, e) \in E'$, $v \in V$ and $e \in E$, exists, if and only if e contains v. We obtain the following result.

Theorem 12 (⋆). *Let $H = (V, E)$ be a hypergraph with the vertex-edge incidence graph $G = (V \cup E, E')$. If G has a planar embedding with vertex set E on the outerface, then we can decide whether H admits a crossing-free strict segment representation in polynomial time.*

Proof Sketch. If the girth (i.e., the length of a shortest cycle) of G is 4, then there are two hyperedges e, e' in H that contain at least two vertices in common. Therefore, H cannot have a strict line segment representation. Otherwise, we first construct a planar embedding of the graph where the vertices of E are on the outerface (Fig. 7(a)), and then give an incremental construction by repeatedly removing some vertices on the unbounded region (Fig. 7(b)). □

The representation that we construct in Lemma 12 can be seen as a contact system of segments, and hence this identifies a class of graphs for which the existence of segment contact representation can be tested in polynomial time without using the conditions of [10,11] that check some properties over all subsystems of at least two paths.

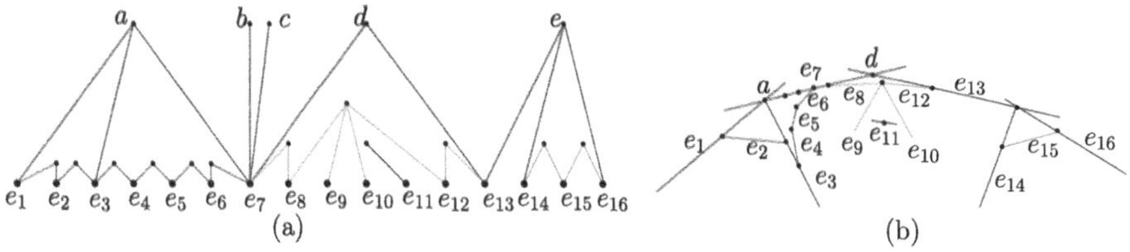

Fig. 7. Construction of a crossing-free strict segment representation in Lemma 12

6 Conclusions and Open Problems

We studied representations of hypergraphs by arrangements of lines/segments. While we answered several complexity questions, many open problems remain:

- What is the time complexity of deciding whether a hypergraph has a crossing-free line representation?
- We have almost characterized the complexity of deciding whether a max-degree-2 hypergraph has a segment representation (hardness for rank at least five and a polynomial algorithm for rank at most three). What about max-degree-2 rank-4 hypergraphs?
- As most straight-line variants are hard, it is a natural question to explore the space where curves can have bends.

References

1. Alper, B., Riche, N.H., Ramos, G.A., Czerwinski, M.: Design study of linesets, a novel set visualization technique. IEEE Trans. Vis. Comput. Graph. **17**(12), 2259–2267 (2011). https://doi.org/10.1109/TVCG.2011.186
2. Alsallakh, B., Micallef, L., Aigner, W., Hauser, H., Miksch, S., Rodgers, P.J.: The state-of-the-art of set visualization. Comput. Graph. Forum **35**(1), 234–260 (2016). https://doi.org/10.1111/CGF.12722
3. Brandes, U., Cornelsen, S., Pampel, B., Sallaberry, A.: Path-based supports for hypergraphs. J. Discrete Algorithms **14**, 248–261 (2012). https://doi.org/10.1016/J.JDA.2011.12.009
4. Buchin, K., van Kreveld, M.J., Meijer, H., Speckmann, B., Verbeek, K.: On planar supports for hypergraphs. J. Graph Algorithms Appl. **15**(4), 533–549 (2011). https://doi.org/10.7155/JGAA.00237
5. Castermans, T., van Garderen, M., Meulemans, W., Nöllenburg, M., Yuan, X.: Short plane supports for spatial hypergraphs. J. Graph Algorithms Appl. **23**(3), 463–498 (2019). https://doi.org/10.7155/jgaa.00499
6. Coxeter, H.S.M., Greitzer, S.L.: Geometry revisited, vol. 19. Mathematical Association of America (1967)
7. Dobler, A., Kobourov, S., Mondal, D., Nöllenburg, M.: Representing hypergraphs by point-line incidences. CoRR (2024). arXiv:2411.13985
8. Flowers, G.: Embeddings of configurations. Ph.D. thesis, University of Victoria (2015)

9. Frank, F., et al.: Using the metro-map metaphor for drawing hypergraphs. In: Bures, T., Dondi, R., Gamper, J., Guerrini, G., Jurdzinski, T., Pahl, C., Sikora, F., Wong, P.W.H. (eds.) Proc. Conference on Current Trends in Theory and Practice of Computer Science (SOFSEM 2021). LNCS, vol. 12607, pp. 361–372. Springer (2021). https://doi.org/10.1007/978-3-030-67731-2_26
10. de Fraysseix, H., de Mendez, P.O.: Representations by contact and intersection of segments. Algorithmica **47**(4), 453–463 (2007)
11. de Fraysseix, H., de Mendez, P.O., Rosenstiehl, P.: Representation of planar hypergraphs by contacts of triangles. In: Hong, S., Nishizeki, T., Quan, W. (eds.) Proceedings of the Graph Drawing (GD 2007). LNCS, vol. 4875, pp. 125–136. Springer (2007) https://doi.org/10.1007/978-3-540-77537-9_15
12. Gonçalves, D.: A planar linear hypergraph whose edges cannot be represented as straight line segments. Eur. J. Comb. **30**(1), 280–282 (2009). https://doi.org/10.1016/J.EJC.2007.12.004
13. Gropp, H.: The drawing of configurations. In: Brandenburg, F. (ed.) Proceedings of the Graph Drawing (GD 1995). LNCS, vol. 1027, pp. 267–276. Springer (1995) https://doi.org/10.1007/BFB0021810
14. Gropp, H.: Configurations and their realization. Discret. Math. **174**(1–3), 137–151 (1997). https://doi.org/10.1016/S0012-365X(96)00327-5
15. Grünbaum, B.: Configurations of Points and Lines. American Mathematical Society (2009)
16. Jacobsen, B., Wallinger, M., Kobourov, S.G., Nöllenburg, M.: Metrosets: visualizing sets as metro maps. IEEE Trans. Vis. Comput. Graph. **27**(2), 1257–1267 (2021). https://doi.org/10.1109/TVCG.2020.3030475
17. Johnson, D.S., Pollak, H.O.: Hypergraph planarity and the complexity of drawing Venn diagrams. J. Graph Theory **11**(3), 309–325 (1987). https://doi.org/10.1002/JGT.3190110306
18. Kim, E., de Mesmay, A., Miltzow, T.: Representing matroids over the reals is $\exists\mathbb{R}$-complete. CoRR arXiv:abs/2301.03221 (2023). https://doi.org/10.48550/arXiv.2301.03221
19. Matousek, J.: Intersection graphs of segments and $\exists\mathbb{R}$. CoRR (2014). arXiv:1406.2636
20. Mäkinen, E.: How to draw a hypergraph. Int. J. Computer Math. **34**(3–4), 177–185 (1990). https://doi.org/10.1080/00207169008803875
21. Schaefer, M.: Complexity of some geometric and topological problems. In: Eppstein, D., Gansner, E.R. (eds.) Proceedings of the Graph Drawing (GD 2009). LNCS, vol. 5849, pp. 334–344. Springer (2009) https://doi.org/10.1007/978-3-642-11805-0_32
22. Shor, P.W.: Stretchability of pseudolines is NP-hard. In: Gritzmann, P., Sturmfels, B. (eds.) Proc. Applied Geometry And Discrete Mathematics (DIMACS 1090). DIMACS Series in Discrete Mathematics and Theoretical Computer Science, vol. 4, pp. 531–554. DIMACS/AMS (1991) https://doi.org/10.1090/DIMACS/004/41
23. Steinitz, E.: Über die Construction der Configurationen n_3. Ph.D. thesis, Breslau (1894)
24. Swaminathan, R., Wagner, D.K.: On the consecutive-retrieval problem. SIAM J. Comput. **23**(2), 398–414 (1994). https://doi.org/10.1137/S0097539792235487
25. Wallinger, M., Jacobsen, B., Kobourov, S.G., Nöllenburg, M.: On the readability of abstract set visualizations. IEEE Trans. Vis. Comput. Graph. **27**(6), 2821–2832 (2021). https://doi.org/10.1109/TVCG.2021.3074615

Reachability in Temporal Graphs Under Perturbation

Jessica Enright, Laura Larios-Jones(✉), Kitty Meeks, and William Pettersson

School of Computing Science, University of Glasgow, Glasgow, Scotland
{jessica.enright,laura.larios-jones,kitty.meeks, william.pettersson}@glasgow.ac.uk

Abstract. Reachability and other path-based measures on temporal graphs can be used to understand spread of infection, information, and people in modelled systems. Due to delays and errors in reporting, temporal graphs derived from data are unlikely to perfectly reflect reality, especially with respect to the precise times at which edges appear. To reflect this uncertainty, we consider a model in which some number ζ of edge appearances may have their labels perturbed by $\pm\delta$ for some δ. Within this model, we investigate temporal reachability and consider the problem of determining the maximum number of vertices any vertex can reach under these perturbations. We show this problem to be intractable in general but efficiently solvable when ζ is sufficiently large. We also give algorithms which solve this problem in several restricted settings. We complement this with some contrasting results concerning the complexity of related temporal eccentricity problems under perturbation.

Keywords: Parameterized Algorithms · Temporal Graphs · Reachability

1 Introduction

Temporal graphs are widely used to model movement and spread in time-sensitive networks [3,6,20,23]. However, the algorithmic tools used in solving problems on these graphs typically assume that the temporal graphs as given are correct and without uncertainty—unfortunately this may not be the case for real-world networks (e.g. [5,27]), and thus this assumption significantly limits the applicability of temporal graph methods.

For example, reachability is used as a measure to assess risk of outbreaks in epidemiologically-relevant graphs such as livestock trading networks or human or animal contact networks [18,21]. It is clear that a calculated reachability may be incorrect if the temporal graph itself is incorrect. From an application perspective, the robustness of the temporal graph's reachability to incorrect timings is important: if a small number of incorrectly-recorded edge appearances can result in a huge increase in reachability, then this temporal graph is more epidemiologically risky than one in which a small number of incorrect appearance times do not significantly increase reachability.

R. Královič and V. Kůrková (Eds.): SOFSEM 2025, LNCS 15538, pp. 255–269, 2025.
https://doi.org/10.1007/978-3-031-82670-2_19

In this work we address the question: what is the complexity, given a specification of the number and magnitude of errors in the timings of the temporal graph's edge appearances, of determining if the reachability of the temporal graph could be above some given threshold? Put another way, we ask if there exists a perturbation of the times allocated to edges that increases the reachability above a target threshold. This second formulation makes clear a link to temporal graph modification questions—an active area of research studying the minimum number of modifications required to a temporal graph that allow it to satisfy some specified property.

Following the definition given by Kempe, Kleinberg and Kumar [22], we say that temporal graphs (G, λ) consist of an underlying graph $G = (V, E)$ and a temporal assignment $\lambda : E(G) \to 2^{\mathbb{N}}$ which describes when each edge is active. We introduce the concept of a (δ, ζ)-perturbation where up to ζ time-edges of a temporal graph are changed by at most $\pm\delta$. We ask, given a temporal graph (G, λ), if there is a perturbation of (G, λ) such that there is a temporal path from some vertex in the graph to at least h vertices. With the framing of a transport network, this asks if we can reach h places from some starting point by connections which occur chronologically. In an epidemiological framework, this asks how many nodes could, in the worst case, be infected by a single initially-infected node.

In previous work, Deligkas and Potapov [8] consider delaying and merging operations to optimise maximum, minimum and average reachability where merging of time-edges assigns all edges which are active in a set of consecutive times the latest time in that set. They show that it is NP-hard to find a set of merging operations that maximises the maximum reachability of a temporal graph, even when the underlying graph is a path. In contrast, they show that finding a set of delaying operations to minimize the maximum reachability is tractable if the number of delays made is unbounded. In a similar way, we find that allowing a large number of perturbations makes our problem easier.

In more recent work, Deligkas et al. [7] investigate the problem which asks how *long* it takes to reach every vertex in a temporal graph from a set of sources following some delays. They consider variations of this problem where the number of delayed edges and total sum of delays must be at most k for some integer k, and show that this problem is `W[2]`-hard with respect to the number of edges changed and prove tractability when the underlying graph has bounded treewidth. Delaying appearances of edges to optimise reachability is also studied by Molter et al. [26], who show that the problem becomes fixed-parameter tractable with respect to the number of delayed edges. Other work on temporal graph modification to optimise reachability has considered deleting edges or edge appearances [12,26] and assigning times to edges in a static graph [13].

Some related work exists on computing structures and properties which are robust to perturbation. Fuchsle et al. [19] ask if there is a path between two given vertices which is robust to delays. In network design, fault-tolerance describes a network's robustness to uncertainty [1,2,4,9,10,24]. Other models containing uncertainty allow for queries to find true values and aim to optimise the number of queries needed to find a solution [14–17].

1.1 Our Contributions

Motivated by the idea of uncertainty in temporal graphs, we work with perturbations of temporal graphs, and focus on the computational complexity of a problem that asks whether there exists a perturbation of an input temporal graph such that there exists a source vertex that has at least a minimum specified reachability. We:

- show that this problem is NP-hard in general, and `W[2]`-hard with respect to the number of edges perturbed (Sect. 2),
- show that the problem is tractable when the number of the perturbations allowed is at least one minus the number of vertices we aim to reach (Sect. 3), and
- show that the problem is tractable using a knapsack-based dynamic program when the underlying graph is a tree, or using a more complex tree decomposition-based algorithm when the underlying graph has bounded treewidth (Sect. 4).

To provide contrast with our results on the reachability problem, we briefly investigate (in Sect. 5) related eccentricity-based problems on temporal graphs with perturbations—in these problems we require not only that vertices are reached from a source, but that they are reached within a given number of edges (shortest eccentricity) or within a given travel duration (fastest eccentricity). In contrast to our findings for our reachability question, for fastest and shortest eccentricity we show that the problem remains hard when we are allowed to perturb all of the edges.

Due to space constraints, many proofs are postponed to an appendix. Statements of results with postponed proofs are indicated with a ($\star$).

1.2 Preliminaries

We give some notation and definitions, as well as including preliminary results relating to reachability in temporal graphs without perturbation.

A *temporal graph* is a pair (G, λ), where $G = (V, E)$ is an underlying (static) graph and $\lambda : E \to 2^{\mathbb{N}}$ is a time-labelling function which assigns to every edge of G a set of discrete-time labels. We call a pair (e, t) where $t \in \lambda(e)$ a *time-edge* and denote by $\mathcal{E}((G, \lambda))$ the set of all time-edges in (G, λ). When the graph in question is clear from context, we denote the number of vertices in the graph $|V(G)|$ by n and the number of edges in the underlying graph $|E(G)|$ by m. We let $\tau(G, \lambda) := \max_{e \in E(G)} |\lambda(e)|$ denote the maximum number of time labels assigned to any one edge in (G, λ), and call this quantity *temporality* following Mertzios et al. [25]. The *maximum lifetime* of a temporal graph, denoted $T(G, \lambda)$, is the largest $t \in \mathbb{N}$ such that there exists an edge $e \in E(G)$ with $t \in \lambda(e)$.

To discuss reachability, we first need a notion of a temporal path. Here we work with *strict temporal paths*, where a *strict temporal path* from v_0 to v_ℓ in (G, λ) is a sequence of time-edges $((v_0v_1, t_1), (v_1v_2, t_2) \dots, (v_{\ell-1}v_\ell, t_\ell))$ where, for each i, $t_i \in \lambda(v_{i-1}v_i)$ and $t_1 < t_2 < \dots < t_\ell$. In this paper, all temporal graphs discussed are strict, and so we will simply refer to them as temporal paths for brevity. We say that the length of a temporal path $((v_0v_1, t_1), \dots, (v_{\ell-1}v_\ell, t_\ell))$ is ℓ. Given a temporal path $P{:=}((v_0v_1, t_1), \dots, (v_{\ell-1}v_\ell, t_\ell))$, for each $1 \leq i < \ell$ the path $P'{:=}((v_0v_1, t_1), \dots, (v_{i-1}v_i, t_i))$ is a *prefix* of P.

If there exists a temporal path from v_s to v_t in (G, λ), we say that v_t is *reachable* from v_s. For a given vertex v_s in a temporal graph (G, λ), we define $\text{reach}((G, \lambda), v_s)$ to be the set of all vertices reachable from v_s in (G, λ). The *maximum temporal reachability* of a graph $R_{\max}(G, \lambda){:=}\max_{v \in V(G)} |\text{reach}((G, \lambda), v)|$ is the cardinality of the largest reachability set in (G, λ).

We say that the *arrival time* of a path $P := ((v_0v_1, t_1), \dots, (v_{\ell-1}v_\ell, t_\ell))$ is t_ℓ. A path P is *foremost* from a set of paths $\mathcal{P}$ if there is no other path $P' \in \mathcal{P}$ with a strictly earlier arrival time. We will write $t_{\text{Fo}}((G, \lambda), v_s, v_t)$ to be the arrival time of a foremost temporal path from v_s to v_t, where $t_{\text{Fo}}((G, \lambda), v_s, v_s) := 0$ and, if there is no temporal path from v_s to v_t, we define $t_{\text{Fo}}((G, \lambda), v_s, v_t){:=}\infty$.

Bui-Xuan et al. [29] introduce a notion that we call a *ubiquitous foremost path* (they refer to this as a ubiquitous foremost journey). Let u and v be two vertices in a temporal graph (G, λ). A foremost temporal path from u to v is a *ubiquitous foremost path* if all of its prefixes are themselves foremost temporal paths. Given a source vertex v_s in a temporal graph (G, λ), a *ubiquitous foremost temporal path tree* on v_s is defined to be a temporal tree (G_T, λ_T) with the following properties:

- G_T is a subtree of G,
- for each $e \in E(G_T)$, $\lambda_T(e) \subseteq \lambda(e)$ with $|\lambda_T(e)| = 1$, and
- for each $v_t \in \text{reach}((G, \lambda), v_s)$, (G_T, λ_T) contains a ubiquitous foremost temporal path from v_s to v_t that arrives at the same time as a ubiquitous foremost temporal path from v_s to v_t in (G, λ).

In particular, note that while there may be multiple ubiquitous foremost temporal paths from v_s to v_t in (G, λ), (G_T, λ_T) will contain exactly one ubiquitous foremost temporal path from v_s to v_t, and the arrival time of a foremost temporal path from v_s to any v_t can be trivially inferred as the earliest active time over all temporal edges incident to v_t. We note that a ubiquitous foremost temporal path tree can be computed efficiently.

Theorem 1.1 (Theorem 2 in [29]). *Given a temporal graph (G, λ) and a source vertex $v_s \in V(G)$, a ubiquitous foremost temporal path tree on v_s can be calculated in $O(m(\log n + \log \tau(G, \lambda)))$ time.*

By applying Theorem 1.1 for each source vertex v_s in $V(G)$, we get the following corollary.

Corollary 1.1. *Given a temporal graph (G, λ) we can calculate $R_{\max}$ in $O(nm(\log n + \log \tau(G, \lambda)))$ time.*

Note that we can reduce the complexity of this to $O(n(n+m))$ by using Algorithm 1 of Wu et al. [28] which calculates, for a given source v_s, foremost arrival times (but not actual temporal paths) for each vertex v_t in $O(n+m)$ time. This, however, does need the input data to be in a specified streaming format which allows their algorithm to calculate foremost arrival times in one pass. For our purposes, knowing foremost arrival times is not as useful as knowing actual foremost temporal paths, so we omit further details and instead refer the reader to [28].

1.3 Problems Considered

Here we introduce the notion of perturbing the appearances of time-edges. To that end, we define two kinds of perturbation as follows.

Definition 1.1. A temporal assignment λ' is a *δ-perturbation* of λ if there is a bijection f from $\mathcal{E}(G,\lambda)$ to $E(G) \times [1,\dots,T(G,\lambda)+\delta]$ such that, for all time-edges, $f((e,t)) = (e,t')$ where t' is an integer in the interval $[\max(1,t-\delta),t+\delta]$. Here we say λ has been *perturbed* by δ, or λ' is a δ-perturbation of λ. We also refer to any time-edge such that $(e,t) \neq f(e,t)$ as perturbed from λ to λ'.

$$|\{(e,t) \mid (e,t) \in \mathcal{E}(G,\lambda) \wedge (e,t) \notin \mathcal{E}(G,\lambda')\}|.$$

Definition 1.2. A temporal assignment λ' is a *(δ,ζ)-perturbation* of λ if it is a δ-perturbation of λ and $|\{(e,t) \mid (e,t) \text{ perturbed from } \lambda \text{ to } \lambda'\}| \leq \zeta$ where f is a bijection from $\mathcal{E}(G,\lambda)$ to $E(G) \times [1,\dots,T(G,\lambda)+\delta]$ such that, for all time-edges, $f((e,t)) = (e,t')$ where t' is an integer in the interval $[\max(1,t-\delta),t+\delta]$.

We can now define the main problem we consider in this paper.

Temporal Reachability with Limited Perturbation (TRLP)
Input: A temporal graph (G,λ) and positive integers ζ, $h \leq n$, and δ.
Question: Is there a (δ,ζ)-perturbation λ' of λ such that $R_{\max}(G,\lambda') \geq h$?

We also consider a straightforward special case of this problem, in which the number of perturbations is unrestricted.

Temporal Reachability with Perturbation (TRP)
Input: A temporal graph (G,λ) and positive integers $h \leq n$ and δ.
Question: Is there a δ-perturbation λ' of λ such that $R_{\max}(G,\lambda') \geq h$?

2 Intractability of TRLP

We begin by stating the main result of this section.

Theorem 2.1 **(⋆).** *TRLP is* NP*-complete. Moreover:*

- *For some $\alpha > 0$ it is* NP*-hard to approximate to within a factor $\alpha \log n$ the minimum number ζ of δ-perturbations required to transform the input temporal graph (G, λ) into a temporal graph with maximum reachability at least h.*
- *TRLP is* W[2]*-hard parameterised by ζ, the number of edges that may be perturbed.*
- *Assuming ETH, there is no $f(\zeta)n^{o(\zeta)}$-time algorithm for TRLP for any computable function f.*

We complement this with an algorithmic result which shows that our ETH lower bound is close to best possible. This uses an algorithm that considers every possible subset of edges of size ζ, and then for each such subset performs a Dijkstra-like exploration of a temporal graph using any edge in the selected subset as early as possible. Repeatedly calling this algorithm for each possible source gives us the desired result.

Theorem 2.2 **(⋆).** *TRLP is solvable in $O\left(n^{2\zeta+3}\log(\tau(G,\lambda))\right)$ time.*

Theorem 2.1 is proved by means of a reduction from Dominating Set, using the following construction.

Construction 2.1. *Given a graph G, let $\mathfrak{C}(G) = (G', \lambda)$ be the temporal graph constructed as follows. Create a vertex v_s in $V(G')$, and for each vertex $v_i \in V(G)$ create two vertices v_i^1 and v_i^2 in $V(G')$. For each $v_i \in V(G)$, add the edges $v_s v_i^1$ and $v_i^1 v_i^2$ to (G'). Then, for each edge $v_i v_j \in E(G)$, add the edges $v_i^1 v_j^2$ and $v_j^1 v_i^2$ to (G'). Let $\lambda(e) = 2$ for each $e \in E(G')$. A sample construction is given in Fig. 1.*

It is clear that we can construct $\mathfrak{C}(G)$ in time polynomial in the size of G. We show in the appendix that, given an instance (G, r) of Dominating Set with $r \geq 1$, G has a dominating set of size r if and only if $(\mathfrak{C}(G), 1, r, |V(G')|)$ is a yes-instance of TRLP.

3 Tractability with Many Perturbations

In this section we show that allowing many perturbations makes our problem easier to solve. In particular, we prove that TRP is solvable in polynomial time, and also that TRLP is solvable in polynomial time if the number ζ of allowed perturbations is at least the target reachability.

As a warm-up, we begin by considering the case in which each perturbation may also be very large: suppose δ is at least the maximum of the lifetime of (G, λ) and the diameter of G. We claim that in this case, if h is the target reachability

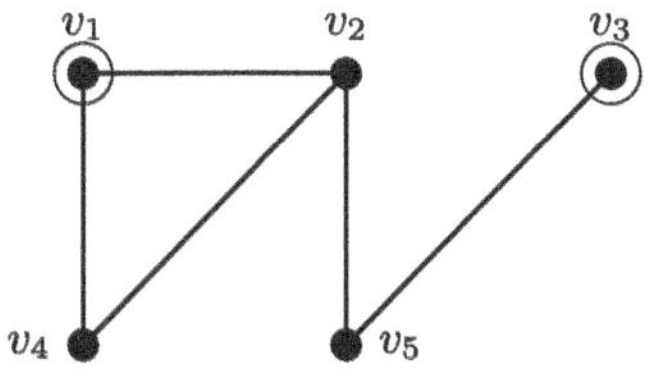

(a) The original graph. The circled vertices form a dominating set.

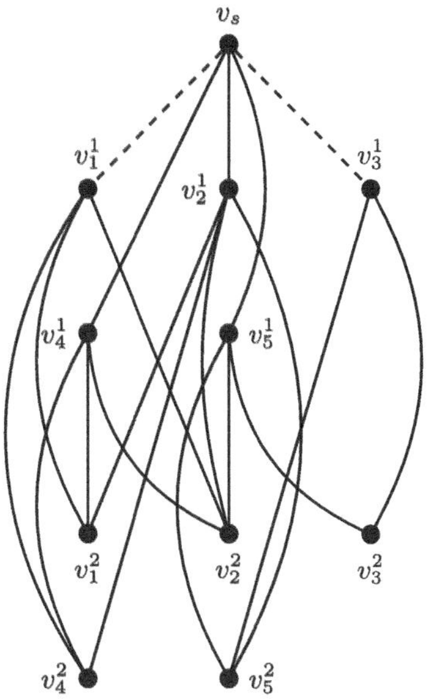

(b) The constructed temporal graph. For readability, times are not listed, but every edge is active only at time 2. The dashed edges are perturbed to time 1 to give a perturbation of the construction where v_s reaches every vertex.

Fig. 1. Diagram of the construction from Construction 2.1

and $\zeta \geq h - 1$, we have a yes-instance. To see this, fix an arbitrary subtree H of G with exactly h vertices (noting that such a subtree contains $h - 1 \leq \zeta$ edges, and if no such tree exists we have a no-instance of TRP), and further pick an arbitrary vertex $r \in V(H)$ which we shall call the root. The permitted perturbations are large enough that we can perturb one appearance of each edge in H so that, for each leaf ℓ of H, the times of these edge appearances are strictly increasing along the path from r to ℓ. It follows that, in the perturbed graph, r has reachability at least h, as required.

The first step towards our tractability results is to show that, given a source vertex in a temporal graph, and an integer δ, we can identify a single δ-perturbation that minimises foremost arrival times to all target vertices: one perturbation achieves such a result exactly because a ubiquitous foremost temporal path tree is a tree with edges active at exactly one time step. Recall that we denote the arrival time of a foremost temporal path from v_s to v_t in (G, λ) by $t_{\text{Fo}}((G, \lambda), v_s, v_t)$.

Lemma 3.1 (⋆). *Given a temporal graph* (G, λ)*, vertex* $v_s \in V(G)$*, and integer* δ*, we can identify in* $O(m(\log n + \log \delta\tau(G, \lambda)))$ *time a* δ*-perturbation* λ' *such that for each* $v_t \in V(G)$*,* $t_{Fo}((G, \lambda'), v_s, v_t) \leq t_{Fo}((G, \lambda''), v_s, v_t)$ *for all* δ*-perturbations* λ''*.*

We use this result to show the following, full details are given in the appendix. Recall that $\tau(G, \lambda) = \max_{e \in E(G)} |\lambda(e)|$.

Theorem 3.1 (⋆). *We can solve TRP in* $O(nm(\log n + \log \delta\tau(G, \lambda)))$ *time.*

Now turning our attention to TRLP, we start with some further notation. To consider a perturbation of fewest time-edges such that the maximum reachability is at least h, we introduce the following notion of worthwhile perturbations.

Definition 3.1. A perturbation λ' of λ is *worthwhile* if reverting any time-edge that has been perturbed from λ to λ' back to the time assigned by λ decreases the maximum reachability of the temporal graph. We call any time-edge (e, t) of (G, λ') worthwhile if λ' is a worthwhile perturbation of λ, $t \in \lambda'(e)$ and $t \notin \lambda(e)$.

We prove the following results about worthwhile perturbations in the appendix.

Lemma 3.2 (⋆). *Let* λ' *be a perturbation of* (G, λ) *such that the maximum reachability of* (G, λ') *is* h*. Suppose there is no* λ'' *which perturbs fewer edges with maximum reachability of* (G, λ'') *at least* h*. Then* λ' *is a worthwhile perturbation of* λ*.*

Lemma 3.3 (⋆). *Suppose that* λ' *is a worthwhile perturbation of* (G, λ)*. Then, for every vertex* v *with maximum reachability in* (G, λ')*, the subgraph induced by* $reach((G, \lambda'), v)$ *contains all edges perturbed by* λ'*.*

Lemma 3.4 (⋆). *The perturbed time-edges in a worthwhile perturbation* λ' *of* (G, λ) *form a forest.*

We can use the above lemmas to show that if the budget of perturbations is large enough, TRP and TRLP are equivalent problems.

Lemma 3.5 (⋆). *If* $\zeta \geq h - 1$*, there is a solution for TRLP on* (G, λ) *and integers* h*,* δ *if and only if there is a solution for TRP on the same temporal graph with* h *and* δ*.*

Combining Theorem 3.1 and Lemma 3.5 gives us the following.

Theorem 3.2. *We can solve TRLP in* $O(nm(\log n + \log \delta\tau(G, \lambda)))$ *time when* $\zeta \geq h - 1$*.*

4 Structural Restrictions on the Underlying Graph

We now consider TRLP when the underlying graph G has restricted structure: in particular when G is a tree, or when has bounded treewidth. When the input underlying graph is a tree, we solve TRLP by a dynamic program which uses Multiple Choice Knapsack to choose the perturbations which maximise the reachability from the vertex we are considering in time $O(n^2\zeta^3(T(G,\lambda)+\delta))$. When G has constant bounded treewidth and the lifetime and size of perturbations allowed are also bounded, we use a dynamic program over a tree decomposition to solve TRLP in polynomial time.

4.1 Knapsack-Based Dynamic Program for Trees

The formal definition of the knapsack problem is as follows:

Multiple Choice Knapsack Problem (MCKP)
Input: An integer c and q mutually disjoint classes $N_1,\ldots,N_q$ of elements such that each element $j \in N_i$ has a profit p_{ij} and weight w_{ij}.
Output: A function $\phi : \bigcup_i N_i \to \{0,1\}$ such that $\sum_{j\in N_i}\phi(j)=1$ for all N_i, $\sum_{i=1}^{q}\sum_{j\in N_i} w_{ij}\phi(j) \leq c$, and $\sum_{i=1}^{q}\sum_{j\in N_i} p_{ij}\phi(j)$ is maximised.

Dudziński and Krzysztof [11] show that this can be solved by a dynamic program in time $O(c\sum_{i=1}^{q} n_i) = O(nc)$ where n_i is the size of the class N_i and n is the total number of elements. In our case, the size of each class is bounded by ζ.

We solve TRLP by a bottom-up dynamic programming argument. The algorithm checks whether there is a (δ,ζ)-perturbation λ' such that a given source vertex v_s has reachability at least h. We root the tree at the source, giving us an orientation so that we can refer to the parents and children of vertices. Since we are asking if there exists a (δ,ζ)-perturbation such that $R_{\max}(G,\lambda) \geq h$, we can repeatedly run the algorithm for different choices of source vertex. Once we reach the root, if there exists a state which corresponds to reachability at least h, we accept. Otherwise, we reject and try another source vertex.

For any vertex v in the rooted tree G, we define a state as the triple (ζ_v, r_v, t_v) where $0 \leq \zeta_v \leq \zeta$, $0 < r_v \leq h$ and $0 \leq t \leq T(G,\lambda)+\delta$. We say that a state (ζ_v, r_v, t_v) is *valid* if and only if there exists a (δ,ζ_v)-perturbation of the temporal subtree G_v rooted at v where v reaches r_v vertices after time t_v. By "after time t_v" we mean that the vertices are reachable by a temporal path which departs v at time t_v at the earliest. We say that the state is *supported* by this perturbation. We further say that a state (ζ_v, r_v, t_v) is *maximally valid* if, for every other valid state (ζ_v, r', t_v) of v, we have $r' < r_v$. Given all maximally valid states of the children of a vertex v, for each time t we can calculate all maximally valid states (ζ_v, r_v, t) of v in $O(d_c\zeta^2)$ time where d_c is the number of children of v. To do this we use the algorithm of Dudziński and Krzysztof for MCKP [11].

The algorithm starts by assigning every leaf the set of states $\{(0,1,t) \colon t \in [0, T(G,\lambda)+\delta]\}$. For any non-leaf vertex v and every time $t \in [0, T(G,\lambda)+\delta]$, we check if the state (ζ_v, r_v, t) is valid and then easily remove states that are not maximally valid. The intuition behind our method is that we use MCKP to find an allocation of perturbations to the subtrees rooted at each child of v which maximises the reachability of v.

Theorem 4.1 (⋆). *TRLP is solvable in $O(n^2\zeta^3(T(G,\lambda)+\delta))$ time on temporal graphs where the underlying graph G is a tree.*

4.2 Tree Decomposition Algorithm

In this section, we describe an algorithm that, given a nice tree decomposition of G, solves TRLP on G. Our strategy is to perform dynamic programming over a nice tree decomposition (see the appendix for the formal definition). Without loss of generality, we assume that the root of the tree decomposition is a bag containing only a source vertex x. The algorithm finds the maximum reachability of the source vertex under any (δ, ζ)-perturbation. To find the maximum reachability of the graph under any (δ, ζ)-perturbation, we can run the algorithm with each possible source vertex. The dynamic program operates by computing a set of states for each bag. Calculation of states requires us to restrict departure and arrival times of temporal paths. Recall that the *arrival time* of a temporal path is the time of the last time-edge in the path. Similarly, the time at which a temporal path *departs* is the time t if first time-edge in the path occurs at $t' \geq t$. If there is no temporal path from a vertex v to a vertex u, we use the convention that the foremost arrival time of a path from v to u is ∞. Our states record:

- the times at which edges in the bag are active following perturbation;
- for each vertex v and time t, the foremost arrival time of a path in the subgraph induced by vertices introduced in the subtree rooted at s (G_s) to each vertex in the bag from v which departs at time t;
- for each set of vertices where each vertex v has a specified (possibly different) departure time t, the total number of distinct vertices reached below the bag by vertices by a path departing each v at time t; and
- the number of edges perturbed below the bag.

Theorem 4.2 (⋆). *Given a temporal graph (G, λ) where the treewidth of G is at most ω, we can solve TRLP in time*

$$O\left(n\omega^4\zeta^3(T(G,\lambda)+\delta)^\omega) \cdot (2\delta h T(G,\lambda))^{4\omega^2(T(G,\lambda)+\delta)^\omega}\right).$$

In particular, we can solve TRLP in polynomial time when the treewidth of the underlying graph, the permitted perturbation size δ and the lifetime of (G, λ) are all bounded by constants.

5 Eccentricity

We now consider two closely related problems under the same model of perturbation. Instead of asking about the number of reachable vertices, we are interested in whether it is possible to reach *all* vertices in the graph by paths whose length or duration is at most some upper bound. It is unsurprising that the problem of determining whether there exists a set of perturbations achieving either goal is NP-complete; in contrast to TRLP, however, these problems do not become easier when we allow an appearance of every edge to be perturbed. If all these perturbations can also be sufficiently large we do still recover tractability.

We have previously discussed strict temporal paths, as well as their lengths. We will use the terms *shortest* and *longest* to be specifically in relation to the number of edges in the path. However, there are other measures of a temporal path that may be relevant under various real-world scenarios. Earlier, we defined foremost paths and the arrival time of a foremost path. In addition, we say that the *duration* of a path $P := ((v_0v_1, t_1), \ldots, (v_{\ell-1}v_\ell, t_\ell))$ is $t_\ell - t_1$. A path P is *fastest* of a set of paths $\mathcal{P}$ if there is no other path $P' \in \mathcal{P}$ of strictly smaller duration.

We begin with formal definitions of the problems we are considering. In both definitions, we are given a temporal graph (G, λ) and a source vertex v_s. This vertex v_s has a *temporal shortest eccentricity* of at most k if, for every vertex $v_t \in V(G)$, a shortest temporal path from v_s to v_t contains at most k edges. The vertex v_s has a *temporal fastest eccentricity* of at most k if, for every vertex $v_t \in V(G)$, a fastest temporal path from v_s to v_t has a duration of at most k. We use the notation $\mathrm{ecc}^{\mathrm{S}}((G, \lambda), v)$ (respectively $\mathrm{ecc}^{\mathrm{Fa}}((G, \lambda), v)$) to mean the temporal shortest (respectively fastest) eccentricity of v in the temporal graph (G, λ). We note that, given a temporal graph (G, λ) and a vertex $v \in V(G)$, both $\mathrm{ecc}^{\mathrm{S}}((G, \lambda), v)$ and $\mathrm{ecc}^{\mathrm{Fa}}((G, \lambda), v)$ can be computed in polynomial time [29].

We now define the problems of determining whether there is a perturbation achieving some given shortest eccentricity; Temporal fastest eccentricity under perturbation (TFaEP) is defined analogously.

Temporal shortest eccentricity under perturbation (TSEP)
Input: A temporal graph (G, λ), source vertex $v_s \in V(G)$, and integers k, δ, and ζ.
Question: Is there a (δ, ζ)-perturbation λ' of λ with $\mathrm{ecc}^{\mathrm{S}}((G, \lambda'), v_s) \leq k$?

Here we have considered two of the three standard notions of optimality for temporal paths. One could naturally define a third problem based on temporal *foremost* eccentricity (requiring the earliest arrival time of a path from v to every other vertex to be at most k), but this problem has previously been studied by Deligkas et al. [7] under another name. The problem ReachFast(k) is to minimise the foremost eccentricity of some *set* S of sources (i.e. minimise the maximum eccentricity over all $s \in S$) when at most k edges can be perturbed, while ReachFastTotal(k) asks the same question when the sum of the sizes of all perturbations must be bounded by k. The behaviour of these problems is similar to those of TSEP and TFaEP: they are intractable in general, but can be solved in polynomial time if both the number and size of the perturbations are unrestricted.

We begin by observing that TSEP and TFaEP are both NP-complete. Inclusion in NP is trivial (the set of perturbations acts as a certificate), and NP-hardness – as well as W[2]-hardness parameterised by ζ and inapproximability (where the goal is to find the smallest number of perturbations required) – follows from the proof of Theorem 2.1.

Recall that TRLP became polynomial-time solvable when we were allowed to perturb at least one appearance of every edge. In contrast with this result, we now show that both TFaEP and TSEP remain NP-hard even when an appearance of every edge can be perturbed, for any constant δ, and for constant values of k. Note that this is also a contrast to the behaviour for foremost eccentricity: Deligkas et al. [7, Algorithm 1] show that this problem is efficiently solvable when ζ is large. We give analogous hardness results for TSEP and TFaEP via reductions from SAT.

Theorem 5.1 ($\star$). *For any integers $k \geq 4$ and any $\delta \geq 1$, TSEP is NP-hard even if $\zeta = m$.*

Theorem 5.2 ($\star$). *For any integers $k \geq 2$ and any $\delta \geq 1$, TFaEP is NP-hard even if $\zeta = m$.*

We conclude this section with one positive result for TSEP and TFaEP. Recall that TRLP becomes trivial when both ζ and δ are sufficiently large (every instance is a yes-instance). While neither TSEP nor TFaEP becomes trivial under the same conditions, both problems can be solved in polynomial time when both ζ and δ are large.

Theorem 5.3 ($\star$). *Each of TSEP and TFaEP can be solved in time $O(m(\log n + \log \delta_T(G,\lambda)))$ on instances $((G,\lambda), v_s, k, \delta, \zeta)$ if $\delta \geq T(G,\lambda)$ and $\zeta \geq n-1$.*

6 Conclusion and Open Questions

Motivated by the uncertainty and errors intrinsic in recording real-world temporal graphs, we have investigated temporal graph reachability problems under perturbation, in particular asking whether there exists a perturbation of time-edges of a temporal graph that results in a minimum required reachability. We find that when arbitrarily many perturbations are allowed, this can be resolved efficiently, but that the problem is, in general, NP-hard and $W[2]$-hard with respect to ζ. Furthermore, under the assumption of ETH, we show that there is no $f(\zeta)n^{o(\zeta)}$-time algorithm for our problem. We have tractability results when the underlying graph is a tree or when it has bounded treewidth and the lifetime and size of permitted perturbations are also bounded by constants. For contrast, we show that allowing arbitrarily many perturbations is not sufficient to give tractability of related eccentricity problems, where the question is not just whether there is some source vertex that reaches some specified number of vertices, but whether it reaches vertices within some travel duration (fastest) or in few steps (shortest).

An obvious avenue for further research is on other temporal graph problems with the same style of perturbations. However, inspired by the motivation to make temporal algorithms more suitable for real-world temporal contact data, we suggest exploring temporal graph modification problems that are robust to perturbations: for example, removing (by, e.g. vaccination) vertices from a temporal network to limit reachability in a way that has guarantees under possible bounds of perturbation.

Acknowldgement. Jessica Enright, Kitty Meeks, and William Pettersson are supported by EPSRC grant EP/T004878/1. William Pettersson is additionally supported by EPSRC grant EP/X013618/1.

Disclosure of Interest. The authors have no competing interests.

References

1. Baswana, S., Choudhary, K., Roditty, L.: Fault tolerant subgraph for single source reachability: generic and optimal. In: Proceedings of the Forty-Eighth Annual ACM Symposium on Theory of Computing. STOC '16, pp. 509–518. Association for Computing Machinery, New York, NY, USA (2016). https://doi.org/10.1145/2897518.2897648,
2. Biló, D., D'Angelo, G., Gualá, L., Leucci, S., Rossi, M.: Blackout-tolerant temporal spanners. In: Erlebach, T., Segal, M. (eds.) ALGOSENSORS 2022. LNCS, vol. 13707, pp. 31–44. Springer, Cham (2022). https://doi.org/10.1007/978-3-031-22050-0_3
3. Casteigts, A.: A journey through dynamic networks (with excursions). Thesis, Université de Bordeaux (2018). https://hal.science/tel-01883384
4. Casteigts, A., Dubois, S., Petit, F., Robson, J.M.: Robustness: a new form of heredity motivated by dynamic networks. Theor. Comput. Sci. **806**, 429–445 (2020). https://doi.org/10.1016/j.tcs.2019.08.008, https://www.sciencedirect.com/science/article/pii/S0304397519304979
5. Chaters, G.L., et al.: Analysing livestock network data for infectious disease control: an argument for routine data collection in emerging economies. Philos. Trans. Roy. Soc. B: Biol. Sci. **374**(1776), 20180264 (2019) https://doi.org/10.1098/rstb.2018.0264, https://royalsocietypublishing.org/doi/full/10.1098/rstb.2018.0264, publisher: Royal Society
6. Deligkas, A., Döring, M., Eiben, E., Goldsmith, T.L., Skretas, G.: Being an influencer is hard: the complexity of influence maximization in temporal graphs with a fixed source (2023). arXiv:2303.11703 [cs]
7. Deligkas, A., Eiben, E., Skretas, G.: Minimizing reachability times on temporal graphs via shifting labels. In: Proceedings of the Thirty-Second International Joint Conference on Artificial Intelligence. IJCAI '23, pp. 5333–5340 (2023). https://doi.org/10.24963/ijcai.2023/592,

8. Deligkas, A., Potapov, I.: Optimizing reachability sets in temporal graphs by delaying. Inf. Comput. **285**, 104890 (2022). https://doi.org/10.1016/j.ic.2022.104890, https://www.sciencedirect.com/science/article/pii/S0890540122000323
9. Dinitz, M., Krauthgamer, R.: Fault-tolerant spanners: better and simpler. In: Proceedings of the 30th Annual ACM SIGACT-SIGOPS Symposium on Principles of Distributed Computing. PODC '11, pp. 169–178. Association for Computing Machinery, New York, NY, USA (2011). https://doi.org/10.1145/1993806.1993830
10. Dubois, S., Feuilloley, L., Petit, F., Rabie, M.: When should you wait before updating? - toward a robustness refinement. In: Doty, D., Spirakis, P. (eds.) 2nd Symposium on Algorithmic Foundations of Dynamic Networks (SAND 2023). Leibniz International Proceedings in Informatics (LIPIcs), vol. 257, pp. 7:1–7:15. Schloss Dagstuhl- Leibniz-Zentrum für Informatik, Dagstuhl, Germany (2023). https://doi.org/10.4230/LIPIcs.SAND.2023.7, https://drops.dagstuhl.de/opus/volltexte/2023/17943, ISSN 1868-8969
11. Dudzinski, K., Walukiewicz, S.: Exact methods for the knapsack problem and its generalizations. Eur. J. Oper. Res. **28**(1), 3–21 (1987). https://doi.org/10.1016/0377-2217(87)90165-2, https://www.sciencedirect.com/science/article/pii/0377221787901652
12. Enright, J., Meeks, K., Mertzios, G.B., Zamaraev, V.: Deleting edges to restrict the size of an epidemic in temporal networks. J. Comput. Syst. Sci. **119**, 60–77 (2021). https://doi.org/10.1016/j.jcss.2021.01.007, https://www.sciencedirect.com/science/article/pii/S0022000021000155
13. Enright, J., Meeks, K., Skerman, F.: Assigning times to minimise reachability in temporal graphs. J. Comput. Syst. Sci. **115**, 169–186 (2021). https://doi.org/10.1016/j.jcss.2020.08.001, https://www.sciencedirect.com/science/article/pii/S0022000020300799
14. Erlebach, T., Hoffmann, M.: Minimum spanning tree verification under uncertainty. In: Kratsch, D., Todinca, I. (eds.) WG 2014. LNCS, vol. 8747, pp. 164–175. Springer, Cham (2014). https://doi.org/10.1007/978-3-319-12340-0_14
15. Erlebach, T., Hoffmann, M., Krizanc, D., Mihal'ák, M., Raman, R.: Computing minimum spanning trees with uncertainty (2008). https://doi.org/10.48550/arXiv.0802.2855, arXiv:0802.2855 [cs]
16. Erlebach, T., Hoffmann, M., de Lima, M.S.: Round-competitive algorithms for uncertainty problems with parallel queries. Algorithmica **85**(2), 406–443 (2022). https://doi.org/10.1007/s00453-022-01035-6
17. Erlebach, T., de Lima, M.S., Megow, N., Schlöter, J.: Learning-augmented query policies for minimum spanning tree with uncertainty. In: Chechik, S., Navarro, G., Rotenberg, E., Herman, G. (eds.) 30th Annual European Symposium on Algorithms (ESA 2022). Leibniz International Proceedings in Informatics (LIPIcs), vol. 244, pp. 49:1–49:18. Schloss Dagstuhl- Leibniz-Zentrum für Informatik, Dagstuhl, Germany (2022). https://doi.org/10.4230/LIPIcs.ESA.2022.49, https://drops.dagstuhl.de/opus/volltexte/2022/16987, ISSN 1868-8969
18. Fielding, H.R., McKinley, T.J., Silk, M.J., Delahay, R.J., McDonald, R.A.: Contact chains of cattle farms in Great Britain. Roy. Soc. Open Sci. **6**, 180719 (2019). https://royalsocietypublishing.org/doi/full/10.1098/rsos.180719
19. Füchsle, E., Molter, H., Niedermeier, R., Renken, M.: Delay-Robust Routes in Temporal Graphs: 39th International Symposium on Theoretical Aspects of Computer Science, STACS 2022. 39th International Symposium on Theoretical Aspects of Computer Science. STACS 2022 (2022). https://doi.org/10.4230/LIPIcs.STACS.2022.30, http://www.scopus.com/inward/record.url?

scp=85126186124&partnerID=8YFLogxK, publisher: Schloss Dagstuhl- Leibniz-Zentrum fur Informatik GmbH, Dagstuhl Publishing
20. Hand, S.D., Enright, J., Meeks, K.: Making life more confusing for firefighters (2022). https://eprints.gla.ac.uk/269008/, Conference Name: 11th International Conference on Fun with Algorithms (FUN 2022) ISBN 9783959772327, ISSN 1868-8969 Meeting Name: 11th International Conference on Fun with Algorithms (FUN 2022) Place: Sicily, Italy
21. Holme, P.: Network reachability of real-world contact sequences. Phys. Rev. E **71**(4), 046119 (2005). https://journals.aps.org/pre/abstract/10.1103/PhysRevE.71.046119
22. Kempe, D., Kleinberg, J., Kumar, A.: Connectivity and inference problems for temporal networks. J. Comput. Syst. Sci. **64**(4), 820–842 (2002). https://doi.org/10.1006/jcss.2002.1829
23. Kutner, D.C., Larios-Jones, L.: Temporal reachability dominating sets: contagion in-temporal graphs. In: Georgiou, K., Kranakis, E. (eds.) ALGOWIN 2023. LNCS, vol. 14061, pp. 101–116. Springer, Cham (2023). https://doi.org/10.1007/978-3-031-48882-5_8
24. Lochet, W., Lokshtanov, D., Misra, P., Saurabh, S., Sharma, R., Zehavi, M.: Fault tolerant subgraphs with applications in kernelization. In: Vidick, T. (ed.) 11th Innovations in Theoretical Computer Science Conference (ITCS 2020). Leibniz International Proceedings in Informatics (LIPIcs), vol. 151, pp. 47:1–47:22. Schloss Dagstuhl-Leibniz-Zentrum fuer Informatik, Dagstuhl, Germany (2020). https://doi.org/10.4230/LIPIcs.ITCS.2020.47, https://drops.dagstuhl.de/opus/volltexte/2020/11732, ISSN 1868-8969
25. Mertzios, G.B., Michail, O., Spirakis, P.G.: Temporal network optimization subject to connectivity constraints. Algorithmica **81**(4), 1416–1449 (2019). https://doi.org/10.1007/s00453-018-0478-6
26. Molter, H., Renken, M., Zschoche, P.: Temporal reachability minimization: delaying vs. deleting. In: Bonchi, F., Puglisi, S.J. (eds.) 46th International Symposium on Mathematical Foundations of Computer Science (MFCS 2021). Leibniz International Proceedings in Informatics (LIPIcs), vol. 202, pp. 76:1–76:15. Schloss Dagstuhl- Leibniz-Zentrum für Informatik, Dagstuhl, Germany (2021). https://doi.org/10.4230/LIPIcs.MFCS.2021.76, https://drops.dagstuhl.de/opus/volltexte/2021/14516, ISSN 1868-8969
27. Myall, A., et al.: Characterising contact in disease outbreaks via a network model of spatial-temporal proximity (2021). https://doi.org/10.1101/2021.04.07.21254497, https://www.medrxiv.org/content/10.1101/2021.04.07.21254497v1, pages: 2021.04.07.21254497
28. Wu, H., Cheng, J., Huang, S., Ke, Y., Lu, Y., Xu, Y.: Path problems in temporal graphs. Proc. VLDB Endow. **7**(9), 721–732 (2014). https://doi.org/10.14778/2732939.2732945,
29. Xuan, B.B., Ferreira, A., Jarry, A.: Computing shortest, fastest, and foremost journeys in dynamic networks. Int. J. Found. Comput. Sci. **14**(02), 267–285 (2003). https://doi.org/10.1142/S0129054103001728, https://www.worldscientific.com/doi/abs/10.1142/S0129054103001728, Publisher: World Scientific Publishing Co

On Computational Completeness of Semi-Conditional Matrix Grammars

Henning Fernau[1], Lakshmanan Kuppusamy[2(✉)], and Indhumathi Raman[3]

[1] Fachbereich 4 – Abteilung Informatikwissenschaften, Universität Trier, 54286 Trier, Germany
fernau@uni-trier.de

[2] School of Computer Science and Engineering, VIT University, Vellore 632014, India
klakshma@vit.ac.in

[3] Department of Computing Technologies, SRM Institute of Science and Technology, Kattankulathur, Chennai 603203, India
indhumar2@srmist.edu.in

Abstract. Matrix grammars are one of the first approaches ever proposed in regulated rewriting, prescribing that rules have to be applied in a certain order. In regulated rewriting, the most interesting case shows up when all rules are context-free. Typical descriptional complexity measures incorporate the number of nonterminals or the matrix length, i.e., the number of rules per matrix. When viewing matrices as program fragments, it becomes natural to consider additional applicability conditions for such matrices. Here, we focus on attaching a permitting and a forbidden string to every matrix in a matrix grammar. The matrix is applicable to a sentential form w only if the permitting string is a subword in w and the forbidden string is not a subword in w. We call such a grammar, where the application of a matrix is conditioned as described, a semi-conditional matrix grammar. We consider (1) the maximal lengths of permitting and forbidden strings, (2) the number of nonterminals, (3) the number of conditional matrices, (4) the maximal length of any matrix and (5) the number of conditional matrices with nonempty permitting and forbidden strings, as the resources (descriptional complexity measures) of a semi-conditional matrix grammar.

We show that certain semi-conditional matrix grammar families defined by restricting resources can generate all recursively enumerable languages.

Keywords: Semi-Conditional & MatrixGrammars · Descriptional Complexity

1 Introduction

A matrix grammar (originally introduced by S. Ábrahám on linguistic grounds in [1]) consists in sequences of context-free rules called matrices; when a matrix

R. Královič and V. Kůrková (Eds.): SOFSEM 2025, LNCS 15538, pp. 270–283, 2025.
https://doi.org/10.1007/978-3-031-82670-2_20

applied to a sentential form, all rules in the sequence are applied in the given order. It is often considered the most important variation of regulated rewriting. Matrix grammars with appearance checking, having three nonterminals, are computationally complete, i.e., they characterize RE [4]. However, the lengths of the matrices are unbounded. It is not clear how to restrict the matrix length while still bounding the number of nonterminals. It is known that matrix grammars without appearance checking are not computationally complete; see [9].

Descriptional complexity (within formal languages) focuses on the influence of syntactic parameters within automata or grammars. For matrix grammars, such parameters of interest could be (e.g.) the number of nonterminals or the number of matrices. For formalisms characterizing RE, it has always been asked how small certain parameters could be while still maintaining computational completeness, see [3] for a survey. We continue this line of research here.

In 1985, Gh. Păun introduced another variant of regulated context-free grammar called semi-conditional grammars [13] where a permitting and a forbidden string, associated to each (context-free) rule, govern the applicability of said context-free rule. In [13], the author also introduced a combination of matrix and semi-conditional grammars, called semi-conditional matrix grammars. To each matrix (containing sequences of possibly erasing context-free rules), a permitting string w_+ of length at most i and a forbidden string w_- of length at most j are associated, and the said matrix is only applicable to the sentential form w if w_+ is a substring of w and if w_- does not occur as a substring of w. The ordered pair (i, j) is called the *degree* of the semi-conditional grammar.

Our Contribution. In this paper, we consider (1) the degree (i, j), (2) the number n of nonterminals, (3) the number m of conditional matrices and (4) the maximal length ℓ of any matrix as descriptional complexity measures of semi-conditional matrix grammars and we denote its corresponding language class by $\mathrm{SCM}(i, j; n; m, \ell)$. Further, in every matrix, if either the permitting or the forbidden string is empty, then we call the semi-conditional grammar as *simple* and its corresponding language class as $\mathrm{SSCM}(i, j; n; m, \ell)$. In [12], it has been proved that $\mathrm{SSCM}(3, 1; 7; 2, 3) = \mathrm{RE}$. In this paper, we improve or complement this result by showing that each of $\mathrm{SSCM}(2, 1; 5; 3, 2)$, $\mathrm{SSCM}(3, 1; 5; 2, 2)$, $\mathrm{SSCM}(3, 1; 4; 3, 3)$ is equivalent to the class RE of recursively enumerable languages. We also show that certain semi-conditional matrix grammars with binary matrices (with matrix length at most 2) can generate RE if we relax on the simplicity condition of the matrices. Our results are surveyed in Table 1.

2 Definitions and Notation

We assume the reader to be familiar with basic notions of formal languages. For a word $w \in V^*$, we call $y \in V^*$ a *subword* (factor) of w if there are $x, z \in V^*$ such that $w = xyz$. Let $\mathrm{sub}(w)$ denote the set of all subwords of w.

Definition 1. *A semi-conditional matrix or* SCM*-grammar [13] is a quadruple* $G = (V, T, M, S)$*, where* V *is an alphabet,* $T \subsetneq V$ *contains the* terminals, $N{:=}V\setminus$

T contains the nonterminals, *and* $S \in N$ *is the* start symbol.
M *is a finite set of* matrices with context conditions *which are of the form* $[(A_1 \to x_1), \ldots, (A_\ell \to x_\ell), P, F]$, *where* $A_i \in N$ *and* $x_i \in V^*$, $P, F \in V^+ \cup \{\emptyset\}$.

Consider the matrix $r = [(A_1 \to x_1), \ldots, (A_\ell \to x_\ell), P, F]$, the sets P and F are called the *permitting* and *forbidden* conditions, respectively; ℓ is known as the *length* of r; $\emptyset \notin V$ is a special symbol, meaning that a condition is missing. If both $P = F = \emptyset$, then we term the matrix r as *unconditional*, otherwise, we call it *conditional*. If all matrices are unconditional, then the degree of G is $(0,0)$ using the standard property that $|\emptyset| = 0$. For brevity, then we simplify $[(A \to x), \emptyset, \emptyset]$ to $A \to x$ hereafter. If at least one condition equals $\emptyset$, then the matrix is termed *simple*. If all matrices are simple, then the SCM grammar itself is termed as a *simple SCM* grammar and is denoted as *SSCM grammar*.

Let $r = [(A_1 \to x_1, \ldots, A_\ell \to x_\ell), P, F] \in M$, with $x, y \in V^*$. Then $x \Rightarrow_r y$ (or simply $x \Rightarrow_G y$ or $x \Rightarrow y$) iff (1) $P \neq \emptyset$ implies $P \in \text{sub}(x)$ and (2) $F \neq \emptyset$ implies $F \notin \text{sub}(x)$, and (3) the matrix $[(A_1 \to x_1), \ldots, (A_\ell \to x_\ell)]$ is applied to x to get y. This means that there are sentential forms $x = y_0, y_1, \ldots, y_\ell = y$ such that y_i is obtained from y_{i-1} by replacing one occurrence of A_i in y_{i-1} by x_i, or, in other words, by applying the context-free rule $A_i \to x_i$ on y_{i-1}, for $i = 1, \ldots, \ell$. Let $\Rightarrow_G^*$ denote the transitive and reflexive closure of $\Rightarrow_G$. The language of G, denoted as $L(G)$, is defined as $L(G) = \{y \in T^* \mid S \Rightarrow_G^* y\}$.

An (S)SCM-grammar is said to be of *degree* (i,j), where $i, j \in \mathbb{N}$, if in every matrix rule $\{[(A_1 \to x_1), \ldots, (A_\ell \to x_\ell)], \alpha, \beta\}$ of M we have $|\alpha| \leq i$ and $|\beta| \leq j$. We denote by (S)SCM$(i, j; n; m, \ell)$, a family of languages generated by (S)SCM-grammars, where

- (i, j) is upper-bounding their degree,
- n is an upper bound on the number of nonterminals,
- m is an upper bound on the number of conditional matrices,
- ℓ is an upper bound on the number of rules in any matrix, i.e., upper-bounding the *length* of any matrix.

The only existing result in the domain of SSCM grammars is (see [12, Theorem 3]) that SSCM$(3, 1; 7; 2, 3)$ = RE. In fact, that result was stated a bit differently, because the authors called $\max(i, j)$ the degree of an SSCM-grammar.

It seems to be very difficult to find computational completeness results for SSCM-grammars with at most four nonterminals and matrix length at most two. As a restriction to binary matrices is (otherwise) a quite common normal form for matrix grammars, we now allow ourselves to drop the simplicity condition. We are not aware of any other descriptional complexity results for SCM grammars. In order to quantify of how much we violate the simplicity condition, we account for the number of non-simple matrices s as a sixth dimension in our notation. In this sense, we note that SSCM$(i, j; n; m, \ell)$ = SCM$(i, j; n; m, \ell, 0)$. In particular, we could refer to Theorem 1 in this new notation as SCM$(2, 1; 5; 3, 2, 0)$ = RE. This way of accounting for non-simple rules was introduced in [5]. The results of this paper are tabulated in Table 1. For comparison, we also include some results from the literature. If no permitting context conditions are present, we arrive at a degree of $(0, j)$, i.e., generalized forbidding matrix grammars [7,11].

Table 1. Results of this paper, plus predecessor results: $\mathrm{SCM}(i,j;n;m,l,s) = \mathrm{RE}$. The $*$ denotes that the respective parameter is unbounded

Degree (i,j)	# Nonterminals n	# Conditional Matrices m	Max. Matrix Length ℓ	# Non-Simple rules s	Reference
(3,1)	7	2	3	0	[12], Thm. 3
(2,1)	5	3	2	0	Thm. 1
(3,1)	5	2	2	0	Thm. 2
(3,1)	4	3	3	0	Thm. 3
(4,3)	4	7	2	6	Thm. 4
(5,2)	4	7	2	4	Thm. 5
(6,3)	4	7	2	3	Thm. 6
(6,3)	3	*	2	4	Thm. 7
(7,2)	3	*	2	3	Thm. 8
(0,2)	*	*	3	0	[13], Thm. 4.4
(0,1)	*	*	2	0	[7], Thm. 2
(0,*)	3	*	*	0	Thm. 9

3 Normal Forms for Type-0 Grammars

In [7,8,10], different normal forms for type-0 grammars have been described. They all present grammars that contain very few nonterminals and are still able to generate every RE language.

Proposition 1. *For each* RE *language* L, $L \subseteq T^*$, *there is a type-0 grammar of the form* $G = (V, T, P \cup \{AB \to \lambda, CD \to \lambda\}, S)$ *with* $L(G) = L$, *where* P *contains only context-free rules and* $N{:=}V \setminus T = \{S, A, B, C, D\}$. P*-rules are:*

1 $S \to uSa$ *with* $u \in \{A, C\}^+$ *and* $a \in T$,
2 $S \to uSv$ *or* $S \to uv$ *with* $u \in \{A, C\}^+$ *and* $v \in \{B, D\}^*$.

As this normal form uses five nonterminals and two non-context-free rules, we call it $(5,2)$-GNF for short. Derivations are performed in three stages:

- In Stage 1, a terminal suffix is created that will be (finally) the terminal word that is produced. Here, only context-free rules from Item 1 are used.
- In Stage 2, only context-free rules from Item 2 are employed.
- Finally in Stage 3, the non-context-free deletion rules come into play.

Sometimes Stages 1 and 2 are together addressed as Phase 1, so that Stage 3 is Phase 2. Unfortunately, it is also possible to mix the first two stages. Geffert [8] could prove that such mixtures can never lead to terminal strings, but when we simulate normal form grammars in the following, we must also consider the corresponding sentential forms. Any sentential form w derivable from S is from

$$L_{(5,2)}{:=}\{A, C\}^*\{S, \lambda\}(\{B, D\}^* \cup T)^* . \quad (1)$$

Based on $(5,2)$-GNF, other normal forms can be derived that basically differ in choosing different encodings of the nonterminal parts. For instance, by applying the morphism defined by $A \mapsto CAA$, $B \mapsto BBC$, $C \mapsto CA$, $D \mapsto BC$, $x \mapsto x$ for $x \in T \cup \{S\}$ to the right-hand side of the context-free rules, one arrives at:

Proposition 2. *For each* RE *language L, $L \subseteq T^*$, there is a type-0 grammar of the form $G = (V, T, P \cup \{AB \to \lambda, CC \to \lambda\}, S)$ with $L(G) = L$, where P contains only context-free rules and $N{:=}V \setminus T = \{S, A, B, C\}$. P-rules look like*

1. *$S \to uSa$ with $u \in \{CA, CAA\}^+$ and $a \in T$,*
2. *$S \to uSv$ or $S \to uv$ with $u \in \{CA, CAA\}^+$ and $v \in \{BC, BBC\}^*$.*

To differentiate from the previously stated normal form, we refer to this one as $(4,2)$-GNF for short. Sentential forms w of this type of grammar belong to:

$$L_{(4,2)}{:=}\{CA, CAA\}^*\{S, \lambda, CC\}(\{BC, BBC\}^* \cup T)^* . \quad (2)$$

Observe that as long as S is present (i.e., in Stages 1 or 2), the strings CC or AB cannot occur as subwords in any sentential form that is derivable from S. Moreover, BA is never a subsequence of any sentential form derivable from S.

Another popular normal form can be called $(3,2)$-GNF. It is derived from $(5,2)$-GNF by the morphism $A \mapsto ABB$, $B \mapsto BA$, $C \mapsto AB$, $D \mapsto BBA$, $x \mapsto x$ for $x \in T \cup \{S\}$ to the right-hand side of the context-free rules. This yields:

Proposition 3. *For each* RE *language L, $L \subseteq T^*$, there is a type-0 grammar of the form $G = (V, T, P \cup \{AA \to \lambda, BBB \to \lambda\}, S)$ with $L(G) = L$, where P contains only context-free rules and $N{:=}V \setminus T = \{S, A, B\}$.*

1. *$S \to uSa$ with $u \in \{ABB, AB\}^+$ and $a \in T$,*
2. *$S \to uSv$ or $S \to uv$ with $u \in \{ABB, AB\}^+$ and $v \in \{BA, BBA\}^*$.*

Sentential forms w of this type of grammar belong to:

$$L_{(3,2)}{:=}\{ABB, AB\}^*\{S, \lambda, AA, ABA\}(\{BA, BBA\}^* \cup T)^* . \quad (3)$$

Masopust and Meduna had a slightly different encoding idea in [10]: keeping a middle marker \$ in the string allows encoding with only two different symbols "elsewhere". Now, to the deletion rules $AB \to \lambda$ and $CD \to \lambda$, resp., there correspond *shrinking rules* $0\$0 \to \$$ and $1\$1 \to \$$ resp., and (only) finally $\$ \to \lambda$. We call the resulting normal form Masopust-Meduna normal form (MMNF).

Proposition 4. *For each* RE *language L, $L \subseteq T^*$, there is a type-0 grammar of the form $G = (V, T, P \cup \{0\$0 \to \$, 1\$1 \to \$, \$ \to \lambda\}, S)$ with $L(G) = L$, where P contains only context-free rules and $N{:=}V \setminus T = \{S, 0, 1, \$\}$. P-rules look like*

1. *$S \to uSa$ with $u \in \{0,1\}^+$ and $a \in T$,*
2. *$S \to uSv$ or $S \to u\$v$ with $u \in \{0,1\}^+$ and $v \in \{0,1\}^*$.*

Sentential forms w of this type of grammar belong to:

$$L_{\mathrm{MM}}{:=}\{0,1\}^*\{S,\lambda,\$\}(\{0,1\}^*\cup T)^*\,. \tag{4}$$

Other encodings are possible for this strategy, for instance, by requiring $u \in \{10, 100\}^+$ and $v \in \{01, 001\}^*$, one gets the property that the penultimate rule to be applied is $1\$1 \to \$$. Such encodings give additional structure, as actually used in this paper. To differentiate this from MMNF, let us call it *strong* MMNF (sMMNF). Sentential forms w of this type of grammar belong to:

$$L_{\mathrm{sMM}}{:=}\{10,100\}^*\{S,\lambda,\$\}(\{01,001\}^*\cup T)^*\,. \tag{5}$$

However, we are now giving a simplified version of what we called *modified MMNF* in [7] (or MMMNF for short) as follows.

Proposition 5. *For each* RE *language* L, $L \subseteq T^*$, *there is a type-0 grammar of the form* $G = (V, T, P \cup \{0\$1 \to \$, 1\$0 \to \$, \$ \to \lambda\}, S)$ *with* $L(G) = L$, *where* P *contains only context-free rules and* $N{:=}V \setminus T = \{S, 0, 1, \$\}$. *The rules in* P *are as in Propos. 4. Sentential forms of this type of grammar are in* L_{MM}.

4 Main Results for SSCM Grammars

Our first theorem is the first one that deals with SSCM grammars of degree $(2, 1)$.

Theorem 1. $\mathrm{SSCM}(2, 1; 5; 3, 2) = \mathrm{RE}$.

Proof. Let $L \in \mathrm{RE}$ be generated by a grammar in $(4, 2)$-GNF $G = (V, T, P \cup \{AB \to \lambda, CC \to \lambda\}, S)$ such that P contains only context-free rules and $N = V \setminus T = \{S, A, B, C\}$ (see Propos. 2). Next, we define the SSCM-grammar $G' = (V', T, P' \cup P'', S)$, where $V' = V \cup \{\#\}$ (assuming $\# \notin V$), P' contains the (single-rule) unconditional matrices of the form $[(S \to \alpha), \emptyset, \emptyset]$ whenever $S \to \alpha \in P$ and P'' contains the three (multi-rule and simple) conditional matrices

$$\begin{aligned} r1 &= [(A \to \#), (B \to \#), \emptyset, \#] \\ r2 &= [(C \to \#), (C \to \#), \emptyset, \#] \\ r3 &= [(\# \to \lambda), (\# \to \lambda), \#\#, \emptyset] \end{aligned}$$

Clearly, G' has degree $(2, 1)$, 5 nonterminals, and 3 conditional binary matrices.

We now show that $L(G') = L(G)$. Trivially, $r1$ (or $r2$, resp.) and then $r3$ can simulate $AB \to \lambda$ (or $CC \to \lambda$, resp.), so that $L(G) \subseteq L(G')$ is clear. For the reverse inclusion, we first make some observations concerning any sentential form w that is derivable in G'; they can be shown by easy inspection or induction and will be used in the following without special mentioning.

1. If w contains no $\#$, then all matrices but $r3$ may apply.
2. If w contains any $\#$, then only unconditional matrices or $r3$ may apply.
3. If w contains no S, then no sentential form w' derivable from w in G' will ever contain any S.

4. If w contains any S, then there is no second occurrence of S in w.

To prove $L(G) \supseteq L(G')$, we will actually show the following claim by induction on the length of the derivation of w: If $S \Rightarrow_{G'}^* w$, then either (1) $S \Rightarrow_G^* w$ or (2) there exist two occurrences of # in w, i.e., $w = w_1\#w_2\#w_3$ and there exists a sentential form w' such that $S \Rightarrow_G^* w'$ and either $w' = w_1Aw_2Bw_3$ or $w' = w_1Cw_2Cw_3$. The claim is trivially true if $w = S$ (derivation length 0).

Consider some sentential form w such that $S \Rightarrow_{G'}^n w$ for some $n > 0$. Then, there is some v such that $S \Rightarrow_{G'}^{n-1} v \Rightarrow_{G'} w$. By induction, we know that either (1) $S \Rightarrow_G^* v$, i.e., $v \in L_{(4,2)}$ as defined in Eq. (2), or (2) there exist two occurrences of # in v, i.e., $v = v_1\#v_2\#v_3$ and there exists a sentential form v' such that $S \Rightarrow_G^* v'$ and either (a) $v' = v_1Av_2Bv_3$ or (b) $v' = v_1Cv_2Cv_3$.

Consider Case (1). We might apply $r1$ to get w. Then, we get $w = w_1\#w_2\#w_3$. As v is also a sentential form of G, BA is not a subsequence of v. Therefore, $v = w_1Aw_2Bw_3$ as claimed. Similarly, we might apply $r2$ to get w, arriving again at $w = w_1\#w_2\#w_3$. Now, $v = w_1Cw_2Cw_3$ as claimed. Finally, if S occurs in v, we can also apply an unconditional matrix. As this (trivially) corresponds to applying a context-free rule, $S \Rightarrow_G^* w$.

Consider Case (2). Hence, $v = v_1\#v_2\#v_3$. If S does not occur in v, then we must apply matrix $r3$. This is only possible if $v_2 = \lambda$, i.e., if $v_2 \neq \lambda$, the derivation is stuck. We mark this observation as [*] which we would recall again later. Hence, $v' = v_1ABv_3$ or $v' = v_1CCv_3$ are derivable in G by induction hypothesis. Now, $v \Rightarrow_{r3} w$ yields $v' \Rightarrow_G w$ or $v' \Rightarrow_G w$ by applying the deletion rules $AB \to \lambda$ or $CC \to \lambda$, respectively. Hence, $S \Rightarrow_G^* w$.

In the following, when considering Cases (2a) and (2b) separately, we assume that S occurs in v. Recall that the position of S is then unique.

Consider Case (2a) first, i.e., there is sentential form v' such that $S \Rightarrow_G^* v'$ and $v' = v_1Av_2Bv_3$. As S occurs in v, it will occur in v_2 by the structure of rules of (4, 2)-GNF. Hence, none of the conditional matrices is applicable on v. Rather, we have to apply an unconditional matrix (with context-free rule $S \to \alpha$) to v in order to get w. Hence, $v = v_1\#v_2'Sv_2''\#v_3$, as $v_2 = v_2'Sv_2''$ for some v_2', v_2''. Then, $w = v_1\#v_2'\alpha v_2''\#v_3$. Now, $v' \Rightarrow_G w'$ with $w' = v_1Av_2'\alpha v_2''Bv_3$ by applying the context-free rule $S \to \alpha$. This shows the claim also in this subcase of Case (2a).

Now, consider Case (2b). This means that there is sentential form v' such that $S \Rightarrow_G^* v'$ and $v' = v_1Cv_2Cv_3$. If S is contained in v_2, then the argument is analogous to the one of the previous paragraph. If S is contained in v_1, then (with $v_1 = v_1'Sv_1''$) applying an unconditional matrix (with context-free rule $S \to \alpha$) to v would result in $w = v_1'\alpha v_1''\#v_2\#v_3$. Now, if one would apply $S \to \alpha$ on v', we get w', with $w' = v_1'\alpha v_1''Cv_2Cv_3$. This shows the claim also in this subcase of Case (2b). As a side-remark: Clearly, one can continue this argument until α does not contain S anymore. But then, the derivation will be stuck according to the analysis of [*], because $v_2 \neq \lambda$ as CC cannot be a substring of any sentential form derivable in Stages 1 or 2 in G. The case when S is contained in v_3 is analogous to the case when S is contained in v_1.

By induction, the claim follows. As for terminal strings w, only Case (1) can happen, $L(G) \supseteq L(G')$ can be inferred. □

The following is a trade-off result of the parameters degree and number of nonterminals compared to the previous theorem. Moreover, it returns to the consideration of degree $(3, 1)$, a case previously considered in [12].

Theorem 2. $\mathrm{SSCM}(3, 1; 5; 2, 2) = \mathrm{RE}$.

Proof. Let $L \subseteq T^*$ be any RE language. By Propos. 5, there is a type-0 grammar of the form $G = (V, T, P \cup \{0\$1 \to \$, 1\$0 \to \$, \$ \to \lambda\}, S)$ with $L(G) = L$ such that P contains only context-free rules with left-hand side S and $N{:=}V \setminus T = \{S, 0, 1, \$\}$, i.e., G is in MMMNF. We need one more additional nonterminal $\#$ in the SSCM grammar G' that we describe next. As before, we take over all context-free rules of G as single-rule unconditional matrices (that we do not count into our numbers). Also, we have the matrix $r_\$ = [(\$ \to \lambda), \emptyset, \emptyset]$. The two non-context-free rules are simulated by the following conditional matrices.

$$r1 = [(0 \to \#), (1 \to \#), \emptyset, \#]$$
$$r2 = [(\# \to \lambda), (\# \to \lambda), \#\$\#, \emptyset]$$

It should be clear how this simulation works, so that $L(G) \subseteq L(G')$ is obvious.

For the converse inclusion $L(G) \supseteq L(G')$, we can actually show the following claim by induction on the length of the derivation of w: If $S \Rightarrow_{G'}^* w$, then either (1) $S \Rightarrow_G^* w$ or (2) there exist two occurrences of $\#$ in w, i.e., $w = w_1\#w_2\#w_3$ and there exists a sentential form w' such that $S \Rightarrow_G^* w'$ and either $w' = w_1 0 w_2 1 w_3$ or $w' = w_1 1 w_2 0 w_3$. Details can be found in the long version of this paper [6]. □

The next result is a trade-off result to the previous theorem and also to [12].

Theorem 3. $\mathrm{SSCM}(3, 1; 4; 3, 3) = \mathrm{RE}$.

Proof. Let $L \in \mathrm{RE}$ be generated by a grammar in (3, 2)-GNF of the form $G = (V, T, P \cup \{AA \to \lambda, BBB \to \lambda\}, S)$ such that P contains only context-free rules and $V \setminus T = \{S, A, B\}$ (see Propos. 1). Next, we define the SSCM-grammar $G' = (V', T, P' \cup P'', S)$, where $V' = V \cup \{\#\}$ (assuming that $\# \notin V$), P' contains the (single-rule) unconditional matrices of the form $[(S \to \alpha), \emptyset, \emptyset]$ whenever $S \to \alpha \in P$ and P'' contains the three (multi-rule) matrices

$$r1 = [(B \to \#), (B \to \#), (B \to \#), \emptyset, \#]$$
$$r2 = [(A \to \#), (A \to \#\#), \emptyset, \#]$$
$$r3 = [(\# \to \lambda), (\# \to \lambda), (\# \to \lambda), \#\#\#, \emptyset]$$

For the formal correctness proof, we refer to the long version [6]. □

5 Non-simple Semi-Conditional Matrix Grammars

In the previous section, we arrived at some computational completeness results for simple SCM grammars. We only got one result with four nonterminals, but in this case, we could not get down to matrix length two. Therefore, we are now relaxing the simplicity condition, rather measuring non-simplicity as an additional parameter, to be able to get several results with four nonterminals and matrix length two. Several trade-offs will be observed.

Theorem 4. SCM(4, 3; 4; 7, 2, 6) = RE.

Table 2. Simulating (3, 2)-GNF in two (related) ways: Theorem 4 for SCM(4, 3; 4; 7, 2, 6) on the left and Theorem 5 for SCM(5, 2; 4; 7, 2, 4) on the right.

$r1 = [(B \to \#), (B \to \#), BBB, \#]$	$r1 = [(B \to \#), (B \to AA), ABBBA, AA]$
$r2 = [(\# \to \lambda), (B \to \#\#\#), A\#\#B, \#\#\#]$	$r2 = [(B \to \#\#\#), A\#AAB, \#\#]$
$r3 = [(A \to \#), (A \to \#), AA, \#]$	$r3 = [(\# \to \lambda), (\# \to \lambda), \#AA\#, \emptyset]$
$r4 = [(\# \to \lambda), (\# \to \lambda), \#\#\#\#, \#B]$	$r4 = [(A \to \#), (A \to \#\#\#), AA, \#]$
$r5 = [(\# \to \lambda), (\# \to \lambda), A\#\#A, BBB]$	$r5 = [(\# \to \lambda), (\# \to \lambda), \#\#\#\#, \emptyset]$
$r6 = [(\# \to \lambda), (\# \to \lambda), B\#\#B, \emptyset]$	$r6 = [(\# \to \lambda), (\# \to \lambda), B\#\#B, \emptyset]$
$r7 = [(\# \to \lambda), (\# \to \lambda), \#\#, A]$	$r7 = [(\# \to \lambda), (\# \to \lambda), \#\#, A]$

Proof. Let $L \in$ RE. We start with a type-0 grammar G in (3, 2)-GNF that generates L. Hence, in Phase 1, when we only apply context-free rules which carry over as unconditional single-rule matrices, we can derive some string from

$$\{ABB, AB\}^*\{S, \lambda\}(\{BA, BBA\} \cup T)^* \subseteq L_{(3,2)}.$$

The two non-context-free erasing rules $BBB \to \lambda$ and $AA \to \lambda$ are simulated by 7 conditional matrices; here, we employ a fourth nonterminal #, see Table 2.

The intended simulation works as follows:

$$\begin{aligned}\alpha ABBBA\beta &\Rightarrow_{r1} \alpha A\#\#BA\beta \Rightarrow_{r2} \alpha A\#\#\#\#A\beta \Rightarrow_{r4} \alpha A\#\#A\beta \\ &\Rightarrow_{r5} \alpha AA\beta \Rightarrow_{r3} \alpha\#\#\beta \Rightarrow_{r6/7} \alpha\beta\end{aligned}$$

Whether to apply matrices $r6$ or $r7$ in the last step depends on whether B's are ending α and starting β or whether $\alpha = \beta = \lambda$.

The formal correctness proof of this construction is in the long version [6]. □

Now we get a trade-off result, as the construction above has uncomparable degree but less non-simple matrices compared to Theorem 5.

Theorem 5. SCM(5, 2; 4; 7, 2, 4) = RE.

Proof. Let $L \in$ RE. We again start with a type-0 grammar G in (3, 2)-GNF that generates L. Compared to the previous construction, only the matrices simulating the two non-context-free erasing rules $BBB \to \lambda$ and $AA \to \lambda$ change as can be seen in Table 2.

Table 3. Simulating sMMNF in two (related) ways: Theorem 6 for SCM(6, 3; 4; 7, 2, 3) on the left and Theorem 7 for SCM(6, 3; 3; ∗, 2, 4) on the right

$r1 = [(S \to \$\$), 0S, \emptyset]$	$r1 = [(S \to SS), 0S, SS]$
$r2 = [(0 \to \$\$\$), (0 \to \$), 0\$\$0, \$\$\$]$	$r2 = [(0 \to SSS), (0 \to S), 0SS0, SSS]$
$r3 = [(1 \to \$\$\$), (1 \to \$), 1\$\$1, \$\$\$]$	$r3 = [(1 \to SSS), (1 \to S), 1SS1, SSS]$
$r4 = [(\$ \to \lambda), (\$ \to \lambda), \$^6, \emptyset]$	$r4 = [(S \to \lambda), (S \to \lambda), S^6, \emptyset]$
$r5 = [(\$ \to \lambda), (\$ \to \lambda), 0\$^4 0, \emptyset]$	$r5 = [(S \to \lambda), (S \to \lambda), 0S^4 0, \emptyset]$
$r6 = [(\$ \to \lambda), (\$ \to \lambda), 1\$^4 1, 0\$]$	$r6 = [(S \to \lambda), (S \to \lambda), 1S^4 1, 0S]$
$r7 = [(\$ \to \lambda), (\$ \to \lambda), \emptyset, 1]$	$r7 = [(S \to \lambda), (S \to \lambda), \emptyset, 1]$

The intended simulation works as follows:

$$\alpha ABBBA\beta \Rightarrow_{r1} \alpha A\#AABA\beta \Rightarrow_{r2} \alpha A\#AA\#\#\#A\beta \Rightarrow_{r3} \alpha\#AA\#\beta$$
$$\Rightarrow_{r3} \alpha AA\beta \Rightarrow_{4} \alpha\#\#\#\#\beta \Rightarrow_{r5} \alpha\#\#\beta \Rightarrow_{r6/7} \alpha\beta$$

Whether to apply matrices $r6$ or $r7$ in the last step depends on whether B's are ending α and starting β or whether $\alpha = \beta = \lambda$. The formal correctness proof of this construction is in the long version [6]. □

We are now presenting yet another trade-off result, now based on sMMNF.

Theorem 6. SCM(6, 3; 4; 7, 2, 3) = RE.

Proof. We start with an sMMNF grammar G for a given arbitrary RE language $L \subseteq T^*$, i.e., G is a type-0 grammar of the form $G = (V, T, P \cup \{0\$0 \to \$, 1\$1 \to \$, \$ \to \lambda\}, S)$ with $L(G) = L$ such that P contains only context-free rules and $N{:=}V \setminus T = \{S, 0, 1, \$\}$. We suggest the simulation rules for $0\$0 \to \lambda$ and for $1\$1 \to \lambda$ in the SCM grammar G' as given in Table 3.

The context-free rules of G are simulated by unconditional matrices in G' except for (a) the rules $S \to u\$v$ in G that will be simulated by $S \to uSv$ and then $r1$, as well as for (b) $\$ \to \lambda$ in G that is simulated by $r7$. We briefly explain the idea behind. Notice that for each rule $S \to u\$v$ in G, there is also a rule $S \to uSv$ in G. The reason why Geffert (and hence also MM) normal forms have this specific feature (and not, as one might expect, directly the rule $S \to \$$) is that otherwise, $\lambda \in L(G)$ would be always true. Another way to avoid the trivial derivation $S \Rightarrow \$ \Rightarrow \lambda$ in an MMNF grammar would be to check, when $S \to \$$ is applied, that indeed some of the rules $S \to uSv$ have been applied before. This is the purpose of the permitting context $0S$ in matrix $r1$. Therefore, we find $S \Rightarrow_G^* \alpha S\beta \Rightarrow_G \alpha\$\beta$ if and only if $S \Rightarrow_{G'} \alpha S\beta \Rightarrow_{G'} \alpha\$\$\beta$. (b) $\$ \to \lambda$ is meant to be the very last rule applied in G; in $r7$, this is explicitly checked with the help of the forbidden context 1. This check makes use of the fact that we started with a grammar G in strong MMNF. We now present intended derivations.

$$\alpha 0\$\$0\beta \Rightarrow_{r2} \alpha\$^6\beta \Rightarrow_{r4} \alpha\$^4\beta \Rightarrow_{r5/r6} \alpha\$\$\beta\,,$$
$$\alpha 1\$\$1\beta \Rightarrow_{r3} \alpha\$^6\beta \Rightarrow_{r4} \alpha\$^4\beta \Rightarrow_{r5/r6/r7} \alpha\$\$\beta\,,$$

Which matrix to apply in the last step depends on the prefix α and suffix β. The formal correctness proof of this construction is in the long version [6]. □

Remark 1. The reader might have wondered if one could not save one of the matrices by opting for a strong version of MMMNF instead of sMMNF. However, then the forbidden context check in $r6$ seems to be impossible. Therefore, this idea would not work.

6 SCM Grammars: When Three Nonterminals Suffice

In this section, we ask ourselves if we could try to obtain computational completeness results with three nonterminals only. Indeed, we can achieve this, as we show in the following, but at the expense of having an unbounded number of (simple) conditional matrices. Otherwise, the idea is simply to modify the previous construction and to re-use the nonterminal S also in Phase 2. More precisely,\$ the role of the central is now taken by the substring SS whose absence must hence be checked in each simulation of a context-free rule. A bit surprisingly, this idea does not increase the degree of the simulating grammar in comparison with the previous Theorem 6.

It is interesting to compare this result with that on (classical) matrix grammars where it is known that three nonterminals suffice to characterize RE, see [4] where the lengths of the matrices are arbitrarily big.

Theorem 7. $\mathrm{SCM}(6,3;3;*,2,4) = \mathrm{RE}$.

Proof. We again start with a grammar G in sMMNF that generates an arbitrary RE language $L \subseteq T^*$. This means that G a type-0 grammar of the form $G = (V,T,P \cup \{0\$0 \to \$, 1\$1 \to \$, \$ \to \lambda\}, S)$ with $L(G) = L$ such that P contains only context-free rules and $N{:=}V \setminus T = \{S,0,1,\$\}$. Any context-free rule $S \to \gamma$ of G is simulated by the simple conditional matrix $r_\gamma = [(S \to \gamma), \emptyset, SS]$. The simulation rules for $0\$0 \to \lambda$ and for $1\$1 \to \lambda$ in the SCM grammar G' are presented in Table 3. None of the matrices $r2$ through $r7$ can be applied already when simulating context-free rules, because each matrix requires the presence of (at least) two occurrences of S one way or the other. Therefore, the arguments from the proof of Theorem 6 nearly literally translate to this case. □

We now present another trade-off result, with uncomparable degrees.

Theorem 8. $\mathrm{SCM}(7,2;3;*,2,3) = \mathrm{RE}$.

Proof. Once more, we start with a grammar G in sMMNF that generates an arbitrary given RE language $L \subseteq T^*$. This means that G a type-0 grammar of the form $G = (V,T,P \cup \{0\$0 \to \$, 1\$1 \to \$, \$ \to \lambda\}, S)$ with $L(G) = L$ such that P contains only context-free rules and $N{:=}V \setminus T = \{S,0,1,\$\}$. Any context-free rule $S \to \gamma$ of G is simulated by the simple conditional matrix

$r_\gamma = [(S \to \gamma), \emptyset, SS]$. We suggest the following simulation rules for $0\$0 \to \lambda$ and for $1\$1 \to \lambda$ in the SCM grammar G', with $\eta \in \{0,1\}$.

$$r1 = [(S \to S1S), 0S0, 1S] \qquad r5_\eta = [(S \to \lambda), (S \to \lambda), \eta 1S1SSS, \emptyset]$$
$$r2 = [(0 \to 11), (0 \to SSSS11), 0S1S0, 11] \qquad r6 = [(S \to \lambda), (S \to \lambda), 0S1SSS0, \emptyset]$$
$$r3 = [(1 \to S), (1 \to SSS), 1S1S1, SS] \qquad r7 = [(S \to \lambda), (S \to \lambda), \emptyset, 0]$$
$$r4 = [(S \to \lambda), (S \to \lambda), SS1SSSS, \emptyset] \qquad r8 = [(1 \to \lambda), \emptyset, S]$$

The intended simulations are:

$$\alpha 0S1S0\beta \Rightarrow_{r2} \alpha 11S1SSSSS11\beta \Rightarrow^2_{r5_1} \alpha 11S1S11\beta,$$
$$\alpha 1S1S1\beta \Rightarrow_{r3} \alpha SS1SSSS\beta \Rightarrow_{r4} \alpha S1SSS\beta \Rightarrow_{r5_0/6} \alpha S1S\beta,$$

(with $\alpha, \beta \neq \lambda$, with α ending with the same symbol as β starts) and

$$1S1S1t \Rightarrow_{r7} 111t \Rightarrow^3_{r8} t \text{ for } t \in T^*.$$

From these three cases, $L(G) \subseteq L(G')$ follows easily by induction. The converse direction is unfortunately a quite intricate case analysis within an inductive argument and is hence deferred to the long version of this paper [6]. □

If one likes to highlight this, we can also deduce from Theorem 7 that is actually slightly better than Theorem 8 in the "long-matrices-count":

Corollary 1. *For each RE language L, there is a SCM grammar generating L with only three nonterminals and only six matrices of length two; all other matrices have length 1.*

Our last result concerning the theme telling that sometimes, three nonterminals are enough to generate all RE languages, is the following one. Its proof is significantly different from the ones previously presented in this paper, as it uses a simulation of graph-controlled grammars with only two nonterminals A and B (not formally introduced in this paper so far). Thus, we the whole argument can be found in the long version [6]. The argument itself is a non-trivial adaptation of the proof of Cor. 6 in [4]. More precisely, in the simulation of the given graph-controlled grammar, we use a third nonterminal C to encode the current vertex (state, viewed as a number) of the control graph explicitly in unary in the sentential form. We can use a (long) sequence of rules $C \to \lambda$ in a matrix to lower-bound the state number and the (long) forbidden context together to make sure that a state transition together with a successful rule application are properly simulated. The failure case is more tricky, as we have to split the tests (absence of left-hand side and upper bound on state number) over two matrices.

Theorem 9. SSCM$(0,*;3;*,*) = $ RE.

7 Conclusions

In this paper, we have tried to delineate the Pareto frontier of descriptional complexity for SCM grammars concerning a number of descriptional complexity parameters. The most natural question here is to complement these results by lower bounds. Or, can we further lower the upper bounds derived in this paper? Should we (also) consider alternative parameters, like the number of matrices that actually match the maximum matrix length? We have discussed this through the paper a bit already, but are far from a systematic study.

It is also interesting to observe that in none of the simulations that we propose (but the last one), the sequence in which the matrix rules are applied within a matrix matters. In other words, literally the same descriptional complexity results hold for SCUM grammars, where UM (not formally introduced) should be read as unordered matrix, a model introduced by Cremers and Mayer in [2].

References

1. Ábrahám, S.: Some questions of phrase-structure grammars. I. Comput. Linguistics **4**, 61–70 (1965)
2. Cremers, A., Mayer, O.: On matrix languages. Information and Control (now Information and Computation) **23**, 86–96 (1973)
3. H. Fernau. Parsimonious computational completeness. In N. Moreira and R. Reis, editors, Developments in Language Theory - 25th International Conference, DLT, volume 12811 of LNCS, pages 12–26. Springer, Cham (2021)
4. Fernau, H., Freund, R., Oswald, M., Reinhardt, K.: Refining the nonterminal complexity of graph-controlled, programmed, and matrix grammars. J. Autom. Lang. Comb. **12**(1/2), 117–138 (2007)
5. H. Fernau, L. Kuppusamy, and I. Raman. Counting simple rules in semi-conditional grammars is not simple. In L. L. Patey and E. Pimentel, editors, Twenty Years of Theoretical and Practical Synergies, 20th Conference on Computability in Europe, CiE, volume 14773 of LNCS, pages 192–204. Springer, Cham (2024)
6. H. Fernau, L. Kuppusamy, and I. Raman. On computational completeness of semi-conditional matrix grammars. Technical Report 2411.15338, ArXiv, Cornell University, USA, November 2024
7. Fernau, H., Kuppusamy, L., Raman, I.: On the computational completeness of generalized forbidding matrix grammars. Theoretical Computer Science **999**, 114539 (2024)
8. Geffert, V.: Normal forms for phrase-structure grammars. RAIRO Informatique théorique et Applications/Theoretical Informatics and Applications **25**, 473–498 (1991)
9. Hauschildt, D., Jantzen, M.: Petri net algorithms in the theory of matrix grammars. Acta Informatica **31**, 719–728 (1994)
10. Masopust, T., Meduna, A.: Descriptional complexity of generalized forbidding grammars. In: Geffert, V., Pighizzini, G. (eds.) 9th International Workshop on Descriptional Complexity of Formal Systems - DCFS, pp. 170–177. University of Kosice, Slovakia (2007)
11. Meduna, A.: Generalized forbidding grammars. International Journal of Computer Mathematics **36**, 31–39 (1990)

12. Meduna, A., Kopeček, T.: Simple semi-conditional versions of matrix grammars with a reduced regulating mechanism. Computing and Informatics **23**, 287–302 (2004)
13. Păun, Gh.: A variant of random context grammars: semi-conditional grammars. Theoretical Computer Science **41**, 1–17 (1985)

Outer-(ap)RAC Graphs

Henry Förster[1,2(✉)], Julia Katheder[1], and Giacomo Ortali[3]

[1] Wilhelm-Schickard-Institut für Informatik, Universität Tübingen, Tübingen, Germany
{henry.foerster,julia.katheder}@uni-tuebingen.de
[2] Campus Heilbronn, Technische Universität München, Heilbronn, Germany
henry.foerster@tum.de
[3] Department of Engineering, University of Perugia, Perugia, Italy
giacomo.ortali@unipg.it

Abstract. An *outer-RAC drawing* of a graph is a straight-line drawing where all vertices are incident to the outer cell and all edge crossings occur at a right angle. If additionally, all crossing edges are either horizontal or vertical, we call the drawing *outer-apRAC* (*ap* for *axis-parallel*). A graph is outer-(ap)RAC if it admits an outer-(ap)RAC drawing. We investigate the class of outer-(ap)RAC graphs. We show that the outer-RAC graphs are a proper subset of the planar graphs with at most $2.5n - 4$ edges where n is the number of vertices. This density bound is tight, even for outer-apRAC graphs. Moreover, we provide an SPQR-tree based linear-time algorithm which computes an outer-RAC drawing for every given series-parallel graph of maximum degree four. As a complementing result, we present planar graphs of maximum degree four and series-parallel graphs of maximum degree five that are not outer-RAC. Finally, for series-parallel graphs of maximum degree three we show how to compute an outer-apRAC drawing in linear time.

Keywords: RAC · beyond planarity · density · series-parallel graphs

1 Introduction

Crossings in graph drawings are well-known to impede readability. This fact was experimentally verified by Purchase [41] in 2000. Follow-up works showed that the topology and geometry of local crossing configurations are deciding factors in how large the impact of crossings on readability actually is. Crossings at larger crossing angles reduce readability to a lesser extent than those at smaller crossing angles [37–39]. These results gave rise to the research field *graph drawing beyond planarity* where graph drawings with specific requirements towards local crossing configurations have been considered. Substantial research has deepened our understanding of beyond-planar graphs; see [24,36] for an overview.

In this paper, we consider *right-angle-crossing drawings*, or *RAC drawings* for short, which are straight-line drawings of graphs where every crossing occurs at a right angle. The RAC drawing model is directly motivated by empirical studies that gave rise to a deeper study of graph drawing beyond planarity [37–39].

R. Královič and V. Kůrková (Eds.): SOFSEM 2025, LNCS 15538, pp. 284–299, 2025.
https://doi.org/10.1007/978-3-031-82670-2_21

Hence, it comes at no surprise, that RAC drawings have been thoroughly investigated. More precisely, a first theoretical study by Didimo et al. [23] established a linear edge density bound shortly after Huang's initial eye tracking study [37]. In addition, they showed that every graph admits a RAC drawing with 3 bends per edge [23] which was shown to be the tight number of bends by Arikushi et al. [7]. Subsequent works on RAC drawings considered edge density [1,40,45], area [29,42], variants where edges are drawn as circular arcs [15], simultaneous RAC drawings [5,11] and algorithms for restricted input graphs [2,10,16]. The complexity of the RAC drawing problem has first been shown to be NP-hard [6] and later to be $\exists\mathbb{R}$-complete [43]; on the other hand, there are FPT algorithms parameterited by feedback edge number and by vertex cover number [14]. Recently, a variant of RAC drawings called *axis-parallel RAC drawings*, or *apRAC drawings* for short, was introduced, in which each crossing edge has slope ± 1 [3][1].

In beyond-planar graph drawing, a classical topic is to consider additional *constraints* for the drawings. One of these constraints is the *outer drawing* model where each vertex must be located on the outer cell of the drawing. Outer drawings may be utilized to visualize highly connected clusters in graphs [4,33] which in real-world networks are often only sparsely connected to each other; see e.g. [30,31]. Previous research has considered outer-k-planar [8,12,34,35], outer-fanplanar [9,13] and outer-confluent graphs [27]. Surprisingly however, the existing literature only considered outer-(ap)RAC drawings with additional constraints on the placement of vertices [18,21] and for outer-1-planar graphs [17].

Our Contribution. We initiate the study of more general outer-(ap)RAC drawings in which vertices can be arbitrarily placed as long as they are incident to the outer cell. In the process, we prove that the outer-RAC graphs are a proper subfamily of the planar graphs in Sect. 3. Moreover, we show that certain planar graphs of low maximum degree do not admit outer-(ap)RAC drawings in Sect. 4. In contrast, we provide efficient outer-(ap)RAC drawing algorithms for series-parallel graphs of low maximum degree in Sect. 5. Finally, we conclude the paper with intriguing open questions.

2 Preliminaries

We assume familiarity with standard notation from graph theory, as found in [25] and basic graph drawing concepts, cf. [44]. In this paper, we consider all graphs to be simple. Let $G = (V, E)$ be a graph. A graph is said to be *cubic*, if all of its vertices have exactly degree 3. In a *subcubic* graph, every vertex has degree at most 3. The terms *(sub)quartic* are defined analogously for vertex degree 4. We call a drawing Γ *planar* if in Γ no two edges intersect except at a common endpoint. We say that a graph G is *planar* if it admits a planar drawing. The connected regions of the plane in a planar drawing Γ are called *faces*, the

[1] Originally, the slopes where defined as 0 and ∞ in [3]. We use the rotated version ± 1 which will allow us to simplify our discussion in Sect. 5.

unbounded face is called *outer face.* A planar drawing Γ in which each vertex is incident to the outer face is called *outerplanar.* If a graph G admits an outerplanar drawing we call G an *outerplanar graph.* In the weak dual graph H of a planar drawing Γ, each face except the outer face is represented by a vertex and faces f_1, f_2 are connected in H if and only if f_1 and f_2 share an edge in Γ. For an embedded outerplanar graph, the weak dual graph is a forest.

Similarly, the connected regions of the plane in a *non-planar* drawing Γ are called *cells* and the unbounded cell is called *outer cell.* Consider a straight-line drawing Γ, i.e., each edge is represented by a single segment. If in Γ, all crossings occur at a right angle and all vertices are located on the outer cell, we call Γ *outer-RAC.* If additionally, all crossing edges have slope ± 1, we call Γ *outer-apRAC.* Moreover, we call a graph *outer-(ap)RAC* if it admits an outer-apRAC drawing.

An *SPQR-tree* $\mathcal{T}$ of a graph G describes a uniquely defined decomposition of G according to its separation pairs [19,20], which can be computed in linear time [32]. A node μ in $\mathcal{T}$ is associated with a graph $skel(\mu)$ called the *skeleton* of μ which consists of *virtual edges* between the vertices of separation pairs in G and at most one edge of G. Each virtual edge corresponds to at least one path between its endpoints in G. The pair of vertices s_μ, t_μ separating the component represented by μ in G are called the *poles* of μ. The vertices s_μ, t_μ are connected by the *parent virtual edge*, which corresponds to a virtual edge in the parent of μ in $\mathcal{T}$. Based on the structure of its skeleton, a node $\mu \in \mathcal{T}$ is of one of four types:

- *S-node*: $skel(\mu)$ forms a cycle of at least three virtual edges, including the parent virtual edge between s_μ and t_μ
- *P-node*: $skel(\mu)$ is comprised of at least three parallel virtual edges between s_μ and t_μ one of which is the parent virtual edge
- *Q-node*: $skel(\mu)$ contains the parent virtual edge and another edge between s_μ and t_μ, representing an actual edge in G. If μ is the root of $\mathcal{T}$, the virtual edge of the skeleton instead corresponds to its unique child node ν.
- *R-node*: $skel(\mu)$ is triconnected and contains the poles s_μ and t_μ. All its edges are virtual, the virtual edge between s_μ and t_μ is its parent virtual edge.

The SPQR-tree $\mathcal{T}$ of a graph G is by definition rooted at a Q-node. Moreover, all its leaves are Q-nodes. Also observe that two S-nodes are never connected to each other and the same holds for two P-nodes in $\mathcal{T}$. The subtree rooted at a node μ induces the so-called *pertinent graph* $pert(\mu)$, which is a subgraph in G obtained from merging the parent virtual edge in the skeleton of each node in the subtree of $\mathcal{T}$ rooted at μ with the corresponding virtual edge in the skeleton of its respective parent node. A *series-parallel graph,* or short *SP-graph,* is a biconnected graph[2] whose SPQR-tree contains no R-nodes. Since the skeleton of both P- and S-nodes are always planar, SP-graphs are always planar.

[2] In the literature there exists another recursive definition for series-parallel graphs [26]. However, one can obtain biconnectivity by a parallel composition with a single edge.

Lemma 1. *Let G be an SP-graph with $n \geq 3$ vertices and let $\mathcal{T}$ be its SQPR-tree rooted at any Q-node. Then, $\mathcal{T}$ contains an S-node μ such that all children of μ are Q-nodes.*

Proof. If G contains no P-node the statement follows immediately. Otherwise, since all leaves of $\mathcal{T}$ are Q-nodes, we find a P-node μ_p by traversing $\mathcal{T}$ top-down, such that the subtree of $\mathcal{T}$ rooted at μ_p contains no other P-node. By simplicity, μ_p has at most one Q-node child and hence at least one S-node child μ_s whose children are all Q-nodes. □

We will make use of Lemma 1 to simplify the discussion of our algorithmic results in Sect. 5 as follows. Let μ_s be an S-node such that all its children are Q-nodes in $\mathcal{T}$. We can now root $\mathcal{T}$ at one of the Q-node children of μ_s, denoted as μ_r. After rerooting, we have that μ_s is the unique child of μ_r.

Corollary 1. *Let G be an SP-graph and $\mathcal{T}$ be its SPQR-tree. $\mathcal{T}$ can be rooted at a Q-node μ_r with unique child μ_s such that μ_s is an S-node and at most one child μ_p of μ_s is a P-node.*

3 Topological Results

In this section, we provide some topological results. A useful tool for these results will be the notion of *blocks* of crossing edges in a given outer-RAC drawing Γ of a graph $G = (V, E)$. To this end, consider the *crossing graph* $C(\Gamma)$ of Γ which contains a vertex for each edge in E and an edge $(e_1, e_2) \in E \times E$ if and only if e_1 and e_2 cross in Γ. A *block* is any maximal connected vertex set in the transitive closure of $C(\Gamma)$; see also Fig. 1. In particular, since in RAC drawings only edges drawn with perpendicular slopes cross, each block contains edges of only two perpendicular slopes. Thus, we can partition the edges of block B obtaining *slope sets* $B_1, B_2 \subset B$ with $B_1 \cap B_2 = \emptyset$ such that each pair of edges of B_i for $i \in 1, 2$ does not cross. Moreover, since we consider the outer-RAC setting, all endpoints of edges of the same block are located on the outer face. Since in addition, edges assigned to different blocks do not cross by definition, we can cover the interior of a RAC drawing with regions called *outlines* of blocks that contain all crossing edges as follows. Consider a block $B \subset E$. Sort the endpoints of the edges in B according to their occurrence in a clockwise cyclic walk along the outer face $f_\circ$ of Γ and enumerate them by $v_1, \ldots, v_k$. Then, there necessarily exists a closed cycle $O(B)$, called the *outline* of B, such that 1.$O(B)$ is disjoint from the interior of $f_\circ$ (in other words, each point $O(B)$ is either inside a bounded cell of Γ or on the boundary of $f_\circ$), 2.$O(B)$ does not cross any edges except at their endpoints, 3.$O(B)$ contains all edges of B but no other edge of Γ, and 4.$O(B)$ traverses $v_1, \ldots, v_k$ in order. More precisely, when traversing along $O(B)$ from v_i to v_{i+1}, one can follow the clockwise last[3] edge of B incident to v_i up to its first crossing

[3] When considering the edges incident to v_i in clockwise order starting from the direction in which we reach v_i along $O(B)$ coming from v_{i-1}.

and then following the edge encountered up to its next crossing where we again follow the crossing edge; see also Fig. 1a. Necessarily, due to the fact that the drawing is outer-RAC, a repeated application of this method will finish at v_{i+1}.

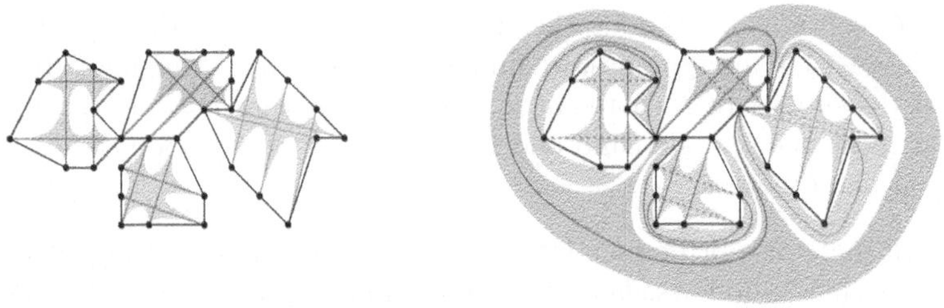

(a) (b)

Fig. 1. (a) An outer RAC graph with its crossing edges being decomposed into blocks (colored edges) and the outlines of the blocks (colored regions). (b) Illustration for the proof of Theorem 1 (Color figure online)

Theorem 1. *Let G be an outer-RAC graph. Then G is planar.*

Proof. Consider any outer-RAC drawing Γ of G. We will describe how to obtain a planar drawing of G. First, extend Γ by all the outlines of blocks obtaining a drawing Γ'. By construction, Γ' is still bounded by the outer cycle C which contains all vertices. Moreover, the outlines of blocks do not cross each other by construction. We now copy the interior of C to the outside. Observe that in this copying operation we disregard the geometry but maintain the topology of the drawing. We now have two copies of each edge, each staying within its respective block outline either in the interior or the exterior of C. For each block B, consider now the slope sets B_1 and B_2. We will remove the copies of B_1 in the interior of C and the copies of B_2 in the exterior of C. Since edges of the same slope set do not cross, we obtain a planar drawing of G by removing the block outlines; see Fig. 1b for an illustration. □

One may wonder if in fact every planar graph is also outer-RAC. We will show that this is not the case by providing a density upper bound for outer-RAC graphs. As an intermediate step, we consider a special subclass of outer-RAC graphs. We call a graph *bounded block graph* if it can be partitioned into a Hamiltonian cycle H and a set of edges E_c and admits a *bounded block drawing* Γ, that is, a drawing where 1. each edge of E_c is drawn straight-line and crosses only at right angles, 2. H forms the crossing-free outer boundary of Γ and is not necessarily drawn straight-line, and 3. all edges of E_c belong to a single block.

Lemma 2. *Let G be a bounded block graph with n vertices. Then, G has at most $2n-2$ edges.*

Proof. Let Γ be a bounded block drawing of G. The crossing edges of G form a block B that can be partitioned into two slope sets B_1 and B_2. In addition, G contains only the plane cycle C which is topologically equivalent to the outline $O(B)$. We now consider for $i \in 1, 2$ the subgraph $B'_i = B[C \cup B_i]$ of G induced by the edges of C and B_i. Recall, that all edges of B_i are parallel, i.e., B'_i is an embedded outerplanar subgraph of Γ. We simplify B'_i obtaining B^*_i by replacing each maximal path p of edges of C on the boundary of the same internal face by a single edge unless this creates a parallel edge with an edge of B_i in which case we replace p by a path of length 2. Hence, each internal face of B^*_i contains exactly 2 edges not belonging to B_i. We claim that for each internal face f of length $4 + k$, we can assign k triangular faces. To this end, recall that the weak dual graph T that has a vertex for each internal face and an edge between faces sharing an edge is in fact a forest. Since each edge of B_i necessarily is part of two faces, f has degree $2 + k$ in T. Moreover, each triangular face has only one edge of B_i, i.e., degree 1 in T. Thus, such an assignment is possible in such a way that two triangular faces t_1, t_2 remain unassigned. We now root T at t_1 and count the number of vertices and edges in an in-order processing of T.

At the root, we encounter 3 vertices and 3 edges. If we encounter another face f, it shares 2 vertices and 1 edge with its parent. That is, if the face has length k with $k \geq 3$, we find $k - 2$ additional vertices and $k - 1$ additional edges. Thus, writing f_j for the number of faces of length j, we obtain the number of vertices n^*_i and edges m^*_i of B^*_i:

$$m^*_i = 3 + 2 + \sum_{j=4}^{\infty} \big((j-1) \cdot f_j + f_j \cdot 2(j-4)\big) = 5 + \sum_{j=4}^{\infty} (3j-9) f_j \qquad (1)$$

$$n^*_i = 3 + 1 + \sum_{j=4}^{\infty} \big((j-2) \cdot f_j + f_j \cdot (j-4)\big) = 4 + \sum_{j=4}^{\infty} (2j-6) f_j \qquad (2)$$

The additive constants refer to faces t_1 and t_2 whereas in the sum expressions we add both the vertices of faces of length j as well as the assigned triangular faces.

We can now transfer back to an accounting for graph B'_i by subdividing the edges replacing paths of non-crossed edges suitably. Each such subdivision creates another edge and another vertex. Writing s for the number of such subdivisions and m'_i and n'_i for the number of vertices and edges of B'_i, respectively, we can refine (1) and (2) as follows:

$$n'_i = n^*_i + s = 4 + s + \sum_{j=4}^{\infty} (2j-6) f_j \Leftrightarrow \sum_{j=4}^{\infty} (2j-6) f_j = n'_j - 4 - s \qquad (3)$$

$$\begin{aligned} m'_i &= m^*_i + s = 5 + s + \sum_{j=4}^{\infty} (3j-9) f_j = 5 + s + \tfrac{3}{2} \sum_{j=4}^{\infty} (2j-6) f_j \\ &= 5 + s + \tfrac{3}{2}(n'_i - 4 - s) \leq \tfrac{3}{2} n'_i - 1 \end{aligned} \qquad (4)$$

Now we compute the number of edges m of G. Necessarily, we have $n = n'_1 = n'_2$. Moreover, both B'_1 and B'_2 contain the outer cycle on n vertices. Thus,

$$m = m'_1 + m'_2 - n \leq \frac{3}{2}n - 1 + \frac{3}{2}n - 1 - n = 2n - 2. \qquad (5)$$

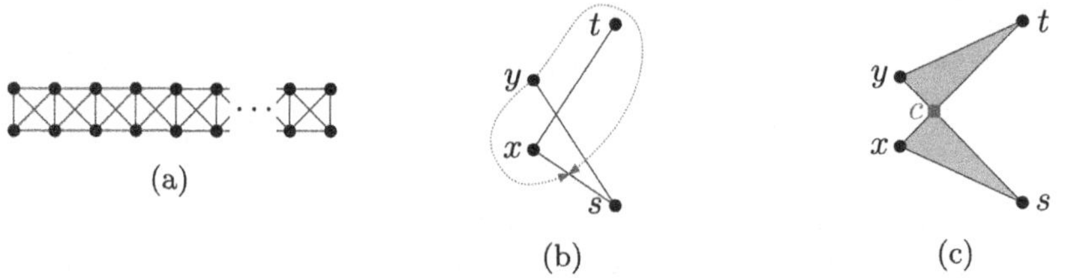

Fig. 2. Illustrations for the proof of (a) Theorem 3 (b) Lemma 3 and (c) Lemma 4

Theorem 2. *Let G be an outer-RAC graph with n vertices. Then, it has at most $m \leq 2.5n - 4$ edges. Moreover, if G has exactly $2.5n - 4$ edges, it can be decomposed into bounded blocks $B_1, \ldots, B_k$ such that for $i \in \{1, \ldots, k\}$ 1.B_i shares exactly one edge with the subgraph induced by $B_1, \ldots, B_{i-1}$ 2.B_i is isomorphic to K_4.*

Proof (Sketch of Proof). We investigate any outer-RAC drawing Γ of G. We first augment G by inserting all outlines of blocks defined by Γ and by triangulating the faces bounded only by crossing-free edges. We obtain a supergraph G' with an associated drawing Γ', in which all crossing edges are straight-line. Then, we observe that G' can be obtained starting from a bounded block graph G'_0 by an iterative procedure. In each step, we have a graph G'_{i-1} and obtain G'_i by merging G'_{i-1} with a new bounded block graph B_i either at a single vertex or at an edge. Finally, we show inductively that in this procedure the upper bound on the number of edges holds. See our extended preprint [28] for details. □

Theorem 2 already also describes a potential matching density lower bound. In fact, the existing literature already establishes that it is indeed outer-RAC. Namely, Dehkordi and Eades proved that every outer-1-planar graph is also outer-RAC [17] whereas Didimo [22] and Auer et al. [8] independently found a lower bound of $2.5n - 4$ edges for outer-1-planar graphs. Here we strengthen their result by explicitly noting that it generalizes to outer-apRAC.

Theorem 3 ([8,17,22]). *There is an infinitely large family of outer-apRAC graphs with n vertices and $2.5n - 4$ edges.*

Proof. K_4 is clearly outer-apRAC, e.g., place the four vertices at coordinates $(0,0)$, $(0,1)$, $(1,0)$ and $(1,1)$. Since the outer cycle of this drawing is square-shaped, we can form a chain of such K_4's by identifying the left edge of a copy with the right edge of another one; see Fig. 2a. In this drawing, only edges of slopes 1 and -1 cross and all vertices are on the outer face, i.e., it is outer-apRAC. □

Also note that the edge density bound differs from a tight bound of $2n-2$ for *circular RAC drawings*, in which all vertices are constrained to lie on a circle [18].

4 Obstructions for Outer-RAC Graphs

Theorem 2 already provides examples of planar graphs that admit no outer-RAC drawings. In this section, we provide planar graphs that admit no outer-RAC drawing despite being of low enough density. To this end, we first observe that parallel short paths admit only a few specific topologies in outer-RAC drawings.

Lemma 3. *Let G be a graph consisting of two paths $p_1 = (s, x, t)$ and $p_2 = (s, y, t)$ with $x \neq y$. Then, in any outer-RAC drawing Γ, p_1 and p_2 are not self-intersecting and one of the following holds:*

P2.1 s, x, t and y occur in this order along the outer cycle of Γ. Moreover, p_1 and p_2 do not intersect.
P2.2 s, x, y and t occur in this order along the outer cycle of Γ. Moreover, p_1 and p_2 intersect exactly once at edges (s, y) and (t, x).
P2.3 s, t, x and y occur in this order along the outer cycle of Γ. Moreover, p_1 and p_2 intersect exactly once at edges (s, x) and (t, y).

Proof. First, observe that each of p_1 and p_2 cannot cross itself as p_1 and p_2 each contain only two straight-line edges that share an endpoint.

In the following, assume that we already have an outer-RAC drawing of p_1. We investigate how p_2 can be added while maintaining that the drawing is outer-RAC. First, if p_1 and p_2 do not cross, we arrive at the configuration described in Case P2.1. Second, assume that p_1 and p_2 cross. Since the pairs of edges $(s, x), (s, y)$ and $(t, x), (t, y)$ share a common endpoint, they cannot cross each other. Hence, only edge pairs $(s, x), (t, y)$ and $(s, y), (t, x)$ may cross. If exactly one of these edge pairs cross, we arrive at one of Cases P2.2 and P2.3.

Hence, it remains to consider the case where (s, x) and (t, y) as well as (s, y) and (t, x) cross. Assume that this is possible. Since (s, x) and (t, y) cross, we have that s, x, y and t either occur in this order around the outer cycle of Γ or in the reversed order s, t, x and y. We assume w.l.o.g. that the order is s, x, y and t. Now, observe that (x, t) and y are separated by at least one vertex in both orientations of the outer cycle of Γ. Since (s, y) must cross (x, t), adding it removes at least one vertex from the outer cycle (see Fig. 2b); a contradiction. □

For three such short paths, we obtain yet a different result:

Lemma 4. *Let G be a graph consisting of three paths $p_1 = (s, x, t)$, $p_2 = (s, y, t)$ and $p_3 = (s, z, t)$ with $x \neq y$, $y \neq z$ and $x \neq z$. Then, in any outer-RAC drawing Γ, p_1, p_2 and p_3 are not self-intersecting, two paths, say p_1 and p_2, cross, whereas p_3 is crossing-free, and one of the following holds:*

P3.1 s, x, y, t and z occur in this order along the outer cycle of Γ. Moreover, p_1 and p_2 intersect exactly once at edges (s, y) and (t, x).
P3.2 s, z, t, x and y occur in this order along the outer cycle of Γ. Moreover, p_1 and p_2 intersect exactly once at edges (s, x) and (t, y).

Moreover, if G is subgraph of an outer-RAC graph G', p_1 and p_2 are not crossed by any edge not belonging to G.

Proof. Consider any outer-RAC drawing Γ of G. Using Lemma 3, we know that each pair of paths can only be realized in three different ways. First note that necessarily two paths must cross as otherwise p_1 and p_2 form a cycle that w.l.o.g. contains p_3; i.e., z is not on the outer cycle.

Thus, w.l.o.g. p_1 and p_2 cross according to Case P2.2 of Lemma 3 (Case P2.3 is symmetric) and let c denote the point where (s, y) and (t, x) cross. Assume for a contradiction that any edge e^* not belonging to p_1 or p_2 crosses $p \in \{p_1, p_2\}$. Note that e^* may be part of p_3 or of a supergraph G' of G. First, since Γ is RAC, e^* cannot cross p at c. Next, observe that p_1 and p_2 induce two triangular regions $T_1 = \triangle sxc$ and $T_2 = \triangle cyt$ which have a right angle at c; see Fig. 2c. Moreover, all edges of p_1 and p_2 are entirely on the boundary of T_1 and T_2. That is, if e^* intersects an edge e_p of p_1 or p_2, it is partially located inside $T \in \{T_1, T_2\}$. In fact, the edge e^* that intersects e_p must cross T twice as none of its endpoints can be located in T as then it would not be on the outer face. However, because T is triangular, it contains no two parallel bounding segments; a contradiction.

Since p_3 crosses neither p_1 nor p_2, Case P3.1 and Case P3.2 arise if p_1 and p_2 cross according to Case P2.2 and Case P2.3, respectively. □

Theorem 4. *There is a SP-graph of maximum degree* 5 *that is not outer-RAC.*

Proof. Consider the graph $K_{2,5}$. It is a parallel composition of five paths of length 2, i.e., it is series-parallel. By Lemma 4, it follows directly that it cannot admit an outer-RAC drawing. □

Theorem 5. *There is a planar triconnected graph of maximum degree* 4 *that is not outer-RAC.*

Proof. Consider the octahedral graph G. It consists of a $K_{2,4}$ composed of four parallel short paths $p_1 = (s, x_1, t)$, $p_2 = (s, x_2, t)$, $p_3 = (s, x_3, t)$ and $p_4 = (s, x_4, t)$ and a cycle on x_1, x_2, x_3, x_4. By Lemma 4, in any of the outer-RAC drawings Γ of G, w.l.o.g. we have that its vertices occur in the order s, x_1, x_2, t, x_3, x_4 along the outer cycle of Γ. Since there is a cycle on x_1, x_2, x_3, x_4 present in G, x_1 is adjacent to $x^* \in \{x_3, x_4\}$. Clearly, edge (x_1, x^*) must cross the drawing of the $K_{2,4}$ as otherwise s or t cannot be on the outer cycle. But then, (x_1, x^*) crosses (x_2, s) which, by Lemma 4, is already crossed by edge (x_1, t); a contradiction. □

In particular, Theorems 4 and 5 motivate us to study SP-graphs of maximum degree four. Namely, SP-graphs are exactly the planar graphs that contain no subdivision of triconnected graphs, whereas our counterexamples for SP-graphs are of maximum degree five. In Sect. 5, we will provide drawing algorithms for such graphs that draw the graph according to its SPQR-tree in a top-down fashion. In particular, for S-nodes, we will realize the skeleton without crossings. In an extended preprint [28], we show that this may not be guaranteed for maximum degree four SP-graphs if we restrict the drawings to be outer-apRAC.

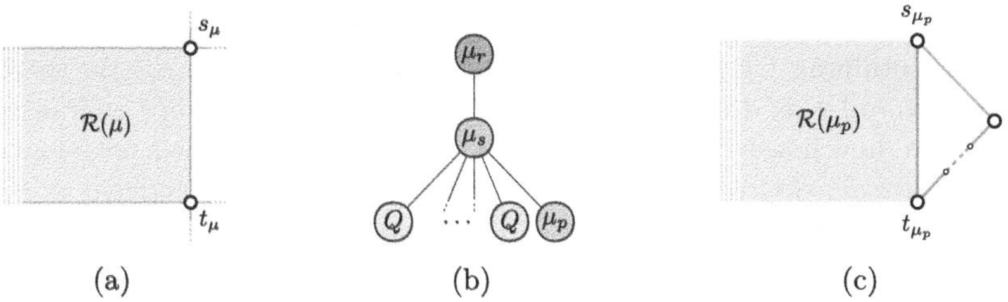

Fig. 3. (a) Invariants I.1 and I.2. (b) and (c) initialization of Theorem 6

5 Outer-RAC Drawings for Bounded Degree SP-Graphs

Theorem 6. *Let G be a biconnected SP-graph with maximum degree 3. An outer-apRAC drawing of G can be computed in $\mathcal{O}(n)$ time.*

Proof. Let G be a subcubic biconnected SP-graph and $\mathcal{T}$ be its SPQR-tree, hence $\mathcal{T}$ contains no R-nodes. In order to construct the outer-apRAC drawing Γ of G, we perform a top-down pre-order visit of $\mathcal{T}$ and draw the skeleton of each visited node depending on its type. Let μ be the vertex currently processed. We draw the virtual edges of $skel(\mu)$ corresponding to child nodes $\nu_1, \dots, \nu_k$ of μ. In this process, we also place the poles of its children, s_{ν_i} and t_{ν_i}, $1 \le i \le k$. Also, when we draw the parent virtual edge corresponding to a node μ, we define a *reserved region* $\mathcal{R}(\mu)$, in which we will draw the skeleton of μ (aside from s_μ and t_μ) later. More precisely, $\mathcal{R}(\mu)$ is defined as the intersection of three half planes, one being the closed half plane left of a line through the poles s_μ and t_μ of μ with s_μ lying above t_μ, the second being the open half plane below a horizontal line through s_μ and the third being the open half plane above a horizontal line through t_μ; see Fig. 3a for an illustration. Further, we maintain the following invariants:

I.1 Virtual edges in $skel(\mu)$ of an already processed node μ that correspond to a not yet processed child node of μ are drawn vertically.
I.2 Let μ be a not yet processed non-Q-node whose parent node in $\mathcal{T}$ has been processed. Then $\mathcal{R}(\mu)$ is free, i.e., it contains only the edge (s_μ, t_μ).
I.3 All crossing edges cross at right angles and have either slope 1 or -1.
I.4 Every already drawn vertex is incident to the outer cell.

We describe how to compute an initial partial drawing adhering to I.1–I.4

Initialization. Observe that if $\mathcal{T}$ contains no P-node, G is a cycle and therefore outerplanar and thus also outer-apRAC. Hence in the following, we can assume that G contains at least one P-node. We then root $\mathcal{T}$ according to Corollary 1; see Fig. 3b. Now, the root μ_r has as a child an S-node μ_s which in turn has exactly one P-node child μ_p. In G, according to the skeleton of μ_s, we have that μ_r and the Q-node children of μ_s form a path P connecting the poles of μ_p. As P contains at least two edges (μ_r and at least one Q-node child of μ_s) we draw the path P such that the poles of μ_p are vertically aligned as depicted in Fig. 3c, maintaining I.1 for the virtual edge corresponding to μ_p. As there are no crossings and P is drawn outside of $\mathcal{R}(\mu_p)$, I.2, I.3 and I.4 are guaranteed.

Next, we show how to handle a non-root node μ in the top-down traversal of $\mathcal{T}$.

μ is a Q-node. $skel(\mu)$ consists of the poles s_μ and t_μ and an edge $e = (s_\mu, t_\mu)$ which is an edge in G. We draw e as a vertical line connecting its endpoints, guaranteeing I.4. As there are no crossing edges and as μ is a leaf in $\mathcal{T}$, I.2 and I.3 are also maintained. Since μ contains no further virtual edge, I.1 is ensured.

μ is a P-node. Since G is subcubic, μ has exactly two child nodes in $\mathcal{T}$, where each is either of type S or type Q. Moreover, according to I.1, (s_μ, t_μ) is a vertical segment and $\mathcal{R}(\mu)$ is free according to I.2. While processing μ, we will partially draw the pertinent graph of each S-node child and remove the drawn edges from its skeleton. The remaining edges of the respective S-node child are then drawn in the recursive case. First, assume that μ has a single S-node child ν_1 and a Q-node child ν_2. Then the virtual edges $e_1 = (s_\mu, s'_\mu) \in skel(\nu_1)$ and $e_2 = (t_\mu, t'_\mu) \in skel(\nu_1)$ are Q-nodes, due to the maximum vertex degree. We delete e_1 and e_2 from $skel(\nu_1)$ and reassign the poles s_{ν_1} to s'_μ and t_{ν_1} to t'_μ. If the modified $skel(\nu_1)$ is empty, we simply draw e_1 and e_2 inside the free region $\mathcal{R}(\mu)$. Otherwise, the edges e_1 and e_2 are then drawn as depicted in Fig. 4a within $\mathcal{R}(\mu)$. Further, we add a virtual edge $e = (s'_\mu, t'_\mu)$ to $skel(\mu)$ which represents the modified node ν_1. The reserved region of ν_1 is defined as $\mathcal{R}(\nu_1)$ and free as it is a subset of $\mathcal{R}(\mu)$. For ν_2, the reference edge is (s_μ, t_μ) which is already drawn vertically. Hence, we maintain I.1, I.2, I.3 and I.4.

Second, consider the case that μ has two S-node children ν_1 and ν_2. By the maximum vertex degree, the virtual edges in $skel(\nu_1)$ and $skel(\nu_2)$ incident to s_μ and t_μ correspond to Q-node children ξ_1, ξ_2 of ν_1 and ξ_3, ξ_4 of ν_2. Further, let $e_i = (u_i, v_i)$ be the virtual edge in $skel(\nu_i)$ corresponding to ξ_i, such that $s_\mu = u_1 = u_3$ and $t_\mu = u_2 = u_4$. We remove the edges e_1, e_2 from $skel(\nu_1)$ and redefine the poles s_{ν_1} as v_1 and t_{ν_1} as v_2. Similarly, we remove e_3, e_4 from $skel(\nu_2)$ such that $s_{\nu_2} = v_3$ and $t_{\nu_2} = v_4$. Next, we draw the edges $e_1, \dots, e_4$ inside $\mathcal{R}(\mu)$ such that e_1 crosses e_4 while e_2, e_3 are drawn crossing-free. Moreover, e_1 is drawn at slope 1 and e_4 at slope -1. If the skeleton of ν_1 is not empty after its modification, we insert a vertical virtual edge $e = (v_1, v_2)$ for the remainder of ν_1 and define the reserved region $\mathcal{R}(\nu_1)$ which is free as it is a subset of $\mathcal{R}(\mu)$. Similarly, we add a vertical virtual edge $e' = (v_3, v_4)$ if $skel(\nu_2)$ is non-empty and define its reserved region $\mathcal{R}(\nu_2)$ which again is free as it also is a subset of $\mathcal{R}(\mu)$. See Fig. 4b for the construction. Otherwise, the crossing-free edges e_2, e_3 are drawn as shown in Fig. 4c inside $\mathcal{R}(\mu)$. The inserted virtual edges and the respective reserved regions clearly fulfill I.1 and I.2, while all other drawn edges correspond to real edges in G and are not considered in the subsequent processing of $\mathcal{T}$. The only crossing is the one of e_1 and e_4, which maintains I.3. Due to this crossing, all vertices are incident to the outer cell, guaranteeing I.4.

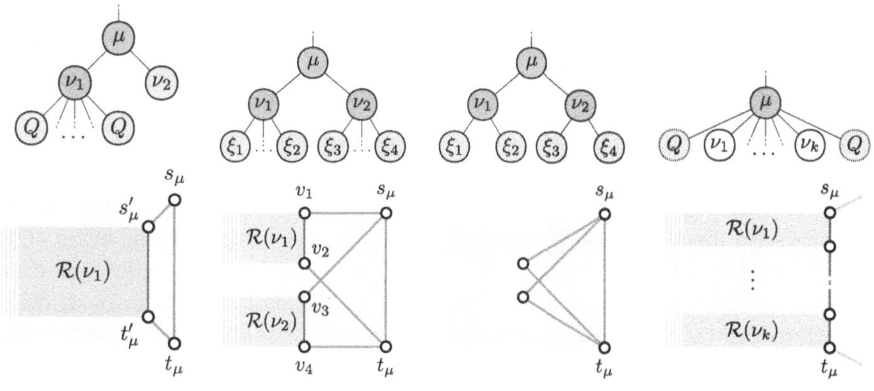

Fig. 4. (a–c) Treatment of P-nodes in the algorithm in the proof of Theorem 6. (d) Treatment of S-nodes in the algorithm in the proof of Theorem 6

μ is a S-node. $skel(\mu)$ consists of a path of virtual edges from s_μ to t_μ. s_μ and t_μ were already placed when processing the parent node in $\mathcal{T}$ such that (s_μ, t_μ) is a vertical segment by I.1. Let $e_i = (u_i, v_i)$ be the virtual edge corresponding to child node ν_i. We draw $e_1, \ldots, e_k$ as equally sized, consecutive vertical lines between s_μ and t_μ, maintaining I.1 and I.4. It follows that $\mathcal{R}(\mu)$ which is free by I.2 is vertically divided into k equally sized sub-regions; see Fig. 4d. As the reserved regions $\mathcal{R}(\nu_1), \ldots, \mathcal{R}(\nu_k)$ only share common borders and are located inside $\mathcal{R}(\mu)$, I.2 is guaranteed. As there are no crossings, I.3 is maintained.

Correctness. For a parent virtual edge corresponding to a not yet drawn node μ, the endpoints are the poles s_μ, t_μ. Thus, I.1 ensures that the reserved region $\mathcal{R}(\mu)$ is well defined. Further, I.2 guarantees that there is no overlap between different parts of the drawing, as the skeleton of each node μ is drawn within $\mathcal{R}(\mu)$ and each child of μ gets a free area that is part of $\mathcal{R}(\mu)$ as described above. Since reserved regions of sibling nodes do not overlap, all occurring crossings in Γ are drawn explicitly in our algorithm and I.3 ensures that they are right-angled and have the same slope. Finally, all vertices in G are incident to the outer face by I.4. As I.1 to I.4 are maintained, the resulting drawing is outer-apRAC.

Running Time. We construct $\mathcal{T}$ in linear time due to [32]. As there are $\mathcal{O}(1)$ operations per node, the top-down visit is in $\mathcal{O}(n)$, resulting in $\mathcal{O}(n)$ time. □

Using similar, more sophisticated techniques we prove in an extended preprint [28]:

Theorem 7. *Let G be a biconnected SP-graph with maximum degree 4. An outer-RAC drawing of G can be computed in $\mathcal{O}(n)$ time.*

6 Open Problems

We conjecture that all subcubic planar graphs and all subquartic SP-graphs are outer-apRAC. For the first conjecture, one needs to draw triconnected subcubic planar graphs whereas for the second one we need a technique for S-nodes that introduces crossings in certain cases. Also, we believe that a high vertex degree at any vertex is also an obstruction for outer-RAC; such a property may be useful for a full characterization. Moreover, an efficient recognition algorithm for general SP- or planar graphs is of interest as well as an area-efficient drawing algorithm for subquartic SP-graphs. To this end note that our drawing algorithms produce exponential-area drawings. Finally, Theorem 1 motivates to study outer-RAC drawings where edges have bends or are drawn with circular arcs [15]. To this end, also note that every graph is outer-apRAC with three bends per edge [23].

References

1. Angelini, P., Bekos, M.A., Förster, H., Kaufmann, M.: On RAC drawings of graphs with one bend per edge. Theor. Comput. Sci. **828–829**, 42–54 (2020). https://doi.org/10.1016/J.TCS.2020.04.018
2. Angelini, P., Bekos, M.A., Katheder, J., Kaufmann, M., Pfister, M.: RAC drawings of graphs with low degree. In: Szeider, S., Ganian, R., Silva, A. (eds.) 47th Int. Symposium on Mathematical Foundations of Computer Science. MFCS 2022. LIPIcs, vol. 241, pp. 11:1–11:15. Schloss Dagstuhl - Leibniz-Zentrum für Informatik (2022). https://doi.org/10.4230/LIPICS.MFCS.2022.11
3. Angelini, P., Bekos, M.A., Katheder, J., Kaufmann, M., Pfister, M., Ueckerdt, T.: Axis-parallel right angle crossing graphs. In: Gørtz, I.L., Farach-Colton, M., Puglisi, S.J., Herman, G. (eds.) 31st Annual European Symposium on Algorithms. ESA 2023. LIPIcs, vol. 274, pp. 9:1–9:15. Schloss Dagstuhl - Leibniz-Zentrum für Informatik (2023). https://doi.org/10.4230/LIPICS.ESA.2023.9
4. Angori, L., Didimo, W., Montecchiani, F., Pagliuca, D., Tappini, A.: Hybrid graph visualizations with chordlink: algorithms, experiments, and applications. IEEE Trans. Vis. Comput. Graph. **28**(2), 1288–1300 (2022). https://doi.org/10.1109/TVCG.2020.3016055
5. Argyriou, E.N., Bekos, M.A., Kaufmann, M., Symvonis, A.: Geometric RAC simultaneous drawings of graphs. J. Graph Algorithms Appl. **17**(1), 11–34 (2013). https://doi.org/10.7155/JGAA.00282
6. Argyriou, E.N., Bekos, M.A., Symvonis, A.: The straight-line RAC drawing problem is NP-hard. J. Graph Algorithms Appl. **16**(2), 569–597 (2012). https://doi.org/10.7155/JGAA.00274
7. Arikushi, K., Fulek, R., Keszegh, B., Moric, F., Tóth, C.D.: Graphs that admit right angle crossing drawings. Comput. Geom. **45**(4), 169–177 (2012). https://doi.org/10.1016/J.COMGEO.2011.11.008
8. Auer, C., et al.: Outer 1-planar graphs. Algorithmica **74**(4), 1293–1320 (2016). https://doi.org/10.1007/S00453-015-0002-1
9. Bekos, M.A., Cornelsen, S., Grilli, L., Hong, S., Kaufmann, M.: On the recognition of fan-planar and maximal outer-fan-planar graphs. Algorithmica **79**(2), 401–427 (2017). https://doi.org/10.1007/S00453-016-0200-5

10. Bekos, M.A., Didimo, W., Liotta, G., Mehrabi, S., Montecchiani, F.: On RAC drawings of 1-planar graphs. Theor. Comput. Sci. **689**, 48–57 (2017). https://doi.org/10.1016/J.TCS.2017.05.039
11. Bekos, M.A., van Dijk, T.C., Kindermann, P., Wolff, A.: Simultaneous drawing of planar graphs with right-angle crossings and few bends. J. Graph Algorithms Appl. **20**(1), 133–158 (2016). https://doi.org/10.7155/JGAA.00388
12. Biedl, T.: Drawing outer-1-planar graphs revisited. J. Graph Algorithms Appl. **26**(1), 59–73 (2022). https://doi.org/10.7155/JGAA.00581
13. Binucci, C., et al.: Fan-planarity: properties and complexity. Theor. Comput. Sci. **589**, 76–86 (2015). https://doi.org/10.1016/J.TCS.2015.04.020
14. Brand, C., Ganian, R., Röder, S., Schager, F.: Fixed-parameter algorithms for computing RAC drawings of graphs. In: Bekos, M.A., Chimani, M. (eds.) GD 2023. LNCS, vol. 14466, pp. 66–81. Springer, Cham (2023). https://doi.org/10.1007/978-3-031-49275-4_5
15. Chaplick, S., Förster, H., Kryven, M., Wolff, A.: Drawing graphs with circular arcs and right-angle crossings. In: Albers, S. (ed.) 17th Scandinavian Symposium and Workshops on Algorithm Theory. SWAT 2020. LIPIcs, vol. 162, pp. 21:1–21:14. Schloss Dagstuhl - Leibniz-Zentrum für Informatik (2020). https://doi.org/10.4230/LIPICS.SWAT.2020.21
16. Chaplick, S., Lipp, F., Wolff, A., Zink, J.: Compact drawings of 1-planar graphs with right-angle crossings and few bends. Comput. Geom. **84**, 50–68 (2019). https://doi.org/10.1016/J.COMGEO.2019.07.006
17. Dehkordi, H.R., Eades, P.: Every outer-1-plane graph has a right angle crossing drawing. Int. J. Comput. Geom. Appl. **22**(6), 543–558 (2012). https://doi.org/10.1142/S021819591250015X
18. Dehkordi, H.R., Eades, P., Hong, S., Nguyen, Q.H.: Circular right-angle crossing drawings in linear time. Theor. Comput. Sci. **639**, 26–41 (2016). https://doi.org/10.1016/J.TCS.2016.05.017
19. Di Battista, G., Tamassia, R.: On-line maintenance of triconnected components with SPQR-trees. Algorithmica **15**(4), 302–318 (1996). https://doi.org/10.1007/BF01961541
20. Di Battista, G., Tamassia, R.: On-line planarity testing. SIAM J. Comput. **25**(5), 956–997 (1996). https://doi.org/10.1137/S0097539794280736
21. Di Giacomo, E., Didimo, W., Eades, P., Liotta, G.: 2-layer right angle crossing drawings. Algorithmica **68**(4), 954–997 (2014). https://doi.org/10.1007/S00453-012-9706-7
22. Didimo, W.: Density of straight-line 1-planar graph drawings. Inf. Process. Lett. **113**(7), 236–240 (2013). https://doi.org/10.1016/J.IPL.2013.01.013
23. Didimo, W., Eades, P., Liotta, G.: Drawing graphs with right angle crossings. Theor. Comput. Sci. **412**(39), 5156–5166 (2011). https://doi.org/10.1016/J.TCS.2011.05.025
24. Didimo, W., Liotta, G., Montecchiani, F.: A survey on graph drawing beyond planarity. ACM Comput. Surv. **52**(1), 4:1–4:37 (2019). https://doi.org/10.1145/3301281
25. Diestel, R.: Graph Theory. Graduate Texts in Mathematics, 5th edn, vol. 173. Springer, Heidelberg (2012). https://doi.org/10.1007/978-3-662-53622-3
26. Eppstein, D.: Parallel recognition of series-parallel graphs. Inf. Comput. **98**(1), 41–55 (1992). https://doi.org/10.1016/0890-5401(92)90041-D
27. Förster, H., Ganian, R., Klute, F., Nöllenburg, M.: On strict (outer-)confluent graphs. J. Graph Algorithms Appl. **25**(1), 481–512 (2021). https://doi.org/10.7155/JGAA.00568

28. Förster, H., Katheder, J., Ortali, G.: Outer-(ap)RAC graphs. CoRR abs/2411.17565 (2024). https://doi.org/10.48550/ARXIV.2411.17565
29. Förster, H., Kaufmann, M.: On compact RAC drawings. In: Grandoni, F., Herman, G., Sanders, P. (eds.) 28th Annual European Symposium on Algorithms. ESA 2020. LIPIcs, vol. 173, pp. 53:1–53:21. Schloss Dagstuhl - Leibniz-Zentrum für Informatik (2020). https://doi.org/10.4230/LIPICS.ESA.2020.53
30. Fortunato, S.: Community detection in graphs. Phys. Rep. **486**(3), 75–174 (2010). https://doi.org/10.1016/j.physrep.2009.11.002
31. Girvan, M., Newman, M.E.J.: Community structure in social and biological networks. Proc. Natl. Acad. Sci. **99**(12), 7821–7826 (2002). https://doi.org/10.1073/pnas.122653799
32. Gutwenger, C., Mutzel, P.: A linear time implementation of SPQR-trees. In: Marks, J. (ed.) GD 2000. LNCS, vol. 1984, pp. 77–90. Springer, Heidelberg (2001). https://doi.org/10.1007/3-540-44541-2_8
33. Henry, N., Fekete, J., McGuffin, M.J.: Nodetrix: a hybrid visualization of social networks. IEEE Trans. Vis. Comput. Graph. **13**(6), 1302–1309 (2007). https://doi.org/10.1109/TVCG.2007.70582
34. Hong, S., Eades, P., Katoh, N., Liotta, G., Schweitzer, P., Suzuki, Y.: A linear-time algorithm for testing outer-1-planarity. Algorithmica **72**(4), 1033–1054 (2015). https://doi.org/10.1007/S00453-014-9890-8
35. Hong, S.-H., Nagamochi, H.: Testing full outer-2-planarity in linear time. In: Mayr, E.W. (ed.) WG 2015. LNCS, vol. 9224, pp. 406–421. Springer, Heidelberg (2016). https://doi.org/10.1007/978-3-662-53174-7_29
36. Hong, S., Tokuyama, T. (eds.): Beyond Planar Graphs, Communications of NII Shonan Meetings. Springer, Cham (2020). https://doi.org/10.1007/978-981-15-6533-5
37. Huang, W.: Using eye tracking to investigate graph layout effects. In: Hong, S., Ma, K. (eds.) APVIS 2007, 6th International Asia-Pacific Symposium on Visualization 2007, pp. 97–100. IEEE Computer Society (2007). https://doi.org/10.1109/APVIS.2007.329282
38. Huang, W., Eades, P., Hong, S.: Larger crossing angles make graphs easier to read. J. Vis. Lang. Comput. **25**(4), 452–465 (2014). https://doi.org/10.1016/J.JVLC.2014.03.001
39. Huang, W., Hong, S., Eades, P.: Effects of crossing angles. In: IEEE VGTC Pacific Visualization Symposium 2008. PacificVis 2008. pp. 41–46. IEEE Computer Society (2008). https://doi.org/10.1109/PACIFICVIS.2008.4475457
40. Kaufmann, M., Klemz, B., Knorr, K., Reddy, M.M., Schröder, F., Ueckerdt, T.: The density formula: one lemma to bound them all. CoRR abs/2311.06193 (2023). https://doi.org/10.48550/ARXIV.2311.06193
41. Purchase, H.C.: Effective information visualisation: a study of graph drawing aesthetics and algorithms. Interact. Comput. **13**(2), 147–162 (2000). https://doi.org/10.1016/S0953-5438(00)00032-1
42. Rahmati, Z., Emami, F.: RAC drawings in subcubic area. Inf. Process. Lett. **159–160**, 105945 (2020). https://doi.org/10.1016/J.IPL.2020.105945
43. Schaefer, M.: RAC-drawability is $\exists\mathbb{R}$-complete and related results. J. Graph Algorithms Appl. **27**(9), 803–841 (2023). https://doi.org/10.7155/JGAA.00646

44. Tamassia, R. (ed.): Handbook on Graph Drawing and Visualization. Chapman and Hall/CRC, Boca Raton (2013). https://www.crcpress.com/Handbook-of-Graph-Drawing-and-Visualization/Tamassia/9781584884125
45. Tóth, C.D.: On RAC drawings of graphs with two bends per edge. In: Bekos, M.A., Chimani, M. (eds.) GD 2023. LNCS, vol. 14465, pp. 69–77. Springer, Cham (2023). https://doi.org/10.1007/978-3-031-49272-3_5

Forest Covers and Bounded Forest Covers

Daya Ram Gaur[1], Barun Gorain[2], Shaswati Patra[2], and Rishi Ranjan Singh[2(✉)]

[1] University of Lethbridge, Alberta, Canada
gaur@cs.uleth.ca
[2] Indian Institute of Technology Bhilai, Raipur, Chattisgarh, India
{barun,shaswatip,rishi}@iitbhilai.ac.in

Abstract. We study approximation algorithms for the forest cover and bounded forest cover problems. A probabilistic $2+\epsilon$ approximation algorithm for the forest cover problem is given using the method of dual fitting. A deterministic algorithm with a 2-approximation ratio that rounds the optimal solution to a linear program is given next. The 2-approximation for the forest cover is then used to give a 6-approximation for the bounded forest cover problem. The use of the probabilistic method to develop the $2 + \epsilon$ approximation algorithm may be of independent interest.

Keywords: Vertex cover · Forest Cover · Approximation Algorithm · Randomized Algorithm · Linear Programming · Dual Fitting · LP Rounding

1 Introduction

Let $G = (V, E)$ be an undirected graph with positive weights in the interval $[0, 1]$ on the edges given by $w : E \rightarrow [0, 1]$. A vertex cover $C \subseteq V$ is a subset of vertices such that every edge in E is incident on some vertex in C. A minimum vertex cover is a vertex cover of minimum cardinality. A tree is a connected component without any cycles. A collection of trees is called forest. The weighted index (wi) of a forest $F = \{T_1, T_2, \ldots T_k\}$ is calculated as follows: $wi(F) = \sum_{e\in E(F)} w_e + k$, where w_e is the weight of edge e, k is the number of connected components in F and $E(F)$ is the set of edges in F. The number of connected components $k = \sum_{i=1}^{k} |T_i| - |E(F)|$ where $|T_i|$ is the number of vertices in tree T_i. So, $wi(F) = \sum_{e\in E(F)} w_e + \sum_{i=1}^{k} |T_i| - |E(F)|$ which can be rewritten as

$$wi(F) = \sum_{i=1}^{k} |T_i| - \sum_{e\in E(F)} (1 - w_e).$$

Full version of this paper is available at https://arxiv.org/abs/2411.16578.

R. Královič and V. Kůrková (Eds.): SOFSEM 2025, LNCS 15538, pp. 300–313, 2025.
https://doi.org/10.1007/978-3-031-82670-2_22

Our goal is to find a pair (C, F) where $C \subseteq V$ is a vertex cover and $F \subseteq E$ is a forest in C such that $wi(F)$ is the minimum possible. This problem is referred to as the forest cover problem.

This is a generalization of the unweighted vertex cover problem. The unweighted vertex cover problem is one of the original 21 NP-complete problems [18]. Papadimitriou and Steiglitz [22, p 432] attibute a 2-approximation algorithm using maximal matching to Gavril and Yannakakis. Vertex cover in bounded degree graphs was studied by Berman and Fujito [5] who have a $2 - 5/(d + 3) + \epsilon$ algorithm for graphs with maximum degree d, and Hochbaum gave $2-2/d$ approximation algorithm for graphs with degree d [17]. Vertex cover is hard to approximate within $2-(2+o_d(1))\frac{\log\log d}{\log d}$ under an assumption known as the unique games conjecture [4]. This lower bound on the approximability matches the upper bound due to Halperin [14] up to $o_d(1)$ factor. An unconditional lower bound of 1.36 on the approximation ratio is due to Dinur and Safra [7]. No approximation algorithm with approximation ratio $2 - \epsilon$ for a constant ϵ is known for unweighted vertex cover. Han, Punnen and Ye [15] have $3/2 + \chi$ approximation algorithm for a parameter χ; no examples where $\chi > 0$ were discovered in their extensive empirical evaluation. Parameterized algorithms for vertex cover have been studied for a while. Very recently, Harris and Narayanswamy [16] gave a $O^*(1.25284^k)$ algorithm to find a vertex cover of size k. This beats the previous long-standing bound of $O^*(1.2738^k)$ of Chen, Kanj and Xia [6] since 2010.

Given an unweighted vertex cover instance, we can create an instance of the forest cover problem by assuming a weight function that assigns a weight of 1 to every edge. This is an approximation preserving reduction. Therefore, the problem is NP-complete, and the best approximation ratio that we can hope for is 2 for the forest cover problem as an approximation ratio of α gives the same approximation ratio for vertex cover.

We study another related graph covering problem; the bounded forest cover problem (BFC). Given a graph $G = (V, E)$ with positive weights on the edges and a parameter $\lambda \geq 0$. The goal is to find a minimum-sized collection of trees $T_1, T_2, \ldots T_k$ such that the total weight of the edges in each tree T_i is at most *lambda* and the vertices in $\cup_{i=1}^{k} T_i$ form a vertex cover of the graph. By minimum-sized collection, we mean k should be the smallest possible. We call this forest cover as opposed to tree cover because tree cover is already used with a different meaning.

The forest cover problem is inspired by graph covering problems and min-max vehicle routing problems [2,3,9,11,12,20]. It is also motivated by the need to dig tunnels or create crossing paths in mine-ridden areas [2]. Similarly, the forest cover problem aims to establish tree-shaped facilities, considering a more general cost function with two components. One part evaluates the cost of travel, while the other accounts for the capital cost of deploying vehicles/robots, etc. The main goal is to minimize deployment and operating costs.

The bounded forest cover problem [11] addresses the gap between tree cover, minimum tree cover, and min-max tree cover problems. It focuses on a variant where the cost function is the same as in the minimum tree cover problem [3], and the goal of graph covering is similar to the tree cover problem in [2]. This problem is motivated by the applications of the tree cover problem [2], with an additional constraint on the weights of trees that need to be selected for covering tasks. This constraint is inspired by modern technologies such as drones and electric vehicles, which can travel limited distance on one full charge. The bounded forest cover problem is relevant to applications of the tree cover problem where the coverage task must be completed using devices with limited operational duration before refueling or recharging.

There are two popular variants of graph covering problems. The first variant seeks covering of the edges, while the second requires a covering of the vertices [2,3,20]. Vertex cover belongs to the former category, while minimum spanning tree and traveling salesman problem are of the latter type. Given a weighted graph and a positive λ, the bounded tree cover problem is to find a collection of trees, each with weight at most λ, such the union of the vertices in the trees is the vertex set itself. Khani and Salvatipour gave a 2.5 approximation algorithm for the bounded tree cover [19]. The special case of when the tree is a path was studied by Levin et al. [3] who gave a 3-approximation algorithm. The bounded path cover problem is a vehicle routing problem where each vehicle travels at a distance of at most λ and the set of vehicles serves all the nodes. A variant occurs when λ is unbounded, and there is a restriction on the number of vehicles (at most k). Here, given a weighted graph, the goal is to find a collection of k paths that cover all the vertices; here, Wu et al. [23] gave a 3/2 approximation algorithm under the assumption that the edge weights satisfy triangle inequality and each vertex is visited exactly once. Suppose we choose the covering subgraph to be a cycle. In that case, we obtain the bounded cycle cover problem, where the vertices of a given graph have to be covered by cycles. The objective is to minimize the number of cycles subject to a maximum length. A 32/7 approximation for the bounded cycle cover problem is due to Yu et al. [24]. The literature on covering graphs by subgraphs is vast, and its numerous approximation algorithms are known. Results in the following papers are the closest to the forest cover problem [2,9,11,19,21]. The main difference between the problems listed in this paragraph and the bounded forest cover problem considered here is the coverage constraint. The coverage constraint in bounded forest cover is on edges; each edge needs to be covered by some vertex in the vertex cover as opposed to covering all the vertices.

1.1 Contributions

Both the forest cover and the bounded forest cover problems are NP-complete. We study approximation algorithms for them. For the forest cover problem, we give a probabilistic algorithm with $2 + \epsilon$ approximation ratio in Theorem 3. We give a deterministic algorithm with 2 approximation ratio in the full version of the paper. For the bounded forest cover problem we give a 6 approximation algorithm. This is the first study on the forest cover problem, and the approximation ratio in this paper is best expected given the conditional hardness results for vertex cover [4,7]. The bounded forest cover problem was first studied in [11], where an 8-approximation algorithm was given. The result in Theorem 5 improves the approximation ratio to 6.

We use the probabilistic method [1] to obtain the $2+\epsilon$-factor approximation algorithm for the forest cover problem. First, we show using the method of dual fitting, in Theorem 2, that the restriction of the forest cover problem admits a 2-approximation. Now, given a graph with weights in the interval $[0,1]$ on the edges, we create a family of LP relaxations that are easy to solve using the result in Theorem 2. We show in Theorem 3 that the average solution to the family satisfies the LP dual of the problem. Each dual solution in the LP family has a corresponding 2-approximate integral solution. We pick the primal integral solution with the smallest value. Such a solution satisfies the primal constraint (to the original problem) and is guaranteed to exist, and is a $2+\epsilon$-approximate solution (Theorem 4). The novel use of the probabilistic method is an essential contribution to this paper, and it might have independent applications.

We present a deterministic algorithm with a 2-approximation ratio for the forest cover problem using LP rounding which is placed in Sect. ?? due to space limit. The algorithm rounds the variables in the solution of the relaxed LP formulation to obtain a forest cover. Each connected component in the subgraph induced by the non-zero variables in the LP optimal solution is treated separately. A minimum spanning tree (MST) is constructed in each component; pendent vertices of MST with low fractional values and edges incident on such vertices are discarded without violating the covering constraint.

2 Forest Cover Problem (FC)

FC Problem: Consider an undirected weighted graph $G = (V, E, w)$ where $w : E \to [0,1]$. A forest is an acyclic subgraph of a graph G. We denote the edges in a forest F as $E(F)$. A vertex cover $C \subseteq V$ is a set of vertices such that for every edge $e = (u, v) \in E$, at least one of u, v is in C. A forest cover of a graph G is a forest in G such that the vertices in the forest form a vertex cover. The weighted index (WI) of a forest $F = \{T_1, T_2, \dots T_k\}$ is calculated $wi(F) = \sum_{i=1}^{k} |T_i| - \sum_{e \in E(F)} (1 - w_e)$, where $|T_i|$ is the number of vertices in tree T_i. The objective is to find a forest cover for a given graph with a minimum weighted index. The decision version asks whether a given graph has a forest cover with WI at most d for some non-negative real number d. See the full version of the paper for an approximation preserving reduction.

Theorem 1. *The forest cover (FC) problem is NP-complete.*

2.1 ILP Formulation for Forest Cover

In this section, we give an integer linear programming formulation for forest cover. This formulation is similar to the ones in [10,13] for the Steiner tree problem.

We use binary variables x_i for each vertex $i \in V$ and y_{ij} for each edge $(i,j) \in E$. The variable x_i is set to 1 if vertex i is present in the forest cover. Otherwise, x_i is 0. Similarly, the variable y_{ij} is set to 1 if edge (i,j) is in the forest cover, otherwise, $y_{ij} = 0$. For $S \subseteq V$, we use $E(S)$ to refer to the edges with both endpoints in S.

$$\min \sum_{i \in V} x_i - \sum_{(i,j) \in E} y_{ij}(1 - w_{ij})$$

$$\begin{aligned} x_i + x_j &\geq 1 && \forall (i,j) \in E \\ x_i &\geq y_{ij} && \forall i \in V, \forall (i,j) \in E \\ \sum_{i \in S} x_i - \sum_{(i,j) \in E(S)} y_{ij} &\geq 1 && \forall S \subseteq V,\ s.t.\ E(S) \neq \emptyset \\ x_i &\in \{0,1\} && \forall i \in V \\ y_{ij} &\in \{0,1\} && \forall (i,j) \in E \end{aligned}$$

The first constraint ensures that at least one end vertex of each edge must be present in the solution. The second constraint ensures that an edge is present in the solution only if both end vertices are present. The third constraint ensures that cycles are absent. The objective function is the weighted index of the forest determined by the values of the variables x, y. The number of constraints in the above ILP is exponential. The following lemma establishes that the optimal solution of the corresponding relaxed LP can be obtained in polynomial time. The proof is available in the full version of the paper.

Lemma 1. *The relaxed linear programming problem of the above ILP can be solved optimally using the ellipsoid method.*

3 Probabilistic Algorithm for Forest Cover

In this section, we propose a probabilistic algorithm for forest cover with approximation factor arbitrarily close to 2. The algorithm is described in two steps. In the first step, we present a deterministic 2-approximation algorithm for forest cover where the weights on the edges are either 0 or 1. In the second step, we use the algorithm for binary weights as a subroutine and give a probabilistic $(2+\epsilon)$-factor approximation algorithm for forest cover, where ϵ is a positive real close to 0.

3.1 Binary Weights

Let $G = (V, E)$ be a graph with binary weights on the edges, i.e., the edge weights are either 0 or 1. We say edge $e \in E(S)$, for $S \subseteq V$ if both the endpoints of e are in the vertex set S. We also use $u \in e$ to refer to the fact that edge e is incident on a vertex u. Let us recall the primal integer program for the forest cover but this time we use different labels for the indices. We call this linear program as P.

$$\min \sum_{u \in V} x_u - \sum_{e \in E} y_e(1 - w_e) \tag{1}$$

$$x_u + x_v \geq 1 \qquad \forall e = (u, v) \in E \tag{2}$$

$$x_u - y_e \geq 0 \tag{3}$$

$$x_v - y_e \geq 0 \qquad \forall e = (u, v) \in E \tag{4}$$

$$\sum_{u \in S} x_u - \sum_{e \in E(S)} y_e \geq 1 \qquad \forall S \subseteq V, \text{ s.t. } E(S) \neq \emptyset \tag{5}$$

$$x_u, y_e \in \{0, 1\} \qquad \forall u \in V, e \in E \tag{6}$$

The dual variables associated with the first set of constraints is z_e, z_{ue}, z_{ve} are dual variables associated with the next two constraints, and the dual variable associated with the last set of constraints is z_S.

The linear programming dual of the integer program above is

$$\max \sum_{e \in E} z_e + \sum_{S \subseteq V} z_S \tag{7}$$

$$\sum_{e:u \in e} z_e + \sum_{e:u \in e} z_{ue} + \sum_{S:u \in S} z_S \leq 1 \qquad \forall u \in V \tag{8}$$

$$\sum_{S:e \in E(S)} z_S + z_{ue} + z_{ve} \geq (1 - w_e) \qquad \forall e \in E \tag{9}$$

$$z_e, z_{ue}, z_{ve}, z_S \geq 0 \qquad \forall e \in E,\ \forall u \in V,\ S \subseteq V \tag{10}$$

Let E_i be the set of edges of weight $i \in \{0, 1\}$. Let V_0 be the set of vertices incident on some edge in E_0 and $V_1 = V \setminus V_0$. Let $E_i(V')$ be the set of edges in E_i where $i \in \{0, 1\}$ with both the endpoints in $V' \subseteq V$. Notice that there are edges of weight 1 with both the endpoints in V_0, and any such edge is not in $E_0(V_0)$. The only edges with both endpoints in V_1 are of weight 1. The edges with one endpoint in V_0 and the other in V_1 are all of weight 1.

$G_0(V_0, E_0(V_0))$ is the subgraph with vertices in V_0 and all the edges in G_0 are of weight 0. Similarly, we define $G_1(V_1, E_1(V_1))$. Stated otherwise, $G_0(V_0, E_0(V_0))$ is the subgraph of G with edges with weight 0, and all the vertices in V_0 are incident on some edges in E_0. The subgraph $G_1(V_1, E_1(V_1))$ contains vertices that are not incident on any weight 0 edges, and all edges in this subgraph have weight 1. Let the number of connected components $C_1, C_2, \ldots, C_k$ in $G_0(V_0, E_0(V_0))$ be k. For each connected component C_i we identify a tree T_i with $|C_i| - 1$ edges. In the subgraph $G_1(V_1, E_1(V_1))$ we find a maximum cardinality matching M.

We will show that $|M| + k$ is a lower bound on the value of the optimal solution to the forest cover problem. We will construct a feasible solution to the dual with-value $k + |M|$.

Lemma 2. *$|M| + k$ is a lower bound on the value of the optimal solution to the forest cover problem.*

Proof. Recall, the k connected components of $G_0(V_0, E_0(V_0))$ are $C_1, C_2, \ldots, C_k$. For each connected component C_i, we have a tree T_i with $|C_i| - 1$ edges. Starting with an initial solution in which all the dual variables are 0, compute a feasible solution as follows:

- For each set of vertices S equal to some C_i, set $z_S = 1$.
- For each edge $e = (u, v) \in M$, set $z_S = 1$, where $S = \{u, v\}$.

Note that the sets $S \subseteq V$ for which $z_S = 1$ are pairwise disjoint. The objective function value is $k + |M|$. What remains to be shown is that the solution is feasible.

First we show that constraints given by (8) are satisfied. Each vertex $u \in V_0$ is in one connected component. Each vertex $u \in V_1$ is incident on at most one edge $e \in M$. In both the cases

$$\sum_{S:u \in S} z_S \leq 1 \text{ and } \sum_{e:u \in e} z_e + \sum_{e:u \in e} z_{ue} = 0$$

Therefore, the first constraint is satisfied for all the vertices. Constraint (9) is interesting only for edges with 0 as it is trivially satisfied for edges with weight 1. Each edge weight 0 is in some connected component C_i (and only one). Therefore, $\sum_{S:e \in E(S)} z_S = 1$ for any edge with weight 0.

Since the solution is a feasible one, the LP relaxation of the primal has a value at least $k + |M|$ (by weak duality). The optimal value for the LP relaxation of the primal is a lower bound on the optimal value of forest cover. □

This lower bound gives us a simple 2-approximation algorithm (Algorithm 1) for the forest cover problem. There are two stages. In the first stage, we compute the connected components in $G_0(V_0, E_0(V_0))$ and create an assignment to the primal variables based on the connected components. In the second stage, we compute a maximum matching $G_1(V_1, E_1(V_1))$ and determine the values of the primal variables. The solution that we construct will be feasible.

The following lemma shows that the solution obtained according to Algorithm 1 is a feasible solution to the primal integer linear program.

Lemma 3. *The solution (x, y) returned by Algorithm 1 is a feasible solution of the primal integer linear program for forest cover.*

Proof. Every edge is incident on some vertex in C_i or incident on an edge in M, therefore, constraint (2) is satisfied. Take any edge $e = (u, v)$ with $y_e = 1$, both the variables $x_u.x_v$ are set to 1. So, (3) and (4) is satisfied. Finally, for any $S \subseteq V$, if $e \in S$ and $y_e = 1$ then both the endpoints of edge have x_u, x_v set to 1: either the y_e was set to 1 in the matching or in the construction of the connected components. Edges in S with $y_e = 1$ form a forest, therefore, constraint (5) is satisfied. □

Algorithm 1: ForestCoverBinary(G)

Let the k connected components of $G_0(V_0, E_0(V_0))$ be $C_1, C_2, \ldots, C_k$. For each connected component C_i we identify a tree T_i with $|C_i| - 1$ edges.
for *each vertex* $u \in \cup_{i=1}^{k} C_i$ **do**
  set $x_u = 1$.
for *each edge* $e \in \cup_{i=1}^{k} T_i$ **do**
  Set $y_e = 1$. The edges that are in this subgraph but not in any tree are assigned a value of 0.
Find a maximum matching M in $G_1(V_1, E_1(V_1))$.
for *each edge* $e = (u, v) \in M$ **do**
  Set $x_u = 1$, $x_v = 1$, and $y_e = 1$.
For all the other vertices, set $x_u = 0$.

The solution constructed above is a feasible integer solution to the primal, and the objective function value is $k + 2|M|$. Therefore, we have the following theorem.

Theorem 2. *Algorithm 1 is a 2-factor approximation algorithm for the forest cover problem on graphs with binary weights.*

3.2 Real Weights

Now we consider the case when the weights on the edges are in the closed interval $[0, 1]$. Let ϵ be a very small positive real close to 0 and $\delta = \epsilon^2$. Let us introduce a very small error δ to the objective function of the linear program P as follows.

$$\sum_{u \in V} x_u - \sum_{e \in E} y_e(1 - w_e - \delta) \tag{11}$$

where δ is a small real positive arbitrarily close to 0. With Eq. 11 as the objective function and the same set of constraints as in P, we call this linear program P'. For arbitrarily small δ, P and P' admits the same optimal solution (X^*, Y^*). Let OPT and OPT' be the optimal values of the objective functions of P and P', respectively. Then $OPT' = OPT + \delta \sum_e y_e^*$.

For ϵ very small, $\epsilon \sum_e y_e^*$ is smaller than OPT, therefore $\delta \sum_e y_e^* \leq \epsilon OPT$. This implies that $OPT' \leq (1 + \epsilon)OPT$. Let D' be the dual of P'. Then D' has the same objective function and the constraint 8 as D. The constraint 9 of D is changed to the following inequality.

$$\sum_{S: e \in E(S)} z_S + z_{ue} + z_{ve} \geq 1 - w_e - \delta \quad \forall e \in E \tag{12}$$

With the definition of D', we are now ready to explain the algorithm in this section. For any edge $e \in E$, let W_e be an indicator variable which is 1 with probability $(1 - w_e)$ (0 with probability w_e). We replace the RHS in the last constraint of the D with this indicator variable. This gives a family of linear programs in which each edge has a weight of 0 or 1.

$$\max \sum_{e \in E} z_e + \sum_{S \subseteq V} z_S \tag{13}$$

$$\sum_{e:u \in e} z_e + \sum_{e:u \in e} z_{ue} + \sum_{S:u \in S} z_S \leq 1 \qquad \forall u \in V \tag{14}$$

$$\sum_{S:e \in E(S)} z_S + z_{ue} + z_{ve} \geq W_e \qquad \forall e \in E \tag{15}$$

$$z_e, z_{ue}, z_{ve}, z_S \geq 0 \qquad \forall e \in E, S \subseteq V \tag{16}$$

Similarly, the last constrain of D' becomes

$$\sum_{S:e \in E(S)} (z_S + z_{ue} + z_{ve}) \geq W_e - \delta \quad \forall e \in E \tag{17}$$

For an experiment, we randomly generate the values $W_e \in \{0, 1\}$ for all edges. This gives us an instance with binary edge weights in $\{0, 1\}$. We can compute a lower bound for this instance and also an upper bound using the results in the previous section.

Suppose we run m such experiments $E_1, \ldots, E_m$. Each of these experiments gives us a feasible solution z^i to the dual LP where $i \in \{1, \ldots, m\}$. Let $\overline{z}$ be the average value of the variables in all the solutions.

Theorem 3. *The average solution $\overline{z}$ over m experiments, where n is the number of edges, $m = n/(2\delta^2)$ and $0 < \delta < 1$, is a feasible solution to D' with a high probability, and the objective function value given by the average solution of all the experiments, $\sum_{e \in E} \overline{z}_e + \sum_{S \subseteq V} \overline{z}_S$ is a lower bound on OPT', the minimum value of the objective function of P'.*

Proof. Since z^i is a feasible solution of D corresponding to the i-th experiment, we have the following sequence of inequalities.

$$\sum_{e:u \in e} z_e^i + \sum_{e:u \in e} z_{ue}^i + \sum_{S:u \in S} z_S^i \leq 1$$

$$\sum_{i=1}^{m} \sum_{e:u \in e} z_e^i + \sum_{i=1}^{m} \sum_{e:u \in e} z_{ue}^i + \sum_{i=1}^{m} \sum_{S:u \in S} z_S^i \leq m$$

$$\sum_{e:u \in e} \sum_{i=1}^{m} z_e^i + \sum_{e:u \in e} \sum_{i=1}^{m} z_{ue}^i + \sum_{S:u \in S} \sum_{i=1}^{m} z_S^i \leq m$$

$$\sum_{e:u \in e} \overline{z}_e + \sum_{e:u \in e} \overline{z}_{ue} + \sum_{S:u \in S} \overline{z}_S \leq 1$$

This shows that $\overline{z}$ satisfies the first constraint (14) of D', as the first constrain is same for D and D'. For each i, the last constraint in D is satisfied, summing it over all i and after taking the average we get the following.

$$\sum_{S:e\in E(S)} z_S^i + z_{ue}^i + z_{ve}^i \geq W_e^i$$

$$\sum_{S:e\in E(S)} \sum_{i=1}^{m} z_S^i + \sum_{i=1}^{m} z_{ue}^i + \sum_{i=1}^{m} z_{ve}^i \geq \sum_{i=1}^{m} W_e^i$$

$$\sum_{S:e\in E(S)} \overline{z}_S + \overline{z}_{ue} + \overline{z}_{ve} \geq \frac{1}{m}\sum_{i=1}^{m} W_e^i$$

From Chernoff-Hoeffding bound [8], we know that

$$Pr[\frac{1}{m}\sum_{i=1}^{m} W_e^i \leq (1-w_e)-\delta] \leq \frac{1}{e^{2m\delta^2}}$$

If we choose $m = n/(2\delta^2)$ where n is the number of edges then $Pr[\frac{1}{m}\sum_{i=1}^{m} W_e^i > (1-w_e)-\delta] \leq 1-\frac{1}{e^n}$. Each edge is set to 0 or 1 with probability w_e independent of other edges. Since there are n constraints, the probability that the above inequality satisfies for every edge is close to 1. Hence, $\overline{z}$ is a feasible solution of D' with very high probability. This implies that the objective function value of D' for the solution $\overline{z}$ is a lower bound of the optimal value of P' with high probability. □

Next, we describe the algorithm for real weights. For each experiment E_i we have a feasible solution D_i to the dual and a solution P_i to the primal problem (for integer weights) given the algorithm (Algorithm 1) in the previous section where W_e^i be the indicator variable with value in $\{0,1\}$ which takes the value 1 with probability $1-w_e$ in the ith experiment. The algorithm returns the solution P_j, which has the minimum objective function value among all P_i's. The following theorem proves that the above algorithm is an approximation algorithm with factor close to 2 with very high probability.

Theorem 4. *The proposed algorithm is a $(2+\epsilon)$ factor approximation algorithm for Forest cover problem with high probability.*

Proof. Let $\overline{P}(\overline{D})$ be the average value of the primal (dual) solutions. Then, $\overline{P} = \sum_{i=1}^{m} P_i/m$, where

$$P_i = \sum_{u\in V} x_u^i - \sum_{e\in E} y_e^i W_e^i + 2|M_i|$$

and the value of the dual solution D_i is

$$D_i = \sum_{e\in E} z_e^i + \sum_{S\subseteq V} z_S^i$$

and M_i is the maximum matching in $G_1(V_1, E_1(V_1))$; in the graph obtained in the ith experiment. The average value $\overline{D} = \sum_{i=1}^{m} D_i/m$.

Since each primal solution is a two approximate solution (to the instance) $P_i \leq 2D_i$, the following inequality holds:

$$m\overline{P} = \sum_{i=1}^{m}\sum_{u\in V} x_u^i - \sum_{i=1}^{m}\sum_{e\in E} y_e^i W_e^i + \sum_{i=1}^{m} 2|M_i| \leq 2\,m\overline{D}$$

Let us rewrite the middle term. The binary value of y_e^i is given by the Algorithm in the previous section and it is fixed. The expected value of $y_e^i W_e^i$,

$$\mathbb{E}[y_e^i W_e^i] = y_e^i \mathbb{E}[W_e^i] = y_e^i(1 - w_e).$$

Therefore, we can rewrite the previous equation (in expectation) as

$$m\overline{P} = \sum_{i=1}^{m}\sum_{u\in V} x_u^i - \sum_{i=1}^{m}\sum_{e\in E} y_e^i(1 - w_e) + \sum_{i=1}^{m} 2|M_i| \leq 2\,m\overline{D}$$

$$\overline{P} = \sum_{u\in V} \overline{x}_u - \sum_{e\in E} \overline{y}_e(1 - w_e) + 2|\overline{M}| \leq 2\overline{D}$$

Since D and D' have the same objective function, the value of the objective functions of both D and D' are same for the average solution $\overline{z}$. Therefore $\overline{P} \leq 2\overline{D'}$. Also, $\overline{D'}$ is a lower bound of Opt' with high probability. Hence, with high probability, $\overline{D'} \leq Opt' \leq Opt(1+\epsilon)$. Hence, with high probability, $\overline{P} \leq 2 \cdot Opt \cdot (1+\epsilon)$.

Each x_u^i, y_u^i for $i \in \{1, \ldots, m\}$ is a feasible solution, so there is a feasible primal solution with value at most twice the average value of the dual solution. Since we have selected the solution which gives the minimum objective function value among all x_u^i, y_u^i for $i \in \{1, \ldots, m\}$, our proposed solution is an integer solution to the primal with value at most the average value of the primal solutions. Therefore, we have a $(2+\epsilon)$-approximate solution with probability close to 1. □

4 Improved Approximation for Bounded Forest Cover (BFC)

Bounded Forest Cover (BFC) was studied in [11], where they show that the problem is NP-complete and gave an 8-approximation algorithm. We improve this factor to 6 using the 2-factor approximation algorithm for the forest cover problem given in the full version of the paper. We define the bounded forest cover problem next.

Bounded Forest Cover(BFC): We are given a graph G with positive weights on the edges and a parameter $\lambda \geq 0$. The goal is to find a minimum-sized collection of trees $T_1, T_2, \ldots T_k$ such that the weight of each $T_i \leq \lambda$ and the vertices in $\cup_{i=1}^{k} T_i$ form a vertex cover of the graph. By minimum-sized collection, we mean k should be the smallest possible. We can assume that the weight on any edge is at most λ.

We will construct a new graph G' from G with weights on the edges in the interval $[0, 1]$ as follows: the vertex and edge set remains unchanged, if w_e is the weight on edge $e \in G$ then w'_e the weight on $e \in G'$ is defined as follows:

$$w'_e = \begin{cases} 1, & if w_e > \lambda/2 \\ 2w_e/\lambda, & otherwise. \end{cases}$$

Lemma 4 (Lemma 2 in [19]). *Given a tree T with weight $w(T)$ and a parameter $\beta > 0$ such that all the edges of T have weight at most β, we can edge-decompose T into trees $T_1, \ldots T_k$ with $k \leq \max\{w(T)/\beta, 1\}$ such that $w(T_i) \leq 2\beta$ for each $1 \leq i \leq k$.*

Next, we solve the forest cover problem approximately using the results in the previous section. This approximate solution gives us a collection C of k trees in G'. Each tree T_i in the collection C is edge-decomposed into a collection S_i of tree $T_1, T_2, \ldots, T_m$ using Lemma 4 (with $\beta = 1$). Each tree in collection S_i has weight at most 2 in G'.

Observation 1. *Let OPT' be the value of the optimal solution to the forest cover problem. Let the optimal solution (trees; each with weight at most λ) to BFC be $\Lambda_1, \ldots, \Lambda_{OPT}$. Then, the value of the optimal solution to bounded forest cover is OPT. Since the optimal solution to BFC is a feasible solution the forest cover problem and $w'(\Lambda_i) \leq 2$ (feasible solution to BFC), we get $OPT' \leq \sum_{i=1}^{OPT} w'(\Lambda_i) + OPT \leq 3OPT$.*

Next, we need to convert the solution in G' to a solution to the bounded forest cover problem and analyze the performance ratio. All the vertices that are in the solution to the 2-approximate solution to the forest cover problem forms a vertex cover in the bounded forest cover problem. The trees in S_i obtained using the edge-decomposition in Lemma 4 (for all i) are the trees in the approximate solution to the BFC.

Observation 2. *If any of trees use a weight of edge 1, then we simply remove the edge from the solution; this does not change the objective function value of the forest cover problem. Therefore, without loss of generality, we assume that all the trees use only edges with weights < 1 in G'. Equivalently, in G, the weight of the edges of all the trees is at most $\lambda/2$.*

The number of trees in BFC, by Lemma 4 are $\sum_{i=1}^{k} \max\{w'(T_i), 1\} \leq \sum_{i=1}^{k}(w'(T_i) + 1)$ where each T_i is a connected component in some G_0 (the graph with weights 0/1 used in the previous section) and $w'(T_i) = \sum_{e \in T_i} w'_e$ and k is the number of connected connected components. $\sum_{i=1}^{k}(w'(T_i) + 1)$ is by definition the value of the 2-approximate solution to the forest cover problem. Therefore, using Observation 1, we get $\sum_{i=1}^{k}(w'(T_i)+1) \leq 2OPT' \leq 6OPT$. Therefore, the number of trees is at most a multiplicative factor of 6 of the minimum possible. Each tree has weight $w(T) = \sum_{e\in T} w_e = \sum_{e\in T} w'_e\lambda/2$ (by Observation 2). Also, $\sum_{e\in T} w'_e \leq 2$, therefore $w(T) \leq \lambda$. Therefore, we have the following theorem.

Theorem 5. *There exists a 6-factor approximation algorithm for the bounded forest cover problem.*

5 Conclusions

We suggest an improved method to approximate the bounded component forest cover problem using LP-rounding. This method is based on the 2-approximation algorithm for the forest cover problem also developed here. We also give a probabilistic approximation algorithm for forest cover with an approximation factor of $2+\epsilon$. The probabilistic technique may be of independent interest. It would be valuable to explore ways to improve the approximation factor and consider covering graphs with other types of subgraphs. Additionally, we could investigate bidirectional linear programming formulations for the forest cover problem.

Acknowledgements. DRG was supported in part by NSERC discovery grant. BG was supported in part by ANRF (Grant Number: MTR/2021/000118). RRS was supported in part by ANRF(Grant Number: CRG/2023/007610).

References

1. Alon, N., Spencer, J.H.: The Probabilistic Method. Wiley, Hoboken (2016)
2. Arkin, E.M., Halldórsson, M.M., Hassin, R.: Approximating the tree and tour covers of a graph. Inf. Process. Lett. **47**(6), 275–282 (1993)
3. Arkin, E.M., Hassin, R., Levin, A.: Approximations for minimum and min-max vehicle routing problems. J. Algorithms **59**(1), 1–18 (2006)
4. Austrin, P., Khot, S., Safra, M.: Inapproximability of vertex cover and independent set in bounded degree graphs. In: 2009 24th Annual IEEE Conference on Computational Complexity, pp. 74–80. IEEE (2009)
5. Berman, P., Fujito, T.: On approximation properties of the independent set problem for degree 3 graphs. In: Akl, S.G., Dehne, F., Sack, J.-R., Santoro, N. (eds.) WADS 1995. LNCS, vol. 955, pp. 449–460. Springer, Heidelberg (1995). https://doi.org/10.1007/3-540-60220-8_84
6. Chen, J., Kanj, I.A., Xia, G.: Improved upper bounds for vertex cover. Theoret. Comput. Sci. **411**(40–42), 3736–3756 (2010)
7. Dinur, I., Safra, S.: On the hardness of approximating minimum vertex cover. Ann. Math. 439–485 (2005)
8. Doerr, B.: Probabilistic tools for the analysis of randomized optimization heuristics. In: Theory of Evolutionary Computation: Recent Developments in Discrete Optimization, pp. 1–87 (2020)
9. Even, G., Garg, N., Könemann, J., Ravi, R., Sinha, A.: Min-max tree covers of graphs. Oper. Res. Lett. **32**(4), 309–315 (2004)

10. Goemans, M.X., Myung, Y.S.: A catalog of Steiner tree formulations. Networks **23**(1), 19–28 (1993)
11. Gorain, B., Patra, S., Singh, R.R.: Graph covering using bounded size subgraphs. In: Bagchi, A., Muthu, R. (eds.) CALDAM 2023. LNCS, vol. 13947, pp. 415–426. Springer, Cham (2023). https://doi.org/10.1007/978-3-031-25211-2_32
12. Guttmann-Beck, N., Hassin, R.: Approximation algorithms for min-max tree partition. J. Algorithms **24**(2), 266–286 (1997)
13. Hajiaghayi, M.T., Jain, K.: The prize-collecting generalized Steiner tree problem via a new approach of primal-dual schema. In: SODA, vol. 6, pp. 631–640 (2006)
14. Halperin, E.: Improved approximation algorithms for the vertex cover problem in graphs and hypergraphs. SIAM J. Comput. **31**(5), 1608–1623 (2002)
15. Han, Q., Punnen, A.P., Ye, Y.: A polynomial time 3/2-approximation algorithm for the vertex cover problem on a class of graphs. arXiv preprint arXiv:0712.3335 (2007)
16. Harris, D.G., Narayanaswamy, N.S.: A faster algorithm for vertex cover parameterized by solution size. In: Beyersdorff, O., Kanté, M.M., Kupferman, O., Lokshtanov, D. (eds.) 41st International Symposium on Theoretical Aspects of Computer Science. STACS 2024, 12–14 March 2024, Clermont-Ferrand, France. LIPIcs, vol. 289, pp. 40:1–40:18. Schloss Dagstuhl - Leibniz-Zentrum für Informatik (2024). https://doi.org/10.4230/LIPICS.STACS.2024.40,
17. Hochbaum, D.S.: Approximation algorithms for the set covering and vertex cover problems. SIAM J. Comput. **11**(3), 555–556 (1982)
18. Karp, R.: Reducibility among combinatorial problems (1972) (2021)
19. Khani, M.R., Salavatipour, M.R.: Improved approximation algorithms for the min-max tree cover and bounded tree cover problems. Algorithmica **69**(2), 443–460 (2014)
20. Kim, T.U., Lowe, T.J., Ward, J.E., Francis, R.L.: A minimum length covering subgraph of a network. Ann. Oper. Res. **18**, 245–259 (1989)
21. Nguyen, V.H.: Approximating the minimum tour cover with a compact linear program. In: van Do, T., Thi, H.A.L., Nguyen, N.T. (eds.) Advanced Computational Methods for Knowledge Engineering. AISC, vol. 282, pp. 99–104. Springer, Cham (2014). https://doi.org/10.1007/978-3-319-06569-4_7
22. Papadimitriou, C.H., Steiglitz, K.: Combinatorial Optimization: Algorithms and Complexity. Courier Corporation, North Chelmsford (1998)
23. Wu, J., Cheng, Y., Yang, Z., Chu, F.: A 3/2-approximation algorithm for the multiple Hamiltonian path problem with no prefixed endpoints. Oper. Res. Lett. **51**(5), 473–476 (2023)
24. Yu, W., Liu, Z., Bao, X.: New approximation algorithms for the minimum cycle cover problem. Theor. Comput. Sci. **793**, 44–58 (2019)

Multi-agent Search-Type Problems on Polygons
(Extended Abstract)

Konstantinos Georgiou(✉), Caleb Jones, and Jesse Lucier

Department of Mathematics, Toronto Metropolitan University, Toronto, ON, Canada
{konstantinos,caleb.w.jones,jesse.lucier}@torontomu.ca

Abstract. We present several advancements in search-type problems for fleets of mobile agents operating in two dimensions under the wireless model. Potential hidden target locations are equidistant from a central point, forming either a unit-radius disk (infinite possible locations) or regular polygons (finite possible locations) inscribed in a unit-radius disk. Building and extending on the foundational disk evacuation problem [23], the disk priority evacuation problem with k Servants [21,27], and the disk w-weighted search problem [49], we make improvements on several fronts. *First*, we establish new upper and lower bounds for the n-gon priority evacuation problem with 1 Servant for $n \leq 13$, and for n_k-gons with $k = 2, 3, 4$ Servants, where $n_2 \leq 11$, $n_3 \leq 9$, and $n_4 \leq 10$, offering tight or nearly tight bounds. The only previous results known were a tight upper bound for $k = 1$ and $n = 6$ in [27] and lower bounds for $k = 1$ and $n \leq 9$ in [49]. *Second*, our work improves the best lower bound known for the disk priority evacuation problem with $k = 1$ Servant from 4.46798 to 4.64666 and for $k = 2$ Servants from 3.6307 of [27] to 3.65332. *Third*, we improve the best lower bounds known for the disk w-weighted group search problem, significantly reducing the gap between the best upper and lower bounds for w values where the gap was largest. These improvements are based on nearly tight upper and lower bounds for the 11-gon and 12-gon w-weighted evacuation problems, while the previous study of [49] was limited only to lower bounds and only to 7-gons.

Keywords: Mobile Agents · Evacuation · Priority Evacuation · Disk · Polygons · Wireless Model

1 Introduction

Search theory is a branch of operations research that focuses on determining optimal strategies for locating targets amid uncertainty and limited information, with search and rescue missions being a prime example. Traditional *search* requires an agent to identify the target quickly, while *evacuation* missions need

Research supported in part by NSERC Discovery grant.
A full version of the paper is available on arXiv [43].

R. Královič and V. Kůrková (Eds.): SOFSEM 2025, LNCS 15538, pp. 314–332, 2025.
https://doi.org/10.1007/978-3-031-82670-2_23

the entire fleet to gather at the hidden target. In some scenarios, only a specific agent must reach the target, or agents with different significance factors influence the solution's quality. Theoretical foundations for these problems date back to the 1960s and have been revitalized by recent advancements in robotics. The field now integrates tools from online algorithms, mobile agent distributed computing, operations research, discrete mathematics, and optimization.

Theoretically, autonomous mobile agents are modeled as volumeless entities navigating discrete domains (networks or graphs) or continuous domains (2D or 3D Euclidean spaces). In continuous domains, possible target locations shape the geometric search areas, including lines, disks, circles, squares, rectangles, and triangles. Significant progress has been made, yielding elegant distributed algorithms for mobile agent computing and technical arguments for impossibility results. However, achieving tight upper and lower bounds remains challenging, with many problems still lacking optimal solutions.

This work is the first to systematically explore a continuous search space with discrete target locations. We study *priority evacuation*, an asymmetric search problem where a distinguished agent, the Queen, must reach the hidden item, aided by other agents, the Servants. The possible target locations form regular n-gons, and we provide upper and lower bounds, often tight, for various n values and numbers of Servants. This problem has been previously studied for the line with 1 Servant and the unit-radius disk with any number of Servants. As by-products of our n-gon results, we improve the best lower bounds for the disk evacuation problem with 1 and 2 Servants. Additionally, we improve the best lower bounds known for the so-called w-weighted search problem on a disk. These improvements are based on new, and nearly tight bounds for the w-weighted search problem on 11-gons and 12-gons. For the formal definitions of all the problems we study, along with a precise quantification of our results, the reader may refer directly to Sect. 2.2.

1.1 Related Work

The field of search problems dates back to the 1960s and has since developed a robust theoretical framework, detailed in various books and surveys [1–3,30, 40,50]. Initially, the focus was on optimizing objectives for single searchers [9, 51], but with the rise of robotic fleets, multi-searcher problems have gained prominence [23,53].

The linear search problem is a classic example studied extensively for both single [6,13] and multiple searchers [16]. Variations aim at minimizing the weighted average of search completion times [48]. Other one-dimensional search settings include searching along rays [14], anomalous terrains [35], graphs [5], and for multiple objects [12,22].

In the past decade, attention has shifted to two-dimensional search problems, exploring domains like polygons [39], disks [23], planes [38], regular polygons [27], equilateral triangles and squares [7,18,34], arbitrary triangles [42], and ℓ_p unit disks [47]. Variations involve different communication models and searcher specifications, such as face-to-face communication [15,29,36], varying searcher speeds [8], searching for multiple exits [53], and diverse communication capabilities [31,41]. Fault tolerance has also been studied, addressing issues like faulty agents and Byzantine faults in [10,11,24,28] as well as in [32,33,46,54].

Research has also explored non-standard objectives, including multi-objective search problems [17], competitive algorithmic approaches [4], and trade-offs between information and cost [52]. Studies have also considered time and energy trade-offs [25,26], and search-and-fetch problems in two dimensions [44,45].

1.2 Closely Related Work, Improvements and Significance of New Contributions

Our work extends and improves a series of studies on search-type problems where possible hidden target locations lie on a unit-radius disk, assuming instantaneous communication between agents under the wireless model. The initial study designed search trajectories to minimize the time for the last agent to reach the target, known as the disk evacuation problem [23]. This foundational study yielded mostly optimal results.

Subsequent research focused on designing search trajectories for a distinguished agent to evacuate quickly, with other agents, termed Servants, assisting. These disk priority evacuation problems were examined for $k = 1, 2, 3$ Servants [27] and for $k \geq 4$ Servants [21]. Notably, significant gaps existed between the upper and lower bounds. The lower bound of 4.38962 for $k = 1$ Servant relied on lower bounds for evacuating the significant agent from a 6-gon, with the n-gon (as n approaches infinity) approximating the disk priority evacuation problem.[1] A tight upper bound for the 6-gon priority evacuation was also provided.

An extension introduced in [49] considered the w-weighted group search on a unit-radius disk with two agents ($w \in [0, 1]$), where $w = 0$ corresponds to the disk priority evacuation problem with $k = 1$ Servant. The paper provided upper and lower bounds, with gaps diminishing as $w \to 0$. Their lower bound relied on w-weighted group search bounds on n-gons with $n \leq 7$, but no upper bounds for the discrete domain were reported. They also provided lower bounds for the n-gon priority evacuation problem with $n \leq 9$, improving the [27] lower bound to 4.46798 for the disk priority evacuation problem with $k = 1$ Servant.

We derive a number of improved results: (i) We provide upper and lower bounds for the n-gon priority evacuation problem with one Servant for $n \leq 13$, achieving nearly tight or tight results (previously only $n = 6$ upper bounds and $n \leq 9$ lower bounds were known). (ii) We study priority evacuation on n_k-gons for $k = 2, 3, 4$ Servants, with $n_2 \leq 11$, $n_3 \leq 9$, and $n_4 \leq 10$, offering tight or nearly tight bounds. No prior results were known. (iii) We improve the best lower bound for the disk priority evacuation problem with 1 Servant from 4.46798 [49] to 4.64666. (iv) We present the first improved lower bound for the

[1] The possible target placements on a disk are uncountable many, whereas for every n, the vertices of an n-gon are finite many. However, for every $\epsilon > 0$, there is large enough n so that every target placement on the disk is no more than ϵ away from a target placement on the n-gon.

disk evacuation problem with 2 Servants from 3.6307 [27] to 3.65332. (v) We improve the lower bounds for the disk w-weighted group search problem of [49] for all $w \in [0, 0.7]$. (vi) Motivated by the fact that the previous improved lower bounds were obtained as corollaries to the w-weighted search problem on n-gons for $n = 12$, we are the first to provide upper and lower bounds for the 11-gon and the 12-gon w-weighted evacuation problem for all $w \in [0, 1]$, achieving tight or nearly tight results.

Our priority evacuation upper bounds are accompanied with search trajectories that resemble the best trajectories known for disk evacuation problems. Despite remaining gaps, this gives further indication that the known search trajectories may be optimal.

1.3 Paper Organization

In Sect. 2.1, we introduce key notation and terminology. The formal definition of the problems we study, along with a formal description of our results can be found in Sect. 2.2. These results pertain to the n-gon priority evacuation problem with k Servants PE_k^n, the disk priority evacuation problem with k Servants DE_k, the n-gon w-weighted search problem $PE^n(w)$, and the disk w-weighted search problem $DE(w)$. The proofs of our results for all these problems appear in the subsequent sections. Indeed, in Sect. 3 we present formulations and relaxations for solving PE_k^n (the same machinery is also applied later to $PE^n(w)$). Then, in Sect. 4 we discuss lower bounds on PE_k^n, and in Sect. 4 upper bounds on the same problem. Equipped with the lower bounds on PE_k^n, we show improved lower bounds on DE_k, $k = 1, 2$ in Sect. 6. Then, in Sect. 7 we justify the reported upper and lower bounds for problems $PE^n(w)$ and $DE(w)$. Finally, in Sect. 8, we conclude with some open questions. Due to space limitations, many proofs are omitted. A full version of the paper, including all omitted proofs, is available on arXiv [43].

2 Preliminaries

2.1 Notation and Terminology

We use $\|\cdot\|_2$ to denote the Euclidean norm over $\mathbb{R}^2$. For $n \in \mathbb{N}$, let $\mathcal{P}_n$ denote the set of all permutations of $\{1, \ldots, n\}$. We define $\mathcal{B}_k^n$ as the set of all n-dimensional $(k+1)$-ary strings, i.e., $\{0, \ldots, k\}^n$, so $\mathcal{B}_1^n$ is the set of n-dimensional binary strings.

We use the term *unit speed trajectory* to refer to a continuous and differentiable function $\tau : \mathbb{R}_+ \to \mathbb{R}^2$ that induces a speed of at most 1. Specifically, if $\tau(t) = (\tau_1(t), \tau_2(t))$, then $\tau(t)$ is 1-Lipschitz continuous, meaning it satisfies $\|\tau(t_1) - \tau(t_2)\|_2 \le |t_1 - t_2|$ for all $t_1, t_2 \in \mathbb{R}_+$.

In this work, n-gons exclusively refer to regular n-polygons inscribed in circles of radius 1. For convenience, we consider n-gons and the circle as embedded in the 2-dimensional Euclidean space. Thus, for a fixed n, the vertices of the n-gon are represented as $V_i^n := (\cos(2i\pi/n), \sin(2i\pi/n)), i = 1, \ldots, n$. It follows that for all $i, j \in \{1, \ldots, n\}$, we have $\|V_i^n - V_j^n\|_2 = 2\sin\left(\frac{\pi}{n} \cdot \mathrm{mod}(|i-j|, n)\right)$, and that the edges of the n-gon have length $2\sin\left(\frac{\pi}{n}\right)$.

Parameter n will denote the number of vertices of the n-gon, and k will denote the number of Servants in our search problem involving $k+1$ mobile agents. The following set will be useful in our later formulations: $\mathcal{X}_k^n := \{0, \ldots, k\} \times \{1, \ldots, n\}$, which corresponds to pairs of agents and n-gon vertices in the multi-agent search problem we are considering.

2.2 Problem Definition and Main Contributions Made Formal

The purpose of this section is to formally define the problems under study and to quantify our results. The search problems are divided into two categories: those where the search domain is a regular polygon inscribed in a unit disk, and those where the search domain is the disk itself. A notable distinction between the two is that, in the polygon search problem, agents start from arbitrary vertices of the polygon (an algorithmic choice), whereas in the disk search problem, agents start from the center of the disk. This is done purely for technical convenience and follows the conventions established in previous literature on the same search problems.

Polygon Priority Evacuation: We study the *Polygon Priority Evacuation Problem* PE_k^n on n-gons with k Servants, a multi-agent search problem where the hidden target lies in a discrete domain, specifically n-gons. In this search problem, the host space is the Euclidean 2-dimensional space (modeled for convenience as a Cartesian plane), and the searchers are $k+1$ unit speed *agents*. Among these agents, one is labeled 0 and is distinguished as the *Queen*, while the other agents, labeled $1, 2, \ldots, k$, are called *Servants*. We consider a regular n-gon with vertices V_i^n, $i = 1, \ldots, n$, inscribed in a unit-radius disk and centered at the origin $O = (0, 0)$. The orientation of the n-gon is known to the agents, i.e. to the algorithm.

Agent movements are determined by unit speed trajectories $\tau_i : \mathbb{R}_+ \to \mathbb{R}^2$, where the initial placements of the agents $\tau_i(0)$ are algorithmic choices, $i = 0, \ldots, k$.These initial positions do not necessarily have to be at the center of the disk, unlike in the disk evacuation problem, which we will study next. The agents' movements are *feasible* if for each polygon vertex $j \in \{1, \ldots, n\}$, there exists an agent $i \in \{0, \ldots, k\}$ and time $t' = t'(j) \in \mathbb{R}_+$ such that $\tau_i(t') = V_j^n$ (i.e., every polygon vertex is eventually visited by some agent). For each polygon vertex $j \in \{1, \ldots, n\}$, we denote by T_j the smallest such t', calling it the *visitation time of vertex* j. The cost of the feasible solution $\{\tau_i\}_{i \in \{0,\ldots,k\}}$, referred to as the *priority evacuation cost*, is defined as $\max_{j \in \{1,\ldots,n\}} \{T_j + \|V_j^n - \tau_0(T_j)\|_2\}$ and the objective is to minimize that cost.

Next, we provide a high-level explanation of the above model. By definition, the agents' specifications correspond to the so-called wireless communication model, which allows them to share information instantaneously. All $k+1$ agents contribute to searching for a hidden target, often referred to as the *exit*, located at one of the vertices of the n-gon. In this online problem, the hidden exit is identified only when any of the $k+1$ agents visit the corresponding n-gon vertex (unknown to the agents), and subsequently, all agents are notified accordingly. For an exit placement, the cost of a solution is given by the time that the Queen *evacuates*, i.e., when she reaches the hidden exit, ignoring thereafter the whereabouts of the k Servants. Overall, this is quantified by the time the hidden exit is discovered, say T_j, plus the time the unit-speed Queen needs to reach the exit, i.e. $\|V_j^n - \tau_0(T_j)\|_2$. Compatible with worst-case analysis, the performance of a feasible solution is defined as the worst-case Queen evacuation time over all exit placements.

In this work, we provide upper and lower bounds on PE_k^n with $k = 1, 2, 3, 4$ Servants over n-gons, and for various values of n. All our results are summarized in the next theorem, which is our first main contribution.

Theorem 1. [2]*For $k = 1, 2, 3, 4$, and for various values of n, Priority Evacuation Problem PE_k^n can be solved in time u_k^n. Also, no algorithm for the problem has evacuation cost less than l_k^n. The upper bounds u_k^n, and the lower bounds l_k^n appear in Table 1. Moreover, for n-gons $n = 12, 13$, we have $u_1^{12} = 3.38511$, $l_1^{12} = 3.38486$, and $u_1^{13} = 3.36362$, $l_1^{13} = 3.36361$.*

Disk Priority Evacuation: Our second contribution pertains to *improved* lower bounds for the *disk priority evacuation problem with k-Servants*, which we denote by DE_k. The problem DE_k was first considered in [27] for $k = 1, 2, 3$ Servants, and in [21] for $k \geq 4$ Servants. For completeness, we include the definition of the problem by adapting the description of PE_k^n. The primary difference between the two problems is that in the disk evacuation problem, the hidden exit can lie *anywhere* on the perimeter of the unit radius disk. Moreover, the agents' starting positions are at the origin, i.e., for the feasible unit speed trajectories $\tau_i : \mathbb{R}_+ \to \mathbb{R}^2$, we have $\tau_i(0) = O = (0, 0)$, for $i = 0, \ldots, k$. We quantify the positioning of the exit by some $\theta \in [0, 2\pi]$, corresponding to the exit placement $P_\theta := (\cos(\theta), \sin(\theta))$. Then, for a feasible solution $\{\tau_i\}_{i=0,\ldots,k}$, we require that for each θ, there exists an agent $i \in \{0, \ldots, k\}$ and a time $t' \in \mathbb{R}_+$ such that $\tau_i(t') = P_\theta$. For each exit placement θ, we denote by $T(\theta)$ the smallest corresponding time t'. The evacuation cost of the feasible solution $\{\tau_i\}_{i\in\{0,\ldots,k\}}$ is then defined as $\sup_{\theta\in[0,2\pi)} \{T(\theta) + \|P_\theta - \tau_0(T(\theta))\|_2\}$ and the objective is to minimize this cost.

Our second main contribution are new lower bounds for DE_k which are summarized in the next statement.

Theorem 2. *No algorithm for DE_k has evacuation cost less than* 4.64666 *for $k = 1$, and less than* 3.65332 *for $k = 2$.*

[2] Our positive and negative results provide only the first 5 digits of our computations, even though our numerical evaluations extend to at least 10 digits of accuracy. Often, we also have closed-form expressions, involving algebraic and trigonometric operations, that describe these numbers. However, for larger values of n, these expressions become too extensive to be informative, and hence we omit them.

Table 1. Summary of our upper and lower bound results for PE_k^n. Every entry is populated with the upper bound u_k^n, the lower bound known l_k^n, and the corresponding optimality gap, i.e. u_k^n/l_k^n. Therefore a reported gap of 1.0 corresponds to an optimal result. For $n = 12, 13$, we only have results for PE_1^n, and we report them separately. However, for these n-gons, the optimality gaps are 1.00007 and 1.0, respectively. The only values known before were u_1^6, and l_1^n for $n \leq 9$.

k \ n	3	4	5	6	7	8	9	10	11
	1.73205	2.14626	2.71441	2.86603	2.97391	3.02649	3.21891	3.21549	3.35919
1	1.73205	2.12132	2.71441	2.86602	2.95125	3.00320	3.21891	3.18712	3.35577
	1.	**1.01175**	**1.**	**1.**	**1.00768**	**1.00776**	**1.**	**1.0089**	**1.00102**
	1.00000	1.70711	1.90211	2.00000	2.14027	2.25951	2.37176	2.38956	2.50211
2	1.00000	1.70710	1.90211	2.00000	2.08348	2.23784	2.35288	2.37810	2.46291
	1.	**1.**	**1.**	**1.**	**1.02726**	**1.00968**	**1.00802**	**1.00482**	**1.01592**
	1.00000	1.00000	1.55017	2.00000	1.86777	1.91342	1.91362	NA	NA
3	1.00000	1.00000	1.53884	1.86602	1.84269	1.91341	1.85083	NA	NA
	1.	**1.**	**1.00736**	**1.0718**	**1.01361**	**1.**	**1.03393**	**NA**	**NA**
	1.00000	1.00000	1.00000	1.5000	1.64960	1.76537	1.68404	1.65153	NA
4	1.00000	1.00000	1.00000	1.50000	1.64959	1.68924	1.66884	1.61803	NA
	1.	**1.**	**1.**	**1.**	**1.**	**1.04507**	**1.00911**	**1.02070**	**NA**

For an informative perspective, we summarize in Table 2 all previous best upper and lower bounds known for DE_k.

Table 2. Summary of best upper and lower bound results known for DE_k, prior to our work. Contrast the upper and lower bound values to the improved lower bounds of Theorem 2

k	1	2	3	4
Upper Bound	4.81854 [27]	3.8327 [27]	3.3738 [27]	3.30129 [21]
Lower Bound	4.56798 [49]	3.6307 [27]	3.2017 [27]	2.91322 [21]

Polygon and Disk w-Weighted Search. Our third contribution pertains to a search problem where the objective is the arithmetic weighted average of the termination times of 2 mobile agents, first considered in [49]. In this problem, the hidden item lies either in the unit radius disk and the 2 agents start from the center of the disk, or it lies on a vertex of an n-gon, and the initial placement of the 2 agents is an algorithmic choice. Having the feasible trajectories identified exactly as in PE_1^n and DE_1, the cost of the solution is instead defined as the arithmetic weighted average of the times that the 2 agents reach the hidden item. Indeed, for each $w \in [0, 1]$ and feasible trajectories τ_0, τ_1, and for the exit placement x, let $T_i(x)$ denote the time that agent $i = 0, 1$ reaches x. The w-weighted search cost for input x is defined as $(T_0(x) + w \cdot T_1(x))/(1 + w)$.

In *n-Gon w-Weighted Search Problem* $PE^n(w)$, the objective is to minimize $\max_x \frac{T_0(x)+w\cdot T_1(x)}{1+w}$, where x ranges over all n vertices of the n-gon. Similarly, for the *Disk w-Weighted Search Problem* $DE(w)$, the objective is to minimize $\sup_x \frac{T_0(x)+w\cdot T_1(x)}{1+w}$ and the supremum is considered over all points x on the perimeter of the disk. From these definitions, it is immediate that setting $w = 0$ results in the priority evacuation objective with 1 Servant, i.e. that $PE^n(0)$ is equivalent to problem PE_1^n, and that $DE(0)$ is equivalent to problem DE_1.

The main contribution pertaining to the w-weighted search objective is an improvement to the previously best lower bound known for the disk. The result is obtained numerically, and is quantified in the next statement.

Theorem 3. *Each of the purple and brown curves in Fig. 1 is a lower bound to the* $DE(w)$*, for all* $w \in [0,1]$*.*[3]

As indicated before, the lower bounds on $DE(w)$ are obtained via reductions to $PE^n(w)$. The best lower bound achieved as w ranges in $[0,1]$ were obtained for $n = 11, 12$. We are therefore motivated to report upper and lower bounds for $PE^{11}(w)$ and $PE^{12}(w)$, demonstrating that our analysis is (nearly) tight. This is quantified in the following statement.

Theorem 4. *For values of* $w \in [0,1]$*, Fig. 2 shows upper and lower bounds to the* $PE^n(w)$ *problem, for* $n = 11, 12$*.* [4]

3 Formulations and Relaxations to PE_k^n

Our contributions are based on the observation that the optimal solution to the PE_k^n problem can be found using a Non-Linear Program (NLP), assuming we know the order in which the polygon vertices are visited and the identities of the agents who first visit those vertices. This idea was first implemented in [49] for $k = 1$ Servants, primarily to provide lower bounds for DE_1, the disk priority evacuation problem with 1 Servant, and for the disk w-weighted search problem. Here, we extend this approach to provide upper and lower bounds also for PE_k^n for $k = 2, 3, 4$ and various values of n, improving the known lower bounds for DE_1 and introducing the first lower bound improvements for DE_k for $k = 2$.

First, we introduce some terminology to clarify the notation. For any feasible solution $\{\tau_i\}_{i\in\{0,\dots,k\}}$ to PE_k^n, each vertex V_i^n of the n-gon is visited by some agent. Let permutation $\rho \in \mathcal{P}_n$ be the ordered list of vertices by their visitation times, where vertex $V_{\rho_i}^n$ has the i'th smallest visitation time (ties are broken arbitrarily). Define $s \in \mathcal{B}_k^n$ as the $(k+1)$-ary string where s_i is the label of the agent visiting ρ_i first. Thus, the corresponding feasible solution to PE_k^n is called an (s,ρ)-*algorithm*. We begin with the following observation.

[3] The lower bounds were computed for values of w from 0 to 1 with step size 0.01.

[4] The lower bounds were computed for values of w from 0 to 1 with step size 0.01. The upper bounds were computed for values of w from 0 to 1 with step size 0.02, hence we only depict them as points.

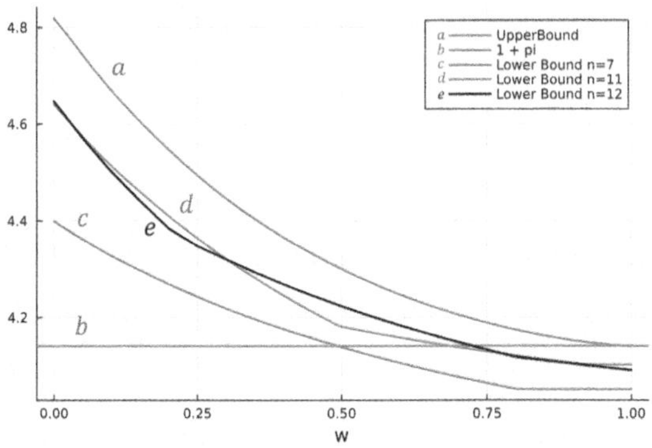

Fig. 1. Known upper bounds compared to our improved lower bounds for $DE(w)$. The blue curve (labeled a) depicts the best upper bound known [49]. The green curve (labeled c) depicts the previously best lower bound known, by a reduction to $PE^7(w)$. The orange line (labeled b) depicts a universal lower bound of $1 + \pi$ also proved in [49]. The purple and black curves (labeled d, e, respectively) are new lower bounds on $DE(w)$ by a reductions to $PE^{11}(w)$ and $PE^{12}(w)$, respectively (hence the maximum of them applies). The lower bounds were calculated for values of $w \in [0, 1]$ starting from 0 and with step size 0.01. (Color figure online)

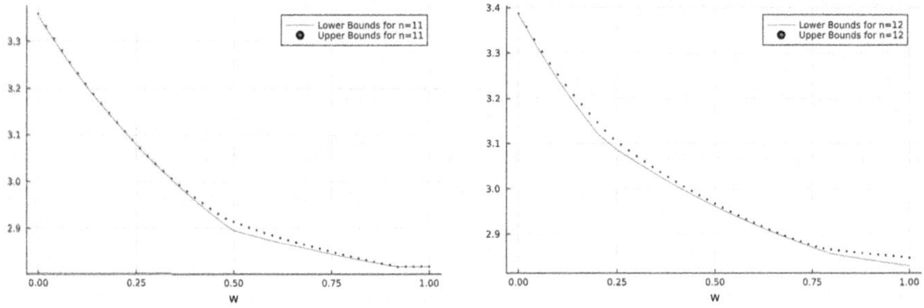

Fig. 2. Upper and lower bound obtained for $PE^{11}(w)$ (left hand-side) and $PE^{12}(w)$ (right hand-side). Starting from $w = 0$, the lower bounds were obtained with w step size of 0.01 and the upper bounds with w step size of 0.02

Observation 1. *The optimal solution to PE_k^n is an (s, ρ)-algorithm, for some $s \in \mathcal{B}_k^n$ and $\rho \in \mathcal{P}_n$.*

For fixed s, ρ, we show next how to find the optimal (s, ρ)-algorithm to PE_k^n. To that end, consider the following Non Linear Program (NLP) on variables $t_1, \ldots, t_n$ and d_{r_1, r_2}, where $r_1, r_2 \in \mathcal{X}_k^n$. The NLP seeks to find the optimal algorithm that visits the polygon vertices according to s, ρ. More specifically, the variable t_i represents the time at which the i-th vertex, i.e., vertex ρ_i, is visited by agent s_i in the algorithm. Similarly, let $r \in \mathcal{X}_k^n$, where r takes the form (i, j), with i being an agent and j corresponding to vertex ρ_j. For such an r, we consider the position of agent i when vertex ρ_i is visited according to the search algorithm. For two such points corresponding to $r_1, r_2 \in \mathcal{X}_k^n$, the variable d_{r_1,r_2} represents their Euclidean distance.

$$\min \max_{j\in\{1,\ldots,n\}} \{t_j + d_{(0,j),(s_j,j)}\} \qquad (\mathrm{NLP}_k^n(s,\rho))$$

$$\begin{aligned}
s.t.:\ & t_{j+1} - t_j \geq d_{(i,j+1),(i,j)},\quad j \in \{0,\ldots,n\}, i \in \{0,\ldots,k\} && (1)\\
& d_{(i,0),(s_l,l)} = 1,\quad l \in \{1,\ldots,n\}, i \in \{0,\ldots,k\} && (2)\\
& d_{(s_j,j),(s_l,l)} = 2\sin\left(mod(|\rho_j - \rho_l|, n)\cdot \tfrac{\pi}{n}\right),\quad j,l \in \{1,\ldots,n\} && (3)\\
& t_0 = -1, t_1 = 0 && (4)\\
& (\mathcal{X}_k^n, d) \text{ is a metric space} && (5)\\
& (\mathcal{X}_k^n, d) \text{ is isometrically embeddable to } \left(\mathbb{R}^2, \|\cdot\|_2\right) && (6)
\end{aligned}$$

Lemma 1 (Introduced in [49], for $k = 1$). *Fix $s \in \mathcal{B}_k^n$ and $\rho \in \mathcal{P}_n$. Then the optimal value to $\mathrm{NLP}_k^n(s,\rho)$ equals the cost of the optimal (s,ρ)-algorithm to PE_k^n.*[5]

We note that while formulation $\mathrm{NLP}_k^n(s,\rho)$ can be effectively coded as an NLP in any programming language, solving it optimally remains a challenge. Due to the non-convex nature of the program, unless a sophisticated method specifically tailored to the NLP is implemented, it is impossible to guarantee a global solution. Indeed, solvers can only ensure that the returned solution is a local optimizer. Motivated by that, we relax $\mathrm{NLP}_k^n(s,\rho)$ into a tracktable optimization problem. This idea is borrowed again from [49].

Lemma 2. *The relaxation of $\mathrm{NLP}_k^n(s,\rho)$ where constraint* (6) *is omitted* (denoted as $\mathrm{LP}_k^n(s,\rho)$)*, is a Linear Program.*

The advantage of considering $\mathrm{LP}_k^n(s,\rho)$ is that the optimization problem can be solved efficiently, and moreover, every solution (even numerical and computer-based) comes with a certificate of global optimality, unlike what can be accomplished for the NLP. For given s, ρ, let $obj_k^n(s,\rho)$ be the optimal solution to $\mathrm{LP}_k^n(s,\rho)$. Since the LP is a relaxation of $\mathrm{NLP}_k^n(s,\rho)$, we see that $obj_k^n(s,\rho)$ is a lower bound on the optimal solution to $\mathrm{NLP}_k^n(s,\rho)$, which is also a lower bound on the evacuation cost of the optimal (s,ρ) algorithm for PE_k^n. Since also the optimal solution to PE_k^n is some (s,ρ)-algorithm, we obtain the following.

Corollary 1. *The optimal cost for solving PE_k^n is at least $\min_{s,\rho} obj_k^n(s,\rho)$, where $obj_k^n(s,\rho)$ denotes the optimal solution to $\mathrm{LP}_k^n(s,\rho)$.*

[5] For our lower bound arguments, we only need that the the optimal value to $\mathrm{NLP}_k^n(s,\rho)$ is a lower bound to the cost of the optimal (s,ρ)-algorithm.

4 Lower Bounds on PE_k^n

The section presents our findings on the lower bounds for the polygon priority evacuation problem PE_k^n for various values of n, and $k = 1, 2, 3, 4$. The results serve as the proof of the lower bounds mentioned in Theorem 1. Due to space limitations, we present a high level overview of our technical work.

For each n and k, the results for PE_k^n are obtained by applying Corollary 1, specifically by solving $\text{LP}_k^n(s, \rho)$ for all $s \in \mathcal{B}_k^n$ and $\rho \in \mathcal{P}_n$, and reporting the smallest value. This work extends the results of [49], which covered only $k = 1$ and $n \leq 9$. A brute force approach to Corollary 1 suggests solving $c_{n,k} := n!(k+1)^n$ linear programs, each with $\Omega\left(n^2 k^2\right)$ variables and $\Omega\left(n^3 k^3\right)$ constraints. Our contribution includes strategies to efficiently handle these configurations, such as assuming initial conditions for vertex visits, omitting provably suboptimal configurations using naive algorithms, and leveraging the symmetry of identical Servants.

These strategies were implemented using Julia's JuMP Clp [19,37], ensuring high accuracy with optimality proofs. Specific variables were preset to known optimal values, reducing runtime. Parallel threading was employed to process LPs, solving only those configurations where the cost was lower than the tentative minimum, which further expedited calculations. Our methods allowed us to handle up to $(n, k) = (13, 1)$ or $(10, 4)$. However, the fewer configurations of $(n, k) = (10, 3)$ $(c_{10,3} = 3.8 \cdot 10^{12})$ proved challenging, indicating that naive bounds used to omit certain LPs were not stringent enough.

Finally, many lower bounds matched our upper bounds, indicating no gap between the LP relaxation and the NLP solution for many cases. Since the relaxation was derived by a metric embedding relaxation, one may expect that optimal LP solutions induce metrics embeddable in $(\mathbb{R}^2, \ell_2)$. However, we report that this is not the case, something we confirmed by solving the appropriate Semidefinite Programs.

5 Upper Bounds on PE_k^n

For each n, k, we prove Theorem 1 by presenting feasible search trajectories to PE_k^n for the $k + 1$ agents and performing a worst case analysis. In this section we demonstrate our techniques by discussing the solution only to PE_1^9, which is also one of the upper bounds that is provably optimal.

When considering a solution to PE_k^n, we fix some $(s, \rho) \in \mathcal{X}_k^n$ and then we describe the (s, ρ)-algorithm together with its analysis based on a figure (depicting the trajectories) and a table giving trajectories' details and performing the worst case analysis. Next we describe the notation that is used in the figures and the tables, e.g. Fig. 3 and Table 3.

Recall that permutation element ρ_i denotes the ith discovered vertex, which is visited by agent $s_i \in \{0, \ldots, k\}$. Servants move always at full speed 1 over the prescribed vertices, and hence no further detail on their movements is required. Their whereabouts are irrelevant after they have visited all their assigned vertices. We denote the length of a regular n-gon edge inscribed in the unit radius disk as $e_n = 2\sin(\pi/n)$, and we denote the origin by $\mathcal{O}$. Vertices of the n-gon where previously denoted as V_j^n, $j = 1, \ldots, n$. In order to ease notation, we drop the superscript, whenever n is clear from the context, and in our figures we simply write V_j.

We use the colour red to denote the trajectory of the Servant(s), while the colour blue is used to denote the trajectory of the Queen. Moreover, we use square vertices to denote the placements of the exits which induce the maximum evacuation time for any given search trajectory. We use a solid line to denote the trajectory the agents follow. We use a dashed line to show when the Queen deviates from the prescribed trajectory to evacuate at an announced exit that yields the maximum evacuation time.

We let t_i denote the first time when vertex ρ_i is discovered by an agent. We denote by $Q^{(i)}$ the location of the Queen at time t_i. We use segments with endpoints of the form $Q^{(i)}, Q^{(i+s)}$ to indicate that the Queen has stayed put at point $Q^{(i+s)}$ from time t_{i+1} and until t_{i+s} Also, the tables include $d(Q^{(i)}, Exit)$ which is the distance between the Queen and exit, when the exit is placed at vertex ρ_i.

The trajectories of the (s, ρ)-algorithms are formally given in tables. For each $i \in \{1, \ldots, n\}$, we specify the location of the Queen at time t_i. Together with the Queen's distance to vertex ρ_i, they define the evacuation cost for the exit placement ρ_i. Whenever the positioning of the Queen is not a vertex, or a self-evident point, we define it formally either as an abstract expression, or as the solution to a formal non-linear system (similarly to how upper bounds have been described for all previous positive results to priority evacuation problems). In the reported trajectory tables, cost entries indicated by $*$ denote the worst case evacuation time for the provided trajectory, over all exit placements. Now we discuss the solution details to PE_1^9.

Example 1 (Upper Bound to PE_1^9). For PE_1^9, we use $s = (1, 0, 1, 0, 1, 0, 1, 0, 1)$ along with $\rho = (1, 2, 9, 3, 8, 4, 7, 5, 6)$, see also Fig. 3 and Table 3.

For the formal description of Queen's positions, we define points $Q^{(5)}$ and $Q^{(7)}$ as convex combinations of the points, V_4 and V_8, and V_4 and V_7, respectively. Thus, $Q^{(5)} = (1-l_1)V_4 + l_1 V_8$ and $Q^{(7)} = (1-l_2)V_4 + l_2 V_7$, for some $l_1, l_2 \in [0, 1]$. l_1 and l_2 are obtained as the solution to the following non-linear system

$$l_1\|V_4 - V_8\| + l_2\|V_4 - V_7\| = e_9 \tag{7}$$

$$(1 - l_1)\|V_4 - V_8\| = l_1\|V_4 - V_8\| + (1 - l_2)\|V_4 - V_7\| + l_2\|V_4 - V_7\| \tag{8}$$

The solution to the above system satisfies $l_1\|V_4 - V_8\| \leq e_9$. (7) is derived by imposing that the time taken by the Queen to travel from $Q^{(5)}$ to V_4 to $Q^{(7)}$ is exactly e_9. Moreover, (8) is derived by equating the evacuation times for when the exit is placed at V_8 and V_7. It follows that $l_1 = 0.06031$ and $l_2 = 0.32635$.

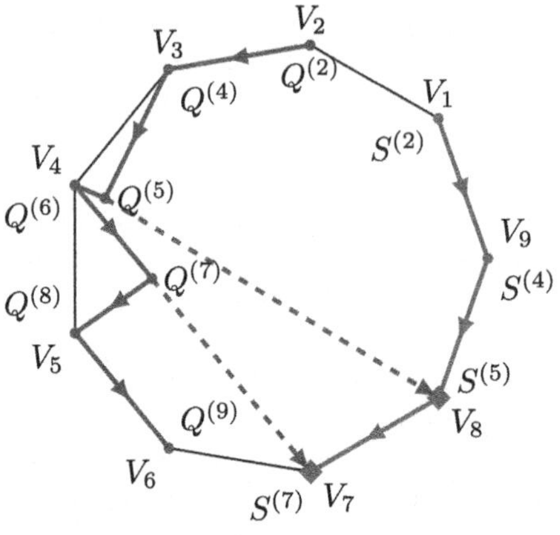

Fig. 3. A search trajectory for PE_1^9. Trajectory details can be found in Table 3. The Queen waits at V_3 for an amount of time such that the Queen arrives at $Q^{(5)}$ at time $2e_9$. The position of $Q^{(5)}$ has been adjusted for clarity

Table 3. Trajectories' details for the upper bound of PE_1^9; see Fig. 3

i	Agent	Vertex	t_i	$Q^{(i)}$	$d(Q^{(i)}, Exit)$	Cost
1	S_1	V_1	0	V_2	e_9	e_9
2	Q	V_2	0	V_2	0	$2\sin(\frac{\pi}{9})$
3	S_1	V_9	e_9	V_3	$\sqrt{3}$	$e_9+\sqrt{3}$
4	Q	V_3	e_9	V_3	0	e_9
5	S_1	V_8	$2e_9$	$(-0.83682, 0.28263)$	1.85083	3.21891*
6	Q	V_4	1.48686	V_4	0	1.48686
7	S_1	V_7	$3e_9$	$(-0.57635, -0.09099)$	1.16679	3.21891*
8	Q	V_5	2.49374	V_5	0	2.49374
9	Q	V_6	3.17778	V_6	0	3.17778

6 Improved Lower Bounds on DE_k for $k = 1, 2$

This section provides the proof of our second main contribution, as stated in Theorem 2. The key element in this proof is the following lemma, first utilized in [27] and restricted to $k = 1$. Notably, it relates the lower bounds for DE_k to those for PE_k^n (through a proper reduction), while also taking into account that, in disk problems, agents start from the center of the disk, unlike in polygon problems where agents can choose their starting positions.

Lemma 3. *The following is true for all $n \geq 3$. Let $l_{n,k}$ be a lower bound on PE_k^n. Then, no algorithm for DE_k has evacuation cost less than $1 + 2\pi/((k+1)n) + l_{n,k}$.*

We can now prove Theorem 2 using Lemma 3 and the lower bounds established by Theorem 1. For this, we utilize the already derived bounds of PE_k^n for $n \geq 6$, and we summarize the induced lower bounds for DE_k in Table 4, which are in agreement with those reported in Theorem 2. The star $*$ next to some lower bounds indicates an improvement upon the previous best results known. Bold text indicates the best result for a fixed k, and over all values of n for which a lower bound to PE_k^n was available. There is no improvement in the state-of-the-art with this method for $k = 3, 4$, but we still report the derived lower bounds for completeness.

Table 4. Lower bounds on DE_k as derived as immediate corollaries of Lemma 3 and Theorem 2. Entries with NA correspond to values of (n, k) for which no lower bound to PE_k^n was derived, and hence these parameters are not applicable to our current argument.

n / k	6	7	8	9	10	11	12	13	Best Previous LB
1	4.38962	4.40005	4.3959	4.56798	4.50128	4.64138	**4.64666***	4.60528	4.56798 [49]
2	3.34907	3.38268	3.49964	3.5856	3.58755	**3.65332***	NA	NA	3.6307 [27]
3	**3.12782**	3.06709	3.10977	3.02537	3.10814	NA	NA	NA	3.2017 [27]
4	2.70944	2.82912	**2.84633**	2.80847	2.74369	NA	NA	NA	2.91322 [21]

7 Upper and Lower Bounds on $PE^n(w)$ and Improved Lower Bounds on $DE(w)$

In this section, we describe the derivation of values in Figs. 1 and 2, leading to Theorems 3 and 4. Setting $k = 1$, we begin with a modification to $\text{NLP}_k^n(s, \rho)$ to model $PE^n(w)$ for all $n \geq 3$ and $w \in [0, 1]$. Note that $PE^n(0)$ coincides with problem PE_1^n. The feasible trajectories of the two agents are characterized by the same series of constraints. The objective of $PE^n(w)$, as defined in Sect. 2.2, is to minimize $\max_{j \in \{1,\ldots,n\}} \left\{ t_j + \frac{d_{(0,j),(s_j,j)} + w \cdot d_{(1,j),(s_j,j)}}{1+w} \right\}$. Importantly, the objective remains linear in variables d. Therefore, similar to our approach in Lemma 2, dropping constraint 6 results in a Linear Program, whose optimal solution provides a lower bound to the optimal (s, ρ)-algorithm for $PE^n(w)$. We denote this Linear Program as $\text{wLP}^n(s, \rho)$, and its optimal solution as $wobj^n(s, \rho)$. The following lemma was used in [49] only for $n \leq 7$.

Lemma 4 ([49]). *No algorithm for $DE(w)$ has a cost less than $1 + \frac{\pi}{n} + \min_{(s,\rho) \in \mathcal{X}_1^n} wobj^n(s, \rho)$, for all $n \geq 3$.*

Figure 1 shows the known upper bound for $DE(w)$ and the known lower bounds obtained by applying Lemma 4 with $n = 7$. As described in Sect. 4, our contribution includes pushing the computational boundaries for computing $\min_{(s,\rho)\in\mathcal{X}_1^n} wobj^n(s,\rho)$ for larger n, reaching $n=12$ for the w-weighted search problem. We compute lower bounds on $PE^n(w)$ for w starting from 0 with a step size of 0.01. These linear programs were solved using Julia's JuMP Clp solver [19,37]. The strongest lower bounds were obtained for $n=11$ and $n=12$, with different n-gon lower bounds dominating for different values of w. Figure 1 depicts the new and stronger lower bounds obtained by applying Lemma 4 for $n=11$ and $n=12$.

The new lower bounds are stronger than the straightforward lower bound of $1+\pi$ reported in [49] for all $w\in[0,0.7]$, and represent an improvement over the previously best lower bound known for the same range of w. More importantly, the gap between the previously best upper and lower bounds is now reduced by more than half for all $w\in[0,0.5]$. This discussion concludes the statements in Theorem 3.

Next, we justify and motivate Theorem 4. The argument used to derive Theorem 3 relies on the computed lower bounds for $PE^n(w)$, which were obtained using LP relaxations to the exact formulations for determining optimal (s,ρ)-algorithms. If these relaxations introduced a significant gap, the lower bound argument would not be tight, modulo the reduction proposed in Lemma 4. Therefore, for $w\in[0,1]$ and $n=11,12$, we aim to find upper bounds for $PE^n(w)$. If these bounds match the derived lower bounds, it would show that the lower bound analysis is tight, and further improvement would require a different technique or higher values of n.

Despite the lower bounds on $\min_{(s,\rho)\in\mathcal{X}_1^n} wobj^n(s,\rho)$ being obtained for various $(s,\rho)\in\mathcal{X}_1^n$ as w ranged over $[0,1]$, computing upper bounds can be done using fixed configurations $(s,\rho)\in\mathcal{X}_1^n$. For $n=11$, we use $s=(1,0,0,1,1,0,1,0,1,0,1)$ and $\rho=(1,2,3,11,10,4,9,5,8,6,7)$, while for $n=12$, we use $s=(1,0,1,0,1,0,1,0,1,0,1,0)$ and $\rho=(1,2,12,3,11,4,10,5,9,6,8,7)$. For these configurations, we compute a numerical solution to the Non-Linear Program $\mathrm{NLP}_k^n(s,\rho)$ using the objective of $PE^n(w)$, from $w=0$ with a step size of 0.02. The numerical solutions were obtained using Julia's JuMP Ipopt [20,37], which uses an interior point method for solving NLPs. Although these solutions are only guaranteed to be locally optimal, they correspond to feasible search trajectories (which may not be optimal). Nevertheless, in Fig. 2, the derived upper and lower bounds are tight or nearly tight, similar to what we observed for PE_k^n.

8 Discussion

In this work, we made several lower bound improvements for well-known multi-agent search problems. Our results are obtained by novel upper and lower bounds, often tight or nearly tight, for searching hidden items where target locations form regular n-gons. We extend existing techniques to larger n-gons and more agents, addressing significant computational challenges. Our contributions also lie in overcoming these challenges. The tightness of our results for n-gons show that further improvements in lower bounds on disk-related search problems require either increasing n or changing the lower bound techniques. The tight bounds are based on solutions to LP relaxations inspired by metric embedding relaxations, which relax Non-Linear Programs (NLPs). We observed that these relaxations often impose no gap, even though the solutions may induce metric-related discrepancies. Investigating this phenomenon and strengthening the LPs to model optimal search algorithms more would be valuable. Finally, our techniques apply to other search problems, especially those with asymmetric objectives. It remains to be seen if these techniques can be adapted for the face-to-face search problems, where strong lower bounds still elude us.

References

1. Ahlswede, R., Wegener, I.: Search Problems. Wiley, Hoboken (1987)
2. Alpern, S., Fokkink, R., Gasieniec, L., Lindelauf, R., Subrahmanian, V.S.: Search Theory: A Game Theoretic Perspective. Springer, New York (2013). https://doi.org/10.1007/978-1-4614-6825-7
3. Alpern, S., Gal, S.: The Theory of Search Games and Rendezvous. Kluwer, London (2003)
4. Angelopoulos, S., Dürr, C., Jin, S.: Best-of-two-worlds analysis of online search. In: Niedermeier, R., Paul, C. (eds.) 36th International Symposium on Theoretical Aspects of Computer Science, STACS 2019, 13–16 March 2019, Berlin, Germany, LIPIcs, vol. 126, pp. 7:1–7:17. Schloss Dagstuhl - Leibniz-Zentrum für Informatik (2019)
5. Angelopoulos, S., Dürr, C., Lidbetter, T.: The expanding search ratio of a graph. Discret. Appl. Math. **260**, 51–65 (2019)
6. Yates, R.B., Culberson, J., Rawlins, G.: Searching in the plane. Inf. Comput. **106**(2), 234–252 (1993)
7. Bagheri, I., Narayanan, L., Opatrny, J.: Evacuation of equilateral triangles by mobile agents of limited communication range. In: Dressler, F., Scheideler, C. (eds.) ALGOSENSORS 2019. LNCS, vol. 11931, pp. 3–22. Springer, Cham (2019). https://doi.org/10.1007/978-3-030-34405-4_1
8. Bampas, E., et al.: Linear search by a pair of distinct-speed robots. Algorithmica **81**(1), 317–342 (2019)
9. Beck, A.: On the linear search problem. Israel J. Math. **2**(4), 221–228 (1964)
10. Behrouz, P., Konstantinidis, O., Leonardos, N., Pagourtzis, A., Papaioannou, I., Spyrakou, M.: Byzantine fault-tolerant protocols for (n, f)-evacuation from a circle. In: Georgiou, K., Kranakis, E. (eds.) ALGOWIN 2023. LNCS, vol. 14061, pp. 87–100. Springer, Cham (2023). https://doi.org/10.1007/978-3-031-48882-5_7
11. Bonato, A., Georgiou, K., MacRury, C., Prałat, P.: Algorithms for p-faulty search on a half-line. Algorithmica, 1–30 (2022)
12. Borowiecki, P., Das, S., Dereniowski, D., Kuszner, Ł: Distributed evacuation in graphs with multiple exits. In: Suomela, J. (ed.) SIROCCO 2016. LNCS, vol. 9988, pp. 228–241. Springer, Cham (2016). https://doi.org/10.1007/978-3-319-48314-6_15
13. Bose, P., De Carufel, J.-L.: A general framework for searching on a line. Theor. Comput. Sci. **703**, 1–17 (2017)
14. Brandt, S., Foerster, K.-T., Richner, B., Wattenhofer, R.: Wireless evacuation on m rays with k searchers. Theor. Comput. Sci. **811**, 56–69 (2020)

15. Brandt, S., Laufenberg, F., Lv, Y., Stolz, D., Wattenhofer, R.: Collaboration without communication: evacuating two robots from a disk. In: Fotakis, D., Pagourtzis, A., Paschos, V.T. (eds.) CIAC 2017. LNCS, vol. 10236, pp. 104–115. Springer, Cham (2017). https://doi.org/10.1007/978-3-319-57586-5_10
16. Chrobak, M., Gasieniec, L., Gorry, T., Martin, R.: Group search on the line. In: Italiano, G.F., Margaria-Steffen, T., Pokorný, J., Quisquater, J.-J., Wattenhofer, R. (eds.) SOFSEM 2015. LNCS, vol. 8939, pp. 164–176. Springer, Heidelberg (2015). https://doi.org/10.1007/978-3-662-46078-8_14
17. Chuangpishit, H., Georgiou, K., Sharma, P.: A multi-objective optimization problem on evacuating 2 robots from the disk in the face-to-face model; trade-offs between worst-case and average-case analysis. Information **11**(11), 506 (2020)
18. Chuangpishit, H., Mehrabi, S., Narayanan, L., Opatrny, J.: Evacuating equilateral triangles and squares in the face-to-face model. Comput. Geom. **89**, 101624 (2020)
19. COIN-OR. Clp: Coin-or linear programming solver. https://github.com/coin-or/Clp. Accessed 04 June 2024
20. COIN-OR. Ipopt: Interior point optimizer. https://github.com/coin-or/Ipopt. Accessed 19 June 2024
21. Czyzowicz, J., et al.: Priority evacuation from a disk: the case of $n \geq 4$. Theor. Comput. Sci. **846**, 91–102 (2020)
22. Czyzowicz, J., Dobrev, S., Georgiou, K., Kranakis, E., MacQuarrie, F.: Evacuating two robots from multiple unknown exits in a circle. In: ICDCN, pp. 28:1–28:8. ACM (2016)
23. Czyzowicz, J., Gasieniec, L., Gorry, T., Kranakis, E., Martin, R., Pajak, D.: Evacuating robots via unknown exit in a disk. In: Kuhn, F. (ed.) DISC 2014. LNCS, vol. 8784, pp. 122–136. Springer, Heidelberg (2014). https://doi.org/10.1007/978-3-662-45174-8_9
24. Czyzowicz, J., et al.: Evacuation from a disc in the presence of a faulty robot. In: Das, S., Tixeuil, S. (eds.) SIROCCO 2017. LNCS, vol. 10641, pp. 158–173. Springer, Cham (2017). https://doi.org/10.1007/978-3-319-72050-0_10
25. Czyzowicz, J., et al.: Energy consumption of group search on a line. In: Baier, C., Chatzigiannakis, I., Flocchini, P., Leonardi, S. (eds.) 46th International Colloquium on Automata, Languages, and Programming (ICALP 2019), Leibniz International Proceedings in Informatics (LIPIcs), vol.132, pp. 137:1–137:15, Dagstuhl, Germany. Schloss Dagstuhl – Leibniz-Zentrum für Informatik (2019). https://doi.org/10.4230/LIPIcs.ICALP.2019.137
26. Czyzowicz, J., et al.: Time-energy tradeoffs for evacuation by two robots in the wireless model. Theor. Comput. Sci. **852**, 61–72 (2021)
27. Czyzowicz, J., et al.: Priority evacuation from a disk: the case of n= 1, 2, 3. Theor. Comput. Sci. **806**, 595–616 (2020)
28. Czyzowicz, J., et al.: Search on a line by byzantine robots. Int. J. Found. Comput. Sci. **32**(04), 369–387 (2021)
29. Czyzowicz, J., Georgiou, K., Kranakis, E., Narayanan, L., Opatrny, J., Vogtenhuber, B.: Evacuating robots from a disk using face-to-face communication. Discret. Math. Theor. Comput. Sci. **22**(4) (2020). https://dmtcs.episciences.org/6732
30. Czyzowicz, J., Georgiou, K., Kranakis, E.: Group search and evacuation. In: Flocchini, P., Prencipe, G., Santoro, N. (eds.) Distributed Computing by Mobile Entities. LNCS, vol. 11340, pp. 335–370. Springer, Cham (2019). https://doi.org/10.1007/978-3-030-11072-7_14
31. Czyzowicz, J., et al.: Group evacuation on a line by agents with different communication abilities. In: ISAAC 2021, pp. 57:1–57:24 (2021)

32. Czyzowicz, J., Killick, R., Kranakis, E., Stachowiak, G.: Search and evacuation with a near majority of faulty agents. In: SIAM Conference on Applied and Computational Discrete Algorithms (ACDA21), pp. 217–227. SIAM (2021)
33. Czyzowicz, J., Kranakis, E., Krizanc, D., Narayanan, L., Opatrny, J.: Search on a line with faulty robots. Distrib. Comput. **32**(6), 493–504 (2019)
34. Czyzowicz, J., Kranakis, E., Krizanc, D., Narayanan, L., Opatrny, J., Shende, S.: Wireless autonomous robot evacuation from equilateral triangles and squares. In: Papavassiliou, S., Ruehrup, S. (eds.) ADHOC-NOW 2015. LNCS, vol. 9143, pp. 181–194. Springer, Cham (2015). https://doi.org/10.1007/978-3-319-19662-6_13
35. Czyzowicz, J., Kranakis, E., Krizanc, D., Narayanan, L., Opatrny, J., Shende, S.: Linear search with terrain-dependent speeds. In: Fotakis, D., Pagourtzis, A., Paschos, V.T. (eds.) CIAC 2017. LNCS, vol. 10236, pp. 430–441. Springer, Cham (2017). https://doi.org/10.1007/978-3-319-57586-5_36
36. Disser, Y., Schmitt, S.: Evacuating two robots from a disk: a second cut. In: Censor-Hillel, K., Flammini, M. (eds.) SIROCCO 2019. LNCS, vol. 11639, pp. 200–214. Springer, Cham (2019). https://doi.org/10.1007/978-3-030-24922-9_14
37. Dunning, I., Huchette, J., Lubin, M.: Jump: a modeling language for mathematical optimization. SIAM Rev. **59**(2), 295–320 (2017)
38. Feinerman, O., Korman, A.: The ants problem. Distrib. Comput. **30**(3), 149–168 (2017)
39. Fekete, S., Gray, C., Kröller, A.: Evacuation of rectilinear polygons. In: Wu, W., Daescu, O. (eds.) COCOA 2010. LNCS, vol. 6508, pp. 21–30. Springer, Heidelberg (2010). https://doi.org/10.1007/978-3-642-17458-2_3
40. Flocchini, P., Prencipe, G., Santoro, N. (eds.) Distributed Computing by Mobile Entities, Current Research in Moving and Computing. LNCS, vol. 11340. Springer, Cham (2019). https://doi.org/10.1007/978-3-030-11072-7
41. Georgiou, K., Giachoudis, N., Kranakis, E.: Evacuation from a disk for robots with asymmetric communication. In: 33rd International Symposium on Algorithms and Computation (ISAAC 2022). Schloss Dagstuhl-Leibniz-Zentrum für Informatik (2022)
42. Georgiou, K., Jang, W.: Triangle evacuation of 2 agents in the wireless model. In: Erlebach, T., Segal, M. (eds.) ALGOSENSORS 2022. LNCS, vol. 13707, pp. 77–90. Springer, Cham (2022). https://doi.org/10.1007/978-3-031-22050-0_6
43. Georgiou, K., Jones, C., Lucier, J.: Multi-agent search-type problems on polygons. arXiv preprint arXiv:2406.19495 (2024)
44. Georgiou, K., Karakostas, G., Kranakis, E.: Search-and-fetch with 2 robots on a disk: Wireless and face-to-face communication models. Discret. Math. Theor. Comput. Sci. **21** (2019)
45. Georgiou, K., Karakostas, G., Kranakis, E.: Treasure evacuation with one robot on a disk. Theor. Comput. Sci. **852**, 18–28 (2021). https://doi.org/10.1016/j.tcs.2020.11.008, https://www.sciencedirect.com/science/article/pii/S030439752030640X,
46. Georgiou, K., Kranakis, E., Leonardos, N., Pagourtzis, A., Papaioannou, I.: Optimal circle search despite the presence of faulty robots. In: Dressler, F., Scheideler, C. (eds.) ALGOSENSORS 2019. LNCS, vol. 11931, pp. 192–205. Springer, Cham (2019). https://doi.org/10.1007/978-3-030-34405-4_11
47. Georgiou, K., Leizerovich, S., Lucier, J., Kundu, S.: Evacuating from ℓ_p unit disks in the wireless model. Theor. Comput. Sci. **944**, 113675 (2023). https://doi.org/10.1016/j.tcs.2022.12.025
48. Georgiou, K., Lucier, J.: Weighted group search on a line & implications to the priority evacuation problem. Theor. Comput. Sci. **939**, 1–17 (2023). https://doi.org/10.1016/j.tcs.2022.10.013

49. Georgiou, K., Wang, X.: Weighted group search on the disk & improved lower bounds for priority evacuation. In: Rescigno, A.A., Vaccaro, U. (eds.) IWOCA 2024. LNCS, vol. 14764, pp. 28–42. Springer, Cham (2024). https://doi.org/10.1007/978-3-031-63021-7_3
50. Hohzaki, R.: Search games: literature and survey. J. Oper. Res. Soc. Jpn. **59**(1), 1–34 (2016)
51. Kleinberg, J.M.: On-line search in a simple polygon. In: SODA, vol. 94, pp. 8–15. Citeseer (1994)
52. Miller, A., Pelc, A.: Tradeoffs between cost and information for rendezvous and treasure hunt. J. Parallel Distrib. Comput. **83**, 159–167 (2015)
53. Pattanayak, D., Ramesh, H., Mandal, P.S., Schmid, S.: Evacuating two robots from two unknown exits on the perimeter of a disk with wireless communication. In: Proceedings of the 19th International Conference on Distributed Computing and Networking, pp. 1–4 (2018)
54. Sun, X., Sun, Y., Zhang, J.: Better upper bounds for searching on a line with byzantine robots. In: Du, D.-Z., Wang, J. (eds.) Complexity and Approximation. LNCS, vol. 12000, pp. 151–171. Springer, Cham (2020). https://doi.org/10.1007/978-3-030-41672-0_9

Generation of Cycle Permutation Graphs and Permutation Snarks

Jan Goedgebeur[1,2] and Jarne Renders[1(✉)]

[1] Department of Computer Science, KU Leuven Kulak, 8500 Kortrijk, Belgium
{jan.goedgebeur,jarne.renders}@kuleuven.be
[2] Department of Applied Mathematics, Computer Science and Statistics, Ghent University, 9000 Ghent, Belgium

Abstract. We present an algorithm for the efficient generation of all pairwise non-isomorphic *cycle permutation graphs*, i.e. cubic graphs with a 2-factor consisting of two chordless cycles, and non-hamiltonian cycle permutation graphs, from which the *permutation snarks* can easily be computed. This allows us to generate all cycle permutation graphs up to order 34 and all permutation snarks up to order 46, improving upon previous computational results by Brinkmann et al. Moreover, we give several improved lower bounds for interesting permutation snarks, such as for a smallest permutation snark of order 6 mod 8 or a smallest permutation snark of girth at least 6. These computational results also allow us to complete a characterisation of the orders for which non-hamiltonian cycle permutation graphs exist, answering an open question by Klee from 1972, and yield many more counterexamples to a conjecture by Zhang.

Keywords: Cycle permutation graphs · Permutation snarks · Exhaustive generation · Canonical construction path method · Orderly generation · Non-hamiltonian

1 Introduction

A *cycle permutation graph* is a cubic graph containing a 2-factor consisting of two chordless cycles. One can easily see that both of these cycles must have length $n/2$, where n is the order of the graph.

Cycle permutation graphs were first introduced by Chartrand and Harary [3] in 1967. They characterised the conditions under which a cycle permutation graph is planar.

In 1972, Klee [8] also studied cycle permutation graphs, where he referred to them as "generalised prisms". He was interested in the hamiltonicity of cycle permutation graphs and asked for which orders there exist non-hamiltonian cycle permutation graphs. Using a computer search Klee determined that up to order 16, the Petersen graph is the only non-hamiltonian cycle permutation graph and he partially solved the question by proving that for every order $n \equiv 2 \bmod 4$ with $n \geq 18$ there exist non-hamiltonian cycle permutation graphs. In this

R. Královič and V. Kůrková (Eds.): SOFSEM 2025, LNCS 15538, pp. 333–346, 2025.
https://doi.org/10.1007/978-3-031-82670-2_24

manuscript, we finish this classification by completely characterising for which orders non-hamiltonian cycle permutation graphs exist.

Cycle permutation graphs were also studied in the 80's by Shawe-Taylor and Pisanski, who asked questions about the girth of such graphs [12,13]. Given a girth g, they construct cycle permutation graphs with girth g or higher.

We are also interested in cycle permutation graphs which are *snarks*, i.e. non-3-edge colourable. In this case we call them *permutation snarks*. We note that in some definitions, snarks are required to be *cyclically* 4*-edge-connected* and/or of *girth* at least 5, however, this is always the case for cycle permutation graphs of order $n > 6$. A graph is *cyclically k-edge connected* if the removal of fewer than k edges cannot separate the graph into two components each containing a cycle and its *girth* is the length of a shortest cycle. Note that a cycle permutation graph with a cycle of length 4 is always hamiltonian and hamiltonian graphs cannot be snarks.

Snarks are particularly interesting since for a lot of open conjectures it can be shown that if the conjecture is false, the smallest possible counterexamples are snarks. This has amongst others been proven for the Cycle Double Cover conjecture [16,17] and Tutte's 5-flow conjecture [18]. A better understanding of the class of snarks can therefore lead to a better understanding of many long-standing open conjectures. Permutation snarks in particular are of interest as their additional structure makes them natural candidates for investigation and this structure in turn allows us to study this class computationally up to significantly higher orders.

It is well known that the Petersen graph is a permutation snark. In 1997 Zhang conjectured that it is the only permutation snark which is cyclically 5-edge-connected [19]. This conjecture was refuted by Brinkmann, Goedgebeur, Hägglund and Markström [1] in 2013, when they developed a new generation algorithm for cubic graphs and snarks and generated all snarks up to 36 vertices and determined that there are 12 cyclically 5-edge-connected permutation snarks on 34 vertices. They obtained counts for permutation snarks by using a filter approach, which given an input graph (given by the generation algorithm) determines whether or not it is a permutation snark. However, this method is not very efficient as in practice very few cubic graphs are cycle permutation graphs and even fewer are permutation snarks. For example on order 32 only 0.24% of the cubic graphs of girth at least 4 (the minimum girth of a cycle permutation graph of order > 6) are cycle permutation graphs. We improve the result by Brinkmann et al. by determining all permutation snarks up to order 46, yielding many more counterexamples to Zhang's conjecture.

We have developed two specialised algorithms for the exhaustive generation of all pairwise non-isomorphic cycle permutation graphs of a given order. The first uses the canonical construction path method [9] to make sure no isomorphic copies are output. The second is based on orderly generation [5,14] and is faster, but only takes care of some of the most common isomorphs. The latter is most useful when the exact number of cycle permutation graphs is not relevant or when the number of output graphs is small enough so that removing isomorphs afterwards is feasible. We describe both approaches in Sect. 2. Their implementations can be found on GitHub [6]. Moreover, we extended both of these algorithms in order to restrict the search to non-hamiltonian cycle permutation graphs or graphs with a given lower bound on the girth in an efficient way. (See Sect. 3.) This allowed us to exhaustively generate all cycle permutation graphs and permutation snarks up to much higher orders.

In 2019, Máčajová and Škoviera [11] gave three methods for constructing permutation snarks and used them to provide permutation snarks of cyclic connectivity 4 and 5 for every large enough order $n \equiv 2 \bmod 8$. They mention that no permutation snark of order $n \equiv 6 \bmod 8$ is known and prove that such a smallest permutation snark – if it exists – must by cyclically 5-edge-connected.

The rest of this paper is organised as follows. In Sect. 2, we explain how the algorithms work and prove their correctness.

In Sect. 3, we use the algorithm to obtain counts for cycle permutation graphs as well as non-hamiltonian cycle permutation graphs and permutation snarks. For the latter, our algorithm allows us to generate all permutation snarks up to order 46. We also increase lower bounds for interesting permutation snarks such as the minimum order for a permutation snark of order $n \equiv 6 \bmod 8$, which was also posed as an open problem in [1].

In Sect. 4, we finish the characterisation of the orders for which non-hamiltonian cycle permutation graphs exist, solving one of the questions by Klee [8].

Due to space constraints we had to omit several proofs and details from this text. A full-length version of this article containing these omissions can be found on ArXiv [7].

2 Generation Algorithms

In this section, we will present our algorithms for exhaustively generating all cycle permutation graphs of a given order n. The basic idea is a backtracking approach, where we start from two cycles of length $n/2$ connected by one edge and add edges between the cycles in all necessary ways. Clearly, this generates all cycle permutation graphs. However, such an approach will give a lot of isomorphic copies. One way solve this is to keep track of all intermediate graphs and to stop adding edges whenever a graph isomorphic to a previously generated graph is found. However, this approach is very memory intensive. For example if one generates all cycle permutation graphs of order 28 using this approach, approximately 160 gigabytes of memory is needed. Therefore, we will use a different approach to avoid the generation of isomorphic copies.

For our first algorithm (see Sect. 2.1), we will use the canonical construction path method introduced by McKay [9], which (if used correctly) guarantees that every graph is generated exactly once without having to store (and compare) all graphs in the memory during the generation process.

For our second algorithm (see Sect. 2.2) – whose idea was suggested to us by Steven Van Overberghe –, we reject isomorphisms using a method based on orderly generation [5,14], which was first used by Rozenfel'd [15] for generating strongly regular graphs. One labelling in each isomorphism class is determined to be canonical and only these labellings will be accepted by the algorithm. For efficiency's sake, we tolerate that some isomorphic graphs have multiple "canonical" labellings, which might lead to isomorphic graphs being output. Again, in this case, every graph is generated at least once without having to store any graphs in memory.

In the next subsections we describe these algorithms in more detail and prove their correctness. Our implementations of both algorithms can be found on GitHub [6] and in [7] we describe how we extensively tested the correctness of the implementations in various ways.

2.1 Canonical Construction Path Method

An *expansion* is an operation which constructs a larger graph from a given smaller one. The reverse operation is called a *reduction*. To use the canonical construction path method we will need to define a *canonical reduction* which is unique up to isomorphism and its inverse operation will then be the *canonical expansion*. Two expansions applied to the same graph are *equivalent* if they lead to isomorphic graphs. This gives us, for each graph to which we apply expansions, classes of equivalent expansions.

When applying an expansion we need to determine whether or not we will accept the newly generated graph and keep adding edges to it or if we discard it and choose another edge to add in the smaller graph. In order to avoid isomorphic copies, we should accept every non-isomorphic intermediate graph exactly once. This can be done by following these two rules:

1. Only accept a graph if it was obtained by canonical expansion.
2. For every graph G to which expansions will be applied, only perform one expansion from each equivalence class of expansions.

In our case there is only one expansion operation, that is: adding an edge between pairs of specific vertices of degree 2.

Consider a subcubic graph G of order n which is a *spanning subgraph*, i.e. containing all vertices, of a cycle permutation graph. The 2-factors of this graph consisting of two chordless cycles of length $n/2$ will be called *permutation 2-factors*. Let C_1 and C_2 be the two induced cycles of a permutation 2-factor F. We say F is *consecutive* if the vertices of degree 3 on C_1 induce a path (or a cycle) or if the vertices of degree 3 on C_2 induce a path (or a cycle).

Let F be a consecutive permutation 2-factor with induced cycles C_1 and C_2. For $i \in \{1, 2\}$, if the degree 3 vertices on C_i induce a path P, let S_i be the set of vertices of degree 2 which are adjacent to P on C_i or let S_i be empty otherwise. We consider a pair of vertices *eligible* if one of the vertices lies in such a set S_i and the other is a vertex of degree 2 on C_{3-i}. We will apply the expansion to any eligible pair of vertices for any consecutive permutation 2-factor of our graph.

Note, that the algorithm would also work if we add an edge between any pair of degree 2 vertices in which one lies on C_1 and the other on C_2. However, restricting the algorithm to eligible pairs drastically reduces the amount of times the recursive algorithm branches and hence makes the implementation a lot more efficient.

High level pseudocode for the recursive method of the algorithm can be found in Algorithm 1.

Algorithm 1. Expand(G)

```
if G is cubic then
    Output G
Determine all eligible pairs of vertices.
Determine the orbits of all such pairs.
for all eligible pairs (u, v) do
    // Cf. Rule 2. of the canonical construction path method.
    if (u, v) is the representative of its orbit then
        if adding edge uv to G is a canonical expansion then // Cf. Rule 1.
            Expand(G + uv)
```

We now explain the algorithm in more detail. Suppose we want to generate the cycle permutation graphs of order n. We start with G_0 consisting of two cycles $C_1 = v_0v_1 \dots v_{n/2-1}$ and $C_2 = w_0w_1 \dots w_{n/2-1}$ and the edge v_0w_0. Note that $C_1 \cup C_2$ forms a consecutive permutation 2-factor. Let G be a (sub)cubic supergraph of G_0 of order n which contains a consecutive permutation 2-factor.

If G is cubic then it is a cycle permutation graph. It should be output and the recursion backtracks. If not, we determine all consecutive permutation 2-factors in order to find the eligible pairs of vertices. In practice, we keep track of all consecutive permutation 2-factors dynamically for efficiency reasons. More specifically: after adding an edge, we remove those which have become non-consecutive or are no longer permutation 2-factors and search for consecutive permutation 2-factors containing the newly added edge. While constructing the induced cycles of such a 2-factor they are stored in a bitset, so that we can use bit operations to determine efficiently whether or not they have chords. This search for new consecutive permutation 2-factors is one of the bottlenecks of the algorithm.

Due to space constraints, we omit a description of further optimisations concerning the computation of permutation 2-factors. It can be found in [7].

Given all consecutive permutation 2-factors, it is straightforward to determine all eligible pairs of vertices. We then need to determine the orbits of these pairs in order to only add an edge between one of the vertex pairs of each orbit, since all pairs in the same orbit would lead to isomorphic graphs. To obtain these orbits we use the program `nauty` [10] for determining all generators of the automorphism group of G. The orbits are then determined using a union-find algorithm. We choose the representative of each orbit to be the first vertex pair of that orbit we encountered while searching for eligible vertex pairs.

Once we have determined that an eligible vertex pair (u, v) is the representative of its orbit, we need to determine whether its addition to G will be a canonical expansion. In practice, we do this by looking at the graph $G + uv$

and analysing its *reducible* edges. These are the edges e for which $G + uv - e$ contains a consecutive permutation 2-factor. These are however easily obtained once one knows all consecutive permutation 2-factors of $G + uv$. Other edges are not reducible and can be ignored since removing them does not give us a graph which needs to be generated.

In order to determine whether or not we should accept $G + uv$ we need to define a canonical reduction which is unique up to isomorphism. The edge whose removal yields the canonical reduction will be the *canonical edge.* To this end, we assign a 10-tuple $(x_0, \ldots, x_9)$ to each reducible edge and let the canonical edge be the one with the lexicographically maximal value for this 10-tuple.

For a reducible edge $e = ab$, the values $x_0, \ldots, x_7$ are invariants of increasing discriminating power and cost, determined empirically. They are defined as follows:

- x_0 is the negative of the number of vertices at distance at most 2 from a or b.
- x_1 (x_2) is the negative of the number of 4-cycles (5-cycles) containing the edge ab.
- x_3 is the number of vertices at distance at most 3 from a or b.
- x_4 is the negative of the number of 6-cycles containing the edge ab.
- x_5 (x_6) is (the negative of) the number of vertices of degree 2 at distance 1 (at most 2) of a and of b.
- x_7 is the number of vertices at distance at most 4 from a or b.

We note that while the discriminating power is negligible for the later invariants, they are necessary for the efficiency of variants of this program, such as the generation of non-hamiltonian cycle permutation graphs or cycle permutation graphs with girth restrictions. See Sect. 3. Moreover, their presence here does not noticeably increase the running time of the algorithm.

While the above values are invariant under isomorphism, their values could be the same for non-isomorphic graphs. Therefore we define $\{x_8, x_9\}$ to be the lexicographically largest label of an edge which is in the same edge orbit as e in the canonical labelling of the graph. We again use `nauty` [10] for determining this canonical labelling of the graph. (This gives us the generators of the automorphism group for free, hence if we accept this expansion, we take care not to call `nauty` a second time.)

Note that in theory we could define the canonical edge using only x_8 and x_9, however, as calling `nauty` is computationally expensive, it is much more efficient to compute the other invariants first. Once such an invariant is able to show that our added edge uv is not canonical, for example if there is a reducible edge with fewer vertices at distance at most 2, then we can reject the expansion immediately and do not need to compute the remaining invariants (including the expensive call to `nauty`). Similarly, the invariants can be sufficient to determine if our added edge is the only reducible edge with maximum value for the tuple and hence it is canonical. The expansion can then immediately be accepted without computing the remaining invariants. As an example, on order 24, invariants $x_0, \ldots, x_7$ are sufficient to discriminate the canonical edge in 94.64% of the cases. If we were to only consider x_8 and x_9 for this order, this would slow down the total computation by a factor of 5.

Theorem 1. *For a given order n, Algorithm 1 will output all pairwise non-isomorphic cycle permutation graphs of order n exactly once.*

Due to page limits we omit the proof. See [7].

2.2 Weak Orderly Generation Method

While the canonical construction path method described in Sect. 2.1 is relatively fast and produces exact counts, one can be even faster by tolerating that some isomorphic copies are output. One of the bottlenecks of the previous approach is the search for new permutation 2-factors in order to obtain all reducible edges. In this section we will present an alternative algorithm based on orderly generation which we called *weak orderly generation*. With this method we will "ignore" the fact that permutation multiple permutation 2-factors can exist. However, in doing so, cycle permutation graphs having multiple permutation 2-factors might be output multiple times.

The idea of this method is based on orderly generation, which was independently introduced by Faradzev [5] and Read [14] in 1978, but had already been used by Rozenfel'd [15] in 1973 for the generation of strongly regular graphs. Orderly generation defines a canonically labelled object for each isomorphism class and generates only that one for each isomorphism class. Ignoring the fact that a cycle permutation graph can have multiple permutation 2-factors, our problem lends itself well to orderly generation and our recursive method needs to branch a lot less, which is why it is a lot faster than the canonical construction path method. The only caveat is that multiple canonically labelled objects can exist for a single isomorphism class, which is why we call it *weak* orderly generation.

We will use the notation $[x] := \{0, 1, \ldots, x-1\}$ for the set of the first x integers (including 0). A *k-permutation* is a permutation $\pi : [k] \to [k]$. A *partial k-permutation* of size l is a permutation from $[l]$ to a subset S of $[k]$ of order l. When $S = [l]$, we say the partial permutation is *restricted*. Let S be a subset of $[l]$, then we define $\pi(S) := \{\pi(s) | s \in S\} \subset [k]$.

We can represent a subcubic graph G containing a consecutive permutation 2-factor F by its order n and a (partial) $n/2$-permutation. Let $C_1 = v_0 v_1 \ldots v_{n/2-1}$ and $C_2 = w_0 w_1 \ldots w_{n/2-1}$ be the cycles of F. Then G is represented by a partial k-permutation π of size l if $w_{\pi(i)}$ is a neighbour of v_i for all $0 \leq i < l$ and G has exactly $n + l$ edges. Vice versa, a partial $(n/2)$-permutation induces a labelled graph of order n.

Certain operations on these partial k-permutations lead to isomorphic graphs. For a given partial k-permutation π of size l with $\pi(0) = 0$, we define the operations:

- $p_1 : \pi \mapsto \pi'$ such that $\pi'(i) = k - \pi(i) \bmod k$ for $i \in [l]$;
- $p_2 : \pi \mapsto \pi'$ such that $\pi'(i) = \pi(l - i) - \pi(l) \bmod k$ for $i \in [l]$.

When π is also restricted, we can define

- $p_3 : \pi \mapsto \pi'$ such that $\pi'(i) = \pi^{-1}(i)$ for $i \in [l]$.

When π is a k-permutation, take $\pi(-1) := \pi(k-1)$. We can define

- $p_4 : \pi \mapsto \pi' : \pi'(i) = (\pi(i-1) - \pi(k-1)) \bmod k$ for $i \in [k]$.

For a given partial k-permutation π of size l with $\pi(0) = 0$, we define $\mathcal{F}(\pi)$ to be the set of all permutations which can be obtained from π by inductively applying p_1, p_2, p_3, p_4 if $l = k$, p_1, p_2, p_3 if π is restricted but $l \neq k$ or p_1, p_2 otherwise.

Let π be a partial k-permutation with $\pi(0) = 0$. Then π is canonical if it is is the lexicographically smallest partial permutation in $\mathcal{F}(\pi)$. The corresponding labelled graph is called *weakly canonical.*

Algorithm 2. Expand(π, l)

---

if $l = k$ **then**
  Output cubic graph represented by π.
for all $x \in [k] \setminus \pi([l])$ **do**
  Extend π to π' by $\pi'(l) = x$.
  if π' is canonical **then**
    Expand(π', $l+1$)

We now explain the algorithm in more detail. A rough outline of the recursive method without any optimisations or technical details can be found in Algorithm 2. Suppose we want to generate the cycle permutation graphs of order n. Let $k := n/2$. We start with G_0 consisting of two disjoint cycles $C_1 = v_0v_1 \ldots v_{k-1}$ and $C_2 = w_0w_1 \ldots w_{k-1}$ and the edge v_0w_0. G_0 is then represented by the partial k-permutation of size 1 which sends 0 to 0. Now let G be a (sub)cubic supergraph of G_0 of order n whose labelling is represented by some partial k-permutation π of size l. If G is cubic, then $k = l$ and we have a cycle permutation graph. It should be output and the recursion backtracks. If not, we recursively add edges from v_l to w_i with $i \in [k] \setminus \pi([l])$. Adding such an edge v_lw_i corresponds to extending the partial k-permutation π by a partial k-permutation π' in which $\pi'(l) = i$ and $\pi'(j) = \pi(j)$ for $j \in [l]$. We must now decide to continue with π' or to discard it. We check if π' is canonical by comparing it to all necessary $f(\pi')$, where f is some composition using p_1, p_2, p_3 and/or p_4.

Note that what we have described is actually (non-weak) orderly generation of the partial k-permutations π with $\pi(0) = 0$, but the translation between k-permutations and cycle permutation graphs is not one-to-one (since a cycle permutation graph can have multiple permutation 2-factors).

Theorem 2. *Algorithm 2 outputs every cycle permutation graph of a given order n at least once.*

Due to page limits we omit the proof. See [7].

The main purpose of this algorithm is to generate non-hamiltonian cycle permutation graphs, in order to extend the exhaustive list of permutation snarks.

As these counts remain relatively low for small orders, it is feasible to perform an isomorphism check afterwards (e.g. using `nauty` [10]) in order to determine the actual counts. We do this via an adaptation described in Sect. 3.2 as well as lookaheads for detecting hamiltonian cycles early in the case of Algorithm 2. See [7].

3 Results

3.1 Cycle Permutation Graphs

Table 1. We omit the single cycle permutation graph of girth 3 on order 6 from this table. The columns with $g \geq k$ indicate the counts of cycle permutation graphs with girth at least k for each order.

Order	$g \geq 4$	$g \geq 5$	$g \geq 6$	$g \geq 7$	$g \geq 8$	$g \geq 9$
8	2	0	0	0	0	0
10	4	1	0	0	0	0
12	10	1	0	0	0	0
14	28	3	0	0	0	0
16	123	11	1	0	0	0
18	667	59	0	0	0	0
20	4 815	402	4	0	0	0
22	41 369	3 602	9	0	0	0
24	411 231	37 178	84	0	0	0
26	4 535 796	424 252	846	1	0	0
28	54 828 142	5 289 603	12 597	0	0	0
30	717 967 102	71 206 645	197 921	1	0	0
32	10 118 035 593	1 027 074 710	3 334 149	6	0	0
34	152 626 831 184	15 800 380 281	58 638 599	190	0	0
36	?	?	1 077 159 843	4 437	1	0
38	?	?	20 642 970 164	147 820	0	0
40	?	?	?	5 166 381	0	0
42	?	?	?	167 517 630	2	0
44	?	?	?	?	33	0
46	?	?	?	?	847	0
48	?	?	?	?	21 294	0
50	?	?	?	?	1 053 289	0
52–58	?	?	?	?	?	0
60	?	?	?	?	?	2
62	?	?	?	?	?	61
64	?	?	?	?	?	1 654

We have used our algorithms to generate all cycle permutation graphs of a given order and also extended them to generate graphs with a lower bound on the girth. This can be done efficiently as the algorithm only adds edges, so we can prune as soon as we have a cycle which is smaller than the desired girth. The results obtained by our implementation of Algorithm 1 can be found in Table 1. These graphs can also be obtained from the House of Graphs [4] at https://houseofgraphs.org/meta-directory/cubic (up to the orders for which it was still feasible to store them). The runtimes needed to obtain these results are summarised in a table in [7].

While the implementation of Algorithm 2 does not give exact counts, it is faster than Algorithm 1 at giving an upper bound. A comparison indicates that the factor with which the second algorithm is faster seems to grow as the order increases and that the ratio of non-isomorphic graphs versus all graphs output by this algorithm seems to increase as well. For example on order 16, 96.85% of the output graphs are non-isomorphic, while on order 34, this is 99.21%. More details are omitted, but can be found in a table in [7].

3.2 Non-hamiltonian Cycle Permutation Graphs

A small adaptation to Algorithms 1 and 2 allows for the efficient generation of all non-hamiltonian cycle permutation graphs of a given order n. This is done by generating cycle permutation graphs of girth at least 5 and by also rejecting an expansion if it leads to a hamiltonian graph. We have adapted the program `cubhamg` included in `nauty` [10] to perform this hamiltonicity check as it is a very fast method for determining the hamiltonicity of (sub)cubic graphs. More details on this algorithm can be found in [2]. For Algorithm 2, we apply lookaheads to avoid performing the hamiltonicity check as much as possible as for this algorithm the hamiltonicity check is the main bottleneck. These lookaheads are described in detail in [7].

Using this adaptation we obtain the counts of non-hamiltonian cycle permutation graphs found in Table 2, and the following propositions, giving a partial answer to the questions asked by Klee [8] concerning the hamiltonicity of cycle permutation graphs. We also increase the previously known counts of permutation snarks from order 34 up to order 46. The graphs can be found in the meta-directory of the House of Graphs [4] at https://houseofgraphs.org/meta-directory/snarks.

Using Algorithm 1 we were able to compute the non-hamiltonian cycle permutation graphs up to order 42. Hence, Algorithm 2 was necessary for obtaining the counts up to order 46. For the counts with girth restrictions, Algorithm 2 was also able to go some orders further. We clarify that these are the exact counts of non-hamiltonian cycle permutation graphs as we filtered the isomorphic ones (using `nauty` [10]) whenever Algorithm 2 was applied. This also means that the counts up to order 42 were independently obtained using two different approaches. Runtimes for the obtained counts are summarised in [7].

In contrast to the general case where the ratio of non-isomorphic graphs versus total graphs output by Algorithm 2 was close to 100% and increasing, in

Table 2. Counts of non-hamiltonian cycle permutation graphs for each order. Columns $g \geq k$ indicate counts with girth at least k for each order, the column $\chi' = 4$ indicates counts of permutation snarks and the column $\lambda_c \geq 5$ indicates counts of cyclically 5-edge-connected permutation snarks.

Order	$g \geq 5$	$\chi' = 4$	$\lambda_c \geq 5$	$g \geq 6$	$g \geq 7$	$g \geq 8$	$g \geq 9$
10	1	1	1	0	0	0	0
12–16	0	0	0	0	0	0	0
18	2	2	0	0	0	0	0
20	0	0	0	0	0	0	0
22	1	0	0	0	0	0	0
24	0	0	0	0	0	0	0
26	64	64	0	0	0	0	0
28	0	0	0	0	0	0	0
30	9	0	0	0	0	0	0
32	0	0	0	0	0	0	0
34	10 778	10 771	12	0	0	0	0
36	4	0	0	0	0	0	0
38	1 848	0	0	0	0	0	0
40	19	0	0	0	0	0	0
42	3 131 740	3 128 893	736	0	0	0	0
44	1 428	0	0	0	0	0	0
46	678 106	0	0	0	0	0	0
48	?	?	?	0	0	0	0
50-54	?	?	?	?	0	0	0
56-58	?	?	?	?	?	0	0
60-70	?	?	?	?	?	?	0

the non-hamiltonian case, the ratio is a lot lower and can vary a lot depending on the order. For example on orders 34, 36 and 38, we get 31.90%, 17.39% and 36.07%, respectively. (See [7] for more details.) An interesting observation is that non-hamiltonian cycle permutation graphs more often than not have multiple permutation 2-factors, which is opposite from the general case. Since isomorphisms occur when output graphs have multiple permutation 2-factors, this explains why there are many more isomorphic graphs output by the algorithm in this case.

It is easy to see that a permutation snark cannot have order 0 mod 4. Otherwise, the induced cycles of a permutation 2-factor are even and we can colour their edges using two colors and give all other edges the third color. Hence, a permutation snark can only exist for orders 2 mod 4. It was shown by Máčajova and Škoviera [11] that there exist (cyclically 5-edge-connected) permutation snarks on every order $n \equiv 2 \bmod 8$ for $n \geq 10$ ($n \geq 34$). However, so far no examples of order $n \equiv 6 \bmod 8$ are known and in fact this was posed as an open problem in [1, Problem 1], where all permutation snarks up to order 34 were determined. Máčajova and Škoviera prove that a smallest example must be cyclically 5-edge-connected.

We show the following.

Proposition 1. *A smallest permutation snark of order* 6 mod 8 *has order at least* 54.

Proof. Using the generation algorithm described in Sect. 2, we generated all non-hamiltonian cycle permutation graphs up to order 46. See Table 2. While there are 1 848 non-hamiltonian cycle permutation graphs of order 38 and 879 828 such graphs of order 46, using two independent algorithms, we verified that all of them are 3-edge-colourable. □

Zhang's conjecture [19] stating that the Petersen graph is the only cyclically 5-edge-connected permutation snark was refuted by Brinkmann et al. [1]. Our search verified their 12 counterexamples on order 34 and found 736 new ones on order 42, cf. Table 2.

While Shawe-Taylor and Pisanski showed in [12] that the girth of cycle permutation graphs can be arbitrarily large, no permutation snarks of girth 6 or higher are known. Our computational results from Table 2 imply the following.

Proposition 2. *The smallest non-hamiltonian cycle permutation graph of girth at least* 6 *has order at least* 50. *The smallest such graph of girth at least* 7 *has order at least* 56. *The smallest such graph of girth at least* 8 *has order at least* 60. *The smallest such graph of girth at least* 9 *has order at least* 72.

Note that as a corollary, we can replace "non-hamiltonian cycle permutation graph" in the above proposition with "permutation snark".

Since we are particularly interested in permutation snarks, one can wonder why we only created a specialised generator for non-hamiltonian cycle permutation graphs and not for permutation snarks. Hamiltonicity is a property which is preserved when adding edges, hence, whenever our algorithm for generating non-hamiltonian cycle permutation graphs encounters a hamiltonian graph, we can prune this branch of the search tree. In contrast, 3-edge-colourability is not preserved under the addition of edges, hence if we encounter a 3-edge-colourable graph, we still need to inspect this branch of the search tree.

One could come up with conditions under which the addition of an edge to a 3-edge-colourable graph can never yield a non-3-edge-colourable graph and use this for pruning in the last step, however, since in the last step of our algorithm there is only one non-edge eligible for addition, this will not speed up the computation time.

4 Orders of Non-hamiltonian Cycle Permutation Graphs

In his 1972 paper [8], Klee studied two questions involving the non-hamiltonicity of cycle permutation graphs, attributed to Ralph Willoughby. The first one asks which permutation graphs admit a hamiltonian cycle. The second asks for which orders n there exist non-hamiltonian cycle permutation graphs. Klee answered this second question partially by proving that for all $n \equiv 2 \bmod 4$ there is a non-hamiltonian cycle permutation graph if and only if n is neither 6 nor 14. We complete the characterisation of the orders for which non-hamiltonian cycle permutation graphs exist and thereby solve Klee's second question. Due to space constraints we omit the proof here. See [7].

Proposition 3. *There is a non-hamiltonian cycle permutation graph of order n if and only if n is even and $n \in \{10, 18, 22, 26, 30\}$ or $n \geq 34$.*

Acknowledgements. The authors thank Steven Van Overberghe for suggesting the idea of Algorithm 2 and Edita Mcajov and Martin koviera for their valuable insights and contributions.

This research was supported by Internal Funds of KU Leuven and by an FWO grant with grant number G0AGX24N. Several of the computations for this work were carried out using the supercomputer infrastructure provided by the VSC (Flemish Supercomputer Center), funded by the Research Foundation Flanders (FWO) and the Flemish Government.

References

1. Brinkmann, G., Goedgebeur, J., Hägglund, J., Markström, K.: Generation and properties of snarks. J. Comb. Theory Ser. B **103**(4), 468–488 (2013). https://doi.org/10.1016/j.jctb.2013.05.001
2. Brinkmann, G., Goedgebeur, J., Mckay, B.D.: The minimality of the Georges-Kelmans graph. Math. Comput. **91**(335), 1483–1500 (2022). https://doi.org/10.1090/mcom/3701
3. Chartrand, G., Harary, F.: Planar permutation graphs. Ann. Inst. H. Poincare **3**(4), 433–438 (1967)
4. Coolsaet, K., D'hondt, S., Goedgebeur, J.: House of graphs 2.0: a database of interesting graphs and more. Discret. Appl. Math. **325**, 97–107 (2023). https://doi.org/10.1016/j.dam.2022.10.013
5. Faradzev, I.A.: Algorithmic Studies in Combinatorics, chap. Generation of nonisomorphic graphs with a given degree sequence, pp. 11–19. Nauka (1978). (Russian)
6. Goedgebeur, J., Renders, J.: Cycle Permutation Graphs, (Version 1) [Computer software] (2024). https://github.com/JarneRenders/Cycle-Permutation-Graphs
7. Goedgebeur, J., Renders, J.: Generation of cycle permutation graphs and permutation snarks (2024). https://arxiv.org/abs/2411.12606
8. Klee, V.: Which generalized prisms admit H-circuits? In: Alavi, Y., Lick, D.R., White, A.T. (eds.) Graph Theory and Applications, pp. 173–178. Springer, Heidelberg (1972). https://doi.org/10.1007/BFb0067368
9. McKay, B.D.: Isomorph-free exhaustive generation. J. Algorithms **26**(2), 306–324 (1998). https://doi.org/10.1006/jagm.1997.0898

10. McKay, B.D., Piperno, A.: Practical graph isomorphism. II. J. Symb. Comput. **60**, 94–112 (2014). https://doi.org/10.1016/j.jsc.2013.09.003
11. Mácajová, E., Škoviera, M.: Permutation snarks of order 2 (mod 8). Acta Math. Univ. Comen. **88**(3), 929–934 (2019)
12. Pisanski, T., Shawe-Taylor, J.: Cycle permutation graphs with large girth. Glas. Mat. **17**, 233–236 (1982)
13. Pisanski, T., Shawe-Taylor, J.: Search for minimal trivalent cycle permutation graphs with girth nine. Disc. Math. **36**(1), 113–115 (1981). https://doi.org/10.1016/0012-365X(81)90179-5
14. Read, R.C.: Every one a winner or how to avoid isomorphism search when cataloguing combinatorial configurations*. In: Alspach, B., Hell, P., Miller, D.J. (eds.) Annals of Discrete Mathematics, Algorithmic Aspects of Combinatorics, vol. 2, pp. 107–120. Elsevier, Amsterdam (1978). https://doi.org/10.1016/S0167-5060(08)70325-X
15. Rozenfel'd, M.Z.: The construction and properties of certain classes of strongly regular graphs (Russian). Uspehi Mat. Nauk **28**(3), 197–198 (1973)
16. Seymour, P.D.: Sums of circuits. In: Bondy, J.A., Murty, U.R.S. (eds.) Graph Theory and Related Topics, pp. 341–355. Academic Press, New York (1979)
17. Szekeres, G.: Polyhedral decompositions of cubic graphs. B. Aust. Math. Soc. **8**(3), 367–387 (1973). https://doi.org/10.1017/S0004972700042660
18. Tutte, W.T.: A contribution to the theory of chromatic polynomials. Can. J. Math. **6**, 80–91 (1954). https://doi.org/10.4153/CJM-1954-010-9
19. Zhang, C.Q.: Integer Flows and Cycle Covers of Graphs. No. 205 in Monographs and Textbooks in Pure and Applied Mathematics. Marcel Dekker, New York (1997)

Expected Density of Random Minimizers

Shay Golan[1,2] and Arseny M. Shur[3](✉)

[1] Reichman University, Herzliya, Israel
golansh1@biu.ac.il
[2] University of Haifa, Haifa, Israel
[3] Bar Ilan University, Ramat Gan, Israel
shur@datalab.cs.biu.ac.il

Abstract. Minimizer schemes, or just minimizers, are a very important computational primitive in sampling and sketching biological strings. Assuming a fixed alphabet of size σ, a minimizer is defined by two integers $k, w \geq 2$ and a total order ρ on strings of length k (also called k-mers). A string is processed by a sliding window algorithm that chooses, in each window of length $w + k - 1$, its minimal k-mer with respect to ρ. A key characteristic of the minimizer is the expected density of chosen k-mers among all k-mers in a random infinite σ-ary string. Random minimizers, in which the order ρ is chosen uniformly at random, are often used in applications. However, little is known about their expected density $\mathcal{DR}_\sigma(k, w)$ besides the fact that it is close to $\frac{2}{w+1}$ unless $w \gg k$.

We first show that $\mathcal{DR}_\sigma(k, w)$ can be computed in $O(k\sigma^{k+w})$ time. Then we attend to the case $w \leq k$ and present a formula that allows one to compute $\mathcal{DR}_\sigma(k, w)$ in just $O(w \log w)$ time. Further, we describe the behaviour of $\mathcal{DR}_\sigma(k, w)$ in this case, establishing the connection between $\mathcal{DR}_\sigma(k, w)$, $\mathcal{DR}_\sigma(k + 1, w)$, and $\mathcal{DR}_\sigma(k, w + 1)$. In particular, we show that $\mathcal{DR}_\sigma(k, w) < \frac{2}{w+1}$ (by a tiny margin) unless w is small. We conclude with some partial results and conjectures for the case $w > k$.

Keywords: Minimizer · Random Minimizer · Expected Density

1 Introduction

The study of length-k substrings (*k-mers*) of long strings dates back to the conjectures of Golomb [7] and Lempel [11], proved by Mykkeltveit in 1972 [14]. He constructed, for each $\sigma \geq 2$ and $k \geq 2$, a minimum-size *unavoidable* set of σ-ary k-mers; "unavoidable" means that every long enough σ-ary string contains a substring from Mykkeltveit's set. The interest to k-mers boosted in 2000s, when sketching of biological sequences was proposed as an alternative approach to full-text indexing with FM-index [4] and similar tools. Minimizers [15,19] provide a very simple way to sample k-mers for sketching and are also used in more involved sampling schemes such as syncmers [3], strobemers [17], and mod-minimizers [8]. For more information, see [1,18,22] and the references therein.

R. Královič and V. Kůrková (Eds.): SOFSEM 2025, LNCS 15538, pp. 347–360, 2025.
https://doi.org/10.1007/978-3-031-82670-2_25

To sample substrings for a sketch of the string S, one fixes integer parameters k and w and chooses one k-mer in every set of w consecutive k-mers in S; this process can be viewed as choosing a k-mer in each "window" of length $(w+k-1)$. By *markup* of S we mean any map assigning to each window its chosen k-mer; we refer to the starting positions of these k-mers in S as *marked positions.* A *(sampling) scheme* is a deterministic algorithm taking a string S and the numbers k, w as the input and computing the markup of S. A scheme must be *local,* which means that the choice of a k-mer in a window depends solely on the window as a string. A *minimizer* is a scheme that fixes a linear order on k-mers and chooses the starting position of the minimal k-mer in each window, breaking ties to the left (breaking ties to the right instead, one will get "dual" minimizers having the same properties as minimizers).

The *density* of a markup is the ratio between the number of marked positions and the length of S. For a scheme $\mathcal{S}$, let $D_{\mathcal{S}}(n)$ be the expected density of the markup of a uniformly random string of length n. The *density of* $\mathcal{S}$ is defined as the limit $D_{\mathcal{S}} = \lim_{n\to\infty} D_{\mathcal{S}}(n)$. Trivially, $D_{\mathcal{S}} \geq 1/w$ by the definition of markup.

Below we consider only minimizers. Given σ and k, $\mathcal{S}_\rho$ denotes the minimizer that uses the order ρ on k-mers (one can view ρ as a permutation of the set of all σ-ary k-mers). To simplify the notation, we write D_ρ instead of $D_{\mathcal{S}_\rho}$. If ρ is chosen uniformly at random from the set of all permutations of σ-ary k-mers, D_ρ becomes a random variable. Its expectation is called the *density of the random order,* denoted by $\mathcal{DR}_\sigma(k,w)$. Note that many schemes [3,8,17,22] make use of minimizers with a (pseudo)random order, and both cases $w \leq k$ and $w > k$ are important for applications [2,12,21]. Typically, a scheme chooses ρ to be the $\leq$ order on hash values of k-mers for some hash function. As the density of the chosen order impacts the size of the obtained sample, computing $\mathcal{DR}_\sigma(k,w)$ has both theoretical and practical interest. Schleimer et al. [19] showed that $\mathcal{DR}_\sigma(k,w) \approx \frac{2}{w+1}$ if "most" windows have no repeated k-mers. Zheng et al. [22] were more precise: $\mathcal{DR}_\sigma(k,w) = \frac{2}{w+1} + o(\frac{1}{w})$ whenever $w < \sigma^{k/(3+\varepsilon)}$, while orders of density $O(\frac{1}{w})$ exist if and only if $w = O(\sigma^k)$. Still, many natural questions about $\mathcal{DR}_\sigma(k,w)$ are open, and we answer some of them in this paper.

We treat σ as a constant and focus on the dependence of density on k and w. In Sect. 2, we describe an algorithm computing $\mathcal{DR}_\sigma(k,w)$ in $O(k\sigma^{k+w})$ time. Then in Sect. 3 we analyse the case $w \leq k$, presenting our main results. We prove a *formula* for $\mathcal{DR}_\sigma(k,w)$, which can be computed in just $O(w \log w)$ time independent of k. Studying this formula, we describe the connection between $\mathcal{DR}_\sigma(k,w)$, $\mathcal{DR}_\sigma(k+1,w)$, and $\mathcal{DR}_\sigma(k,w+1)$; in particular, we show that $\mathcal{DR}_\sigma(k,w) < \frac{2}{w+1}$ for almost all pairs w, k with $w \leq k$. The notion of *major run,* playing the key role in obtaining these results, can be of independent interest. In Sect. 4 we briefly consider the case $w \gg k$, where $\mathcal{DR}_\sigma(k,w)$ approaches its infimum σ^{-k}. The paper ends with a discussion in Sect. 5. The proofs and tables omitted due to space constraints can be found in the full version [6].

Notation and Definitions. In what follows, Σ, σ, k, and w denote, respectively, the alphabet $\{0,\ldots,\sigma-1\}$, its size, the length of the marker substrings (k-mers) and the number of k-mers in a window. A string s over Σ is a sequence of characters $s = s[1]s[2] \cdots s[|s|]$, where $|s|$ denotes the length of s. The *reversal* of s is $\overleftarrow{s} = s[|s|] \cdots s[2]s[1]$. If $\leq i \leq j \leq |s|$, we call $s[i..j] = s[i]s[i+1] \ldots s[j]$ a *substring* of s. It is a *prefix* if $i = 1$ and a *suffix* if $j = |s|$. A *repeat* is a pair of equal substrings in a string. A *k-string* is a string of length k; we similarly write *k-prefix*, *k-suffix*, *k-substring*, and *k-repeat.*

An integer p is a *period* of s if $s[1..|s| - p] = s[p+1..|s|]$. If moreover p is the minimal period of s and $p \leq |s|/2$, then s is called *p-periodic.* A string s is *primitive* if it is *not* p-periodic for every divisor p of $|s|$. A p-periodic substring of s is a *run* (in s), if it is not a part of a longer p-periodic substring of s.

For a set F of strings, its *prefix tree* contains all prefixes of strings from F as vertices and all pairs (u, ua), where $a \in \Sigma$, as directed edges. For a string s, its *suffix tree* is the prefix tree of the set of all suffixes of s, additionally compressed by replacing each maximal non-branching path by a single edge.

By $\mathsf{Perm}(\sigma, k)$ we denote the set of all permutations of all σ-ary k-strings.

Density of Minimizers. Let $\mathcal{S}_\rho$ be a minimizer for certain σ, k, w, and consider computing the markup of a string S. Suppose the scheme processes windows left to right, and in the current window containing a substring ub, where $b \in \Sigma$, a new position is marked. This happens either if the k-suffix of ub is its unique minimal k-mer (recall that the ties are broken to the left) or if the substring au from the previous window has its k-prefix as the minimal k-mer. One can restate this condition as "the minimal k-mer of the substring aub is either its prefix or its unique suffix". A $(k + w)$-substring with this property is called a *gamechanger* (also known as *charged context* [22]). One can see that the density D_ρ of $\mathcal{S}_\rho$, defined in the introduction, equals the fraction of gamechangers among all σ-ary $(k+w)$-strings. This gives us finite and computationally efficient definition of D_ρ.

The *density factor* is the density multiplied by the factor of $(w + 1)$. Such a normalization allows one to compare the density over a range of window sizes. We write $\mathcal{DFR}_\sigma(k, w) = (w + 1)\mathcal{DR}_\sigma(k, w)$.

2 Computing Expected Density in the General Case

In order to compute the value $\mathcal{DR}_\sigma(k, w)$ efficiently, we need a more efficient representation for it. Let $P_{\sigma,k}(v)$ be the probability that a $(w + k)$-string v is a gamechanger for a randomly chosen order $\rho \in \mathsf{Perm}(\sigma, k)$. We need two lemmas.

Lemma 1. *If v contains t distinct k-mers, then $P_{\sigma,k}(v) = \frac{2}{t}$ if the k-suffix of v has no other occurrences in v and $P_{\sigma,k}(v) = \frac{1}{t}$ otherwise.*

Proof. As v has t distinct k-mers, the probability that the k-prefix of v is minimal among these k-mers is $\frac{1}{t}$, and the same applies to the k-suffix. Now the claim follows from the definition of gamechanger. □

Lemma 2. *One has $\mathcal{DR}_\sigma(k, w) = \frac{1}{\sigma^{w+k}} \sum_{v \in \Sigma^{w+k}} P_{\sigma,k}(v)$.*

Proof. Given $v \in \Sigma^{w+k}$, $\rho \in \mathsf{Perm}(\sigma, k)$, let $I(\rho, v)$ be 1 if v is a gamechanger according to ρ and 0 otherwise. From definitions, $D_\rho = \frac{1}{\sigma^{w+k}} \sum_{v\in\Sigma^{w+k}} I(\rho, v)$ and $P_{\sigma,k}(v) = \frac{1}{|\mathsf{Perm}(\sigma,k)|} \sum_{\rho\in\mathsf{Perm}(\sigma,k)} I(\rho, v)$. Then $\mathcal{DR}_\sigma(k, w) = \frac{\sum_{\rho\in\mathsf{Perm}(\sigma,k)} D_\rho}{|\mathsf{Perm}(\sigma,k)|} = \frac{\sum_{\rho\in\mathsf{Perm}(\sigma,k)} \sum_{v\in\Sigma^{w+k}} I(\rho,v)}{|\mathsf{Perm}(\sigma,k)|\cdot\sigma^{w+k}} = \frac{1}{\sigma^{w+k}} \sum_{v\in\Sigma^{w+k}} P_{\sigma,k}(v)$. □

Lemma 2 allows one to compute $\mathcal{DR}_\sigma(k, w)$ without iterating over a huge set $\mathsf{Perm}(\sigma, k)$: indeed, it suffices to compute the probability $P_{\sigma,k}(v)$ for every $(w+k)$-string v. By Lemma 1, for the string v we need just to count distinct k-mers in it. This can be done in $O(wk)$ time naively or in $O(w + k)$ time using, e.g., the suffix tree of v. The following theorem shows that we can do better, spending just $O(k)$ amortized time per string.

Theorem 1. *The density $\mathcal{DR}_\sigma(k, w)$ can be computed in $O(k\sigma^{w+k})$ time and $O(w + k)$ space.*

Proof (sketch). Let $\mathsf{kmers}(v)$ be the number of distinct k-mers in the string v. The idea of fast computation of $\mathsf{kmers}(v)$ for all $(k + w)$-strings is as follows.

Let $\mathcal{T}$ be the prefix tree of the set Σ^{w+k}. We perform the depth-first traversal of $\mathcal{T}$, maintaining the suffix tree of the current node. Descending from a node u to its child ua, we update the suffix tree $\mathcal{ST}(u)$ to $\mathcal{ST}(ua)$ and in the process see whether the k-suffix of ua is a substring of u. Thus we obtain the number $\mathsf{kmers}(ua)$ from $\mathsf{kmers}(u)$. When ascending back from ua to u, we revert the changes, restoring $\mathcal{ST}(u)$. Traversing the whole tree $\mathcal{T}$, we get $\mathsf{kmers}(v)$ for each leaf v, which is exactly what we need. If restoring $\mathcal{ST}(u)$ from $\mathcal{ST}(ua)$ is not slower than updating $\mathcal{ST}(u)$ to $\mathcal{ST}(ua)$, then the time complexity of our scheme is $O(t\sigma^{w+k})$, where t is the worst-case time for one iteration of the suffix tree algorithm. We achieve $t = k$ with a version of Weiner's algorithm [20]. □

Remark 1. *The formula from Lemma 2 can be used to compute the expected density only for small values of k and w. For such values, $(k+w)$-strings can be viewed as σ-ary numbers of $(k+w)$ digits. Then it is faster to stick to computation over integers and avoid string operations and data structures. Instead of the suffix tree, a dictionary of k-mers of the current node can be maintained. Then the traversal of the prefix tree $\mathcal{T}$ can be organized in $O(1)$ amortized time per node (counting dictionary operations as $O(1)$).*

3 The Case $w \le k$

The expected density of a random minimizer depends on windows with k-repeats, as every window v, in which all $(w + 1)$ k-mers are distinct, satisfies $P_{\sigma,k}(v) = \frac{2}{w+1}$. If $w \le k$, then equal k-mers in a window necessarily overlap or touch, and thus create a periodic substring. This simple observation has very strong implications; in particular, it leads to a formula for $\mathcal{DR}_\sigma(k, w)$ and to a deep understanding of the behaviour of this function.

We start with combinatorial lemmas describing the mutual location of all repeated k-mers in a window. Let x be a p-periodic run in a string v. We call x a *major run*, if $|x| \ge |v|/2 + p$.

Lemma 3. *A string contains at most one major run.*

Proof. Aiming at a contradiction, assume that v has ρ_1-periodic major run x_1 and ρ_2-periodic major run x_2. As their total length is at least $|v| + p_1 + p_2$, they overlap in v by a substring y of length at least $p_1 + p_2$. If $p_1 = p_2$, then x_1 and x_2 form a p_1-periodic substring that contains both of them; this contradicts the definition of run. If $p_1 \neq p_2$, then y has periods p_1 and p_2 and then has the period $p = \gcd(p_1, p_2)$ by the Fine–Wilf periodicity lemma [5]. Since p divides p_1, the run x_1 has a p_1-substring that is an integer power of a p-substring. Hence x_1 is p-periodic. By the same argument, x_2 is also p-periodic. As $p < \max\{p_1, p_2\}$, we get a contradiction with our assumption on the periods of x_1 and x_2. □

Lemma 4. *Let $k \geq w$ and let a $(w+k)$-string v have a repeated k-mer. Then*

(i) v has a major run of length at least $p + k$, where p is its period;
(ii) all occurrences of repeated k-mers in v are inside the major run;
(iii) v has $w - i$ distinct k-mers, where the major run is p-periodic and has length $p + k + i$.

Proof. (i) Suppose that a k-mer u occurs in v at positions i and $j > i$. Since $k \geq |v|/2$, these two occurrences either overlap or touch, and thus $v[i..j+k-1]$ is a p-periodic substring of length $j - i + k$, where $p \leq j - i$ (the case $p < j - i$ takes place if there is a third occurrence of u between the two considered). This substring can be extended to a p-periodic run in v; the length of this run is at least $p + k$, so it is major by definition.

(ii) The above procedure (start with any two occurrences of one k-mer and extend the obtained periodic substring to a run) results in a major run, and this run is unique by Lemma 3.

(iii) The word v has $w + 1$ k-mers in total. By (ii), only the k-mers inside its major run x can repeat. By definition of p-periodic, p is the minimum period of x, so x contains exactly p distinct k-mers among its total of $p + i + 1$ k-mers. The statement now follows. □

3.1 The Formula for $\mathcal{DR}_\sigma(k, w)$

Let $\mathsf{Rep}_{\sigma,k,w}$ be the set of all σ-ary $(k+w)$-strings with k-repeats. By Lemma 1, the probability $P_{\sigma,k}(v)$ equals $\frac{2}{w+1}$ for every $v \in \Sigma^{w+k} \setminus \mathsf{Rep}_{\sigma,k,w}$. Then by Lemma 2 we have

$$\mathcal{DR}_\sigma(k,w) = \frac{2}{w+1} + \frac{1}{\sigma^{w+k}} \underbrace{\left(\sum\nolimits_{v \in \mathsf{Rep}_{\sigma,k,w}} P_{\sigma,k}(v) - \frac{2}{w+1} \left|\mathsf{Rep}_{\sigma,k,w}\right| \right)}_{\mathsf{Dev}_\sigma(k,w)} \tag{1}$$

The expression in parentheses in (1) shows how far is the density from the value $\frac{2}{w+1}$. We refer to this expression as *deviation* (of the random order) and denote it by $\mathsf{Dev}_\sigma(k,w)$. Our aim is to design, in the case $w \leq k$, a formula for $\mathsf{Dev}_\sigma(k,w)$ and thus for $\mathcal{DR}_\sigma(k,w)$. Let $\mathsf{Prim}_\sigma(n)$ denote the number of σ-ary primitive words of length n.

Lemma 5. *If $w \le k$, then*

$$\left|\mathsf{Rep}_{\sigma,k,w}\right| = \sum\nolimits_{p=1}^{w} \mathsf{Prim}_\sigma(p)\sigma^{w-p}\left(w-p+1-\tfrac{w-p}{\sigma}\right) \tag{2}$$

Proof. By Lemma 4(i), every string $v \in \mathsf{Rep}_{\sigma,k,w}$ contains a major run, say v' of period q, and $|v'| \ge q+k$. On the other hand, if a $(k+w)$-string v contains a p-periodic substring of length $p+k$, then the k-prefix and k-suffix of this substring are equal, and hence $v \in \mathsf{Rep}_{\sigma,k,w}$. Therefore, to prove the lemma we need to count, for each period $p = 1,\dots,w$, the number of $(k+w)$-strings containing a p-periodic run of length $\ge p+k$. Each such string v can be uniquely represented as $v = v_1v_2v_3v_4$, where $|v_2| = k$, $|v_3| = p$, and v_2v_3 is a suffix of the major run. Note that v_3 is primitive (otherwise, the run would have a smaller period) and v_2 is uniquely determined by v_3 due to periodicity. Further, there is no restrictions on v_1; if $|v_4| = 0$, we have σ^{w-p} options for v_1, to the total of $\mathsf{Prim}_\sigma(p)\cdot\sigma^{w-p}$ options for v. Now let $|v_4| > 0$. As the major run ends with v_3, one has $v_3[1] \ne v_4[1]$; there are no other restrictions for v_4. Since the number of options for a non-zero length of v_4 is $w-p$, we have $\mathsf{Prim}_\sigma(p)\cdot(\sigma-1)\sigma^{w-p-1}(w-p)$ options for v with nonempty v_4. Adding the numbers obtained for empty and nonempty v_4, we obtain exactly the term for p in (2). □

Lemma 6. *If $w \le k$, then*

$$\sum_{v\in\mathsf{Rep}_{\sigma,k,w}} P_{\sigma,k}(v) =$$

$$\sum_{t=1}^{w} \tfrac{1}{t}\cdot\Big(\mathsf{Prim}_\sigma(t) + \sum_{p=1}^{t-1}\mathsf{Prim}_\sigma(p)\sigma^{t-p}\cdot\Big(2t-2p+1-\tfrac{4t-4p-1}{\sigma}+\tfrac{2t-2p-2}{\sigma^2}\Big)\Big) \tag{3}$$

Proof. As in Lemma 5, we view elements of $\mathsf{Rep}_{\sigma,k,w}$ as strings containing a p-periodic run of length $\ge p+k$, for some $p = 1,\dots,w$. But unlike Lemma 5, here we need to count each string v with the weight $P_{\sigma,k}(v)$ computed by Lemma 1. This weight depends on the length of the run (Lemma 4) and its location (whether it is a suffix of v or not). Let us count $(w+k)$-strings with t distinct k-mers. Such a string can be decomposed as $v = v_1v_2v_3v_4$, where v_2v_3 is the major run and $|v_3| = p$. Then $|v_2| = w+k-t$ by Lemma 4(iii). This implies $|v_1| + |v_4| = t-p$.

First consider the case where the major run is a suffix of v and hence $P_{\sigma,k}(v) = \frac{1}{t}$ by Lemma 1. There are $\mathsf{Prim}_\sigma(p)$ options for the run (for every v_3, v_2 is unique due to periodicity). As $|v_4| = 0$, it remains to consider v_1. If $p = t$, then v_1 is empty, and if $p < t$, then there are $\sigma^{t-p-1}(\sigma-1)$ options for v_1, as the last letter of v_1 breaks the period of the run. In total, we have $\mathsf{Prim}_\sigma(t) + \sum_{p=1}^{t-1}\mathsf{Prim}_\sigma(p)\sigma^{t-p-1}(\sigma-1)$ strings of weight $\frac{1}{t}$.

Now let $|v_4| > 0$ and thus $p < t$ and also $P_{\sigma,k}(v) = \frac{2}{t}$ by Lemma 1. If $|v_1| = 0$, we get, symmetric to the above, $\sum_{p=1}^{t-1}\mathsf{Prim}_\sigma(p)\sigma^{t-p-1}(\sigma-1)$ strings of weight $\frac{2}{t}$. Finally, if $|v_1| > 0$, then both the last letter of v_1 and the first letter of v_4 break the period of the run. Then for fixed lengths of v_1 and v_4 we get, similar to the above, $\sum_{p=1}^{t-2} \mathsf{Prim}_\sigma(p)\sigma^{t-p-2}(\sigma-1)^2$ strings of weight $\frac{2}{t}$. This amount should be multiplied by $(t-p-1)$ possible choices of length for v_1 and v_4. Adding up the numbers obtained in all three cases, we get the term for t in (3). □

The definition of $\mathsf{Dev}_\sigma(k,w)$ and Lemmas 5, 6 immediately imply

Proposition 1. *If $w \le k$, then the deviation of the random order is*

$$\mathsf{Dev}_\sigma(k,w) = \sum_{t=1}^{w} \tfrac{1}{t}\cdot\Big(\mathsf{Prim}_\sigma(t) + \sum_{p=1}^{t-1} \mathsf{Prim}_\sigma(p)\sigma^{t-p}\cdot\Big(2t-2p+1-\tfrac{4t-4p-1}{\sigma}+\tfrac{2t-2p-2}{\sigma^2}\Big)\Big)$$
$$-\tfrac{2}{w+1}\cdot\sum_{p=1}^{w}\mathsf{Prim}_\sigma(p)\sigma^{w-p}(w-p+1-\tfrac{w-p}{\sigma}) \tag{4}$$

In particular, $\mathsf{Dev}_\sigma(k,w)$ is independent of k.

Substituting (4) into (1), we obtain the main result of this section.

Theorem 2. *If $w \le k$, the random order has the expected density $\mathcal{DR}_\sigma(k,w) = \frac{2}{w+1} + \frac{\mathsf{Dev}_\sigma(k,w)}{\sigma^{w+k}}$, where $\mathsf{Dev}_\sigma(k,w)$ is given by the formula (4). In particular, $\mathcal{DR}_\sigma(k,w)$ can be computed in $O(w\log w)$ time independently of σ and k.*

Proof. The formula is already proved, so it remains to show the time complexity. Note that if a string is not primitive, then it has the form u^m, where u is primitive and m is an integer greater than 1. Therefore, $\mathsf{Prim}_\sigma(p) = \sigma^p - \sum_{d|p}\mathsf{Prim}_\sigma(d)$. To compute $\mathsf{Prim}_\sigma(1),\ldots,\mathsf{Prim}_\sigma(w)$ by this formula, we initialize a table with the values $\sigma,\sigma^2,\ldots,\sigma^w$ and process it left to right; processing the cell i, we subtract its value from the values in cells $2i, 3i, \ldots$. This clearly takes $O(w\log w)$ time. Then we compute the deviation according to (4) in just $O(w)$ time by memorizing the values of the internal sum for the smaller values of t. □

Given Theorem 2, we can study details of the behaviour of the function $\mathcal{DR}_\sigma(k,w)$ in the half-quadrant defined by the inequality $w \le k$.

3.2 From $\mathcal{DR}_\sigma(k,w)$ to $\mathcal{DR}_\sigma(k+1,w)$

Proposition 2. *If $w \le k$, then $\mathcal{DR}_\sigma(k+1,w) = \frac{2}{w+1} + \frac{1}{\sigma}\cdot(\mathcal{DR}_\sigma(k,w) - \frac{2}{w+1})$.*

Proof. As we know from Proposition 1, $\mathsf{Dev}_\sigma(k,w) = \mathsf{Dev}_\sigma(k+1,w)$. Then the result is immediate from Theorem 2. □

Then we immediately have

Corollary 1. *If w is fixed and k tends to infinity, the density $\mathcal{DR}_\sigma(k,w)$ approaches $\frac{2}{w+1}$ (equivalently, the density factor $\mathcal{DFR}_\sigma(k,w)$ approaches 2) at* exact *exponential rate σ.*

The crucial fact that $\mathsf{Dev}_\sigma(k,w)$ does not depend on k looks unexpected and calls for a better explanation of its nature. Below we explain it establishing a natural bijection between the sets $\mathsf{Rep}_{\sigma,k,w}$ and $\mathsf{Rep}_{\sigma,k+1,w}$.

Consider the following function ϕ defined on $\mathsf{Rep}_{\sigma,k,w}$ ($w \le k$). Given a string $v \in \mathsf{Rep}_{\sigma,k,w}$, let x be its major run (Lemma 4(i)), let p be the period of x, and let $a_1a_2\cdots a_p$ be the p-suffix of x. We write $v = \ell x r$. Then $\phi(v)$ is a $(k+w+1)$-string of the form $\ell x' r'$ such that $x' = xa_1$ and r' is defined as follows. If either r is empty, or $p=1$, or $r[1] \ne a_2$, then $r' = r$; otherwise (i.e., if $r[1] = a_2$), r' is obtained from r by replacing $r[1]$ with a_1.

Theorem 3. *Let $w \le k$. Then ϕ is a bijection of $\mathsf{Rep}_{\sigma,k,w}$ onto $\mathsf{Rep}_{\sigma,k+1,w}$ and $P_{\sigma,k}(v) = P_{\sigma,k+1}(\phi(v))$ for every $v \in \mathsf{Rep}_{\sigma,k,w}$.*

Proof. We first prove the following claim.

Claim. If a string $v \in \mathsf{Rep}_{\sigma,k,w}$ has p-periodic major run x and $a_1a_2\cdots a_p$ is the p-suffix of x, then xa_1 is the major run in $\phi(v)$.

Proof. Let $v = \ell x r$. Since x is p-periodic and ends with $a_1a_2\cdots a_p$, $x' = xa_1$ is also p-periodic. Since x' is preceded in $\phi(v) = \ell x' r'$ by the same string ℓ as x in v, it cannot be extended to the left; it also cannot be extended to the right if r' is empty. For nonempty r' we consider two cases. If $r' = r$, then by the definition of ϕ either $p = 1$ or $r[1] \ne a_2$. In both cases, $r'[1] = r[1]$ breaks the period p in $\phi(v)$. If $r' \ne r$, then $p > 1$ and $r'[1] = a_1$ while $r[1] = a_2$. We know that $r[1] \ne a_1$, because $r[1]$ breaks the period p in v. Hence $a_1 \ne a_2$, and once again $r'[1]$ breaks the period p. Therefore the p-periodic string x' can be extended neither to the left or to the right, and thus is a run. Clearly, x' satisfies the length condition in the definition of major run. Hence, the claim is proved.

Now we proceed with the proof of the theorem. Since $|x| \ge k+p$ by the definition of major run, the k-prefix of x repeats in x. Then the $(k+1)$-prefix of x' repeats in x'. Therefore, $\phi(v) \in \mathsf{Rep}_{\sigma,k+1,w}$. To prove that ϕ is bijective it suffices to consider an arbitrary string $v' \in \mathsf{Rep}_{\sigma,k+1,w}$ and show that it has exactly one preimage by ϕ. By Lemma 4(i), v' has a major run.

Let $v' = \ell x' r'$, where x' is the major run, and let $x' = \hat{x}a_1a_2\cdots a_pa_1$, where p is the period of x'. We denote $x = \hat{x}a_1a_2\cdots a_p$. By the Claim and Lemma 3, every preimage of v' has x as the major run. Therefore, by the definition of ϕ, we can consider as candidate preimages of v' only strings of the form $\ell x r$ where r either equals r' or differs from r' in the first letter only. In particular, if r' is empty, then ℓx is the only candidate preimage and clearly $\phi(\ell x) = \ell x' = v'$. Now let $r' = b'\hat{r}$, $r = b\hat{r}$. The definition of ϕ tells us that either $b = b'$ or $b = a_2$ and $b' = a_1$. Then in the case $b' \ne a_1$ the only candidate for b is b', and one can check that $\phi(\ell x b'\hat{r}) = v'$ by definition. For the case $b' = a_1$ we observe that $b \ne a_1$, because x is a run in $\ell x r$. Then $b = a_2$ is the only candidate, and again, $\phi(\ell x a_2\hat{r}) = v'$ by definition. Thus we proved that v' has exactly one preimage by ϕ, and therefore ϕ is a bijection.

It remains to prove that ϕ preserves probabilities to be a gamechanger. Since the major runs x (of v) and x' (of $\phi(v)$) have the same period, Lemma 4(iii) implies that the number of distinct k-mers in v equals the number of distinct $(k+1)$-mers in $\phi(v)$. Next, note that the k-suffix of x has at least two occurrences in x. Hence, by Lemma 4(ii), the k-suffix of v has another occurrence in v if and only if x is a suffix of v. By a similar argument, the $(k+1)$-suffix of $\phi(v)$ has another occurrence in $\phi(v)$ if and only if x' is a suffix of $\phi(v)$. By the definition of ϕ, x is a suffix of v if and only if x' is a suffix of $\phi(v)$. Therefore, $P_{\sigma,k}(v) = P_{\sigma,k+1}(\phi(v))$ by Lemma 1. □

3.3 From $\mathcal{DR}_\sigma(k, w)$ to $\mathcal{DR}_\sigma(k, w+1)$

Comparing the densities $\mathcal{DR}_\sigma(k, w)$ and $\mathcal{DR}_\sigma(k, w+1)$ we need to compare their deviations. As $\mathsf{Dev}_\sigma(k, w)$ does not depend on k by Proposition 1, below we write $\mathsf{Dev}_\sigma(w)$. Let $\Delta_\sigma(w) = \mathsf{Dev}_\sigma(w+1) - \mathsf{Dev}_\sigma(w)$. We prove the following.

Lemma 7. *For every $\sigma \geq 2$ and $w \leq k-1$, one has $\Delta_\sigma(w) = \frac{S_1+S_2}{(w+1)(w+2)}$, where*

$$\begin{aligned}S_1 = (0-w)\sigma^0\mathsf{Prim}_\sigma(w+1) + (2-w)\sigma\mathsf{Prim}_\sigma(w) + (4-w)\sigma^2\mathsf{Prim}_\sigma(w-1) + \cdots \\ \cdots + (w-2)\sigma^{w-1}\mathsf{Prim}_\sigma(2) + w\sigma^w\mathsf{Prim}_\sigma(1) \quad (5)\end{aligned}$$

$$\begin{aligned}S_2 = (w-0)\sigma^0\mathsf{Prim}_\sigma(w) + (w-2)\sigma\mathsf{Prim}_\sigma(w-1) + (w-4)\sigma^2\mathsf{Prim}_\sigma(w-2) + \cdots \\ \cdots + (4-w)\sigma^{w-2}\mathsf{Prim}_\sigma(2) + (2-w)\sigma^{w-1}\mathsf{Prim}_\sigma(1) \quad (6)\end{aligned}$$

Proof. By the definition of deviation, $\Delta_\sigma(w) = \Delta_1 - \Delta_2$, where

$$\Delta_1 = \sum_{v\in\mathsf{Rep}_{\sigma,k,w+1}} P_{\sigma,k}(v) - \sum_{v\in\mathsf{Rep}_{\sigma,k,w}} P_{\sigma,k}(v),$$
$$\Delta_2 = \tfrac{2}{w+2}\left|\mathsf{Rep}_{\sigma,k,w+1}\right| - \tfrac{2}{w+1}\left|\mathsf{Rep}_{\sigma,k,w}\right|.$$

Note that all terms in the sum (3) are independent of w. Then Δ_1 equals such a term for $t = w+1$, i.e.,

$$\Delta_1 = \tfrac{1}{w+1}\Big(\mathsf{Prim}_\sigma(w+1) + \sum_{p=1}^{w}\mathsf{Prim}_\sigma(p)\cdot\sigma^{w+1-p}\Big(2w-2p+3-\tfrac{4w-4p+3}{\sigma}+\tfrac{2w-2p}{\sigma^2}\Big)\Big).$$

Using (1) and (2), we compute

$$\begin{aligned}\Delta_2 &= \tfrac{2}{w+2}\sum_{p=1}^{w+1}\mathsf{Prim}_\sigma(p)\sigma^{w-p+1}\big(w-p+2-\tfrac{w-p+1}{\sigma}\big)\\ &\quad - \tfrac{2}{w+1}\sum_{p=1}^{w}\mathsf{Prim}_\sigma(p)\sigma^{w-p}\big(w-p+1-\tfrac{w-p}{\sigma}\big)\\ &= \tfrac{2}{w+2}\mathsf{Prim}_\sigma(w+1)\\ &\quad + \sum_{p=1}^{w}\mathsf{Prim}_\sigma(p)\sigma^{w-p}\Big(\tfrac{2(w-p+2)\sigma}{w+2} - \tfrac{2(w-p+1)}{w+2} - \tfrac{2(w-p+1)}{w+1} + \tfrac{2(w-p)}{(w+1)\sigma}\Big)\end{aligned}$$

Grouping the corresponding terms in the expressions for Δ_1 and Δ_2, we get

$$\Delta_\sigma(w) = \Delta_1 - \Delta_2 = \Big(\tfrac{1}{w+1} - \tfrac{2}{w+2}\Big)\mathsf{Prim}_\sigma(w+1)$$
$$+\sum_{p=1}^{w} \mathsf{Prim}_\sigma(p)\sigma^{w-p}\Big(\tfrac{(2w-2p+3)\sigma}{w+1} - \tfrac{(2w-2p+4)\sigma}{w+2} + \tfrac{2w-2p+2}{w+2} - \tfrac{2w-2p+1}{w+1} + 0\Big)$$
$$= \frac{-w\mathsf{Prim}_\sigma(w+1)+\sum_{p=1}^{w} \mathsf{Prim}_\sigma(p)\sigma^{w-p}((w-2p+2)\sigma-w+2p)}{(w+1)(w+2)} \qquad (7)$$

Unwrapping the sum, we get exactly $S_1 + S_2$ in the numerator. □

To estimate $\Delta_\sigma(w)$ from (5) and (6), we need to evaluate Prim_σ. We recall (see, e.g., [13]) that

$$\mathsf{Prim}_\sigma(p) = \sum_{d|p} \mu(d)\sigma^{p/d}, \qquad (8)$$

where the *Möbius function* $\mu(n)$ is defined as follows. If n is *square-free*, i.e., a product of t distinct primes for some t, then $\mu(n) = (-1)^t$, including $\mu(1) = (-1)^0 = 1$. Otherwise, $\mu(n) = 0$. For example, $\mathsf{Prim}_\sigma(1) = \sigma$, $\mathsf{Prim}_\sigma(2) = \sigma^2 - \sigma$, $\mathsf{Prim}_\sigma(4) = \sigma^4 - \sigma^2$, $\mathsf{Prim}_\sigma(60) = \sigma^{60} - \sigma^{30} - \sigma^{20} - \sigma^{12} + \sigma^{10} + \sigma^6 + \sigma^4 - \sigma^2$.

Substituting the values of Prim_σ into (5) and (6), one can see that for any fixed w, $\Delta_\sigma(w)$ is a polynomial in σ; in particular,

$$\Delta_\sigma(1) = \tfrac{\sigma}{3}, \Delta_\sigma(2) = \tfrac{\sigma^2}{6}, \Delta_\sigma(3) = \tfrac{2\sigma^3+3\sigma^2-3\sigma}{20}, \Delta_\sigma(4) = \tfrac{2\sigma^4+2\sigma^3-6\sigma^2+4\sigma}{30} \qquad (9)$$

The following lemma can be proved by direct computation.

Lemma 8. *For every $\sigma \geq 2$ and every $w \leq \min\{10, k-1\}$, $\Delta_\sigma(w) > 0$.*

To the contrast, as w grows, $\Delta_\sigma(w)$ becomes negative and approaches $-\infty$.

Lemma 9. *Let $\sigma \geq 2$ be fixed. There exist constants $C_\sigma, C'_\sigma > 0$ such that $\frac{2\sigma+6-C_\sigma w}{(w+1)(w+2)}\sigma^{w-1} < \Delta_\sigma(w) < \frac{2\sigma+6-C'_\sigma w}{(w+1)(w+2)}\sigma^{w-1}$ whenever $11 \leq w \leq k-1$.*

Proof. Assume $w \geq 11$ and denote $\tilde{\Delta}_\sigma(w) = (w+1)(w+2)\Delta_\sigma(w)$ for convenience. By Lemma 7 and formula (8), we have $\tilde{\Delta}_\sigma(w) = \sum_{t=1}^{w+1} c_t(w)\sigma^t$, where each coefficient $c_t(w)$ is a sum of $O(w)$ numbers, each of absolute value at most w. We first compute the three leading coefficients of $\tilde{\Delta}_\sigma(w)$.

The exponent σ^{w+1} does not appear in S_2 (6), while in S_1 (5) it appears with the coefficient $c_{w+1}(w) = -w+(2-w)+\cdots+(w-2)+w = 0$. Next, σ^w appears in S_2 with the coefficient $w+(w-2)+(w-4)+\cdots+(4-w)+(2-w) = w$; in S_1 it appears only in the term containing $\mathsf{Prim}_\sigma(2) = \sigma^2 - \sigma$, with the coefficient $2-w$. Hence $c_w(w) = 2$. Finally, the exponent σ^{w-1} appears once in S_2 (in the term of $\mathsf{Prim}_\sigma(2)$ with the coefficient $w-4$) and twice in S_1 (in the terms of $\mathsf{Prim}_\sigma(4)$ and $\mathsf{Prim}_\sigma(3)$, with the coefficients $6-w$ and $4-w$ respectively). Hence $c_{w-1}(w) = 6-w$. Therefore, the three leading terms in $\tilde{\Delta}_\sigma(w)$ sum up to $X = (2\sigma+6-w)\sigma^{w-1}$; below we define Y, Z so that $\tilde{\Delta}_\sigma(w) = X + Y + Z$.

All lower terms in $\tilde{\Delta}_\sigma(w)$ are due to the monomials $\pm\sigma^{p-m}$ appearing in (8) (the expansion of $\mathsf{Prim}_\sigma(p)$) for certain $p > m \geq 2$. Every such monomial contributes to the coefficients at σ^{w+1-m} and σ^{w-m} (see (7)). If $\mathsf{Prim}_\sigma(p)$ contains

$\pm\sigma^{p-m}$, then $p = (p-m)d$ for a square-free d. Hence $p = \frac{md}{d-1}$, implying that $d-1$ divides m and the maximum value of p is $2m$. In particular, $p = 3, 4$ for $m = 2$; $p = 6$ for $m = 3$; $p = 5, 6, 8$ for $m = 4$; and $p = 6, 10$ for $m = 5$. From (5), (6) (or from (7)) we see that the monomials $\pm\sigma^{p-m}$ with $m \in \{2, 3, 4, 5\}$ contribute $Y = (w-4)\sigma^{w-2} + (20-2w)\sigma^{w-3} + (3w-30)\sigma^{w-4} - 8\sigma^{w-5}$ to $\tilde{\Delta}_\sigma(w)$ (the contribution of $-\sigma^{p-2}$ to the coefficient of σ^{w-1} is included into X).

We rewrite $Y = (y_1 w - y_2)\sigma^{w-1}$, where $y_1 = \frac{1}{\sigma} - \frac{2}{\sigma^2} + \frac{3}{\sigma^3} > 0$ and $y_2 = \frac{4}{\sigma} - \frac{20}{\sigma^2} + \frac{30}{\sigma^3} + \frac{8}{\sigma^4} > 0$. As $w \geq 11$, we have $(y_1 - \frac{y_2}{11})w\sigma^{w-1} \leq Y < y_1 w\sigma^{w-1}$.

Let Z be the sum of terms in $\tilde{\Delta}_\sigma(w)$ arising from monomials $\pm\sigma^{p-m}$ in (8) with $m \geq 6$. Then $\tilde{\Delta}_\sigma(w) = X + Y + Z$ as desired. As mentioned above, for a fixed m, the values of p satisfy $p = \frac{md}{d-1}$, where d is square-free. So either $p = m+1$ or $d - 1 \leq \frac{m}{2}$ and $d - 1 \neq 3$. Hence the number of options for p is at most $\frac{m}{2}$. The coefficient for σ^{p-m} for one term of (7) is between $-w$ and w. Therefore, we bound the absolute value of Z as $|Z| \leq \sum_{m=6}^{\infty} \frac{m}{2} w\sigma^{w+1-m}$. Factoring $\frac{w\sigma^w}{2}$ out and substituting $x = \sigma^{-1}, t = 6$ into the textbook formula $\sum_{m=t}^{\infty} mx^{m-1} = \frac{tx^{t-1}-(t-1)x^t}{(1-x)^2}$, we obtain $|Z| \leq zw\sigma^{w-1}$, where $z = \frac{6\sigma^{-2}-5\sigma^{-3}}{2(\sigma-1)^2}$.

For any $\sigma \geq 2$, one can easily check that $y_1 + z < 1$ and $y_1 - \frac{y_2}{11} - z > -\infty$. Then $\tilde{\Delta}_\sigma(w) = X+Y+Z$ is between $(2\sigma+6-C_\sigma w)\sigma^{w-1}$ and $(2\sigma+6-C'_\sigma w)\sigma^{w-1}$ for some constants $C_\sigma, C'_\sigma > 0$. The lemma follows from definition of $\tilde{\Delta}_\sigma(w)$. □

Now we describe the main features of the "horizontal" behaviour of the density $\mathcal{DR}_\sigma(k,w)$. As we compare the values for different w, it is convenient to formulate the result in terms of density factor $\mathcal{DFR}_\sigma(k,w) = (w+1)\mathcal{DR}_\sigma(k,w)$.

Theorem 4. *For every $\sigma \geq 2$ there exist integers $w_\sigma, w'_\sigma \geq 11$ such that*

(i) $\mathcal{DFR}_\sigma(k,w) > 2$ if $2 \leq w \leq \min\{w_\sigma, k\}$;
(ii) $\mathcal{DFR}_\sigma(k,w) < 2$ if $w'_\sigma \leq w \leq k$; in this case $2 - \mathcal{DFR}_\sigma(k,w) = \Theta(\frac{1}{w\sigma^k})$.

Proof. Note that if $w = 1$, for any order ρ we have $D_\rho = 1 = \frac{2}{w+1}$ and therefore $\mathcal{DFR}_\sigma(k,1) = 2$. Then (i) follows from Lemma 8. Further, Lemma 9 proves that $\Delta_\sigma(w)$ is negative starting from some value of w and approaches $-\infty$ as w grows; then the deviation becomes negative starting from some $w = w'_\sigma$, implying the first statement of (ii). Now note that $\Delta_\sigma(w) = \Theta(\frac{\sigma^w}{w})$, and hence the same bound works for $\mathsf{Dev}_\sigma(w)$. Formula (1) implies the second statement of (ii). □

Remark 2. *Combining the bounds from Theorem 4 with experiments for small alphabets, we claim a stronger result for those alphabets. Namely, a* single *constant separate the zones where $\mathcal{DFR}_\sigma(k,w) > 2$ and $\mathcal{DFR}_\sigma(k,w) < 2$ (see [6], Tables 1, 2). However, we have no proof of this property for arbitrary alphabets.*

4 The Case $w > k$

A straightforward lower bound for $\mathcal{DR}_\sigma(k,w)$ (and for the expected density of *any* particular order) is σ^{-k}: every position of the minimal k-mer is marked, and the expected density of such positions in a random string is σ^{-k}. If w is very big compared to k, then almost all windows contain the minimal k-mer; thus, $\mathcal{DR}_\sigma(k,w)$ approaches σ^{-k}. The next proposition clarifies what is "very big" in this context.

Proposition 3. *Let $N = \sigma^k$ and let $w = \frac{\sigma}{\sigma-1}N(\ln N + g(N))$ for arbitrary fixed positive function g. Then $\mathcal{DR}_\sigma(k,w) = (1 + O(e^{-g(N)}))\sigma^{-k}$.*

Proof. Let $A_{\sigma,u}(n)$ be the number of σ-ary n-strings having no occurrence of the k-mer u and let $A_{\sigma,k}(n) = \max\{A_{\sigma,u}(n) \mid u \text{ is a } k\text{-mer}\}$. Since some k-mer should be marked in each window having no occurrence of the minimal k-mer, we get the upper bound $\mathcal{DR}_\sigma(k,w) \le \frac{1}{\sigma^k} + \frac{A_{\sigma,k}(w+k-1)}{\sigma^{w+k-1}}$. The function $A_{\sigma,k}(n)$ can be estimated by the method of Guibas and Odlyzko [9,10]; we use the bound based on [16, Sect. 4]: $A_{\sigma,k}(n) \le (1 + \frac{k}{\sigma^k})(\sigma - \frac{\sigma-1}{\sigma^k})^n$. Then we have

$$\frac{A_{\sigma,k}(w+k-1)}{\sigma^{w+k-1}} \le (1 + \tfrac{k}{\sigma^k})(1 - \tfrac{\sigma-1}{\sigma^{k+1}})^{w+k-1} < (1 + \tfrac{k}{\sigma^k})e^{-\frac{\sigma-1}{\sigma^{k+1}}(w+k-1)}.$$

Since $(1+\frac{k}{\sigma^k})e^{-\frac{(\sigma-1)(k-1)}{\sigma^{k+1}}} = O(1)$, by substituting $w = \frac{\sigma}{\sigma-1}N(\ln N + g(N))$ we get

$$\mathcal{DR}_\sigma(k,w) \le \tfrac{1}{\sigma^k} + O(e^{-\frac{\sigma-1}{\sigma^{k+1}}\cdot\frac{\sigma^{k+1}}{\sigma-1}(k\ln\sigma + g(N))}) = \tfrac{1}{\sigma^k} + O(\tfrac{e^{-g(N)}}{\sigma^k}).$$

□

5 Discussion and Future Work

Random minimizer is an object interesting for both theory and practice. Its main characteristic is the density $\mathcal{DR}_\sigma(k,w)$ studied in this paper. We provide a detailed description of density for the case $w \le k$; the only remaining point of interest is to prove that for every σ the density passes the limit value $\frac{2}{w+1}$ only once (see Remark 2 and [6], Tables 1, 2).

The case $w > k$ presents more open problems, which can can be easily seen if we plot $\mathcal{DR}_\sigma(k,w)$ as a function of w for some σ and k (see [6], Fig. 1). We know approximate values of this function for small w (see (4) and Theorem 4) and for very big w (Proposition 3), but to fill the intermediate range is an open problem. The following simple lemma shows that $\mathcal{DR}_\sigma(k,w)$ is monotone in w.

Lemma 10. *$\mathcal{DR}_\sigma(k,w+1) \le \mathcal{DR}_\sigma(k,w)$.*

Proof. Consider an arbitrary minimizer $\mathcal{S}_\rho$ with parameters $k, w+1$ and an arbitrary window $u = u[1..w+k]$. When processing u, the scheme chooses some k-mer $u[i..i+k-1]$. Then $\mathcal{S}_\rho$ with parameters k, w chooses the same k-mer $u[i..i+k-1]$ when processing the window $u[1..w+k-1]$, or $u[2..w+k]$, or both. Hence for any processed string, the set of positions marked by $\mathcal{S}_\rho$ with the parameters $k, w+1$ forms a subset of positions marked by $\mathcal{S}_\rho$ with the parameters k, w. The lemma now follows from definitions. □

In order to describe the behaviour of $\mathcal{DR}_\sigma(k, w)$ for "medium" values of w, it is necessary to describe the ranges of $w = w(\sigma, k)$ where the density is $\frac{2+o(1)}{w}$; $\leq \frac{C}{w}$ for an absolute constant $C > 2$; $\leq \frac{C}{\sigma^k}$ for an absolute constant $C > 1$. From [22] we have the lower bound for the first range: all values $w = O(\sigma^{k/3-\varepsilon})$ are inside it. We believe that this bound is a big underestimate. Another result of [22] implies an upper bound $w = O(\sigma^k)$ for the second range, for every C. Our conjecture for the third range is $w = \Omega(\sigma^k \cdot k \ln \sigma)$.

Acknowledgments. S. Golan is supported by Israel Science Foundation grant no. 810/21. A. Shur is supported by the ERC grant MPM no. 683064 under the EU's Horizon 2020 Research and Innovation Programme and by the State of Israel through the Center for Absorption in Science of the Ministry of Aliyah and Immigration.

References

1. Chikhi, R., Holub, J., Medvedev, P.: Data structures to represent a set of k-long DNA sequences. ACM Comput. Surv. **54**(1) (2021). https://doi.org/10.1145/3445967
2. Deorowicz, S., Kokot, M., Grabowski, S., Debudaj-Grabysz, A.: KMC 2: fast and resource-frugal k-mer counting. Bioinformatics **31**(10), 1569–1576 (2015)
3. Edgar, R.: Syncmers are more sensitive than minimizers for selecting conserved *k*-mers in biological sequences. PeerJ **9**, e10805 (2021). https://doi.org/10.7717/peerj.10805
4. Ferragina, P., Manzini, G.: Opportunistic data structures with applications. In: 41st Annual Symposium on Foundations of Computer Science. FOCS 2000, pp. 390–398. IEEE Computer Society (2000). https://doi.org/10.1109/SFCS.2000.892127
5. Fine, N.J., Wilf, H.S.: Uniqueness theorems for periodic functions. Proc. Am. Math. Soc. **16**(1), 109–114 (1965)
6. Golan, S., Shur, A.M.: Expected density of random minimizers (2024). arxiv:2410.16968
7. Golomb, S.W.: Shift Register Sequences. Holden-Day, San Francisco (1967)
8. Groot Koerkamp, R., Pibiri, G.E.: The mod-minimizer: a simple and efficient sampling algorithm for long k-mers. In: Pissis, S.P., Sung, W. (eds.) 24th International Workshop on Algorithms in Bioinformatics. WABI 2024. LIPIcs, vol. 312, pp. 11:1–11:23. Schloss Dagstuhl - Leibniz-Zentrum für Informatik (2024). https://doi.org/10.4230/LIPICS.WABI.2024.11
9. Guibas, L.J., Odlyzko, A.M.: Maximal prefix-synchronized codes. SIAM J. Appl. Math. **35**, 401–418 (1978)
10. Guibas, L.J., Odlyzko, A.M.: String overlaps, pattern matching, and nontransitive games. J. Comb. Theory A **30**, 183–208 (1981)
11. Lempel, A.: On extremal factors of the de Bruijn graph. J. Comb. Theory B **11**, 17–27 (1971)
12. Li, H.: Minimap2: pairwise alignment for nucleotide sequences. Bioinformatics **34**(18), 3094–3100 (2018)
13. Lothaire, M. (ed.): Combinatorics on Words, 2 edn. Cambridge Mathematical Library. Cambridge University Press, Cambridge (1997)

14. Mykkeltveit, J.: A proof of Golomb's conjecture for the de Bruijn graph. J. Comb. Theory B **13**, 40–45 (1972)
15. Roberts, M., Hayes, W., Hunt, B.R., Mount, S.M., Yorke, J.A.: Reducing storage requirements for biological sequence comparison. Bioinformatics **20**(18), 3363–3369 (2004). https://doi.org/10.1093/bioinformatics/bth408
16. Rubinchik, M., Shur, A.M.: The number of distinct subpalindromes in random words. Fundam. Informaticae **145**(3), 371–384 (2016). https://doi.org/10.3233/FI-2016-1366
17. Sahlin, K.: Effective sequence similarity detection with strobemers. Genome Res. **31**(11), 2080–2094 (2021). https://doi.org/10.1101/gr.275648.121
18. Sahlin, K., Baudeau, T., Cazaux, B., Marchet, C.: A survey of mapping algorithms in the long-reads era. Genome Biol. **24**, 133 (2023). https://doi.org/10.1186/s13059-023-02972-3
19. Schleimer, S., Wilkerson, D.S., Aiken, A.: Winnowing: local algorithms for document fingerprinting. In: Proceedings of the 2003 ACM SIGMOD International Conference on Management of Data. SIGMOD '03, pp. 76–85. Association for Computing Machinery, New York, NY, USA (2003). https://doi.org/10.1145/872757.872770
20. Weiner, P.: Linear pattern matching algorithms. In: 14th Annual Symposium on Switching and Automata Theory, Iowa City, Iowa, USA, 15–17 October 1973, pp. 1–11. IEEE Computer Society (1973). https://doi.org/10.1109/SWAT.1973.13
21. Wood, D.E., Salzberg, S.L.: Kraken: ultrafast metagenomic sequence classification using exact alignments. Genome Biol. **15**, 1–12 (2014)
22. Zheng, H., Kingsford, C., Marçais, G.: Improved design and analysis of practical minimizers. Bioinformatics **36**, i119–i127 (2020). https://doi.org/10.1093/bioinformatics/btaa472

Correction to: Beyond Image-Text Matching: Verb Understanding in Multimodal Transformers Using Guided Masking

Ivana Beňová, Jana Košecká, Michal Gregor, Martin Tamajka, Marcel Veselý, and Marián Šimko

Correction to:
Chapter 7 in: R. Královič and V. Kůrková (Eds.): ***SOFSEM 2025: Theory and Practice of Computer Science*****, LNCS 15538,**
https://doi.org/10.1007/978-3-031-82670-2_7

The original version of the book was inadvertently published with an error related to the author names, text part and alignment in Tables 2 and 3. It has been corrected.

The updated version of this chapter can be found at
https://doi.org/10.1007/978-3-031-82670-2_7

R. Královič and V. Kůrková (Eds.): SOFSEM 2025, LNCS 15538, p. C1, 2025.
https://doi.org/10.1007/978-3-031-82670-2_26

Author Index

R. Královič and V. Kůrková (Eds.): SOFSEM 2025, LNCS 15538, pp. 361–363, 2025.
https://doi.org/10.1007/978-3-031-82670-2

The manufacturer's authorised representative in the EU is Springer Nature Customer Service Centre GmbH, Europaplatz 3, 69115 Heidelberg, Germany. If you have any concerns regarding our products, please contact ProductSafety@springernature.com

Printed and bound by CPI Group (UK) Ltd, Croydon, CR0 4YY

24/04/2026

02096365-0008